HIGH FREQUENCY
OCEAN ACOUSTICS

Related Titles from AIP Conference Proceedings

700 Review of Progress in Quantitative Nondestructive Evaluation: Volume 23
Edited by Donald O. Thompson and Dale E. Chimenti, February 2004, 2 vol. hard cover set,
CD-ROM included, 0-7354-0173-X

676 Experimental Chaos: 7th Experimental Chaos Conference
Edited by V. In, L. Kocarev, T. L. Carroll, B. J. Gluckman, S. Boccaletti, and J. Kurths,
August 2003, 0-7354-0145-4

621 Ocean Acoustic Interference Phenomena and Signal Proceessing
Edited by William A. Kuperman and Gerald L. D'Spain, June 2002, 0-7354-0070-9

524 Nonlinear Acoustics at the Turn of the Millennium: ISNA 15; 15th International
Symposium on Nonlinear Acoustics
Edited by Werner Lauterborn and Thomas Kurz, July 2000, 1-56396-945-9

To learn more about these titles, or the AIP Conference Proceedings Series, please visit
the webpage **http://proceedings.aip.org**

HIGH FREQUENCY OCEAN ACOUSTICS

High Frequency Ocean Acoustics Conference

La Jolla, California   1 – 5 March 2004

EDITORS
Michael B. Porter
Martin Siderius
Science Applications International Corporation
San Diego, California

William A. Kuperman
Scripps Institution of Oceanography
La Jolla, California

SPONSORING ORGANIZATIONS
Office of Naval Research
Acoustical Society of America

Melville, New York, 2004
AIP CONFERENCE PROCEEDINGS ■ **VOLUME 728**

Editors:

Michael B. Porter
Martin Siderius

Science Applications International Corporation
10260 Campus Point Drive
San Diego, CA 92121
USA

E-mail: MikePorter@HLSResearch.com
sideriust@saic.com

William A. Kuperman
Scripps Institution of Oceanography
La Jolla, CA 92093-0238
USA

E-mail: wak@mpl.ucsd.edu

The articles on pp. 12-21, 90-97, 98-105, 106-113, 114-121, 141-148, 149-156, 157-164, 165-172, 393-401, 402-412, 413-419, 420-427, 428-437, and 438-446 were authored by U. S. Government employees and are not covered by the below mentioned copyright.

The article on pp. 32-39 was prepared on behalf of Defence Research and Development Canada (DRDC Atlantic), and therefore the copyright belongs to the Crown, that is to the Canadian Government.

The article on pp. 508-513 is © British Crown copyright - Dstl 2004 - published with the permission of the Controller of Her Majesty's Stationery Office.

Authorization to photocopy items for internal or personal use, beyond the free copying permitted under the 1978 U.S. Copyright Law (see statement below), is granted by the American Institute of Physics for users registered with the Copyright Clearance Center (CCC) Transactional Reporting Service, provided that the base fee of $22.00 per copy is paid directly to CCC, 222 Rosewood Drive, Danvers, MA 01923. For those organizations that have been granted a photocopy license by CCC, a separate system of payment has been arranged. The fee code for users of the Transactional Reporting Service is: 0-7354-0210-8/04/$22.00.

© 2004 American Institute of Physics

Individual readers of this volume and nonprofit libraries, acting for them, are permitted to make fair use of the material in it, such as copying an article for use in teaching or research. Permission is granted to quote from this volume in scientific work with the customary acknowledgment of the source. To reprint a figure, table, or other excerpt requires the consent of one of the original authors and notification to AIP. Republication or systematic or multiple reproduction of any material in this volume is permitted only under license from AIP. Address inquiries to Office of Rights and Permissions, Suite 1NO1, 2 Huntington Quadrangle, Melville, N.Y. 11747-4502; phone: 516-576-2268; fax: 516-576-2450; e-mail: rights@aip.org.

L.C. Catalog Card No. 2004113574
ISBN 0-7354-0210-8
ISSN 0094-243X
Printed in the United States of America

CONTENTS

Preface...xi

SEDIMENT ACOUSTICS

Wave and Material Properties of Marine Sediments:
Theoretical Relationships for Geoacoustic Inversions.........................3
    M. J. Buckingham*
Empirical Predictions of Seafloor Properties Based on Remotely
Measured Sediment Impedance 12
    M. D. Richardson and K. B. Briggs
High-Frequency Geoacoustic Inversion of Ambient Noise Data Using
Short Arrays 22
    M. Siderius and C. Harrison
Using Buried Directional Receivers in High-Frequency
Seafloor Studies..32
    J. C. Osler and A. P. Lyons
Geoacoustic Inversion of Broadband Data from the Florida Straits............40
    R. Chapman and Y. Jiang
High-Frequency Rapid Geo-Acoustic Characterization 47
    K. D. Heaney

ACOUSTIC COMMUNICATIONS

The Impact of Underwater Acoustic Channel Structure and Dynamics
on the Performance of Adaptive Coherent Equalizers 57
    J. Preisig
Spatio-Temporal Focusing for Elimination of Multipath Effects in
High Rate Acoustic Communications.....................................65
    M. Stojanovic
Synthetic Undersea Acoustic Transmission Channels 73
    D. Green and J. Rice
Acoustic Communication Using Time-Reversal Signal Processing:
Spatial and Frequency Diversity..83
    D. Rouseff, J. A. Flynn, J. A. Ritcey, and W. L. J. Fox
Environmental Effects on Phase Coherent Underwater
Acoustic Communications: A Perspective from Several
Experimental Measurements...90
    T. C. Yang
Environmental and Motion Effects on Orthogonal Frequency Division
Multiplexed On-Off Keying..98
    P. J. Gendron and T. C. Yang

*Indicates invited speaker

High-Frequency FH-FSK Underwater Acoustic Communications:
The Environmental Effect and Signal Processing 106
    W.-B. Yang and T. C. Yang
Underwater Acoustic Communication Channel Capacity:
A Simulation Study..114
    T. J. Hayward and T. C. Yang

BOUNDARY SCATTERING AND VOLUME FLUCTUATIONS

Progress and Research Issues in High-Frequency Seafloor Scattering 125
    D. R. Jackson*
Modeling Shallow Water Propagation with Scattering from
Rough Boundaries 132
    E. I. Thorsos, F. S. Henyey, W. T. Elam, S. A. Reynolds,
    and K. L. Williams
Mid-Frequency Sonar Backscatter Measurements from a
Rippled Bottom..141
    J. L. Lopes, R. Lim, and K. W. Commander
The Dependence of Long-Range Reverberation on Bottom Roughness 149
    R. Gauss, D. Fromm, K. LePage, and R. Gragg
The Influence of the Sea Surface and Fish on
Long-Range Reverberation 157
    R. Gauss, D. Fromm, K. LePage, J. Fialkowski, and R. Nero
Environmental Effects of Waveguide Uncertainty on Coherent
Aspects of Propagation, Scattering, and Reverberation 165
    K. D. LePage and B. E. McDonald
Towards a Deterministic High-Frequency Shallow Water
Ray Propagation Model..173
    L. Pautet and E. Pouliquen
Nonlinear Bubble Dynamics and the Effects on Propagation Through
Near-Surface Bubble Layers..180
    T. G. Leighton*
The Sea Surface Bounce Channel: Bubble-Mediated Energy Loss
and Time/Angle Spreading 194
    P. H. Dahl
On the Relationship between Signal Bandwidth and Frequency
Correlation for Surface Forward Scattered Signals........................204
    L. Culver and D. Bradley
Modeling Acoustic Signal Fluctuations Induced by
Sea Surface Roughness 214
    R. M. Heitsenrether and M. Badiey
Mid-Frequency Signal Fluctuations and Target Localization.................222
    W. S. Hodgkiss, G. L. D'Spain, and D. E. Ensberg
HF Doppler Acoustic Imaging of the Ocean Surface and Interior 230
    R. Pinkel* and J. A. Smith

*Indicates invited speaker

Detection of High-Frequency Sources in Random/Uncertain Media........... 237
    L. H. Sibul, C. M. Coviello, and M. J. Roan

MARINE MAMMALS

The Dolphin Sonar: Excellent Capabilities in Spite of
Some Mediocre Properties.. 247
    W. W. L. Au*
Biomimetic Signal Processing Using the Biosonar
Measurement Tool (BMT)... 260
    A. T. Abawi, P. Hursky, M. B. Porter, C. Tiemann, and S. Martin
Active Sonar and the Marine Environment 272
    E. M. Sevaldsen and P. H. Kvadsheim
Predicting the Environmental Impact of Active Sonar 280
    A. J. Duncan, R. D. McCauley, and A. L. Maggi
Acoustic Propagation Studies for Sperm Whale Phonation Analysis
during LADC Experiments.. 288
    N. A. Sidorovskaia, G. E. Ioup, J. W. Ioup, and J. W. Caruthers
Underwater Ambient Noise and Sperm Whale Click Detection during
Extreme Wind Speed Conditions 296
    J. J. Newcomb, A. J. Wright, S. Kuczaj, R. Thames, W. R. Hillstrom,
    and R. Goodman

EXPERIMENTAL AND MEASUREMENT TECHNIQUES

The Kauai Experiment... 307
    M. B. Porter, P. Hursky, M. Siderius, M. Badiey, J. Caruthers,
    W. S. Hodgkiss, K. Raghukumar, D. Rouseff, W. Fox, C. de Moustier,
    B. Calder, B. J. Kraft, K. McDonald, P. Stein, J. K. Lewis, and S. Rajan
Ocean Variability Effects on High-Frequency Acoustic Propagation
in KauaiEx... 322
    M. Badiey, S. E. Forsythe, M. B. Porter, and the KauaiEx Group
Telesonar Testbed Instrument Provides a Flexible Platform
for Acoustic Propagation and Communication Research in the
8–50 kHz Band.. 336
    V. K. McDonald, P. Hursky, and the KauaiEx Group
Channel Effects on Direct-Sequence Spread Spectrum Rake Receiver
during the KauaiEx Experiment 350
    P. Hursky, V. K. McDonald, and the KauaiEx Group
Impact of Thermocline Variability on Underwater Acoustic
Communications: Results from KauaiEx 358
    M. Siderius, M. B. Porter, and the KauaiEx Group
Side-Scan Sonar Survey Operations in Support of KauaiEx 366
    J. W. Caruthers, E. Quiroz, C. Fisher, R. Meredith, N. A. Sidorovskaia,
    and the KauaiEx Group

*Indicates invited speaker

High Frequency Tomography Using Bottom-Mounted Transducers 373
    J. K. Lewis, P. J. Stein, S. Rajan, J. Rudzinsky, A. Vandiver, and the KauaiEx Group

Results from the Elba HF-2003 Experiment 385
    F. Jensen, L. Pautet, M. B. Porter, M. Siderius, V. McDonald, M. Badiey, D. Kilfoyle, and L. Freitag

Panama City 2003 Broadband Shallow-Water Acoustic Coherence Experiments 393
    S. Stanic, E. Kennedy, D. Malley, B. Brown, R. Meredith, R. Fisher, H. Chandler, R. Ray, and R. Goodman

Panama City 2003 Acoustic Coherence Experiments: Environmental Characterization 402
    R. Meredith, R. Fisher, S. Stanic, E. Kennedy, D. Malley, and B. Brown

Broadband Horizontal and Vertical Spatial Coherence Measurements 413
    T. H. Ruppel, S. Stanic, G. V. Norton, R. W. Meredith, E. T. Kennedy, R. R. Goodman, and M. A. Wilson

Broadband Temporal Coherence Results from the June 2003 Panama City Coherence Experiments 420
    H. Chandler, E. Kennedy, R. Meredith, R. Goodman, and S. Stanic

Panama City 2003 Acoustic Coherence Experiments: Low Frequency Bottom Penetration Fluctuation Measurements in a Multipath Environment 428
    R. W. Meredith, E. T. Kennedy, D. Malley, R. A. Fisher, R. Brown, and S. Stanic

A High-Speed, Multi-Channel Data Acquisition System 438
    D. Malley, B. Brown, E. Kennedy, R. Meredith, and S. Stanic

TARGET MODELING, SYSTEMS, AND APPLICATIONS

Navy Applications of High-Frequency Acoustics 449
    H. Cox*

Virtual Source Approach to Scattering from Partially Buried Elastic Targets 456
    H. Schmidt*

A Finite-Element Tool for Scattering from Localized Inhomogeneities and Submerged Elastic Structures 464
    M. Zampolli, D. S. Burnett, F. B. Jensen, A. Tesei, H. Schmidt, and J. B. Blottman III

High-Frequency Material-Dependent Scattering Processes for Tilted Truncated Cylindrical and Disk-Shaped Targets 472
    P. L. Marston

Detection of Direct-Path Arrivals for Multi-Narrowband Sequences (3–30 kHz) in Shallow Water 478
    A. Zoksimovski and C. de Moustier

*Indicates invited speaker

A New Synthetic Aperture Sonar Design with Multipath Mitigation 489
      M. Pinto, A. Bellettini, L. S. Wang, P. Munk, V. Myers, and L. Pautet

Mid- to High-Frequency Ambient Noise Anisotropy and Notch-Filling Mechanisms 497
      P. Ferat and J. Arvelo

Measurements and Predictions of High Frequency Ambient Noise 508
      A. Holden

Ultrasonic Time Reversal Mirrors 514
      M. Fink,* G. Montaldo, and M. Tanter

Time Reversal Ocean Acoustic Experiments at 3.5 kHz: Applications to Active Sonar and Undersea Communications 522
      H. Song, P. Roux, T. Akal, G. Edelmann, W. Higley, W. S. Hodgkiss,
      W. A. Kuperman, K. Raghukumar, and M. Stevenson

Time-Reversal and Spatial Diversity: Issues in a Time-Varying Geometry Test 350
      S. M. Jesus and A. J. Silva

A High-Frequency Active Underwater Acoustic Barrier Experiment Using a Time Reversal Mirror; Model-Data Comparison 539
      A. Tesei, H. C. Song, P. Guerrini, P. Roux, W. S. Hodgkiss, T. Akal,
      M. Stevenson, and W. A. Kuperman

Author Index 547

*Indicates invited speaker

PREFACE

For many researchers, high-frequency acoustics has implied low-interest physics. In this view, it is the domain where the ocean volume consumes the sound energy, and the ocean surface or bottom thoroughly scatters it. The propagation physics is then little deeper than that of a spherical wave and the applications are simple devices such as fathometers and ADCP's. However, in recent years several new applications have motivated interest in the high-frequency band, which in turn have revealed much fascinating unexplored territory.

As every community defines high and low relative to their own center of reference we shall first explicitly identify HF as the 3-50 kHz band. In this band, we find 1) the heart of modern acoustic communications work, 2) sperm whale clicks that allow tracking at long ranges, 3) dolphin whistles and clicks, and 4) an ambient noise spectrum which beautifully illuminates surface and bottom and thereby reveals clearly the bottom type. Sound in this band can refract under the influence of the ocean thermal structure and then propagate to long ranges. To the surprise of many, the sound energy is *not* annihilated by contact with rough, moving boundaries, but instead can reflect many times off the boundary and still yield distinct echoes.

As we have explored these various applications we find a great deal of uncharted ground. The area of acoustic communications provides an interesting example. Much work has been done in the 8-50 kHz band where (as you can see in this volume) one will typically encounter a rich multipath structure. Should we consider the surface as frozen in time or can we look at an ensemble average over the period of ocean swell? Unfortunately the answer is neither, for over the typical symbol processing time, the swell is effectively frozen, while smaller scale surface features such as ocean chop may move significantly. Thus the sound energy provides an acoustic flash that freezes the large-scale features but blurs the short scale features. At the high end of the band, bubbles can completely mask the surface, depending on sea state. The rough ocean bottom is clearly static, but alas, the source or receiver may well be moving leading to another dynamic scattering process. Perhaps we can rely on purely refracted paths? These paths may come and go with the swirling patterns of cold and warm water in the ocean.

In short, there are questions about every aspect affecting sound propagation in this band with broad import across a variety of applications. These questions motivated a conference held 1-5 March 2004 at La Valencia Hotel in La Jolla, California, whose proceedings are introduced here.

Thanks are due to the Office of Naval Research for providing the financial support that made this conference possible. We also gratefully acknowledge Betty Kunowski (with assistance from Barbara Jones) who served as the conference secretariat and was involved in almost all aspects of the conference organization.

M. B. Porter
M. Siderius
W. A. Kuperman

SEDIMENT ACOUSTICS

Wave and Material Properties of Marine Sediments: Theoretical Relationships for Geoacoustic Inversions

Michael J. Buckingham

Marine Physical Laboratory, Scripps Institution of Oceanography
University of California, San Diego, 9500 Gilman Drive, La Jolla, CA 92093-0238, USA
Also affiliated to: Institute of Sound and Vibration Research
The University, Southampton SO17 1BJ, UK

Abstract. In recent years, a theory of wave propagation in marine sediments has been developed, based on the grain-to-grain interactions that occur during the passage of compressional and shear waves. The theory yields a dispersion pair, representing phase speed and attenuation, for each wave. These expressions are functions of frequency and the physical properties of the sediment, that is, the porosity, density, grain size and over-burden pressure (or depth in the medium). The predicted functional dependencies are compared with extensive data sets that have appeared in the literature over the past couple of decades. No adjustable parameters are available to help improve the comparisons. In all cases, the theory shows a high level of agreement with the data. This agreement even extends to both attenuations, in that the theory, which predicts *intrinsic* attenuation, arising from the conversion of wave energy into heat, accurately traces out the lower bound of the widely-distributed measurements. This is physically reasonable, since the data represent *effective* attenuation, which includes additional sources of loss such as scattering from shell fragments and other inhomogeneities in the medium. It is suggested that the set of simple algebraic expressions comprising the theory have application in evaluating the geoacoustic parameters of the seabed, all of which may be computed from knowledge of just one, say the compressional wave speed or the porosity.

INTRODUCTION

Marine sediments are granular materials (sands, silts and clays), saturated with seawater, that support the propagation of a compressional (longitudinal) wave and a shear (transverse) wave. Commonly, sediments are unconsolidated, that is, the grains are unbonded with the facility to move relative to one another. The compressional wave of the second kind, or slow wave, predicted by Biot's theory [1,2] of wave propagation in porous media has not been observed in unconsolidated sediments, despite several attempts to detect it. Either the slow wave is negligible in or absent from such materials and is not further considered here.

The wave properties of sediments are the phase speeds and attenuations of the compressional and shear waves. Over several decades, these wave properties have been carefully measured, both *in situ* and in core samples, by several groups of investigators. It is now well-established that the wave properties depend more or less systematically on the porosity, density and grain size of the medium, along with the measurement frequency and over-burden pressure, which translates into depth in the sediment. The correlations that have been observed to exist between the wave and physical properties of the medium have been expressed by Hamilton *et al.* [3,4] and Richardson and Briggs [5] through empirical regression equations.

An alternative approach to the correlations between the wave and physical properties of sediments has been pursued by Buckingham [6], who developed a theoretical model of wave propagation in saturated granular materials. The basis of his model is a particular form of inter-granular shearing, which yields dispersion relationships for the compressional and the shear wave. These dispersion relationships, in the form of algebraic expressions for the phase speed and the attenuation of both types of wave, depend explicitly on the grain size, porosity, depth in the sediment, and measurement frequency. They are independent of such parameters as the pore-fluid viscosity, the permeability and tortuosity, all of which are present in Biot's dispersion relationships [1,2]. Moreover, Buckingham's theory, unlike Biot's, does not include a bulk or shear frame modulus, since the stiffness of the material (*i.e.*, the elasticity of the mineral "frame") is naturally accounted for by the grain-to-grain interactions.

In this paper, a brief comparison is made between Buckingham's theoretical dispersion relationships and data sets, taken from the literature, on the inter-relationships between the wave and physical properties of marine sediments. As will be demonstrated, the theoretical predictions accurately match the data in practically all regards.

DISPERSION RELATIONS

The expressions for the sound speed, c_p, and attenuation, α_p, from Buckingham's inter-granular shearing theory are as follows:

$$c_p = \frac{c_o}{\text{Re}\left[1 + \frac{\gamma_p + (4/3)\gamma_s}{\rho_o c_o^2}(j\omega T)^n\right]^{-1/2}} \tag{1a}$$

and

$$\alpha_p = -\frac{\omega}{c_o}\text{Im}\left[1 + \frac{\gamma_p + (4/3)\gamma_s}{\rho_o c_o^2}(j\omega T)^n\right]^{-1/2}. \tag{1b}$$

The corresponding expressions for the shear wave speed, c_s, and attenuation, α_s, are

$$c_s = \sqrt{\frac{\gamma_s}{\rho_o}}\frac{(\omega T)^{n/2}}{\cos\left(\frac{n\pi}{4}\right)} \tag{2a}$$

and

$$\alpha_s = \omega \sqrt{\frac{\rho_o}{\gamma_s}} (\omega T)^{-n/2} \sin\left(\frac{n\pi}{4}\right) . \tag{2b}$$

In these dispersion pairs, $j = \sqrt{-1}$, ω is angular frequency, T is an arbitrary time, set equal to one second, and introduced to avoid awkward dimensions arising when frequency is raised to a fractional power, ρ_o is the bulk density of the material, and c_o is Wood's [7] sound speed in the equivalent suspension:

$$c_o = \sqrt{\frac{\kappa_o}{\rho_o}} , \tag{3}$$

where ρ_o is the bulk modulus of the medium.

Both ρ_o and κ_o may be expressed as weighted means of the respective values for the two constituent materials, the mineral grains (ρ_g, κ_g) and pore fluid (ρ_w, κ_w):

$$\rho_o = N\rho_w + (1-N)\rho_g \tag{4}$$

and

$$\frac{1}{\kappa_o} = N\frac{1}{\kappa_w} + (1-N)\frac{1}{\kappa_g} , \tag{5}$$

where N is the fractional porosity. Representative values of the properties of the mineral grains and pore fluid are, respectively: ρ_g = 2730 kg/m^3, κ_g = 3.36 x 10^{10} Pa; and ρ_w = 1005 kg/m^3, κ_w = 2.37 x 10^9 Pa.

The three remaining parameters in the dispersion relationships, (γ_p, γ_s, n), characterize grain-to-grain shearing that occurs during the passage of a wave. Analogous to the Lamé parameters of elasticity theory, the first two, γ_p and γ_s, are (real) compressional and shear moduli, whereas the third, n, is a positive fractional index, which is a measure of the strain hardening that is postulated to occur at grain contacts as inter-granular sliding progresses. From the Hertz theory of elastic spheres in contact [8], the two moduli are expressed in terms of the porosity, N, grain diameter, u_g, and depth in the sediment, d, as follows:

$$\gamma_p = \gamma_{po} \left[\frac{(1-N)u_g d}{(1-N_o)u_{go} d_o}\right]^{1/3} \tag{6a}$$

and

$$\gamma_s = \gamma_{so} \left[\frac{(1-N)u_g d}{(1-N_o)u_{go}d_o} \right]^{2/3} , \qquad (6b)$$

where the compressional and shear coefficients, γ_{po} and γ_{so}, respectively, take numerical values that are independent of the bulk physical properties of the medium. Although (γ_{po}, γ_{so}, n) may vary weakly from one sediment to another, due to microscopic differences between grain-surface properties, their values are held constant throughout the remaining discussion: $\gamma_{po} = 3.9 \times 10^8$ Pa, $\gamma_{so} = 4.65 \times 10^7$ Pa and $n = 0.0851$. (See Buckingham [9] for the derivation of these values). The three reference parameters, porosity N_o, grain size u_{go}, and depth d_o, in Eqs. (6), serve to avoid awkward dimensions appearing when the terms in square brackets are raised to a fractional power. These reference parameters are chosen for convenience ($N_o = 0.37$, $u_{go} = 1000$ μm, $d_o = 0.3$ m) and do not represent additional unknowns.

Equations (1) to (6) specify completely the predicted properties of the compressional and shear wave in a marine sediment. These expressions depend explicitly on grain size and porosity, and when measured values for both are available, they should be used in evaluating the theory. This, however, is not always possible. For instance, if the wave properties were to be evaluated as functions of porosity or of grain size, bearing in mind that the latter two are correlated, then it is necessary to identify a relationship connecting N and u_g. From a random-packing, rough-grain argument, Buckingham [10] has derived such a relationship:

$$N = 1 - P_s \left\{ \frac{u_g + 2\Delta}{u_g + 4\Delta} \right\}^3 , \qquad (7)$$

where $P_s = 0.63$ is the packing factor of a random "close" packing of uniform spheres and the rms roughness, or shape, parameter, Δ, characterizes the degree of non-sphericity exhibited by the mineral grains in the sediment. It is possible that Δ may differ slightly from one sediment to another, but for the purpose of the following discussions it is held constant with the value $\Delta = 1$ μm.

The predicted wave properties as functions of frequency, porosity, grain size, and depth in the sediment, may now be investigated, using Eqs. (1) to (7) with the parameter values that have been cited. For the dispersion relations in Eqs. (1) and (2) to be valid, these theoretical functional dependencies must follow the trends of the corresponding data sets. Note that no adjustable parameters are available to help improve any of the following comparisons.

FREQUENCY DEPENDENCE

Straightforward approximations [6] of the dispersion relationships in Eqs. (1) and (2) indicate that both wave speeds show near-logarithmic dispersion and both attenuations vary almost linearly with the frequency. Recent measurements of the compressional wave properties in sand sediments by Simpson *et al.* [11,12], made at frequencies between 3 and 100 kHz, are consistent with the theory, indicating weak logarithmic dispersion and an attenuation that varies linearly with frequency.

Similar behavior is displayed by compressional-wave data discussed by Buckingham and Richardson [13] from the SAX99 medium sand sediment in the Gulf of Mexico. Fig. 1 shows the SAX99 measurements along with the corresponding theoretical curves from Eqs. (1). Both theory and experiment show logarithmic dispersion, at a level of approximately 1% per decade, whilst the measured and predicted attenuations exhibit a near linear dependence on frequency.

Few data sets are available on the shear wave speed and attenuation as functions of frequency. Brunson and Johnson [14] made laboratory measurements of shear attenuation in a medium sand and found a near-linear dependence on frequency over the frequency range from 0.45 to 7 kHz, consistent with Eq. (2b). Their attenuation data are reproduced in Buckingham's [6] Fig. 2a.

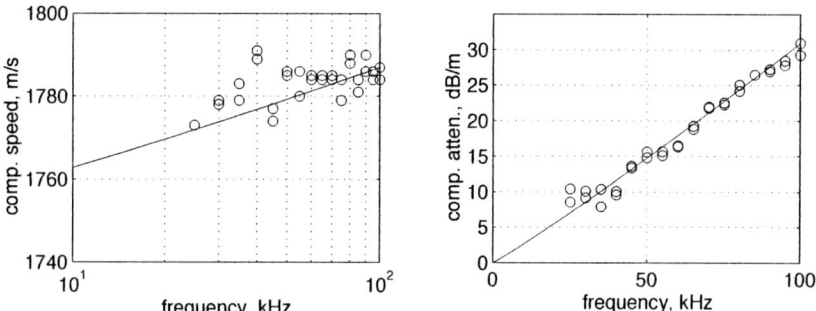

FIGURE 1. Sound speed and attenuation as functions of frequency. Note the semi-logarithmic and linear axes, respectively. Data are for the medium sand sediment at the SAX99 site.

DEPTH DEPENDENCE

Richardson *et al.* [15] have reported 400 kHz measurements of the compressional wave speed and attenuation as functions of depth in diver-collected core samples. Figures 2a and 2b show these data sets and for comparison the curves from Eqs. (1). The compressional speed ratio in Fig. 2a is the sound speed in the sediment normalized to that in seawater at the same temperature.

Below a depth of 10 cm, the predicted wave speed falls within the limits of the data in Fig. 2a, but slightly under-estimates the measurements at shallower depths. This small discrepancy may be due to ducting caused by the steep gradient in the sound speed profile near the interface, which could lead to the measurements being over-estimated.

The theoretical attenuation profile in Fig. 2b does not fall within the range of the widely distributed measurements but instead traces the lower limit of the data points. Such behavior is exactly as expected. The theory predicts the *intrinsic* attenuation, arising from the conversion of wave energy into heat. The data, on the other hand, represent the *effective* attenuation, which includes the intrinsic attenuation plus any additional sources of loss, due, for example, to scattering from inhomogeneities such as shell fragments that may be present in the medium. Thus the intrinsic attenuation should delineate the lower bound to the effective attenuation, as it does in Fig. 2b.

Shear wave speed and attenuation profiles have been reported by Richardson [16,17] for a sand sediment in the North Sea, at a site designated C1. Figures 2c and 2d show the comparisons of the data with the predictions of Eqs. (2). The theoretical shear speed follows the data points satisfactorily and, at depths below 10 cm, the theoretical intrinsic attenuation curve coincides with the lower bound of the effective attenuation data. At shallower depths, Richardson [16] has suggested that the data are under-estimated due to ducting in the steep sound speed profile.

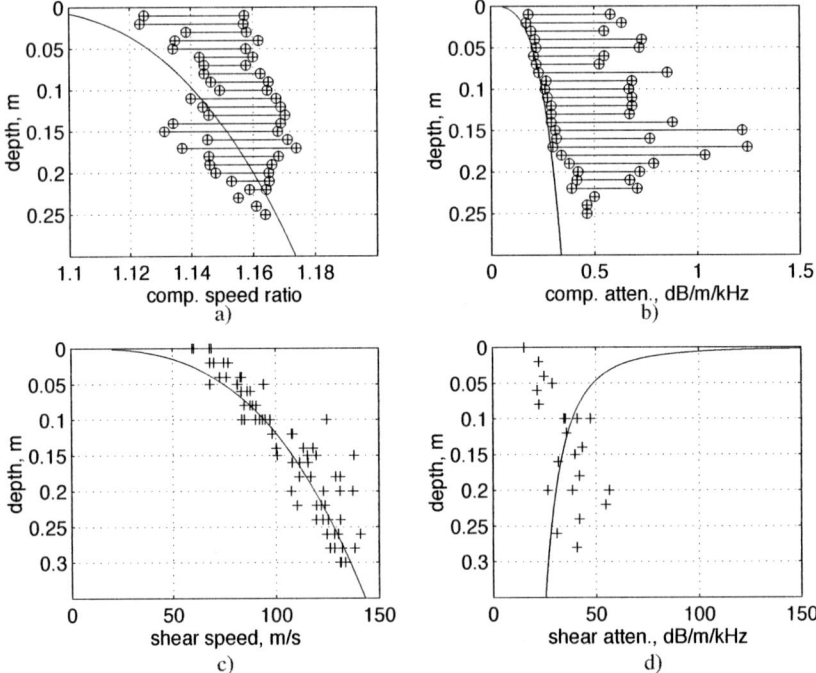

FIGURE 2. Depth dependence of a) the sound speed ratio, b) the sound attenuation, c) the shear speed, and d) the shear attenuation.

POROSITY DEPENDENCE

Richardson and colleagues have made numerous measurements, on core samples and *in situ*, of the phase speeds and attenuations of compressional waves and shear waves in a wide variety of siliciclastic sediments. The materials range from coarse sands to clays, spanning porosities from 0.37 to 0.9. Their core data on compressional wave properties are summarized in [5]. Figures 3a and 3b show mostly *in situ* data for the compressional wave speed (normalized to the sound speed in seawater) and attenuation versus porosity.

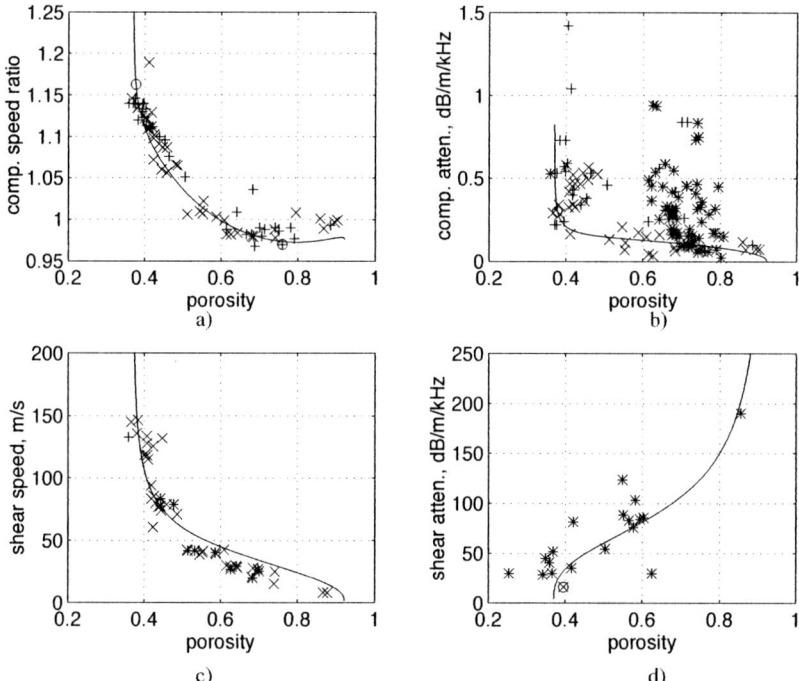

FIGURE 3. Porosity dependence of a) the sound speed ratio, b) the sound attenuation, c) the shear speed, and d) the shear attenuation.

In Fig. 3a, the theoretical dispersion curve accurately follows the trend of the data points throughout the full porosity range. Note the extremely high gradient of both the curve and the data in the low porosity region, $N < 0.4$. The broad theoretical minimum in the vicinity of $N = 0.8$ is primarily due to a similar minimum in the Wood's sound speed, c_o, appearing in the numerator of Eq. (1a).

The effective attenuation data in Fig. 3b are widely scattered, falling mostly above the theoretical curve, which represents the intrinsic attenuation. As in Fig. 2b, the theory traces out the lower bound of the attenuation data, as it should. Once again, a very high gradient appears in the low porosity region, where $N < 0.4$.

Figure 3c shows Richardson's shear-wave speed and attenuation data as functions of porosity. The theoretical curve for the shear speed represents well the trend of the data, with both theory and measurements showing an extremely high gradient when $N < 0.4$. Although fewer data on shear attenuation are available, Richardson [17] has reported a sufficient number of measurements to identify a trend, as can be seen in Fig. 3d. Consistent with the previous attenuation comparisons, the theoretical curve representing the intrinsic attenuation in Fig. 3d delineates the lower bound of the effective attenuation data (apart from one rogue point at $N \approx 0.63$).

GRAIN SIZE DEPENDENCE

Richardson's measurements of wave properties versus porosity usually include an estimate of the mean grain diameter, with the coarser materials corresponding to the lower porosities. As discussed by Buckingham [9], the compressional and shear wave dispersion relationships in Eqs. (1) and (2) match the grain-size data with a quality similar to that shown in the porosity plots of Fig. 3. Both the compressional and shear wave speed follow the trends of the grain-size data accurately; and both of the predicted intrinsic attenuations outline the lower bounds of the widely scattered, effective attenuation data points.

CONCLUDING REMARKS

Two theoretical dispersion pairs, one for the compressional wave and the second for the shear wave in a marine sediment, have been compared with extensive data sets on the wave properties of sediments. The theory, which indicates that the compressional and shear wave properties are coupled, has no adjustable parameters. All the comparisons of the theory with data - wave speeds and attenuations versus frequency, depth in the sediment, porosity and grain size - show a high level of agreement. It has been demonstrated that the theoretical curves accurately follow the trends of the data in every case. This agreement extends even to the attenuation: the theory, which predicts *intrinsic* attenuation, due to the conversion of wave energy into heat, traces out the lower bound to the highly scattered attenuation data points. The latter represent the *effective* attenuation, which includes additional loss mechanisms such as scattering from shell fragments and other inhomogeneities in the medium.

The precision of the theory of wave propagation in marine sediments suggests that it has practical application in characterizing the geoacoustic properties of the seabed. All that is needed is a single measurement of either a wave property, say the compressional wave speed, or a physical property such as the porosity from a core sample. The remaining geoacoustic parameters may then be computed from a few simple algebraic expressions, which include the two sets of dispersion relationships representing the wave properties of the compressional and the shear wave.

ACKNOWLEDGMENTS

This work was supported by Dr. Ellen Livingston, Ocean Acoustics Code, the Office of Naval Research, under grant number N00014-04-1-0063.

REFERENCES

1. M. A. Biot, *J. Acoust. Soc. Am.* **28**, 168-178 (1956).
2. M. A. Biot, *J. Acoust. Soc. Am.* **28**, 179-191 (1956).
3. E. L. Hamilton and R. T. Bachman, *J. Acoust. Soc. Am.* **72**, 1891-1904 (1982).
4. E. L. Hamilton, "Acoustic properties of sediments", in *Acoustics and the Ocean Bottom*, edited by A. Lara-Saenz, C. Ranz Cuierra, and C. Carbo-Fité, Consejo Superior de Investigacions Cientificas, Madrid, 1987, pp. 3-58.

5. M. D. Richardson and K. B. Briggs, "Empirical predictions of seafloor properties based on remotely measured sediment impedance". in *High Frequency Ocean Acoustics*, edited by M. Porter, M. Siderius, and W. A. Kuperman, (2004).
6. M. J. Buckingham, *J. Acoust. Soc. Am.* **108**, 2796-2815, (2000).
7. A. B. Wood, *A Textbook of Sound*, Bell and Sons, London, 1964.
8. S. P. Timoshenko and J. N. Goodier, *Theory of Elasticity*, McGraw-Hill, New York, 1970.
9. M. J. Buckingham, *J. Acoust. Soc. Am.* in preparation (2004).
10. M. J. Buckingham, *J. Acoust. Soc. Am.* **102**, 2579-2596 (1997).
11. H. J. Simpson and B. H. Houston, *J. Acoust. Soc. Am.* **107**, 2329-2337 (2000).
12. H. J. Simpson, *et al.*, *J. Acoust. Soc. Am.* **114**, 1281-1290 (2003).
13. M. J. Buckingham and M. D. Richardson, *IEEE J. Ocean. Eng.* **27**, 429-453 (2002).
14. B. A. Brunson and R. K. Johnson, *J. Acoust. Soc. Am.* **68**, 1371-1375 (1980).
15. M. D. Richardson, *et al.*, *IEEE J. Ocean. Eng.* **26**, 26-53 (2001).
16. Michael D. Richardson, private communication.
17. M. D. Richardson, "Variability of shear wave speed and attenuation in surficial marine sediments," in *Impact of Littoral Environmental Variability on Acoustic Predictions and Sonar Performance*, edited by N. G. Pace and F. B. Jensen, Kluwer, Dordrecht, 2002, pp. 107-114.

Empirical Predictions of Seafloor Properties Based on Remotely Measured Sediment Impedance

Michael D. Richardson and Kevin B. Briggs

Marine Geosciences Division, Naval Research Laboratory, Stennis Space Center MS 39529-5004

Abstract. Numerous acoustic systems have been developed over the past 25 years for remote classification of the seabed. Many systems use inversions of echo returns to estimate seafloor impedance and then use empirical relationships to predict other seabed properties from values of impedance. New regressions are presented, separately for siliciclastic and carbonate sediments, which allow prediction of sediment grain size, porosity, bulk density, percent sand and gravel and sound speed ratio and attenuation from values of an index of impedance (product of sound speed ratio and bulk density). This index is independent of pore water temperature and salinity and water depth. The regressions are based on nearly 800 cores collected from 67 shallow-water sites around the world (12 carbonate and 55 siliciclastic sites). Data are typically restricted to the upper 30 cm of sediment. The regressions based on the nearly 4,500 common data points from core measurements (3,922 for siliciclastic and 621 for carbonate sediments) do not vary significantly from the regressions for siliciclastic sediments first presented by Richardson and Briggs (1993) or between carbonate and siliciclastic sediments suggesting the empirical predictions universally apply to coastal sediments. Sound speed dispersion, sediment disturbance during core collection and measurement, inequalities between sample size (acoustic footprint vs. core diameter), spatial variability, and regression error all affect the accuracy of sediment property predictions.

INTRODUCTION

Many acoustic sediment classification systems use the amplitude of echo returns to estimate seafloor impedance. Empirical relationships between seafloor impedance and sediment physical properties are then used to map seafloor physical properties such porosity, bulk density, percent sand and gravel, or mean grain size and geoacoustic properties such as sound speed and attenuation. In this paper, we provide an update to the empirical relationships first given by Richardson and Briggs [1] to predict values of seafloor properties from a temperature-independent index of acoustic impedance. In the 1993 paper, the authors analyzed 1,243 measurements of impedance and physical properties from 211 cores collected from 22 sediment types at 11 siliciclastic sites. For this paper the data set has been expanded to over 4,500 measurements from 67 shallow water sites, with both siliciclastic and carbonate sediments represented (Tables 1 and 2). Our first objective is to determine if the almost-5-times-larger data set collected over a wider range of sediment types yields differences in empirical relationships between an index of impedance and related sediment physical and geoacoustic properties. The second objective is to determine whether these empirical regressions yield different predictions from carbonate and siliciclastic sites as suggested by Richardson and colleagues from the analyses of sediments collected from the Florida Keys [2].

METHODS FOR SEDIMENT COLLECTION AND LABORATORY DATA ANALYSES

Sediment geoacoustic and physical property measurements were made from sediments collected with 45-cm-long, 5.9-cm-inside-diameter, clear, polycarbonate coring tubes. Most sediments were collected by divers but sediments collected from eight sites (Montauk Point, Quinault Range, Arafura Sea, Russian River, Eel River, North Sea , TOSSEX, and Straits of Juan de Fuca), which were too deep for diving operations, were subsampled from $0.25 m^2$ spade box cores. Cores were capped at both ends immediately after collection to retain the overlying water and kept in an upright position during transport to the laboratory for analysis. Collection, measurement, and handing procedures were designed to minimize sampling disturbance and to maintain an intact sediment-water interface within the coring tube.

Sound speed and attenuation were measured on sediment at 1-cm intervals within the core tubes, usually within 24 hours of collection, using time-of-flight and amplitude of pulsed 400-kHz sine waves transmitted across the core tube [3]. Sediment sound speed is calculated from the differences in time-of-flight between sediment and distilled water within identical core tubes, the measured inside diameter of the core tube (5.9 cm), and the sound speed within the distilled water. Attenuation is measured as 20 log of the ratio of the mean amplitude of the waveform transmitted through water to those transmitted through sediment. Sound speeds are reported as the unitless sound speed ratio (V_p ratio) which is the ratio of measured sound speed to the sound speed of pore water at the same temperature, salinity and pressure. Attenuation is expressed in units of dB $m^{-1} kHz^{-1}$ (k) after Hamilton [4].

Sediments were then extruded from sediment cores and sectioned at 2-cm intervals to determine sediment porosity and grain size distribution. Porosity was determined from weight loss of sediments dried at 105° C for 24 hours and corrected for residual salt. Grain density was determined using a pyncnometer. Sediment bulk density was calculated from the porosity and densities of pore water and sediment grains. Sediment grain size was determined from disaggregated samples by dry sieving for sand-sized particles and by either pipette methods or Micromeritics sedigraph for silt- and clay-sized particles.

Sediment impedance (Z, kg $m^{-1} s^{-1}$) is the product of sediment sound speed and bulk density. Sediment sound speed is dependent on pore water temperature and salinity and pressure (water depth). Furthermore, sound speed in sediment at a single site can vary up to 10% over the range of seasonal conditions expected in coastal waters [1]. Therefore, the pore-water-independent Index of Impedance (IOI), which is the product of the sediment bulk density and velocity ratio, is used to calculate empirical relationships between sediment impedance and other sediment physical properties.

RESULTS AND DISCUSSION

Sediment physical and geoacoustic properties were measured on over 800 cores collected from 67 shallow-water sites around the world (12 carbonate and 55 siliciclastic sites). The total of 4,582 collocated measurements is nearly 5-times the number of measurements used by Richardson and Briggs [1] to determine similar empirical relationships between the index of impedance (IOI) and sediment physical and geoacoustic properties and includes measurements in carbonate sites (609) as well as siliciclastic sites (3973). Sediment at the 55 siliciclastic sites ranged from very-high-porosity clays (such as Eckernförde Bay, Baltic Sea or St. Andrew Bay, Florida) to coarse sands in the northeastern Gulf of Mexico (Table 2). Siliciclastic sampling sites were generally associated with high-frequency acoustic bottom scattering experiments and include sites in the Mediterranean, Baltic and North Seas, and along the entire range of Atlantic, Pacific and Gulf coasts of the US [1,3,5]. Carbonate sampling sites are geographically restricted to tropical waters along the southern coastline of Florida [2] and Hawaii but, nevertheless include several sediment types (Table 1).

The Index of Impedance (*IOI*) provides excellent predictions of sound speed ratio, bulk density, and porosity for both carbonate and siliciclastic sediments (Figures 1 and 2; Tables 3 and 4). This is not surprising as *IOI* is the product of velocity ratio (V_pR) and bulk density, and both sediment bulk density and porosity are determined from the same wet loss measurements. Predictions of V_p, V_pR, bulk density and porosity for carbonate and siliciclastic sediments vary less than 14 m s^{-1}, 0.001, 0.04 g cm^{-3}, 4% respectively, over the full range of values of *IOI* suggesting regressions for each parameter derived from the entire data set is appropriate (Table 5). The coefficients of determination (r^2) between *IOI* and sediment mean grain size and percent sand and gravel are lower for carbonates than siliciclastic sediments. The lower values of r^2 between *IOI* and grain size properties due to scatter in the data justify combined regressions using all carbonate and siliciclastic data in spite of up to 0.6 phi and 17% differences in predicted mean grain size and percent sand and gravel. Based on the data presented, attenuation is poorly predicted from impedance.

TABLE 1. Mean values of sediment physical and geoacoustic properties from carbonate sites located in southern Florida and in Hawaii. Sediment properties include sound speed (V_p, m s^{-1}), sound speed ratio (V_pR, no units), attenuation (α, dB m^{-1}; k, α kHz^{-1}), mean grain size (M_z, phi) porosity (η, %), density (ρ, g cm^{-3}) and the Index of Impedance (*IOI*, g cm^{-3}). Sites are ordered as increasing values of *IOI*.

Site	V_p	V_pR		M_z			k	IOI	Sediment
Hawaii/mud	1495.3	0.977	68.6	8.67	84.02	1.296	0.171	1.267	calc. silty clay
MarqKeys	1555.6	1.017	391.3	6.15	59.66	1.726	0.978	1.755	calc. s-s-clay
SG98-5	1560.8	1.020	322.3	5.85	59.59	1.748	0.806	1.783	calc. s-s-clay
DTortugas	1561.8	1.021	343.0	6.62	59.00	1.755	0.858	1.792	calc. s-s-clay
LFK/fine	1581.3	1.034	365.8	5.40	57.19	1.759	0.914	1.818	calc. s-s-clay
Hawaii-4	1609.7	1.052	246.2	3.88	56.42	1.771	0.615	1.864	calc. silty sand
SG98-2	1669.4	1.091	383.1	1.57	49.47	1.921	0.958	2.096	crse. skel. sand
Hawaii/crse	1639.4	1.072	695.2	0.74	45.18	1.960	1.738	2.100	crse. coral sand
Hawaii-2	1671.6	1.093	438.3	2.33	47.68	1.933	1.096	2.112	calc. med. sand
LFK/crse	1704.7	1.114	488.9	0.54	41.97	2.054	1.222	2.289	crse. coral sand
RebShoal	1733.1	1.133	279.1	1.26	43.85	2.022	0.698	2.290	carbonate sand
SG98-3	1777.3	1.162	236.7	1.66	40.92	2.067	0.592	2.401	ooid/skel. sand

TABLE 2. Mean values of sediment physical and geoacoustic properties from 55 siliciclastic sites world-wide. Sediment properties include sound speed (V_p, m s^{-1}), sound speed ratio (V_pR, no units), attenuation (α, dB m^{-1}; k, α kHz^{-1}), mean grain size (M_z, phi) porosity (η, %), density (ρ, g cm^{-3}) and the Index of Impedance (*IOI*, g cm^{-3}). Sites are ordered as increasing values of *IOI*.

Site	Vp	VpR		M_z			k	IOI	Sediment Type
SABay	1518.9	0.993	38.7	10.94	89.14	1.170	0.097	1.162	clay
Eck93	1515.5	0.991	72.3	9.88	87.40	1.188	0.181	1.177	silty clay
CLBight	1521.9	0.995	114.0	8.10	86.50	1.223	0.285	1.216	silty clay
JDF7	1507.2	0.985	114.2	8.50	83.43	1.313	0.285	1.294	silty clay
LISound	1503.1	0.982	—	7.64	76.64	1.411	—	1.386	clayey silt
Orcas	1511.9	0.988	179.1	8.08	75.22	1.403	0.448	1.387	clayey sand
Diga	1480.4	0.968	58.0	10.05	69.12	1.506	0.145	1.458	silty clay
JDF4	1521.7	0.995	206.2	6.93	74.35	1.470	0.517	1.462	glacial till
Arafura	1511.4	0.988	347.8	5.24	71.63	1.494	0.869	1.476	clayey sand
Portovenere	1501.7	0.982	66.2	9.45	68.30	1.546	0.166	1.518	silty clay
STeresa	1502.4	0.982	122.3	8.78	66.98	1.569	0.306	1.541	silty clay
RussRiver	1545.5	1.010	231.8	6.35	64.35	1.597	0.579	1.613	clayey sand
Viareggio	1511.3	0.988	99.5	8.98	61.74	1.634	0.249	1.615	silty clay
Eck93	1609.7	1.052	210.7	4.59	59.38	1.659	0.527	1.745	sand-silt-clay
EelRiver	1554.6	1.016	190.7	7.17	57.32	1.745	0.477	1.773	clayey silt
ATB/G40	1651.9	1.080	219.8	2.56	56.61	1.716	0.549	1.853	fine sand
JDF1	1617.6	1.057	238.5	4.37	55.37	1.800	0.596	1.903	silty fine sand
Tellaro	1614.4	1.055	184.7	6.08	50.70	1.820	0.462	1.921	sand-silt-clay
Monasteroli	1652.4	1.080	220.2	5.12	46.62	1.891	0.550	2.042	sand-silt-clay
JDF6	1668.2	1.090	314.3	2.94	47.56	1.922	0.786	2.096	fine sand/s-s-c
Tirrenia	1683.1	1.100	127.6	3.72	45.76	1.906	0.319	2.097	v.fine sand
VAzzura	1686.4	1.102	156.5	4.14	45.17	1.911	0.391	2.106	muddy sand
SG98-6	1649.6	1.078	632.5	0.08	43.47	2.001	1.581	2.158	shell/coral hash
JDF5	1701.5	1.112	213.8	2.31	45.44	1.946	0.534	2.164	fine sand/s-s-c
LTB	1716.8	1.122	317.1	2.54	43.57	1.929	0.793	2.165	fine sand
Quinault	1709.3	1.117	177.2	2.94	41.76	1.971	0.443	2.202	fine sand
PC93	1708.5	1.117	404.0	0.98	40.93	2.008	1.010	2.242	coarse sand
PCII	1716.4	1.122	391.2	0.85	41.09	2.000	0.978	2.244	c. sand/sh. hash
TBay/crse	1754.2	1.147	610.2	1.36	44.85	1.966	1.526	2.254	coarse/fine sand
KB/lyn	1709.2	1.117	586.9	0.90	40.14	2.020	1.467	2.256	shell hash
Charl/fine	1728.4	1.130	281.0	1.97	39.94	2.001	0.703	2.260	fine sand
SG98-10	1752.1	1.145	164.1	1.62	40.69	1.979	0.410	2.266	medium sand
Charl/crse	1729.1	1.130	308.1	1.44	39.63	2.006	0.770	2.267	medium sand
PC84	1742.9	1.139	241.7	2.61	40.08	1.998	0.604	2.276	fine sand
SWEAT	1747.6	1.142	213.3	2.23	40.38	2.007	0.533	2.292	fine sand
SG98-9	1747.1	1.142	206.7	1.56	39.45	2.010	0.517	2.295	medium sand
TBay/fine	1746.0	1.141	206.1	2.92	40.16	2.013	0.515	2.297	fine sand
ATB/B14	1752.6	1.146	107.2	2.15	39.52	2.006	0.268	2.298	fine sand
SG98-1	1713.0	1.120	430.2	0.84	40.66	2.053	1.076	2.299	shell hash
IRB	1745.2	1.141	281.2	1.77	40.63	2.023	0.703	2.307	medium sand
SG98-8	1747.1	1.142	265.7	2.14	39.65	2.026	0.664	2.314	shelly fine sand
PCB I&II	1755.1	1.147	176.1	2.34	39.72	2.018	0.440	2.315	fine sand
NS	1735.0	1.134	226.1	1.87	41.07	2.046	0.565	2.320	medium sand
MVCO	1755.1	1.147	154.5	2.52	38.49	2.028	0.386	2.327	fine sand
PCB99	1764.2	1.153	133.5	2.24	39.33	2.020	0.334	2.329	fine sand
MonPt	1744.4	1.140	92.1	2.04	37.21	2.045	0.230	2.332	fine sand
KB/bar	1758.2	1.149	254.4	1.33	37.28	2.047	0.636	2.352	medium sand
Duck	1758.8	1.150	116.2	2.53	39.54	2.051	0.291	2.357	fine sand
JDF2	1771.6	1.158	179.5	2.03	39.10	2.039	0.449	2.361	medium sand
PE99	1770.7	1.157	153.0	1.28	37.08	2.052	0.383	2.375	medium sand
PE00	1774.1	1.160	149.5	1.21	37.32	2.050	0.374	2.377	medium sand
SAX99	1766.3	1.154	177.5	1.27	37.27	2.066	0.444	2.385	medium sand
NoSea	1779.0	1.163	155.7	1.93	37.56	2.054	0.390	2.388	med/fine sand
TOSSEX	1762.7	1.152	161.8	1.93	35.64	2.075	0.404	2.391	med/fine sand
HoodCanal	1767.1	1.155	184.6	1.34	36.46	2.108	0.462	2.435	medium sand

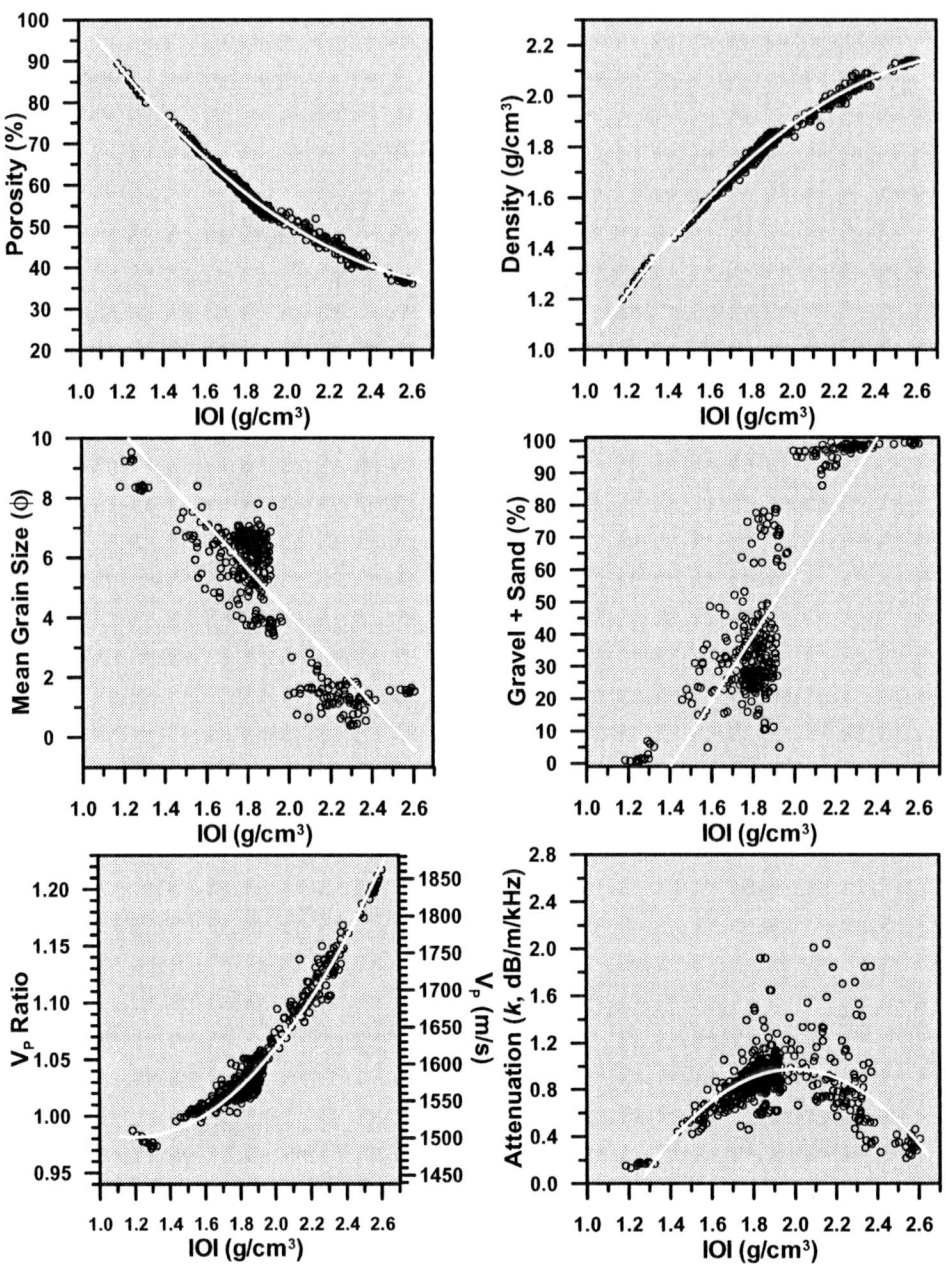

FIGURE 1. Empirical relationships used to predict sediment physical and acoustic properties from the Index of Impedance *(IOI)* for carbonate sediments. Data and regressions (Table 1, Table 3) are based on 69 cores collected from 12 sites around southern Florida and in the Hawaiian Islands.

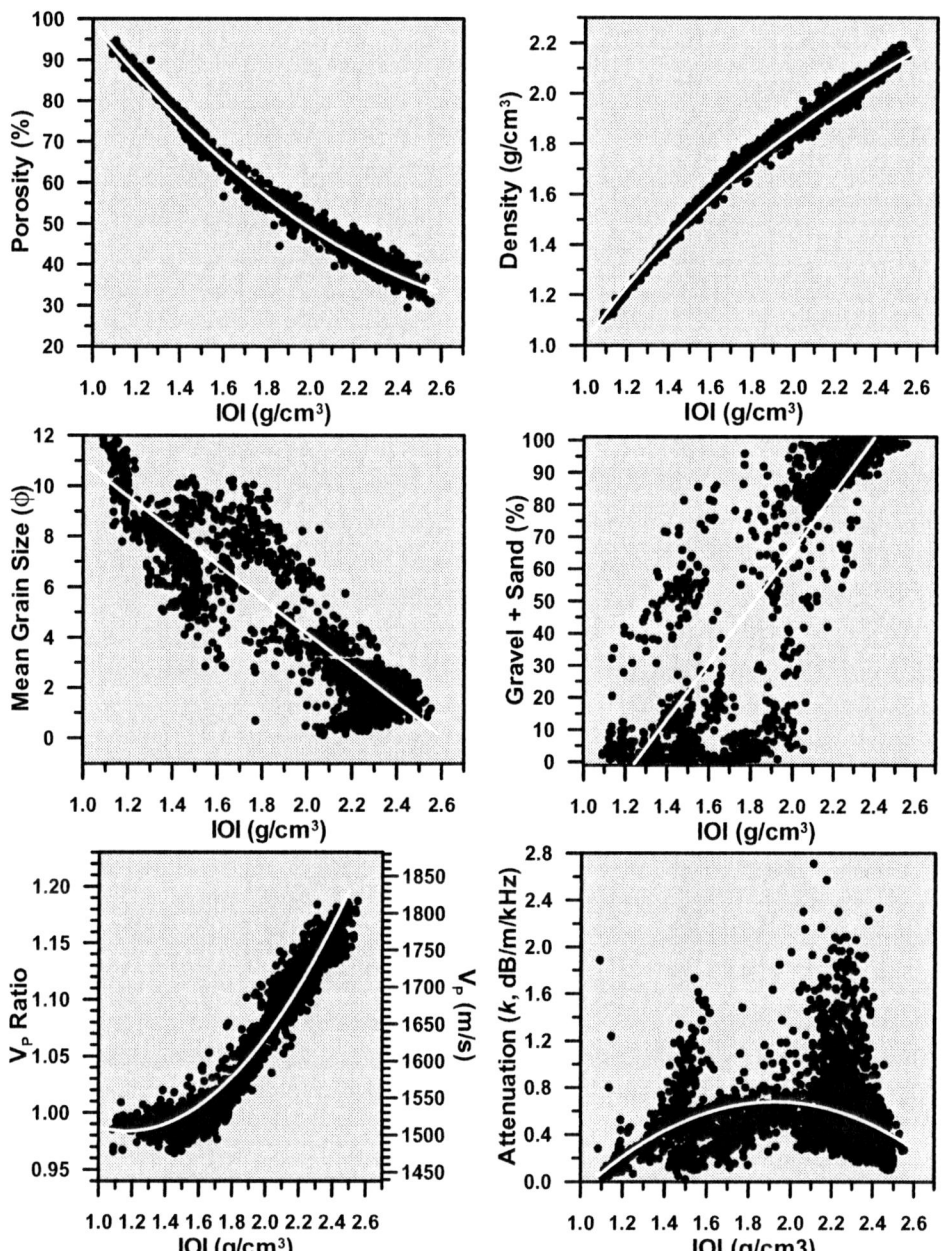

FIGURE 2. Empirical relationships used to predict sediment physical and acoustic properties from the Index of Impedance (*IOI*) for siliciclastic sediments. Data and regressions (Table 2; Table 4) are based on over 3,900 measurements made on cores collected from 55 shallow-water sites world-wide.

TABLE 3. Empirical Predictive Relationships for Sediment Physical and Geoacoustic Properties Based on the Index of Impedance (IOI) for Carbonate Sediments. Coefficient of determination (r^2) is given for each regression.

Parameter	Regression	r^2
Sound Speed Ratio	= 1.164 - 0.3001(IOI) + 0.1253(IOI)^2	0.96
Attenuation (k)	= -5.96 + 6.94(IOI) – 1.174(IOI)^2	0.43
Porosity (%)	= 186.18 – 102.20(IOI) + 17.29(IOI)^2	0.99
Density (g cm^{-3})	= -0.52 + 1.81(IOI) – 0.305(IOI)^2	0.99
Mean Grain Size (θ)	= 19.3 -7.6(IOI)	0.75
Sand and Gravel (%)	= -143.2 + 101.4(IOI)	0.73

TABLE 4. Empirical Predictive Relationships for Sediment Physical and Geoacoustic Properties Based on the Index of Impedance (IOI) for Siliciclastic Sediments. Coefficient of determination (r^2) is given for each regression.

Parameter	Regression	r^2
Sound Speed Ratio	= 1.149 - 0.2821(IOI) + 0.1203(IOI)^2	0.97
Attenuation (k)	= -2.61 + 3.41(IOI) – 0.885(IOI)^2	0.16
Porosity (%)	= 178.60 – 94.60(IOI) + 14.86 (IOI)^2	0.99
Density (g cm^{-3})	= 1.01 + 1.22LN(IOI)	0.99
Mean Grain Size (θ)	= 17.7 -6.8(IOI)	0.85
Sand and Gravel (%)	= -109.6 + 87.7(IOI)	0.82

TABLE 5. Empirical Predictive Relationships for Sediment Physical and Geoacoustic Properties Based on the Index of Impedance (IOI) for Siliciclastic and Carbonate Sediments Combined. Coefficient of determination (r^2) is given for each regression.

Parameter	Regression	r^2
Sound Speed Ratio	= 1.164 - 0.3001(IOI) + 0.1253(IOI)^2	0.97
Attenuation (k)	= -3.31 + 4.33 (IOI) - 1.138 (IOI)^2	0.22
Porosity (%)	= 174.16 - 89.12(IOI) + 13.37 (IOI)^2	0.99
Density (g cm^{-3})	= 1.02 + 1.21LN(IOI)	0.99
Mean Grain Size (θ)	= 17.9 - 6.0(IOI)	0.84
Sand and Gravel (%)	= -113.4 + 89.1(IOI)	0.81

CONCLUSIONS

The Index of Impedance (*IOI*) can be used to predict accurately sound speed, density, and porosity in seafloor sediments and, with a lesser degree of accuracy, predict mean grain size and percent sand and gravel. The lower values of the coefficient of determination (r^2) between *IOI* and mean grain size (percent sand and gravel) compared to sediment bulk density, porosity, or sound speed reflect the lack of fundamental physical relationship between mean grain size and either sediment bulk density or sound speed (Fig. 3). The coefficients of determination (r^2) are 0.84 and 0.64, respectively. In muddy sediments, consolidation (dewatering) lowers porosity and increases density without a change in mean grain size. In sands, porosity can vary up to 10%, depending on packing [7]. Given the same packing a uniform assemblage of spheres would theoretically achieve the same porosity regardless of grain diameter (size). Using values of mean grain size as an index, especially in the silt-size range, may be very misleading because of major differences in sorting (standard deviation of the particle size distribution). Well-sorted sediment composed of wholly silt-size particles may have the same mean grain size as poorly sorted sediment with a mixture of sand- and clay-size particles. The resultant density and sound speed of these two sediments, however, might be very different. Given the aforementioned issues, it is perhaps amazing that empirical regressions between grain size-related parameters and sediment density, porosity, sound speed, or impedance have any predictive value.

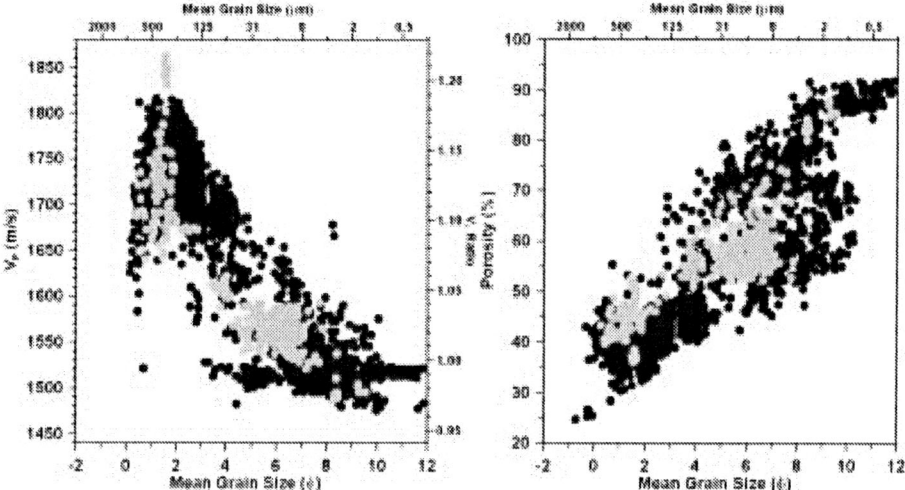

FIGURE 3. Scatter diagrams of mean grain size versus porosity and sound speed (V_p). The lighter colored symbols, which represent carbonate sediments, overlay the darker colored symbols, which represent siliciclastic sediment. Mean grain sizes less than 4 phi are in the sand grain size, 4-8 phi are silt sized particles and greater than 8 phi are clay-sized particles.

The use of different empirical *IOI* regressions for carbonate sediments than for siliciclastic sediments may be justified for some specific carbonate sites where intra-particulate porosity is high [2] but is not justified for more generalized relationships for all coastal sediments, based on the data presented here. Therefore, the regressions based on the combined data set (Table 5) are recommended for general use. The regressions presented by Richardson and Briggs [1] in 1993 do not significantly differ from those regressions developed from the much larger data sets from siliciclastic sediments used here (Tables 4 and 5). Mean absolute differences between the 1993 and 2004 *IOI* regressions were as follows: 1.4% for porosity, 0.0108 g cm^{-3} for bulk density, 0.13 phi for mean grain size, 0.0025 for velocity ratio and 3.8 m s^{-1} for sound speed. A regression between IOI and percent sand and gravel was not calculated by

Richardson and Briggs [1]. Attempts to predict compressional wave attenuation from *IOI* at these high acoustic frequencies (400 kHz) have failed because of the high, but unknown, contribution of scattering to the overall measured attenuation. Intrinsic attenuation probably does not exceed the lower level of the curvilinear fits given in Figures 1 and 2. However, it is notable that attenuation in the carbonate sediments is on average 0.14 dB m^{-1} kHz^{-1} (56 dB m^{-1} @ 400 kHz) higher than for siliciclastic sediments.

Poro-elastic models predict that sound speed is dispersive, especially in sandy sediments [6]. The empirical relationships presented here were developed using sound speeds measured at 400 kHz. Typical echo sounders operate at 3.5 to 30 kHz, where sound speeds and thus impedance values may be lower. This dispersion effect is more pronounced in sand compared to muddy sediments. In the example given by Williams et al [7] for the sand sediment of the SAX99 experiments, measured sound speeds were 25-75 m/s higher at 400 kHz than over the 3.5- to 30-kHz frequency band. The calculated values of IOI, given the mean sediment density of 2.066 g cm^{-3}, would be 2.4 g cm^{-3} at 400 kHz and 2.3 g cm^{-3} at 3.5 kHz. Based on this amount of sound speed dispersion, sediment properties predicted at 3.5 kHz would be different than from measured sound speeds (400 kHz): porosity is 2.6% higher, bulk density is 0.05 g/cm^3 lower, mean grain size is 0.69 phi units higher (finer), and sound speed is 44 m/s higher.

ACKNOWLEDGMENTS

We acknowledge the considerable help of Richard I. Ray, the NORDA/NRL diving officer, who has been our indispensable diving companion for the past 20 years. The data summarized in this paper was collected over the past 25 years with financial support of the Office of Naval Research, Naval Research Laboratory, and the SACLANT Undersea Research Centre. This paper is NRL contribution number NRL/PP/7430-04-6.

REFERENCES

1. Richardson, M. D., and Briggs, K. B., "On the use of acoustic impedance values to determine sediment properties", in *Acoustic Classification and Mapping of the Seabed*, edited by N.G Pace and D.N. Langhorne, Proceedings of the Institute of Acoustics 15(2): University of Bath, Bath UK, 1993, pp. 15-24
2. Richardson, M.D., Lavoie, D.M. and Briggs, K.B. 1997. "Geoacoustic and physical properties of carbonate sediments of the Lower Florida Keys", *Geo-Marine Letters* **17**:316-324 (1997).
3. Richardson, M.D. and Briggs, K.B. "In-situ and laboratory geoacoustic measurements in soft mud and hard-packed sand sediments: Implications for high-frequency acoustic propagation and scattering." *Geo-Marine Letters* **16**:196-203 (1996).
4. Hamilton, E.L. "Prediction of in-situ acoustic and elastic properties of marine sediments", *Geophysics* 36:266-289 (1971).
5. Richardson, M.D., Briggs, K.B., Bentley, S.J. Walter, D.J. and T.H. Orsi, T.H. "The effects of biological and hydrodynamic processes on physical and acoustic properties of sediments off the Eel River, California. *Marine Geology* **182**:121-139 (2002).

6. Stoll, R.D. "Velocity dispersion in water-saturated granular sediment" *Journal Acoustical Society of America* **111(2)**:785-793 (2002).
7. Williams, K.L.., Jackson D.R., Thorsos, E.L., Tang, D. and Schock S.G. "Comparison of sound speed and attenuation measured in a sandy sediment to predictions based on Biot theory of porous media" *IEEE Journal of Oceanic Engineering* **27(3)**:413-428 (2002).

High-Frequency Geoacoustic Inversion of Ambient Noise Data Using Short Arrays

Martin Siderius* and Chris Harrison †

*Center for Ocean Research, SAIC, 10260 Campus Point Drive, San Diego, CA, 92121
†NATO Undersea Research Centre 400 Viale San Bartolomeo, La Spezia Italy, 19138

Abstract. Ocean ambient noise is generated in many ways such as from winds, rain and shipping. A technique has recently been developed (Harrison and Simons, J. Acoust. Soc. Am, Vol. 112 no. 4, 2002) that uses the vertical directionality of ambient noise to determine seabed properties. It was shown that taking a ratio of upward looking beams to downward produces an estimate of the reflection loss. This technique was applied to data in the 200–1500 Hz band using a 16-m vertical array. Extending this to higher frequencies allows the array length to be substantially shortened and greatly reduces interference from shipping. If array lengths can be reduced to about 1 m then it may be possible to hull-mount or tow such an array from a surface ship or submerged vehicle (e.g. an autonomous underwater vehicle). Although this seems attractive the noise is primarily generated by wind which in turn causes a rough sea-surface and bubbles and these factors combined with increased volume attenuation may degrade this type of reflection loss estimate at high frequencies. In this paper, we examine measured noise data from the October 2003 ElbaEx experiment using a 5.5 m array in the 1–4 kHz frequency band. Results indicate the noise field is predictable with modeling and the ratio of upward looking to downward looking beams produces an approximation to the reflection loss which can be inverted for seabed properties. For short arrays (a 1 m aperture is considered here), the beamforming is not ideal over a broad-band of frequencies. The beams are broadened and this leads to an up/down ratio that does not produce a good estimate of reflection loss. This can be especially problematic at low grazing angles which is the part of the reflection loss curve that is often most important to estimate correctly. Techniques will be presented for mitigating the impact of beamwidth and grating lobes on estimating the seabed properties.

INTRODUCTION

Using measurements of ocean ambient noise to produce an estimate of seabed properties is attractive for several reasons. 1) Since ambient noise results from wind and rain interacting with the sea-surface the sound sources exist everywhere. 2) This sheet source provides an angular spread of plane-waves that have interacted with the bottom and therefore contain information about seabed properties. 3) Passive measurements not requiring a sound projector greatly simplify the design of an experiment or survey technique. 4) With concerns over the impact of sound on marine mammals, an environmentally friendly geoacoustic inversion method that does not require a human-made sound source is highly attractive.

Although the dependency of ambient noise on seabed properties has been widely reported, only recently has a method been developed that uses vertical directionality of ambient noise data to produce the bottom power reflection loss. This was demonstrated for several sites using a 16 m vertical aperture for frequencies of 200–1500 Hz [1].

The method uses a ratio between beams traveling from the direction of the surface to those coming from the seabed and in theory this ratio equals the bottom power reflection loss. Actual beamforming introduces beam widths and this can be a big problem as array lengths become short (relative to wavelength) or when hydrophone spacing is larger than half a wavelength and grating lobes are introduced that erroneously mix the up and down going beams. Beamforming and separating up and downgoing beams can be especially problematic near grazing angles which is important for longer range propagation.

For practical applications, it is difficult to cover large areas using long vertical arrays. It is possible to have long vertical arrays drift to determine bottom properties and even a sub-bottom profile [2]. However, it is difficult to tow them in a specific pattern such as during a multibeam bathymetry survey. For this, short arrays are attractive since these can be towed or mounted on the hull of a surface ship, or these could be mounted on an autonomous underwater vehicle (AUV). If short vertical apertures are feasible it may even be possible to use the slight tilt in long towed arrays. As vertical aperture shrinks, it is natural to shift to higher frequencies and this has the added benefit of operating outside the frequency band dominated by shipping. While distant shipping noise does not generally cause a major concern, nearby shipping can interfere with the beamformed output in such a way that the reflection loss estimates are not valid.

In this paper we examine using high-frequency (1–4 kHz) noise data for obtaining reflection loss and geoacoustic properties. Further, we explore the feasibility of using very short vertical apertures (1–5 m). We begin by describing noise modeling— this is useful to illustrate how beamforming ambient noise can produce reflection loss and also is the basis for the inversion technique for determining geoacoustic properties of the seabed. This noise modeling is appropriate for high frequencies and broad band data since it is built on a ray approach. The next section gives a short description of the beamforming issues that, in some cases, hinders obtaining the ambient noise inversion. Finally, data from the recent ElbaEx experiment [3] is used to demonstrate the inversion technique.

HIGH FREQUENCY AMBIENT NOISE MODELING AND ESTIMATING REFLECTION LOSS

A variety of methods exist for modeling the ocean's acoustic ambient noise field [4]. Probably the most widely used methods are normal modes (Kuperman-Ingenito) or wavenumber integration [5, 6]. These wave solutions are not ideal for high frequency, or broadband calculations since they become too computationally intensive. Since we are considering both broad-band and high-frequency data for inversion these methods are not attractive. Here, a simpler ray approach is used which involves far fewer calculations yet provides (arguably) a better solution than full-wave models [7]. A broad-band of frequencies can be computed at arbitrarily high frequency in a fraction of the time needed for wave calculations and this type of calculation is included in the noise coherence model CANARY [8, 9].

A ray based derivation will be presented here for the noise cross-spectral density function that follows Harrison [7] but simplified by only considering vertically separated

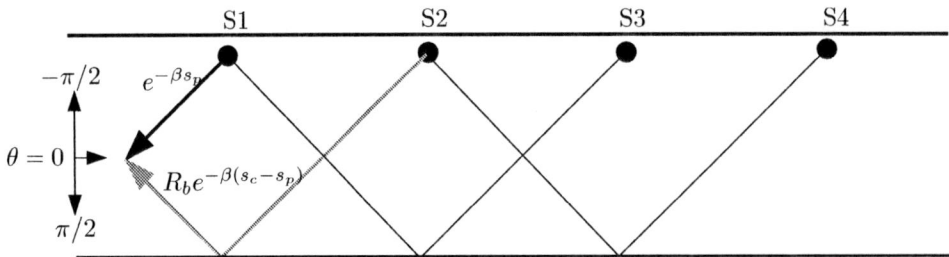

FIGURE 1. The geometry for noise sources near the surface. The sources shown are those that would arrive at a location from equal angles looking toward the surface and towards the seabed. A full cycle distance is indicated by s_c and partial (from S1 to receiver) by s_p and bottom reflection loss by R_b. The coordinate system for up and down looking angles is shown on the left side of the diagram.

hydrophones in an iso-sound speed water column (these assumptions are not required; but making them allows for a more clear derivation to illustrate the method).

The acoustic field at frequency ω can be calculated from ray amplitudes and arrivals at a receiver depth, z_r and range r according to,

$$P(\omega, z_r, r) = \sum_{n=1}^{N} A_n e^{i\omega D_n}, \qquad (1)$$

where A_n are the arrival amplitudes for the nth eigenray and D_n the corresponding delays. The noise cross-spectral density function between two vertically separated hydrophones at depths z_1 and z_2 ($z_2 > z_1$) is,

$$C_\omega(z_1, z_2) = \int_0^{2\pi} \int_0^\infty P(z_1) P^*(z_2) g^2(\theta) r dr d\phi \qquad (2)$$

where $g(\theta)$ is a term that allows for the noise sources to have directionality. If the noise sources are dipoles and we assume azimuthal symmetry,

$$C_\omega(z_1, z_2) = 2\pi \int_0^\infty \sum_n |A|^2 e^{ik(z_2 - z_1)\sin\theta} \sin^2\theta r dr. \qquad (3)$$

In eq. (3), the cross terms of the double sum have been ignored. The final step is to include the amplitude term which for each path is,

$$|A|^2 = \frac{\cos\theta}{r|(dr/d\theta)\sin\theta|} Q P_m \qquad (4)$$

where Q and P_m are needed for the volume and boundary losses. The value of these terms are illustrated using Fig. 1 where four noise sources are shown that contribute to a receiver in the water column from the same angles in the upward and downward directions.

The first source has only volume losses to the receiver, $Q = e^{-\beta s_p}$, where β is the volume loss per distance and s_p is the ray partial cycle distance. The losses from the second source include a longer path and a bottom interaction: $Q = e^{-\beta(s_c-s_p)}R_b$, where s_c is the complete ray cycle distance and R_b is the bottom reflection loss. Each of the next sources also include losses from one or more full cycle distance which results in a geometric series for the compounding losses. That is, $P_m = \prod_{m=1}^{m} R_b e^{-\beta s_c}$, or, for angles between $-pi/2$ to 0 (sources S3, S5, S7 ...),

$$Loss_{-\pi/2 \to 0} = e^{-\beta s_p}\{1 + R_b e^{-\beta s_c} + (R_b e^{\beta s_c})^2 + ...\} = e^{-\beta s_p}\frac{1}{1-R_b e^{-\beta s_c}}. \quad (5)$$

For angles between 0 to $pi/2$ (sources S2, S4, S6 ...),

$$Loss_{0 \to \pi/2} = e^{-\beta(s_c-s_p)}R_b\{1 + R_b e^{-\beta s_c} + (R_b e^{-\beta s_c})^2 + ...\} = e^{-\beta(s_c-s_p)}R_b \frac{1}{1-R_b e^{-\beta s_c}}. \quad (6)$$

Combining these, the cross-spectral density can be written,

$$C_\omega(z_1,z_2) = 2\pi \int_0^{\pi/2} \frac{1}{1-R_b e^{\beta s_c}} \cos\theta \sin\theta \{e^{ik(z_2-z_1)\sin\theta} + R_b e^{-ik(z_2-z_1)\sin\theta}\}d\theta, \quad (7)$$

where the small partial cycle distance volume attenuation terms have been suppressed. Note in eq. (7) the *rdr* term has cancelled and the integration is over angle θ. There are two plane-waves in eq. (7); one traveling from the surface down and one from the seabed up. These differ by exactly the term R_b, the power reflection coefficient. Therefore, if beamforming is ideal and these plane-waves could be extracted and corresponding up and down angles divided, then the reflection loss at each angle would result. Note, that even though several assumption were made (such as dipole sources) these are not important factors in the final result since these terms would cancel out when dividing up and down plane-waves.

Vertically beamforming from short or undersampled arrays

As mentioned, if beamforming is ideal the downward traveling plane-waves can be separated from those traveling upward and taking the ratio exactly produces the power reflection loss of the seabed. If array length and hydrophone spacing are sufficient the beamforming is close enough to ideal and the technique produces a good estimate of reflection loss. In general, we desire to have the seabed properties characterized over a broad-band of frequencies. Collecting the broad-band ambient noise data is not a problem, but beamforming and dividing up and down plane-waves can be. Consider a 32 element array designed for 4 kHz (hydrophone spacing of 0.18 m and length of 5.58 m). In the top two panels in Fig. 2 the beampattern is shown for 4 kHz looking both broadside and endfire. Looking broadside, the beam is narrow, about $\pm 2°$ at 6 dB down, and looking endfire about $\pm 5°$. Either broadside or enfire show sidelobes that are down significantly (around 13 dB). When steering near endfire the up and down beams will be distinguishable when the steer angle is greater than about 2°. Likewise, as the beams move from broadside to endfire the reflection loss curve will be "smudged", but

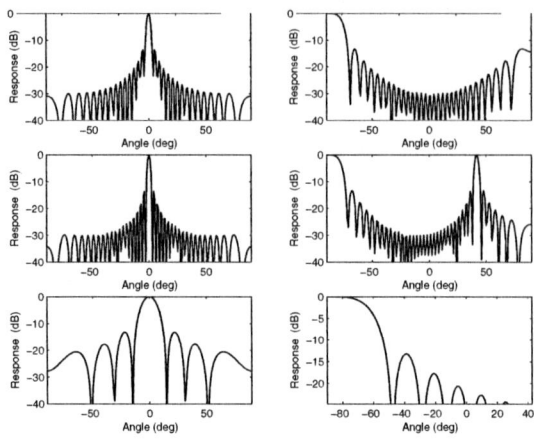

FIGURE 2. Beamformed output for frequencies of 4 kHz (top two panels) 5 kHz (middle) and 1 kHz (bottom). All use an array with 0.18 m hydrophone spacing and length of 5.58 m. Left panels show broadside beamforming and right panels endfire.

only slightly. The smudge is never worse than 10° so, at this frequency, for this array the reflection loss estimate should be good. Now consider beamforming at 5 kHz as shown in the middle panels of Fig. 2. In this case, as we move away from broadside to endfire there is aliasing that forms a beam in the downward direction. This beam will also contribute and these contributions destroy the reflection loss estimate. Next, consider beamforming at 1 kHz shown in the lower panels of Fig. 2. Here, there is no aliasing but the beamwidths are very large, over $\pm 30°$ at endfire. These large beams will cause such a severe smudging that the up/down ratio is no longer a good representation of reflection loss.

GEOACOUSTIC INVERSION OF HIGH FREQUENCY AMBIENT NOISE DATA

ElbaEx 2003

In October 2003, in collaboration with the NATO Undersea Research Centre, a series of experiments took place to the north and south of Elba Island in the Mediterranean Sea. These experiments were designed to study high-frequency acoustic propagation and the performance of underwater acoustic communications. On October 29, 2003, an ambient noise experiment was conducted at the north site near Capraia Island. An array with 32 hydrophones having 0.18 m separation was allowed to drift from an initial position of $42°55.5'N$, $10°5.4'E$ while recording ambient noise. The low end of the spectrum was only filtered to prevent interference from mechanical noise at the high end, an anti-aliasing filter cut the data off at around 3.8 kHz. Only data below 3.5 kHz are considered here. The water column sound speed profile was slightly downward refracting with a late

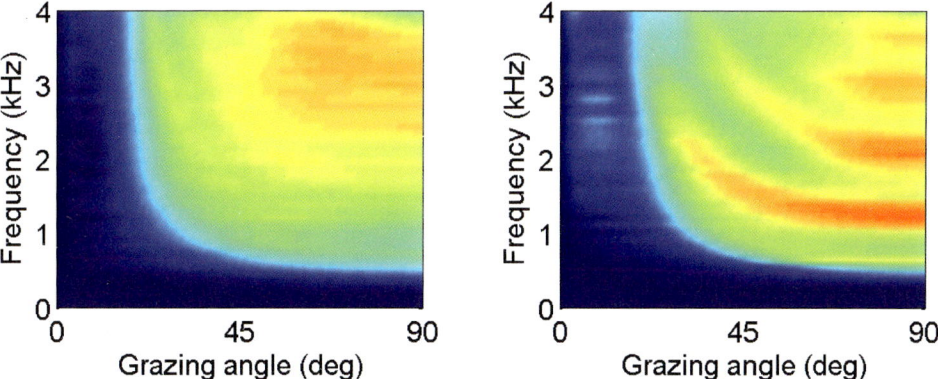

FIGURE 3. Reflection loss inferred from noise measured on a drifting VLA. The left panel shows 4 mnutes of data from 11:30 UTC and the right panel 4 minutes from 13:00 UTC processed by beamforming and dividing the upward and downward looking beams. Data are shown on a 0-15 dB color scale.

summer type profile. The ambient noise data were processed to form the cross-spectral density matrix and in the following sections the reflection loss and seabed properties are estimated from about four minutes of data.

In Fig. 3, two examples are shown taking measured ambient noise, followed by forming the cross-spectral density, vertically beamforming and dividing the beams looking toward the surface by those steered towards the seabed. There are two notable differences between the left panel (11:30 UTC) and the right (13:00 UTC): first, there is an interference pattern evident in the right panel that is absent in the left panel. These type of fringe patterns are formed from layers in the seabed and already by observation the left panel can be represented (for these frequencies) as a homogeneous half-space. The second difference is the slightly higher loss near grazing angles in the data shown in the right hand panel. In the next section we use an inversion scheme to determine the seabed properties from these data.

Inverting reflection loss

One of the most attractive features of the up/down, ambient noise inversion is that it isn't really an inversion. It is an extremely simple processing technique to produce the reflection loss curve. In some cases, this reflection loss curve may be all that is required. For propagation modeling, particularly ray based models, this is a sufficient representation of the bottom. However, as we have seen, actual arrays introduce a beampattern that will smudge the reflection loss and this is not always a good representation of the true reflection loss. Also, for angles near horizontal small errors in correcting for diffraction or array tilt (or other experimental errors) can cause slightly offset beams and therefore errors in the reflection loss estimate for these angles. In addition, we seek to invert the

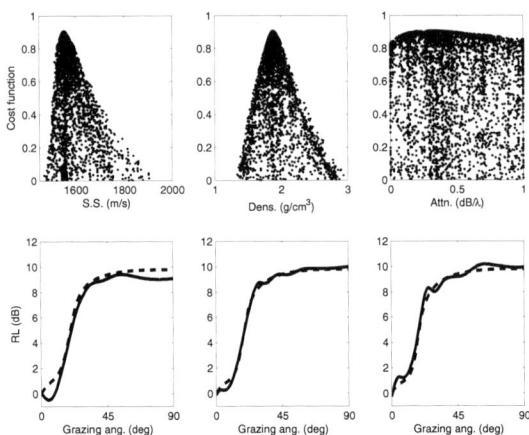

FIGURE 4. Top panels show the cost function output from the genetic algorithm search with the seabed parameterized by sound speed, density and attenuation (x-axes indicate the search bounds). The peaks indicate where the best fit occurs; for the 11:30 UTC data the best fit values are: sound speed: 1553 m/s; density: 1.9 g/cm^3; attenuation 0.3 dB/λ. Lower panels show the measured up/down ratio (solid lines) and the best fit modeled up/down ratio. Left panel is 2 kHz, middle is 2.7 kHz and the right panel 3.5 kHz. These are approximations to the reflection loss curves at these frequencies.

inferred reflection loss to produce geoacoustic properties of the seabed. While these are not always needed, the inversion insures the seabed properties have physical meaning consistent over a broad-band of frequencies and can be used to correct the errors and smudging in the estimated reflection loss. In theory, once the correct geoacoustic properties are determined they can be used over all frequencies.

All the effects introduced into the measured data from the environment, such as water column sound speed profile, surface losses, volume attenuation and physical array parameters (length, number of hydrophones) can be included in the formula for noise cross-spectral density (eq. (7)). This, can then be beamformed and the up/down beam ratio taken in exactly the same way as for the measured data. To produce a true reflection loss and geoacoustic properties the beamformed output and/or the up/down beam ratio can be compared with the model results using a variety of environments and a search algorithm to find the best. In our case, we compare the mean square error between measured and calculated quantities and direct the search using a genetic algorithm. A genetic algorithm is useful since the number of geoacoustic properties can be large depending on the number of layers in the seabed. Exhaustive searching for the best fit quickly becomes unwieldy. The number of layers to include in the model can be estimated by transforming the reflection loss into an impulse response using the sub-bottom profiling technique [2]. The data considered here first is from 11:30 UTC where there is no evidence of structure in the reflection loss curves (see Fig. 3) and, therefore, a simple half-space was used to describe the seabed (i.e. a sound speed, density and attenuation constant).

In Fig. 4, the results are shown from the processing in the 2–3.5 kHz band with the 32 elements of the array (5.58 m length). In the lower figure, the solid line represents

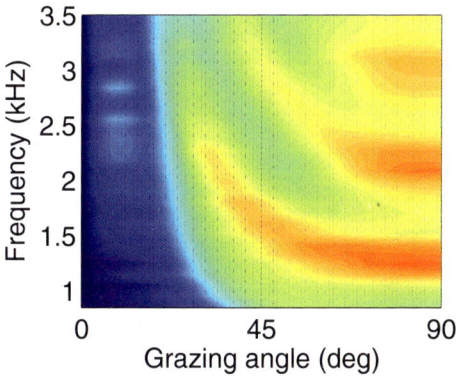 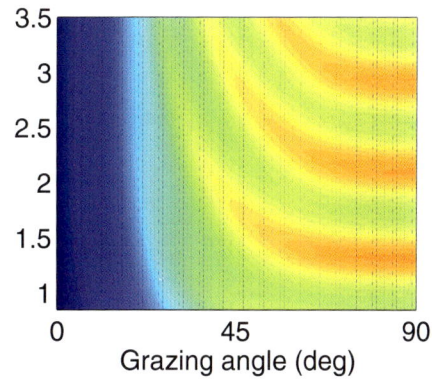

FIGURE 5. Experimental data (left panel) and model (right panel) of the up/down beam ratio. The model has a sediment layer of 0.99 m over a half space. The sediment properties found are: sediment sound speed: 1561 m/s, sediment density: 1.91g/cm^3, sediment attenuation: $0.6 \text{ dB}/\lambda$, half-space sound speed: 1625 m/s, density: 2.1g/cm^3, attenuation: $0.002 \text{ dB}/\lambda$.

the inferred reflection loss at 2, 2.7 and 3.5 kHz determined from taking the ratio of up to downward steered beams. These reflection loss estimates are compared against thousands of possible seabed types that are used to calculate cross-spectral density, beamformed and further processed for the up/down reflection loss estimate. The output values of the cost function for each of the three seabed parameters are plotted against the seabed property value (the x-axes indicate the search bounds for each parameter). The peak indicates the point having best agreement with the data. There are clear indications that the seabed sound speed, density and attenuation are well determined from the data. In the lower panel of Fig. 4, the dashed lines show the output from the model. Note, that the processed measured data has unrealistic values of reflection loss at low angles but realistic values come out of the model. Once the seabed properties are obtained, the "true" reflection loss curves can be generated at any desired frequency without the beamforming "smudging".

In another example, four minutes of data are taken at about 13:00 UTC and the up/down beam processing is used with the results shown in the right side panel of Fig. 3. In this case the fringe pattern indicates the presence of a layer in the seabed. Therefore, a seabed layer over a half-space was the assumed geoacoustic seabed for inversion and this has 7 parameters (sound speed, density and attenuation in the sediment layer and halfspace and the sediment thickness). The final result shows a sediment layer of 1 m and the model/data inferred reflection loss comparison is shown in Fig. 5. While the sound speed in the sediment layer for the 13:00 UTC data is similar to the previous example at 11:30 UTC, the attenuation constant found at 13:00 is higher. This appears consistent with the measurements for the region at low grazing angles where the bottom loss is greater (comparing low grazing angles in the left and right panels of Fig. 3. In this example data from 1–3.5 kHz were used.

TABLE 1. Seabed properties found through inversion of ambient noise (11:00 UTC) data using different array configuration and geoacoustic parameterization of the seabed.

	Sound speed (m/s)	Density (g/cm^3)	Attenuation (dB/λ)
32 phones-5.54 m	1553	1.9	0.30
8 phones-1.44 m	1546	1.9	0.13
16 phones-5.4 m *	1567	2.0	0.35

* Beamformed output used for model/data comparison

Inverting beamformed data from short or undersampled arrays

In the previous examples, the full array which is 32 hydrophones with 0.18 m spacing (total length of 5.58 m) was used. Here, we consider reducing this in two ways, first, 8 hydrophones are used for about 1.4 m aperture. Then, 16 hydrophones with 0.36 m spacing is considered (5.4 m total length). For the case with 1.4 m and 8 hydrophones the beams become large and the up/down ratio of beams smudges the reflection loss to the point where it no longer reasonably represents the reflection loss. However, using the modeling the smudging is done in exactly the same way and, somewhat surprisingly, the final seabed values are nearly the same as those found using the full array. In the second case, the sparse sampling of hydrophones causes grating lobes that destroy the up/down ratio and the results do not even resemble reflection loss curves. However, in this case, the beamformed output is compared (rather than the up/down ratio) with that from the model (i.e. the final step of forming the up/down ratio is skipped). Even though the beamformed output has grating lobes, the model is the same and again the final seabed values found using the search algorithm are nearly the same as for the other examples. All these results are summarized in Table 1 for the data at 11:00 UTC.

CONCLUSIONS

In this paper we examined the possibility of using ocean ambient noise data in the 1–4 kHz frequency band to determine properties of the seabed. A noise model based on a ray approach was used along with a genetic algorithm search to match measured and simulated data. This approach compensates for the beamforming "smudging" of the reflection loss as well as for artifacts that sometimes occur for low grazing angles. A genetic algorithm was used to direct the search and find the best fit between model and data with the best fit assumed to be a good representation of the seabed. The tests showed well determined seabed properties and a good match between measurements and model. The seabed properties are also sensible based on preliminary assessment of grain sizes (from a grab sample) and impulse response measurements taken at various ranges using a vertical array and a controlled sound projector. Ambient noise in this band of frequencies and possibly even higher frequencies appears to have advantages of reduced array length requirements and also less interference from shipping. With the short arrays considered here it may be feasible to use this method for surveys either using surface ships or submerged vehicles such as autonomous underwater vehicles (AUV's).

ACKNOWLEDGMENTS

The authors would like to thank the NATO Undersea Research Centre and the staff that participated in the ElbaEx experiments. In particular, Chief Scientists Finn Jensen (1st half) and Mark Stevenson (2nd half), Engineering Coordinator, E. Michelozzi, and Data Acquisition Coordinator, P. Boni. We would also like to thank the Captain and crew of the R/V Alliance. This research was supported by the Office of Naval Research, Ocean Acoustics Program and the NATO Undersea Research Centre.

REFERENCES

1. Harrison, C. H., and Simons, D. G., *J. Acoust. Soc. Am.*, **112**, 1377–1389 (2002).
2. Harrison, C. H., *J. Acoust. Soc. Am. (to appear)* (2004).
3. Jensen, F. B., "Results from the Elba HF-2003 experiment," in *Proceedings of the High-Frequency Ocean Acoustics Conference*, AIP, 2004.
4. Jensen, F. B., Kuperman, W. A., Porter, M. B., and Schmidt, H., *Computational Ocean Acoustics*, American Institute of Physics, Inc., New York, 1994.
5. Kuperman, W. A., and Ingenito, F., *J. Acoust. Soc. Am.*, **67**, 1988–1996 (1980).
6. Schmidt, H., OASES users guide and reference manual, MIT, Dept. of Ocean Engineering (1999).
7. Harrison, C. H., *J. Acoust. Soc. Am.*, **99**, 2055–2066 (1996).
8. Harrison, C. H., *Applied Acoustics*, **51**, 289–315 (1997).
9. Harrison, C. H., Brind, R., and Cowley, A., *J. Comp. Acoust*, **9**, 327–345 (2001).

Using Buried Directional Receivers in High-Frequency Seafloor Studies

John C. Osler[*1] and Anthony P. Lyons[†1]

[*]*Defence R&D Canada – Atlantic, P.O. Box 1012, Dartmouth, Nova Scotia, Canada*
[†]*The Pennsylvania State University, Applied Research Laboratory, P.O. Box 30, State College, PA, 16804-0030, USA*

Abstract. Knowledge of acoustic arrival angle can be useful for studying penetration mechanisms and for estimating sediment sound speed dispersion. The arrival angle of the acoustic field penetrating the water-sediment interface, however, is difficult to measure using sparsely distributed pressure sensors. The arrival angle, as well as amplitude information, can be unambiguously obtained by measuring particle motion with directional receivers. An experiment was conducted off of Elba Island, Italy, to assess the feasibility of a novel technique that uses high-frequency accelerometers to measure the directionality of acoustic arrivals. Measurements of acoustic penetration into a sandy seafloor were obtained over a wide frequency range (2.5 to 29 kHz) using off-the-shelf accelerometers, adapted for use in marine studies. The sensors that were developed are suitable for the penetration studies for which they were devised, but their angular resolution would limit their application in dispersion studies. Insights from this experiment will guide the design of new directional sensors that are suitable to study dispersion.

INTRODUCTION

There have been repeated experiments [1, 2, 3] demonstrating "anomalous" acoustic penetration into the seabed below the critical angle (i.e. beyond that predicted by elastic wave theory for smooth seabeds). Modeling and penetration experiments have suggested various causes for the anomalous penetration including a Biot "slow" compressional wave [2] and scattering from interface roughness [4]. Chotiros [2] used 20 kHz data obtained on a sparse buried array of hydrophones to estimate the direction and speed of acoustic waves by intensity superposition and interpreted the results as evidence of the Biot slow wave. Thorsos et al. [4] showed that scattering, essentially straight down from the water-sediment interface, could also account for subcritical penetration. For an array of receivers such as that used by Chotiros [2], the scattered field could essentially mimic the slow wave. The fact that one cannot distinguish between the competing mechanisms of a Biot slow wave and interface scattering, points out the difficulty in using sparsely distributed pressure sensors to measure arrival angle. Direct measurements of the arrival angle of acoustic waves penetrating the seafloor using accelerometers (or other types of vector sensors) should be able to conclusively distinguish between the two subcritical penetration mechanisms by illuminating not just *when* the signal arrives but *from which direction*. The refracted

---

[1] Research initiated and field experiment conducted when the authors were colleagues at the NATO Undersea Research Centre, Viale San Bartolomeo, 400, 19138 La Spezia, Italy

Biot slow wave (typical speed 1200-1300 ms^{-1}) would arrive from a grazing angle between approximately 30° and 45°, whereas scattering by the rough water-sediment interface can arrive over a wide range of angles controlled by the ripple wavelength, water/sand speed ratio, and incident angle [4].

In analyzing their sediment acoustic penetration data, Maguer et al. [3] noted that a lower sound speed than that measured on ground truth cores was required to fit the experimental observations to model calculations. Laboratory measurements of sound speed are made by transmitting acoustic signals a known distance through the sediment core and core liner. This is typically done at a much higher frequency, e.g. 200 kHz, than the experimental data under consideration, 4 and 10 kHz in the case of Maguer et al. [3]. The discrepancy between *in situ* and ground truth measurements was attributed to sound speed dispersion [3] and later confirmed experimentally [5]. More recent investigations [6, 7] have also confirmed sound speed dispersion for a sandy site off of Panama City, Florida. Although comparisons between measurements and predictions of dispersion based on Biot theory have been promising, they have not been conclusive to date due to uncertainties both in parameter estimation and acoustic measurement techniques [6].

As a complement to these techniques [6], sediment sound speed could be measured using buried directional receivers. The angle to which sound is refracted when transmitted across the water-sediment interface is a function of the sediment sound speed. Therefore, the angle at which an acoustic pulse arrives at buried directional sensors, measured as a function of frequency, can be used to make estimates of sound speed dispersion. Sediment to water sound speed ratios at the SAX99 experimental site have been estimated to range between 1.09 below ~500 Hz to 1.15 above 50 kHz [6]. These ratios yield an approximate difference in arrival angles at low and high frequency of 10-15° (Fig. 1), depending upon how close one can work to the critical angle and avoid interference between the refracted and evanescent fields. The angular resolution required to estimate sound speed dispersion would be approximately ±1.5° (Fig. 1a). By contrast, the angular resolution required to distinguish penetration mechanisms is less stringent, approximately ±5°, unless one happens to receive arrivals at an angle at which both penetration mechanisms can occur (30-45°).

To establish the feasibility of using buried accelerometers in the aforementioned seafloor studies, the concept was evaluated during a sea-trial in 1999, the Acoustic Penetration Experiment (APEx99). Two fundamental capabilities must be demonstrated. First, the sensors must be able to achieve the respective angular resolutions required for penetration and sound speed dispersion studies. In Fig. 1, the arctangent of the ratio of the on-axis/off-axis response is used to predict the angular resolution capability of the sensor that may in turn be compared with the anticipated requirements for the seafloor studies. Second, the sensors must be sufficiently sensitive in the frequency band of interest and not be limited by self-noise. Uni-axial accelerometers were selected because of their sensitivity at high frequency. They were buried in orthogonal pairs in order to decompose the acoustic arrivals into vertical and horizontal components. The following section concerns the issues relevant to the design of a sensor package and calibrations of the on-axis and off-axis sensitivity. Results from APEx99 are then presented, followed by a discussion of the measured *in-situ* angular response and the viability of this technique for seafloor studies.

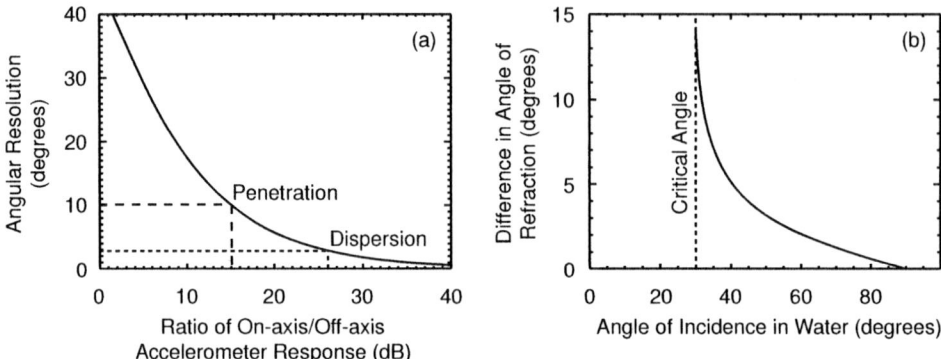

FIGURE 1. (a) Estimated angular resolution as a function of the ratio of on-axis to off-axis accelerometer sensitivity. Penetration (long dash) and dispersion (short dash) measurement requirements are superimposed. (b) Difference in angle of refraction for sediment/water sound speeds ratios of 1.15 and 1.09 as a function of grazing angle.

THEORETICAL CONSIDERATIONS

Several criteria governed the design of the buried directional receivers used in this study. The ideal receiver should have a high sensitivity to forcing on its axis of response, sufficient, for example, to detect arrivals scattered downward from the seabed by a pulse incident at a shallow grazing angle. Meanwhile, in order to be effective at discriminating the angle of arrival, it should be relatively insensitive to off-axis forcing by translational motion, and to rotational motion, as this would manifest itself as a spurious on-axis translational component. It should be rigid within the frequency band of interest and thus free of any internal resonances. It should also be compact such that an orthogonal pair of receivers could be located within a fraction of an acoustic wavelength of each other in order to measure the acoustic field at that point in the seabed.

The directional receiver that has been developed consists of an accelerometer fastened to a thin disk by a mounting stud (Fig. 2). The disk ensures that it is well coupled and oriented in the medium in which it is placed. The mass of the coupling disk is approximately ten times that of the accelerometer, as per the manufacturers recommendation, and such that the accelerometer has a minimal effect on the motion of the object to which it is attached. The accelerometers were manufactured by Endevco Inc. and two versions of the same model were used, 7259A-25 and 7259A-100 with sensitivities of 25 and 100 mvg^{-1} respectively. They have a wide bandwidth with an amplitude response that is flat from 10 Hz to 50 kHz with a deviation of less than 1 dB. As they are not designed for underwater applications, the sensor housing and connection to the 1 mm diameter conducting cable had to be potted. This presented some difficulties, as the potting compound did not bond readily with the Teflon coating of the cable.

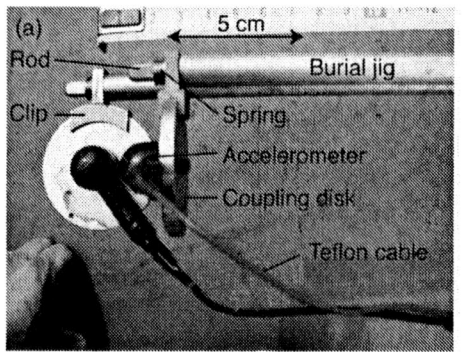 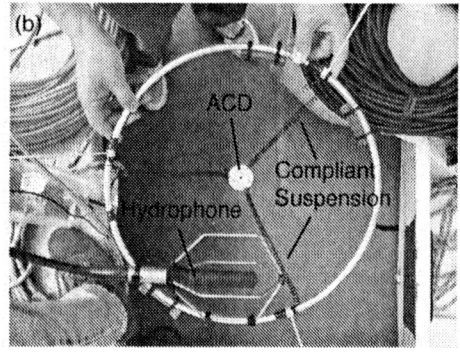

FIGURE 2. (a) Pair of uni-axial accelerometers mounted to coupling disks that are orthogonal to each other. The sensors are attached to the burial jig using clips on the edges of the disks. Springs temporarily fix the clips to the inner, square, rod of the burial jig. The sensors are released when the square rod is retracted. (b) The accelerometers are calibrated individually in water by mounting them on a frame using very soft springs. A hydrophone is also attached to the circular frame.

Specifying the dimensions and composition of a coupling disk is challenging as conflicting design requirements are encountered: maximum on-axis sensitivity requires minimal inertial mass; insensitivity to rotational motion requires maximum disk radius; minimizing internal resonances requires a thick disk with minimum radius; compact pair of sensors requires minimal disk radius. To elucidate these points and find the optimum compromise, the physics of each of these criteria requires consideration. In what follows, all variables (forces, accelerations, etc…) are assumed to be harmonic with angular frequency $\omega = 2\pi f$. Assuming plane wave propagation, differentiation with respect to time becomes a multiplication by $j\omega$, so if a denotes acceleration, then $a/j\omega$ denotes velocity.

In a Newtonian fluid, the equation of motion for the accelerometer coupling disk assembly (hereafter called the ACD) when freely suspended in water is

$$(a_d - a_w)(m_i + m_a) = -a_w(m_i - m_w) - D \tag{1}$$

where a_d and a_w are the accelerations of the coupling disk and the surrounding water respectively, m_i, m_a, and m_w are the inertial mass of the assembly, the virtual (or added) mass of water it entrains, and the mass of water it displaces (values in Table 1). The term $(a_d - a_w)$ represents the acceleration of the disk relative to the water and D is a drag force that is insignificant because the particle velocities are near zero.

The ACD is forced on its axis of sensitivity (perpendicular to the face of the ACD) by an acoustic pressure

$$p = \rho c u = \frac{\rho c}{\omega} a_w, \tag{2}$$

where ρ is the density of water, c is the speed of sound in water, and u is the particle velocity in water. Combining Equations (1) and (2), we find that the ratio of the response of an accelerometer to the incident pressure is

$$\frac{a_d}{p} = \frac{\omega}{\rho c} \frac{(m_w + m_a)}{(m_i + m_a)}. \tag{3}$$

Although the intrinsic frequency response of the accelerometer is uniform as a function of frequency from 10 Hz to 50 kHz, in accordance with the manufacturers specifications, the response of the ACD increases as a function of frequency (Fig. 3), behaving as $20\log_{10}(\omega)$. The sensitivity of the ACD can be increased by reducing its inertial mass. From this standpoint, a coupling disk manufactured with aluminum would be preferable. The theoretical resonance frequency of an aluminum disk would also be higher though still within our measurement band (Table 1). However, the threaded holes for the accelerometer stud mount on the aluminum disks could not withstand the calibrated torque required to fix the accelerometers onto the coupling disk–a crucial and unexpected requirement to minimize the resonance at approximately 13 kHz associated with the stud (Fig. 3).

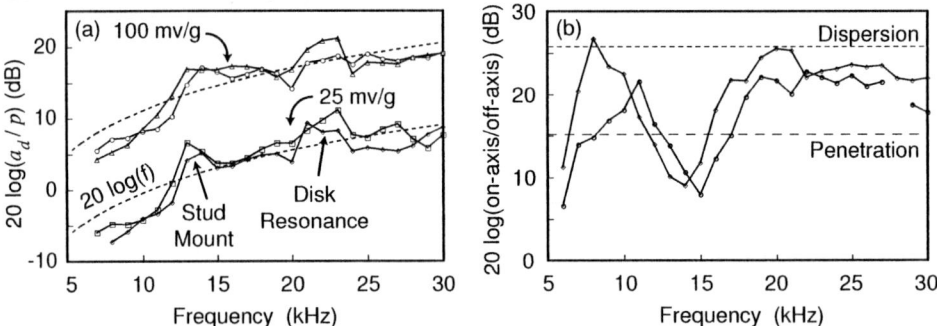

FIGURE 3. (a) On-axis response of four ACDs calibrated individually in water using the setup in Fig. 2b. The overall trend of the data follows 20 log(f) as expected. The resonance around 13 kHz is due to the stud used to mount the accelerometer to the coupling disk. (b) Ratio of on-axis to off-axis response of two ACDs to acoustic pressure in water. The horizontal lines are the minimum ratios required in order to achieve the angular resolution requirements for dispersion and penetration studies (from Fig.1).

For the case of the ACD freely suspended in water, it is straightforward to estimate the received pressure that is required to exceed the self-noise[2] of the accelerometer, 8 x 10^{-3} ms^{-2} for the 7259A-25 and 5x10^{-3} ms^{-2} for the 7259A-100. Using the higher self-noise value for the 7259A-25 and the appropriate values for m_i, m_a, and m_w in Equation (3), an incident pressure of 139 dB re 1 μPa would be required at 3 kHz and an incident pressure of 119 dB re 1 μPa would be required at 30 kHz. These incident pressures are readily attainable using sources used in this experiment (ITC-3013, ITC-2007, and Simrad TOPAS with maximum source levels of 185, 189, and 213 dB re 1 μPa respectively).

When the ACD is buried in the seabed, it is more difficult to estimate its response. For an initial estimate, Equation (3) may still be used, but one must treat the seabed as a higher density fluid. In this case, an incident pressure of 143 dB re 1 μPa would be required at 3 kHz and an incident pressure of 123 dB re 1 μPa would be required at 30

---

[2] More often than not, accelerometers are limited by their self-noise rather than by ambient noise.

kHz. Equation (3) can be extended to include the effects of a reaction force within the seabed. For sake of brevity, these equations are not shown but the effect is to decrease the response of the ACD, thus higher incident pressures are required.

It is necessary for the ACD to be relatively insensitive to off-axis forcing. The manufacturers specification for the accelerometers is 5% of the on-axis sensitivity, or an on-axis/off-axis ratio of -26 dB. Considering the response of the ACD in water, Equation (3) may continue to be used, however, the added mass, m_a, will be considerably lower as the cross sectional area is much smaller. As a consequence, the inertial mass term will dominate, further reducing the transverse sensitivity of the ACD compared to the on-axis sensitivity. However, the ACD may be limited by its response due to rotational motion, rather than by its transverse sensitivity.

TABLE 1. Properties of the accelerometer coupling disk assembly.

Variable	Units	Stainless Steel	Aluminum
r, Radius	m	0.022	0.022
t, Thickness	m	0.004	0.004
m_i, Inertial Mass	kg	0.050	0.016
m_w, Mass of water displaced	kg	0.006	0006
m_a, Added mass, on-axis of a disk $(8/3\rho_w r^3)$	kg	0.029	0.029
m_a, Added mass, off-axis of a sphere $(2/3\pi\rho_w r^3)$	kg	0.0003	0.0003
β factor (see text for details)	Not applicable	0.340	0.195
Inertial response, $(m_w+m_a)/(m_i+m_a)$	dB	-7.24 dB	-2.82 dB
Theoretical resonance frequency of disk	Hz	22121	24442

The importance of rotation may be estimated by considering the value of

$$\beta = md^2 / I \qquad (4)$$

where m is the sum of all mass terms (inertial and added/virtual) and I is the sum of all inertia terms (including added/virtual) multiplied by d^2, the square of the moment arm. To avoid problems with rotational motion [8], the inertial terms must be greater than the mass terms and/or the moment arm must be as small as possible, i.e. β <<1. Since the added moment of inertia of the disk increases as a function of the disk radius raised to the fifth power [8], this term quickly begins to dominate. For a coupling disk made of stainless steel and a radius of 0.22 m (Table 1), a value of β=0.34 is predicted in water. Calibrations in water (Fig. 3b) show that the ratio of on-axis/off-axis response is suitable for penetration studies, except at the frequencies that are effected by the stud mount (Fig. 3b). A β factor may be calculated for an ACD buried in a higher density fluid seabed, however, this does not account for the stiffness of the seabed (angular spring constant). This would serve to further resist any rotational motion of the ACD and increase the on-axis/off-axis response ratio.

SENSOR PERFORMANCE

In April and May 1999, two pairs of ACDs were deployed in a sandy sediment in a water depth of 12 m as part of APEx99 in Biodola Bay, Elba Island, Italy. Driven initially by penetration questions and later by sound speed dispersion issues, the goals of the APEx99 experiment included evaluating the feasibility of using vector sensors such as accelerometers in higher-frequency seafloor studies. Directional sources on a movable, telescopic tower and moored omni-directional sources transmitted pulses over a range of frequencies (2.5-29 kHz) at a variety of grazing angles: above critical for *in situ* calibration; at critical to study interaction of refracted and evanescent waves, and subcritical to measure acoustic penetration into a sandy seafloor. The well-characterized site [3] had a sand layer thickness of approximately 2 m, a bulk density of 1920 kgm^{-3}, and a compressional wave speed estimated at 1720 ms^{-1} (measured at 200 kHz from diver cores). The sand had a mean grain diameter of $\phi = 2.25$ (0.21 mm) with a standard deviation of 0.6 ϕ (~0.1 mm). The water sound speed was measured with a CTD to be 1530 ms^{-1}, thereby creating a critical angle of approximately 29°.

The vertical and horizontal acceleration of the ACDs in response to a pulse of sound can be used to measure the angle of arrival by examining the orientation of the major axis of the elliptical motion (Fig. 4). The analysis can be repeated for different pulses to measure arrival angle as a function of frequency (Fig. 5a). The width of the minor axis of the ellipse (Fig. 4a) indicates that some signal is not arriving in phase. (Phase quadrature between components would yield a circular particle motion.) The physical separation of the ACDs introduces a phase shift that one can correct as a function of frequency, but this requires some *a priori* knowledge about the angle of arrival and the properties of the seabed. As these are the very quantities to be measured, an iterative approach is required in some cases. However, phase shifts do not change the orientation of the major axis; they broaden the minor axis and degrade the ability to discern the major axis (Fig. 5b). The simultaneous arrival of acoustic signals from multiple directions is of greater concern. This may change the amplitude of the components and bias estimates of the angle of arrival.

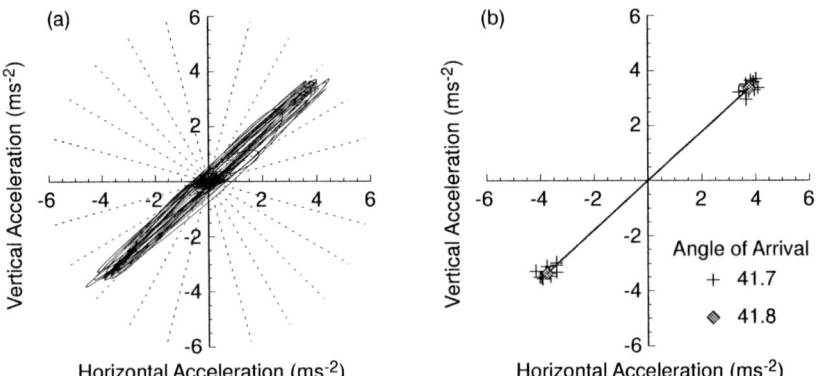

FIGURE 4. (a) Sample hodogram of particle acceleration on ACDs at 50 cm depth for a 2 ms pulse of sound at 16 kHz, incident upon the seabed at a grazing angle of 49.4°. The pulse is refracted to approximately 41.8° as anticipated based on the sound speed ratio (b) Diagonal line represents the best fit to the angle of arrival, following the peak amplitude of both components on cycle-by-cycle basis (plus symbols) and fitting a sine wave, in a least squares sense, on a ping-by-ping basis (diamonds).

CONCLUSIONS

One of the main goals of the APEx99 experiment was to test the feasibility of using vector sensors such as accelerometers to study acoustic penetration into sediment at high frequency. The orthogonal pair of uni-axial accelerometers devised for penetration studies is suitable for that application; however, their angular resolution (as revealed by the scatter in Fig. 5a) is not sufficient for dispersion studies. Design improvements to make vector sensors more amenable to dispersion studies will include: improved angular resolution; co-located pressure for purposes of calibration as well as intensity and impedance processing; and tri-axial accelerometers in a single body to avoid ambiguities due to phase separation between receivers.

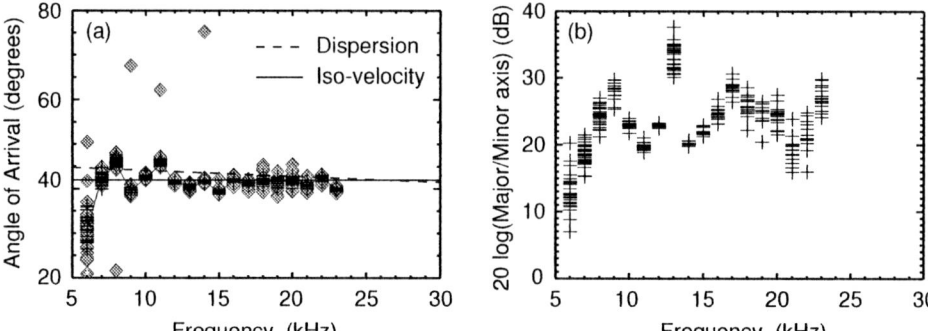

FIGURE 5. (a) Angle of arrival as a function of frequency (25 pings per frequency). The lines are the anticipated angle of arrival for an iso-velocity sand (solid) and for a linear sound speed dispersion from 5-30 kHz for water/sediment sound speed ratios ranging from 1.09-1.15 (dashed). Poor performance at 6 kHz is due to low signal-to-noise. The variability from 8-11 kHz is related to the *in situ* calibration. (b) Ratio of major and minor axes of the hodogram ellipses as a function of frequency.

ACKNOWLEDGMENTS

The authors would like to thank the captain and crew of *RV Manning*, Mr. P. Boni, Mr. R. Stoner, Mr. M. Paoli, Dr. E. Pouliquen, Mr. E. Muzi and Mr. A. Brogini. Analysis of APEx99 data was supported by ONR.

REFERENCES

1. J.L. Lopes, *J. Acoust. Soc. Amer.*, **99**, 2473(A) (1996).
2. N.P. Chotiros, *J. Acoust. Soc. Amer.*, **97**, 199-214 (1995).
3. A. Maguer, W. L. J. Fox, H. Schmidt, E. Pouliquen, and E. Bovio, *J. Acoust. Soc. Amer.*, **107**, 1215-1225 (2000).
4. E. I. Thorsos, D. R. Jackson, and K. L. Williams, *J. Acoust. Soc. Amer.*, **107**, 263-277 (2000).
5. A. Maguer, E. Bovio, and W. L. J. Fox, *J. Acoust. Soc. Amer.*, **108**, 987-996 (2000).
6. K. L. Williams, D. R. Jackson, E. I. Thorsos, D. Tang and S. G. Schock, *IEEE J. Ocean. Eng.*, **27**, 413-428 (2002).
7. M. J. Buckingham and M. D. Richardson, *IEEE J. Ocean. Eng.*, **27**, 429-453 (2002).
8. J. C. Osler and D. M. F. Chapman, *J. Geophys. Res.*, **103**, 9879-9894 (1998).

Geoacoustic Inversion of Broadband Data from the Florida Straits

Ross Chapman and Yongmin Jiang

School of Earth and Ocean Sciences, University of Victoria, PO Box 3055, Victoria, B.C. V8W 3P6 Canada

Abstract. Acoustic propagation experiments have been carried out in the Florida Straits with a multi-frequency broadband source that transmitted M-sequence pulses over a range of 10 km to a sparse-filled vertical line array. The sound source was cycled in octaves from 100 Hz to 3200 Hz, transmitting each octave for one hour. This paper presents results of matched field inversions of the acoustic field data to estimate geoacoustic model parameters for the experimental site. The inversion is very sensitive to the sediment sound speed at the sea floor. The estimated value of 1560 m/s is consistent with fine-grain calcareous sand material that is considered to represent ground truth for the site.

INTRODUCTION

Experiments with broadband sound sources were carried out in the South Florida Straits to study variability and coherence in sound propagation on the continental shelf. Previous work by DeFerrari [1] has concentrated on analysis of the effects of the oceanographic conditions in the strait that give rise to spatial and temporal inhomogeneities of the sound speed in the water column. The sound propagation is bottom limited in the shallow water environment, but there is relatively sparse information about the geoacoustic parameters. This paper describes results of matched field inversion of the acoustic field data from the experiment to estimate a geoacoustic profile for the region.

The experiment is described in the next section. This is followed in the next section by a discussion of the features of sound propagation that affect the design of the inversion method. The results of the inversion are presented and summarized in the final section.

FLORIDA STRAITS EXPERIMENT

The experimental geometry is shown in Fig. 1. A sound source moored on the sea floor transmitted M-sequence signals to a sparse-filled vertical line array at a range of 10 km. The average water depth over the distance between the source and receivers was 145 m. The sequence of events was a 6-hour transmission series that consisted of M-sequence signals at 100 Hz for the first hour, then 200 Hz for the next hour, then 400 Hz, increasing each hour in octave steps to 3,200 Hz for the 6th hour. This series of signals was repeated throughout the experiment. At each frequency, the M-sequence repetition rate was ~2.55 s, and the spectral band width at each frequency was one quarter of the value of the carrier frequency. The received signals were coherently averaged for one minute, and processed to generate complex envelopes [2]. The vertical line array consisted of 32 sensors that were non-uniformly spaced from 39 m to 140 m. The system was bottom moored and stretched nearly vertical by a subsurface float. Data from seven hydrophones were used in the inversions (Fig 2).

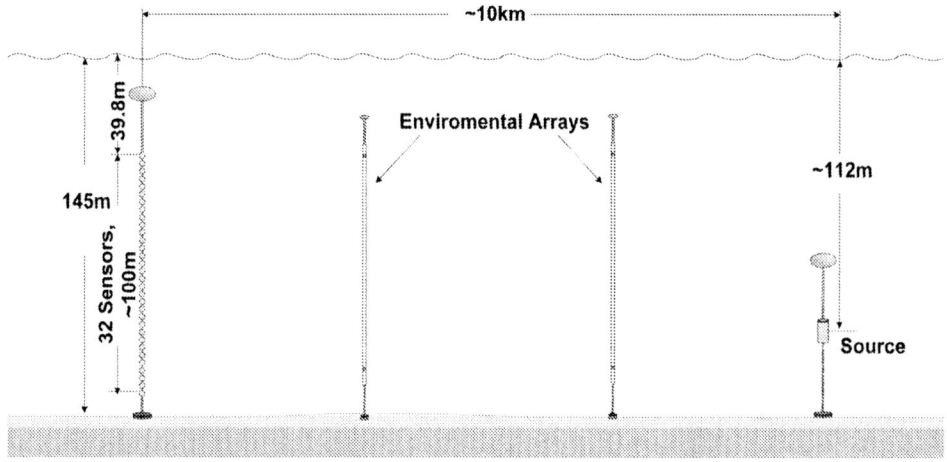

FIGURE 1. Experimental geometry for the South Florida Strait site.

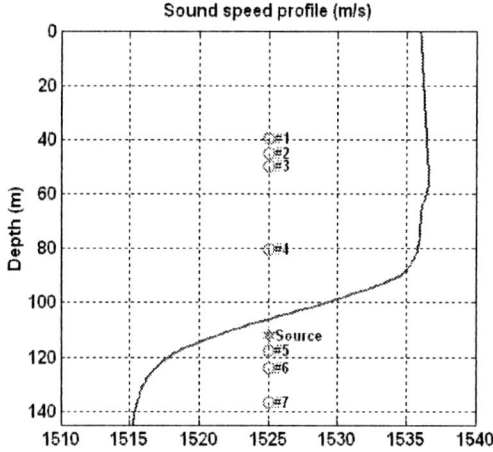

FIGURE 2. Sound speed profile, and locations of source and receivers in the water column.

The sound speed profile in the water was measured throughout the 28-day experiment at 10-element environmental sensor arrays at two locations along the propagation path. Temperature measurements at the two stations generally were correlated for periods of up to three hours, indicating very little range dependence. The measured profile in Fig. 2 shows the sound speed for the time period of the acoustic data that were used in this work. The conditions were uniform in the upper part of the water to a depth of about 80 m. Below this depth, there was a strong thermocline that generated a waveguide in the bottom portion of the channel. The sound source was located at 112 m in the deep waveguide, and the receiver depths relative to the waveguide are shown in Fig. 2.

GEOACOUSTIC INVERSION

Acoustic Propagation

The signal field at all frequencies from a source in the deep waveguide consists of two components that can be described in terms of ray theory (Fig. 3). Steep angle rays propagate by Surface-Reflected/Bottom-Reflected (SRBR) paths that span the entire water column. These paths are received at all sensors in the array, and arrival times span about 300–400 ms. Shallow angle rays less than 7.6° are trapped in the deep waveguide, and arrive as a strongly focused group of refracted/bottom-reflected (RBR) rays. Arrival times of rays in this group are within 10–20 ms, and the intensity of the group is 10–15 dB higher than that of the SRBR components. This signal is observed only on the deeper hydrophones of the vertical array that are within the deep waveguide. In both cases of SRBR and RBR paths, the propagation is bottom limited with 8–10 bottom interactions along the travel path.

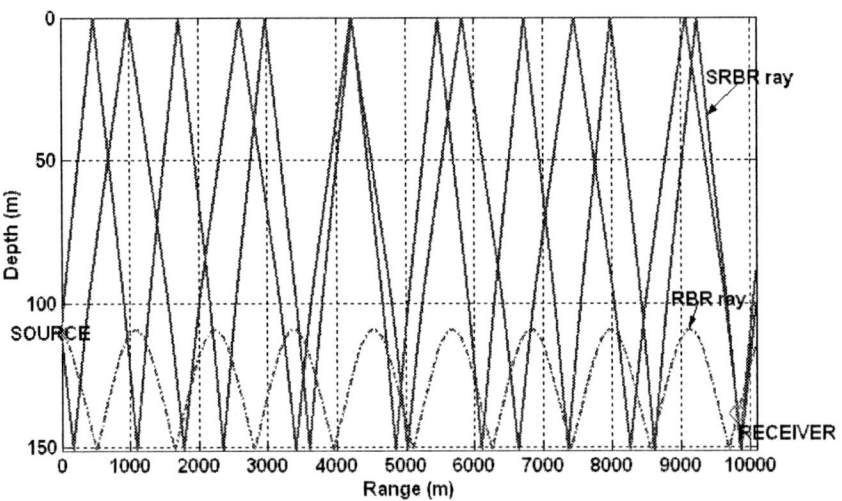

FIGURE 3. Illustration of RBR and SRBR rays from the source at 112 m.

Strategies for Inversion

The overall objective of this work is to invert geoacoustic parameters for a simple gradient layer model of the sound speed profile in the sediment, using data separately from each of the six frequency bands that were transmitted in the experiment. The very long range geometry presents significant challenges for implementing matched field inversion. Conventional matched field processing, which makes use of spatial phase coherence across the array, is the method of choice for the lower frequency signals. However, at higher frequencies, (> 800 Hz) it is more effective to exploit the temporal coherence of the signal, and base the inversion on modeling the signal waveform at single sensors.

Normal mode propagation models are most appropriate for calculating acoustic fields for matched field inversions at the lower frequencies, but at higher frequencies it is more efficient to use ray theory for calculating the waveform [3,4]. However, it is first necessary to benchmark the ray model. We show in Fig. 4 a comparison between transmission loss calculated by ray theory [5] and normal modes [6] for the Florida Straits sound speed profile and a half space geoacoustic model with sediment sound speed of 1700 m/s. The left and right panels show the results for receivers at 40 m and 130 m, respectively, for the source depth of 112 m in the waveguide. The comparison demonstrates that ray theory can model the field accurately for the shallow receiver where the propagation is by SRBR rays. However, the ray model does not perform as well in modeling the RBR waveguide propagation. This result suggests that the inversions at higher frequencies should be restricted to data from the shallow hydrophones above the deep waveguide.

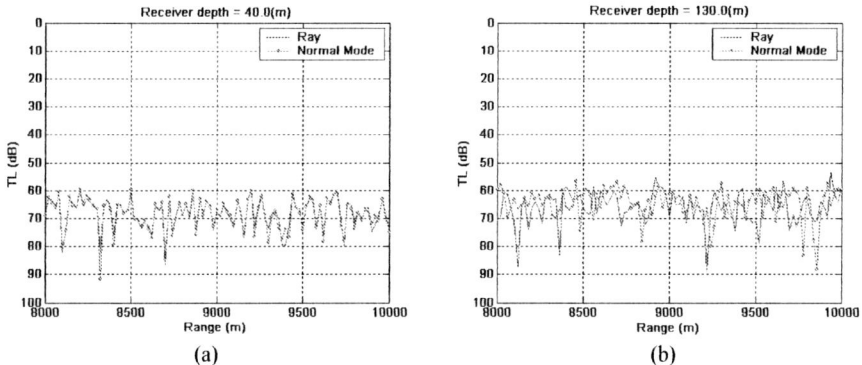

FIGURE 4. Comparison between TL calculated by normal mode and ray theory at different receiver depth (a) Receiver depth at 40 m; (b) Receiver depth at 130 m

Fast Gibbs Sampling

In this paper we describe the application of conventional matched field inversion to spectral components of the transmitted signal. The spatial coherence across the array was first examined for each spectral component in the band to select five frequencies that were suitable for matched field processing. The spatial coherence was not uniformly high across the band, and, generally, poor coherence was associated with low signal strength at several of the sensors.

We assumed that the environment was range independent, and used the normal mode model ORCA [6] to calculate the replica fields for a simple gradient layer geoacoustic model as shown in Fig. 5. This model is similar to the model used in previous work by DeFerrari and Monjo [1] to predict channel pulse response. However, the shear wave effect was not considered here. The complete model consisted of five geoacoustic parameters, including the water depth, and the density, attenuation, sound speed gradient and sound speed at the top of the sediment layer. Both the density and attenuation were assumed to be constant with depth. The inversion also estimated three geometrical parameters of the experiment – the source depth, depth of the topmost receiver in the vertical array, and the range.

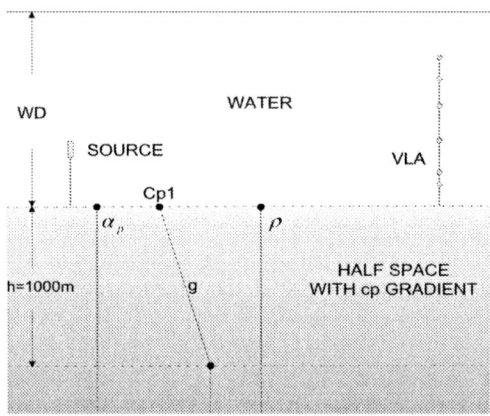

FIGURE 5. Geoacoustic model used for the South Florida Strait environment.

The model parameters were estimated using the Fast Gibbs Sampling method [7,8]. This approach provides an unbiased, asymptotically converging sample of the *a posteriori* probability distribution that represents the complete solution to the inverse problem in the Bayesian formulation. The method uses the same selection criterion as in conventional optimization by simulated annealing, except that all samples are drawn at T = 1. The samples were generated by evaluating a multi-frequency cost function that was based on the single frequency, normalized Bartlett processor.

Inversion Results

The marginal densities that were derived from the Gibbs *a posteriori* probability distribution are shown in Fig. 6 for the eight model parameters. The distributions indicate that all parameters except the sound speed gradient have been well estimated. The estimates for the geometrical parameters are generally consistent with ground truth measurements from the experiment. The range and water depth estimates are strongly correlated, but the inversion prefers deeper depths and shorter ranges than expected from the experimental deployment. Range errors of up to 250 m are not uncommon in localizing the source and array, but the water depth errors are more difficult to interpret. If the impedance contrast at the sea floor is weak, the low frequency 200-Hz signal may be sensing a deeper interface within about a wavelength of the sea floor.

Sediment sound speed is very well estimated, and the maximum *a posteriori* (MAP) value is consistent with expected values for calcareous sand that is thought to be the sea bed material in the region. Density is not well estimated, but the MAP value is also consistent with calcareous sand. These estimates represent at best an average over the multiple bottom interactions along the propagation paths.

Overall correlations between the measured signal and signal envelopes calculated using the MAP estimates are about 80 % over the array. The estimated parameters support SRBR and RBR modes that model the dominant first arrival components of the signal very well. However, a group of weaker arrivals that are delayed by about 300 ms are not predicted by the gradient sound speed geoacoustic model. These later arrivals in the 200-Hz signal are likely due to interactions with the subbottom structure. This interaction is described by the sediment sound speed gradient in the working geoacoustic model. However, the inversion indicated that there is little information about the gradient parameter in the data. At higher frequencies, the inversion sensitivity to the details of the deeper structure is likely to remain low.

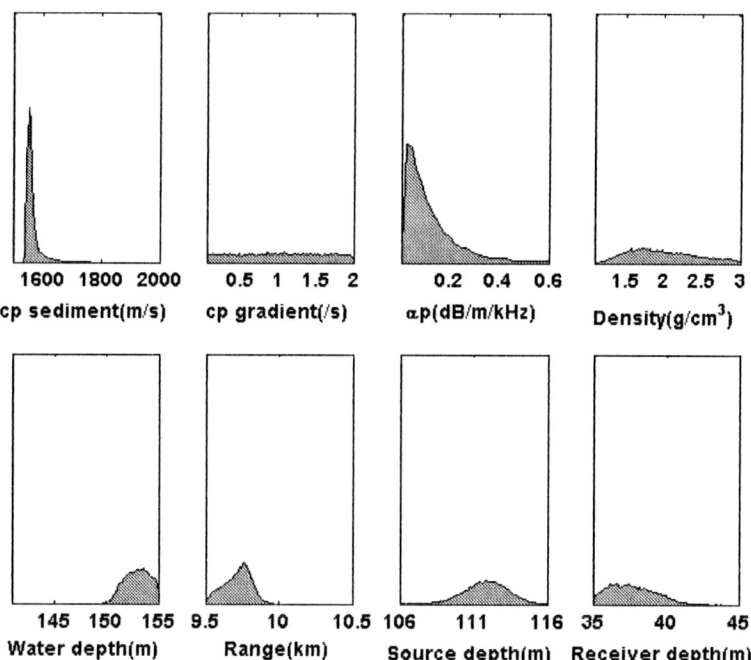

FIGURE 6. Marginal densities generated from the Gibbs sampling inversion for the geoacoustic and geometric model parameters.

SUMMARY

Strategies for inversion of low and high frequency signals that were transmitted in the experiment in the South Florida Straits are discussed in this paper. The inversion strategy for the low frequency data is based on conventional matched field processing, whereas the inversions at high frequencies are based on waveform matching of the received signals. The inversion results for the 200-Hz data provide a baseline geoacoustic model for comparing with the results from inversions at higher frequencies. The inverted geoacoustic profile is sensitive to the sediment parameters at the sea floor, and the estimates for sound speed and density are consistent with expected values for calcareous sediment material.

ACKNOWLEDGMENTS

This work was supported by the Ocean Acoustics Team at the Office of Naval Research. Discussions with Harry DeFerrari and Neil Williams about the experiment were greatly appreciated.

REFERENCES

1. H. A. DeFerrari, N. J. Williams and H. B. Nguyen., "Variability, coherence and predictability of shallow water acoustic propagation in the Straits of Florida," *Impact of Littoral Environmental Variablility on Acoustic Predictions and Sonar Performance*, edited by Nicholas G. Pace and Finn B. Jensen, Dordrecht/Boston/London: KLUWER ACADEMIC PUBLISHERS, 2002, pp.245-254.
2. Hien B. Nguyen, Harry A. DeFerrari, and Neil J. Williams, "Ocean Acoustic Sensor Installation at the South Florida Ocean Measurement Center," *IEEE J. Oceanic Eng.*, vol. 27, NO.2, 2002, pp.235-244.
3. Evan. K. Westwood, "Broadband Matched Field Source Localization", *J. Acoust. Soc. Am*, **91**, 2777-2789, (1992).
4. N. R. Chapman, J. Desert, A. Agarwal, Y. Stephan and X. Demoulin, "Estimation of Seabed Models by Inversion of Broadband Acoustic Data", *Acta Acustica*, 88, 756-759, (2002).
5. M. B. Porter and H. Bucker, "Gaussian Beam Tracing for Computing Ocean Acoustic Fields", *J. Acoust. Soc. Am*, **82**, 1349-1359, (2002).
6. Evan K. Westwood, C. T. Tindle and N. R. Chapman, "A Normal Mode Model for Acousto-elastic Ocean Environments," *J. Acoust. Soc. Am*, **100**, 3631-3645, (1996).
7. Stan E. Dosso, "Quantifying Uncertainty in Geoacoustic Inversion. I. A fast Gibbs sampler approach," *J. Acoust. Soc. Am*, **111**, 129-142, (2002).
8. Stan E. Dosso, and Peter L. Nielsen, "Quantifying Uncertainty in Geoacoustic Inversion. II. Application to Broadband, Shallow-water Data," *J. Acoust. Soc. Am*, **111**. 143-159, (2002).

High-Frequency Rapid Geo-acoustic Characterization

Kevin D. Heaney

Lockheed-Martin ORINCON Corporation, 4350 N. Fairfax Dr., Arlington VA 22203

Abstract. The Rapid Geo-acoustic Characterization (RGC) algorithm was developed to perform geo-acoustic characterization using the interference patterns of surface ships of opportunity in shallow water. It has been applied successfully at low frequencies to synthetic inversion workshop data and at sea in near real-time. The RGC algorithm determines an effective sediment, which matches the slope (waveguide invariant) and spacing (reciprocal of the time-spread) of interference patterns as well as the fall off in range of the received level. In this paper, we evaluate the extension of the algorithm to mid and high-frequency acoustic signals in shallow water. The issues explored are the robustness of the waveguide invariant, time-spread and TL slope at frequencies above 2 kHz. The algorithm will be applied to a synthetic data example with a known bottom in a range-independent environment.

INTRODUCTION

An approach has been developed to generate effective geo-acoustic parameters using the received low frequency (f<800Hz) acoustic energy from passing surface ships of opportunity. The Rapid Geo-acoustic Characterization (RGC) algorithm uses a set of acoustic observables to compare pre-computed predictions with the measured data. The pre-computation of these observables from a reduced set of sediment parameters provides the near real-time (after the data has been taken) performance. The RGC has been applied to range-dependent inversion workshop data[1, 2], surface ship of opportunity data[3] and active receptions from a seismic survey transmission[4]. The RGC has been used for low frequency acoustics because of the focus on geo-acoustic parameters, surface ship noise as sources, and low frequency bottom loss model upgrades. This paper is an exploration into the utility of the RGC algorithm for higher frequencies (2-6 kHz).

The first issue is the presence of striations at mid to high frequencies. There is no particular reason to presume that striations would not exist, given the coherent interaction of a finite number of acoustic paths, yet at the higher frequencies the fields are known to exhibit much finer scale structure. We begin this paper by looking at some measured data taken by the NATO Undersea Research Centre during their BOUNDARY 2003 experiment. In Fig. 1, the spectrogram for a passing surface ship is shown from 0 to 1500 Hz. The top panel is the broadband Bearing Time Record (BTR). The ship that is being tracked passes from 90 degrees at the beginning of the data, through end-fire CPA at 19:00. The Scissorgram (beam-steered spectrogram) is shown in the bottom panel. The striations are clear out to a range of 6 km and up to a frequency of 1200 Hz.

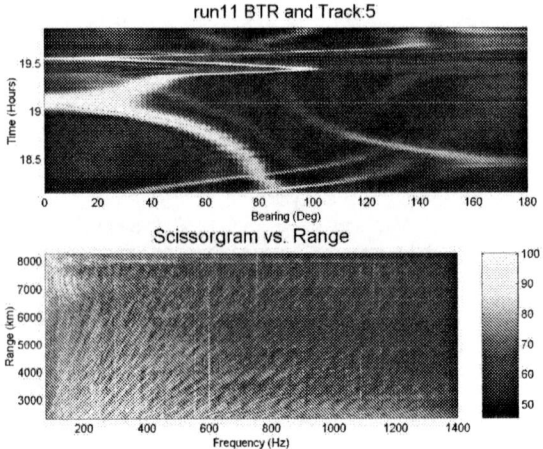

FIGURE 1. At sea measurements during BOUNDARY 2003 of low and mid-frequency striations.

This paper is organized as follows. In section 2 the RGC algorithm is presented. The issue of whether we can use the low frequency acoustic observables at higher frequencies is then examined. The final section involves the application of the RGC to simulated data for a simple sediment model using the frequency band from 4 to 6 kHz.

THE RAPID GEO-ACOUSTIC CHARACTERIZATION ALGORITHM

Striation Slope – Waveguide Invariant (β)

The persistent feature of striations in a shallow-water waveguide has a concise theoretical explanation. The "waveguide invariant" β, was introduced in the Russian literature and is presented in Brekhovskikh and Lysanov[5]. We present a heuristic argument looking at scales of variation in the range/wavenumber and time/frequency domain.

We begin by looking at the dominant oscillations in the intensity of the field as a function of range, at a fixed frequency. Following the Fourier transform analogy, the minimum spacing in range is related to the maximum spread in horizontal wavenumber

$$\Delta r_{min} \approx \frac{2\pi}{\Delta k_{max}} = \frac{2\pi}{(k_{max} - k_{min})} = -\frac{1}{f}\left(\frac{1}{c_{p0}} - \frac{1}{c_{pmax}}\right)^{-1} \quad (1)$$

where the wavenumber has been written as the frequency divided by the phase velocity (c_p).

The time spread is the reciprocal of the minimum frequency spacing of striations, measured from consecutive peaks in the striation pattern at a specific range:

$$\tau = r\left(\frac{1}{v_{g0}} - \frac{1}{v_{g\,max}}\right) \approx \frac{1}{\Delta f_{min}} \quad (2)$$

where the travel time has been written as the range (r) divided by the group velocity (v_g). The normalized slope is the ratio of the frequency spacing divided by the frequency, to the range spacing divided by the range. Taking the ratio of (1) divided by r and the reciprocal of (2) divided by f yields:

$$\frac{\Delta f / f}{\Delta r / r} = -\frac{\left(\frac{1}{c_{p0}} - \frac{1}{c_{p\,max}}\right)}{\left(\frac{1}{v_{g0}} - \frac{1}{v_{g\,max}}\right)} = \beta \quad (3)$$

where β is the waveguide invariant, or normalized slope. For environments that are dominated by surface-reflecting, bottom-reflecting (SRBR) ray paths, the waveguide invariant is nearly one for any choice of rays or modes; hence, the name invariant. In the Pekeris waveguide, $\beta = \cos^2 \theta_{critical}$, which is nearly one. In shallow-water waveguides with more complex sediments, the waveguide invariant varies with the geo-acoustic parameters of the sediment and the water depth[6].

Time Spread (τ)

Having used the slope of the striations as the first observable, there is freedom to choose either the spacing in range or in frequency as the second. The time spread (τ) (Eq. 2) is chosen as the second acoustic observable rather than the wavenumber spacing, because of its relative ease of physical interpretation.

TL slope (α)

The third acoustic observable is the slope of the incoherent TL as a function of range. This parameter is used rather than the absolute TL, to remove the requirement of knowing the source signature. For many shallow-water environments, the Transmission Loss (either band-averaged, or incoherent) can be written as[7]:

$$TL(r) = A + B \log(r) - \alpha r \quad (4)$$

At ranges beyond several water depths, the high angle propagation will be attenuated and this simple empirical fit to the TL (in dB) will be valid. Cylindrical spreading is assumed ($B = 10$) and the attenuation observable, α, is defined as the coefficient to a linear fit of the band-averaged TL curve after taking out the cylindrical spreading. The units of *TL* and *A* are dB, *r* is in kilometers, and α is in dB/km. In this way, it can be shown that α is directly related to an average of the modal attenuation (measured in dB/km).

Sediment Model

To facilitate the rapid search of a large parameter space, a simplified sediment model is used. The model is a single, homogenous unconsolidated sediment layer overlying a hard, reflective basement (half-space). The sediment properties (sound speed, density and attenuation) as a function of depth are determined using a parametric model based on the work of Hamilton[8] and Bachman[9]. The model is written explicitly in Heaney[1]. In the low-frequency RGC, the two search parameters are the sediment thickness and the sediment grain size (or compressional speed at the water-sediment interface). For higher frequencies, we do not expect the sediment thickness to be a factor. A much more significant issue (given the larger relative bandwidth) is the frequency exponent of the sediment attenuation. To this end, we perform a 2-dimensional exhaustive search for the sediment compressional speed (at the interface) and the frequency exponent of attenuation (γ). The equation governing the attenuation coefficient in the sediment is therefore:

$$\alpha(z) = \alpha_0(z) \left(\frac{f(Hz)}{1000} \right)^{\gamma} \quad (5)$$

HIGH FREQUENCY DATA AND ACOUSTIC OBSERVABLE COMPUTATION

We now look at a synthetic data set generated using a range-independent normal mode code. The downward refracting profile is shown in Fig. 2.

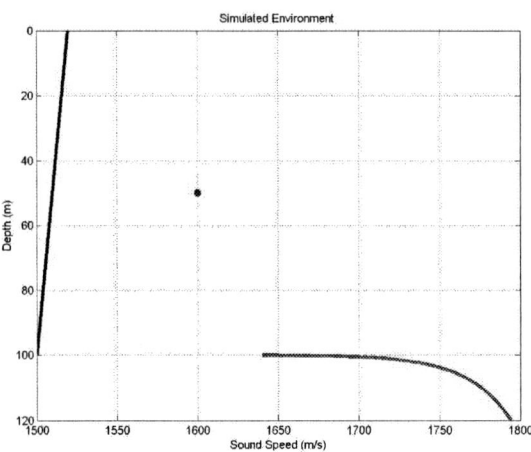

FIGURE 2. Sound speed in the water and sediment for simulations. Source/Receiver depth is 50m.

For this scenario, we model the received signal as if it were from a towed source (at 50 m) to a hydrophone (or beamformed array) at 50 m. The maximum source-receiver range is 3 km, given that we do not expect (because of volume attenuation and source energy spectral levels) to see higher frequencies at longer ranges.

The striation patterns generated from a broadband TL computation are shown in Fig 3. This is a complex pattern exhibiting three distinct interference patterns. The lowest frequency pattern is, we believe, a Lloyds mirror pattern, with a convergence zone at 1800 m. The high frequency oscillations pointing to the upper left (the origin of the plot) correspond to striations with a waveguide invariant value near one. The low frequency oscillations slanting to the upper right correspond to striations with the waveguide invariant value near to -3.

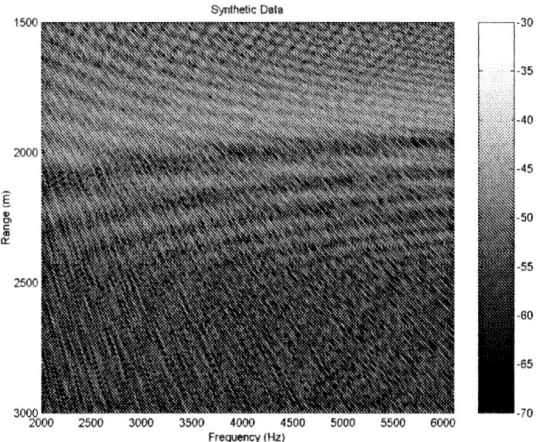

FIGURE 3. Transmission Loss (TL) from a submerged source to a receiver as a function of range/frequency.

Brekhovskikh[5] has shown that for the perfectly reflecting waveguide $\beta = 1$ and for the purely refracted case $\beta = -3$. This computation demonstrates that striations with $\beta = 1$ are visible out to 6 kHz and could, therefore, be used in a geo-acoustic inversion. The high density of striations makes the estimation of the time-spread (reciprocal of the minimum spacing of the dominant striations) difficult so we will utilize a different approach to estimate the time-spread. By computing the autocorrelation of the signal, we can get a (normalized) look at the impulse response of the channel. The results of this computation are shown in Fig. 4.

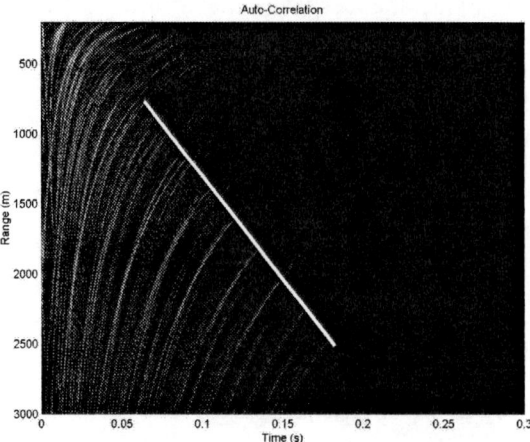

FIGURE 4. Auto-correlation of the signal (corresponding to the spectrum plotted in Fig. 3). The line is hand-drawn to demonstrate the estimation of the time spread as a function of range.

We are now in a position to estimate the acoustic observables for this data-set. The observables are computed at a series of ranges (1000, 1500, 2000, 2500 m) and a frequencies (1000, 2000, …. 6000 Hz). The waveguide invariant (β), the time-spread (τ), and the slope of the incoherent (band-averaged) TL vs. range (α) is shown in Fig. 5 as a function of range and frequency.

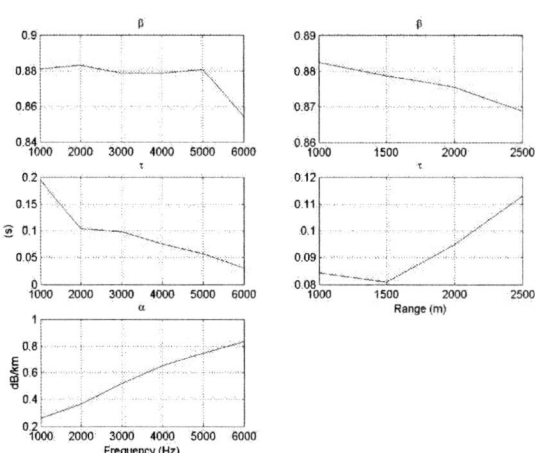

FIGURE 5. Acoustic Observables for simulated dataset. Top panels: waveguide invariant Middle Panels: Time Spread Bottom Panel: TL slope

We see that there is a linear change in the waveguide invariant with range. As the propagation range gets larger, higher angles are stripped out and the effective (average) striation slope decreases. The increase in the time-spread with range indicates that dispersion is dominating over mode attenuation (an indication of a hard-sediment). The TL-slope plot in the lower left indicates that α is a strong function of frequency and is less than 1 dB/km which is also indicative of a hard-sand sediment.

RGC SOLUTION

We now compute determine the RGC solution by computing the acoustic observables for a range of compressional sound speeds at the water-sediment interface (1470 – 1740 m/s) and a range of attenuation coefficients frequency exponents (1 - 2). The approach is quite rapid because we use normal modes at a single frequency to estimate the waveguide invariant (Equation 3), the time-spread and then use the incoherent TL to determine the slope of the TL with range (Equation 5). The full range-dependent model implemented in Heaney[1], used the Parabolic Equation model and required a broadband run for each geo-acoustic estimate. The acoustic observables are computed at each range/frequency for each sediment speed and frequency exponent. The cost function (for each observable) is defined as the RMS error between the predictions and the data over the range/frequency locations of the computation. The results are shown in Fig. 6. For completeness we include using the incoherent TL as a cost function. (This is in the event where the source and receiver are well calibrated.)

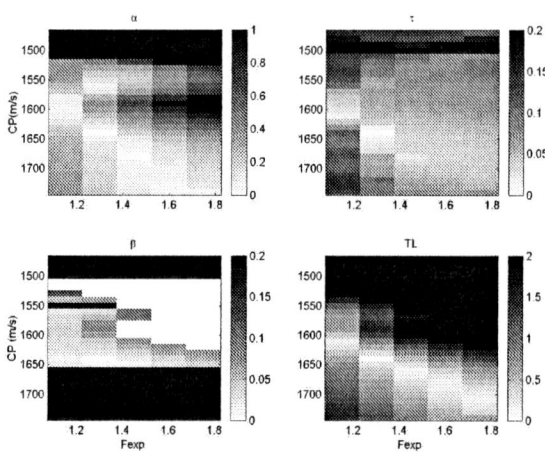

FIGURE 6. Cost functions for the acoustic observables: TL slope, Time-Spread, Waveguide Invariant, and Incoherent TL.

The correct solution (which we are guaranteed to get in this noiseless, simulated example) is with a compressional speed of 1640 m/s and a frequency exponent of 1.3. What is immediately clear from this result is that the correct solution is estimated, but there is degeneracy (ambiguity) between frequency exponent and compressional speed. For the TL inversion in particular (lower-right), sediments with lower compressional speed and lower frequency exponent have similar TLs than those with higher compressional speeds and higher exponents. The combination of the cost functions does reduce the ambiguity because they have slightly different shapes. The waveguide invariant cost function requires further analysis. It is clear that the numerical code for rapidly estimating the waveguide invariant crashes for high values of compressional speed.

CONCLUSION

We have examined whether the Rapid Geo-acoustic Characterization (RGC) algorithm could be successfully applied at mid to high frequencies. From experimental data and numerical modeling we have seen that stable striations are visible at short ranges for passing surface ships in the 1-6 kHz region. These striations do contain information about the geo-acoustics of the region. A synthetic case was run, and it was demonstrated that the acoustic observables (striation slope, striation spacing, and slope of the incoherent TL vs. range) could be used to successfully invert for the geo-acoustic parameters.

ACKNOWLEDGMENTS

This work was sponsored by SPAWAR and the SBIR office, under the GAIT (Geo-Acoustic Inversion Toolbox) contract. The BOUNDARY 2003 test was conducted by NATO's SACLANT Centre for Undersea Research, with Peter N. Nielsen acting as Senior Scientist.

REFERENCES

1. K. D. Heaney, "Rapid Geoacoustic Characterization: Applied to Range-Dependent Environments," *IEEE Journal of Ocean Engineering*, vol. 29, pp. 43-50, 2004.
2. N. R. Chapman, S. Chinbing, D. King, and R. B. Evans, "Special Issue on Geo-acoustic Inversion In Range-Dependent Shallow-Water Environments," *IEEE Journal of Ocean Engineering*, vol. 28, pp. 317-319, 2003.
3. K. D. Heaney, "Rapid Geoacoustic Characterization Using a Ship of Opportunity," *IEEE Journal of Ocean Engineering*, vol. 29, pp. 88-99, 2004.
4. K. D. Heaney, D. D. Sternlicht, A. Teranishi, B. Castille, and M. Hamilton, "Active Rapid Geoacoustic Characterization using a Seismic Survey Source," *IEEE Journal of Ocean Engineering*, vol. 29, pp. 100-109, 2004.
5. L. M. Brekhovskikh and Y. P. Lysanov, in *Fundamentals of Ocean Acoustics*, 2nd ed. New York: Springer-Verlag, 1991, pp. 140-145.
6. G. L. D'Spain and W. A. Kuperman, "Application of waveguide invariants to analysis of spectrograms from shallow water environments that vary in range and azimuth," *Journal of the Acoustical Society of America*, vol. 106, pp. 2454-2470, 1999.
7. R. J. Urick, *Principles of Underwater Sound for Engineers*. New York: McGraw-Hill Book Company, 1967.
8. E. L. Hamilton, "Geoacoustic Modeling of the Seafloor," *Journal of the Acoustical Society of America*, vol. 68, 1980.
9. R. T. Bachman, "Parameterization of Geoacoustic Properties," 1989.

ACOUSTIC COMMUNICATIONS

The impact of underwater acoustic channel structure and dynamics on the performance of adaptive coherent equalizers

Dr. James Preisig

Dept. of Applied Ocean Physics and Engineering, Woods Hole Oceanographic Institution Woods Hole, MA 02543

Abstract. Channel estimate based equalizers are those for which observations of the received signal are used to estimate the channel impulse response and possibly the statistics of the interfering noise field, and these estimates are used to calculate the equalizer filter coefficients. Channel estimate based decision feedback equalizers (CE-DFE), linear MMSE equalizers (L-MMSE), and Passive Time-Reversal equalizers (P-TR) are all examples of coherent channel estimate based equalizers. Equations are derived for analyzing the performance of these channel estimated based equalizers. The performance is characterized in terms of the mean squared soft decision error (σ_s^2) of each equalizer. This error is decomposed into two components. These are the minimum achievable error (σ_o^2) and the excess error (σ_ε^2). The former is the soft decision error that would be realized by the equalizer if the filter coefficient calculation were based upon perfect knowledge of the channel impulse response and statistics of the interfering noise field. The latter is the additional soft decision error that is realized due to errors in the estimates of these channel parameters. Here, the impact of errors in the channel impulse estimation errors on σ_ε^2 is considered (i.e., the equalizer is assumed to have accurate knowledge of the statistics of the interfering noise field.). The error equations allow for a direct comparison of the performance characteristics of these equalizers as well as an evaluation of the impact of the characteristics of the acoustic channel on equalizer performance.

INTRODUCTION

Adaptive coherent equalizers can be divided into two classes, the first being direct adaptation equalizers and the second being channel estimate based equalizers. Direct adaptation equalizers are those for which the filter coefficients of the equalizer are directly adjusted based upon observations of the received signal. Channel estimate based equalizers are those for which observations of the received signal are used to estimate the channel impulse response and possibly the statistics of the interfering noise field and these estimates are used to calculate the equalizer filter coefficients. Channel estimate based decision feedback equalizers (CE-DFE), linear MMSE equalizers (L-MMSE), and Passive Time-Reversal equalizers (P-TR) [3] are all examples of coherent channel estimate based equalizers. The use of the expressions derived here allows these three equalizers to be evaluated in the following context: that is, given estimates of the channel impulse response and the statistics of the interfering noise fields, how do the three different methods of computing equalizer filter coefficients impact the equalizer performance?

The format of this paper is as follows. The next section outlines notation as well as the expressions for the modeled channel impulse response and the equalizer filter coefficients. This section develops insights that allows the equalizer filtering problem to be thought of in terms of "replica vectors" that are derived from the channel impulse response. The section following that presents expressions for the soft decision error achieved by each of the three types of equalizers. The expressions are used to develop insight into the characteristics of channels that most significantly impact equalizer performance, and the relative performance characteristics of the three types of equalizers are evaluated. Finally, the performance equations for a particular type of channel estimator (the exponentially weighted least squares estimator) are presented and that equation are used to evaluate a required error correlation matrix.

CHANNEL AND EQUALIZER MODEL

The acoustic channel is modeled as a time-varying, discrete time system described by the complex baseband input delay-spread function[1] (IDSF) [1]. The received signal at time n is given by

$$u[n] = \tilde{\mathbf{g}}^h[n]\tilde{\mathbf{d}}[n] + v[n]. \tag{1}$$

where

$$\tilde{\mathbf{g}}[n] \triangleq [g[n, N_c - 1], \cdots, g[n, 0], \cdots, g[n, -N_a]]^t, \text{ and}$$

$$\tilde{\mathbf{d}}[n] \triangleq [d[n - N_c + 1], \cdots, d[n], \cdots, d[n + N_a]]^t$$

are samples of the IDSF and transmitted data symbols, respectively. $d[n]$ is the transmitted data symbol at time n, $v[n]$ is the interfering noise at time n, $g[n,m]$ is the IDSF for delay m at time n, the superscript t denotes transpose, and the superscript h denotes Hermitian. The quantities N_a and N_c denote, respectively, the number of acausal and causal taps in the IDSF[2].

Throughout this paper, lower case letters denote scalar quantities, lower case bold face letters denote vectors (all vectors are assumed to be column vectors), and upper case bold face letters denote matrices. Herein the received signal is assumed to be sampled at the transmit symbol rate. The extension of the analysis to fractionally spaced systems is conceptually straight forward, but the notation is cumbersome. The final results of the analysis are equally applicable to symbol rate and fractionally spaced systems.

The equalizers considered here each consist of a finite impulse response (FIR) *feedforward* filter that filters the received signals and, in the case of the CE-DFE, an FIR feedback filter that filters and feeds back estimates of the transmitted data symbol.[3] The

---

[1] The input delay-spread function is one form of what is more commonly referred to as the time-varying impulse response

[2] Physical underwater acoustic channels are all causal. However, it is sometimes conceptually convenient to think of some point in delay (the variable m) within the IDSF to be the zero delay tap and those points that preceed this point in the delay variable to be acausal taps.

[3] The development herein assumes a single channel equalizer. The extension to a multichannel equalizer utilizing a receive array is straightforward and does not alter the results of the analysis.

output of the filter is the soft decision estimate, $\hat{d}_s[n]$, of the transmitted data symbol, $d[n]$. The estimate $\hat{d}_s[n]$ is the input to a decision device that generates the final estimate, $\hat{d}[n]$, of the transmitted data symbol.

For a linear equalizer (e.g., the L-MMSE and P-TR equalizers) the soft decision estimate of the transmitted data symbol, $\hat{d}_s$, is given by

$$\hat{d}_s[n] = \mathbf{h}^h[n]\,\mathbf{u}[n], \tag{2}$$

where $\mathbf{h}[n]$ is a vector of the feedforward filter coefficients at time n and

$$\mathbf{u}[n] \triangleq [u[n-L_c-1], \cdots, u[n], \cdots, u[n+L_a]]^t. \tag{3}$$

Here, L_c and L_a denote the number of causal and acausal taps, respectively, of the feedforward filter. The notation $\mathbf{h}_{\text{lin}}$ and $\mathbf{h}_{\text{tr}}$ will be used to denote the filter coefficient vectors for the L-MMSE and P-TR equalizers, respectively. For the CE-DFE, $\hat{d}_s$ is given by

$$\hat{d}_s[n] = \mathbf{h}_{\text{ff}}^h[n]\,\mathbf{u}[n] + \mathbf{h}_{\text{fb}}^h[n]\,\hat{\mathbf{d}}_{fb}[n]. \tag{4}$$

Here, $\mathbf{h}_{\text{ff}}$ and $\mathbf{h}_{\text{fb}}$ are vectors of the coefficients of the CE-DFE feedforward and feedback filters, respectively. For a feedback filter of length L_{fb} symbols, $\hat{\mathbf{d}}[n]$ is a vector of estimates of past transmitted data symbols given by $\hat{\mathbf{d}}_{fb}[n] \triangleq [\hat{d}[n-L_{fb}], \cdots, \hat{d}[n-1]]^t$.

For both the linear and decision feedback equalizers, $\mathbf{u}[n]$ is the received signal vector that is used by the feedforward filter to generate the soft decision estimate. Combining (1) and (3) yields

$$\mathbf{u}[n] = \mathbf{G}[n]\,\mathbf{d}[n] + \mathbf{v}[n], \tag{5}$$

where

$$\mathbf{d}[n] \triangleq [d[n-L_c-N_c+2], \cdots, d[n], \cdots, d[n+L_a+N a]]^t$$

and

$$\mathbf{v}[n] \triangleq [v[n-L_c], \cdots, v[n], \cdots, v[n+L_a]]^t.$$

$\mathbf{G}[n]$ is the sampled IDSF matrix with the ith row composed of $\tilde{\mathbf{g}}^h[n-L_c+i]$ packed with leading and trailing zeros to position it the appropriate columns of the matrix with respect to the elements of the vector $\mathbf{d}[n]$. It is instructive to represent $\mathbf{G}$ using its column vectors indexed in the following manner:

$$\mathbf{G}[n] = \left[\mathbf{g}_{(N_c+L_c-2)}, \cdots, \mathbf{g}_1, \mathbf{g}_0, \mathbf{g}_{-1}, \cdots, \mathbf{g}_{-(N_a+L_a)}\right], \tag{6}$$

The dependence of the columns of $\mathbf{G}[n]$ on the time index n is suppressed here for notational convenience. The vector $\mathbf{g}_i$ is a replica vector for the data symbol $d[n-i]$ in the received signal vector $\mathbf{u}[n]$. Partition the transmit data symbols in $\mathbf{d}[n]$ into three groups: $\mathbf{d}_{fb}[n] \triangleq [d[n-L_{fb}], \cdots, d[n-1]]^t$, $d[n]$, and $\mathbf{d}_o[n]$ which is composed of the remaining elements of $\mathbf{d}[n]$. Partition the columns of $\mathbf{G}[n]$ into three similarly defined sets: $\mathbf{G}_{fb}$, $\mathbf{g}_0$, and $\mathbf{G}_o$. Then (5) can be rewritten as

$$\mathbf{u}[n] = \mathbf{g}_0 d[n] + \mathbf{G}_{fb}\mathbf{d}_{fb}[n] + (\mathbf{v}[n] + \mathbf{G}_o\mathbf{d}_o[n]). \tag{7}$$

The first term is the portion of the received signal vector, $\mathbf{u}[n]$, that corresponds to the transmitted data symbol to be estimated, $d[n]$. The second term is the portion of $\mathbf{u}[n]$ that can be cancelled by the output of the feedback filter in a CE-DFE, and the terms in the parenthesis represent an effective observation noise that the feedforward filter must try to eliminate. Assuming that the data sequence is a zero mean, white sequence with a variance of one[4], the data sequence is independent of the channel IDSF and $v[n]$, and that $\mathbf{v}[n]$ is a zero mean sequence with covariance $\mathbf{R}_v$ that is independent of the channel IDSF, the effective noise correlation matrix, $\mathbf{Q}$, can be written as

$$\mathbf{Q} = \mathbf{R}_v + \mathbf{G}_o \mathbf{G}_o^h. \tag{8}$$

With the model and quantities so defined, a number of approaches can be used to calculate the optimal filter coefficients. One such approach is given in [2]. In that paper, the effective noise correlation matrix, denoted with the symbol $\mathbf{R}$, includes the impact of channel estimation errors. Therefore, the calculated filter coefficients and subsequent error analysis are valid for the case where the DFE has accurate knowledge of both the noise statistics and the second order statistics of the channel estimation errors. For the filter calculation and performance analysis presented here, there is no assumption that the DFE knows the statistics of the channel estimation errors.

The filter coefficients for the three equalizers are calculated using estimated quantities for $\mathbf{R}_v$ and $\mathbf{G}$. In the following expressions, these estimated quantities are denoted by the hat (e.g., $\hat{\mathbf{R}}_v$). The filter coefficient vectors for the L-MMSE and CE-DFE equalizers are selected to minimize the mean squared soft decision error ($\mathrm{E}[|\hat{d}_s[n] - d[n]|^2]$ assuming that the estimates of $\mathbf{R}_v$ and $\mathbf{G}$ are accurate and that the statistical assumptions stated in the paragraph before (8) hold. The expressions for these filter coefficient vectors are

$$\mathbf{h}_{\mathrm{ff}} = \frac{\hat{\mathbf{Q}}^{-1} \hat{\mathbf{g}}_0}{1 + \hat{\mathbf{g}}_0^h \hat{\mathbf{Q}}^{-1} \hat{\mathbf{g}}_0}, \quad \mathbf{h}_{\mathrm{fb}} = -\hat{\mathbf{G}}_{fb}^h \mathbf{h}_{\mathrm{ff}}, \text{ and} \tag{9}$$

$$\mathbf{h}_{\mathrm{lin}} = \frac{(\hat{\mathbf{Q}} + \hat{\mathbf{G}}_{fb} \hat{\mathbf{G}}_{fb}^h)^{-1} \hat{\mathbf{g}}_0}{1 + \hat{\mathbf{g}}_0^h (\hat{\mathbf{Q}} + \hat{\mathbf{G}}_{fb} \hat{\mathbf{G}}_{fb}^h)^{-1} \hat{\mathbf{g}}_0}. \tag{10}$$

The P-TR equalizer is a normalized matched filter so its coefficients are given by

$$\mathbf{h}_{\mathrm{tr}} = \frac{\hat{\mathbf{g}}_0}{\hat{\mathbf{g}}_0^h \hat{\mathbf{g}}_0}. \tag{11}$$

EQUALIZER PERFORMANCE

The performance is characterized in terms of the mean squared soft decision error $\left(\sigma_s^2 \triangleq \mathrm{E}[|\hat{d}_s[n] - d[n]|^2]\right)$ of each equalizer. This expectation is conditioned upon the estimate of the channel IDSF. This error is decomposed into two components. These are

---

[4] The assumption of unit variance can be made without any loss of generality of the results.

the minimum achievable error (σ_o^2) and the excess error (σ_ε^2) such that $\sigma_s^2 = \sigma_o^2 + \sigma_\varepsilon^2$. σ_o^2 is the soft decision error that would be realized by the equalizer if the filter coefficient calculation were based upon perfect knowledge of the channel impulse response and statistics of the interfering noise field. σ_ε^2 is the additional soft decision error that is realized due to errors in the estimates of these channel parameters.

The Minimum Achievable Error

For the three different equalizers, the minimum achievable error is given by

$$\sigma_{o_{dfe}}^2 = \frac{1}{1 + \hat{\mathbf{g}}_0^h \hat{\mathbf{Q}}^{-1} \hat{\mathbf{g}}_0}, \qquad (12)$$

$$\sigma_{o_{lin}}^2 = \frac{1}{1 + \hat{\mathbf{g}}_0^h (\hat{\mathbf{Q}} + \hat{\mathbf{G}}_{fb} \hat{\mathbf{G}}_{fb}^h)^{-1} \hat{\mathbf{g}}_0}, \quad \text{and} \qquad (13)$$

$$\sigma_{o_{tr}}^2 = \frac{\hat{\mathbf{g}}_0^h (\hat{\mathbf{Q}} + \hat{\mathbf{G}}_{fb} \hat{\mathbf{G}}_{fb}^h) \hat{\mathbf{g}}_0}{\hat{\mathbf{g}}_0^h \hat{\mathbf{g}}_0}. \qquad (14)$$

Comparing equations (12), (13), and (14), it can be shown that

$$\sigma_{o_{dfe}}^2 \leq \sigma_{o_{lin}}^2 \leq \sigma_{o_{tr}}^2.$$

Furthermore, it can be shown that $\sigma_{o_{dfe}}^2$ and $\sigma_{o_{line}}^2$ will always decrease when the number or received signal channels or the length of the feedforward or feedback filters is increased.

Note that the minimal achievable error for the DFE is completely characterized by the quadratic product of the replica vector associated with the data symbol being estimated, $\mathbf{g}_{(0)}$, and the inverse of the effective noise correlation matrix, $\mathbf{Q}$, which represents the contribution of the observation noise and the acausal data symbols ($d[m]$ for $m > n$) and causal symbols $d[m]$ for which $m < n - L_{fb}$ to the input of the feedforward filter and the soft data estimate. Thus, the minimal achievable error of the DFE is determined by the projection of the replica vector $\mathbf{g}_{(0)}$ on both the observation noise correlation matrix $\mathbf{R}_v$ and the replica vectors corresponding to the acausal data symbols and the causal symbols not canceled by the output of the feedback filter. The structure of the channel IDSF impacts the minimal achievable error through these replica vectors. Note that in non-minimum phase channels where there will be a large number of replica vectors corresponding to acausal channel taps, the minimal achievable error will tend to larger than in minimum phase channel with comparable delay spreads. An example of one such class of non-minimum phase channels is the long range, deep water channel.

The Excess Error

The impact of error in the estimation of channel parameters on the performance of each coherent equalizer is quantified by the excess error, σ_ε^2. Here, the impact of errors in the estimation of the IDSF is considered. Let the true channel IDSF matrix, $\mathbf{G}[n]$ be given by

$$\mathbf{G}[n] = \hat{\mathbf{G}}[n] + \mathbf{E}_G \qquad (15)$$

where $\mathbf{E}_G$ is the error in the estimate of the IDSF matrix. For analytic simplicity, assume further that $\mathbf{E}_G$ is statistically independent of $\hat{\mathbf{G}}[n]$ and has zero mean. Then,

$$\sigma^2_{\varepsilon_{dfe}} = \mathbf{h}^h_{ff} \mathbf{R}_{E_G} \mathbf{h}_{ff}, \qquad (16)$$

$$\sigma^2_{\varepsilon_{lin}} = \mathbf{h}^h_{lin} \mathbf{R}_{E_G} \mathbf{h}_{lin}, \text{ and} \qquad (17)$$

$$\sigma^2_{\varepsilon_{tr}} = \mathbf{h}^h_{tr} \mathbf{R}_{E_G} \mathbf{h}_{tr}, \qquad (18)$$

where $\mathbf{R}_{E_G} \triangleq E[\mathbf{E}_G \mathbf{E}^h_G | \hat{\mathbf{G}}]$. Thus the sensitivity of each equalizer to channel IDSF estimation errors is determined by the magnitude squared of the vector of the equalizer's feedforward filter coefficients and the projection of these coefficient vectors on the eigenstructure of $\mathbf{R}_{E_G}$. For the region of high adaptive processing gain, that is $\hat{\mathbf{g}}^h_0 \hat{\mathbf{Q}}^{-1} \hat{\mathbf{g}}_0 \gg 1$, and for equalizers with the same feedforward filter length, it can be shown that

$$|\mathbf{h}_{ff}|^2 > |\mathbf{h}_{lin}|^2 > |\mathbf{h}_{tr}|^2.$$

Thus, if $\mathbf{R}_{E_G}$ is a scalar times the identity matrix and the equalizers have the same number of taps in their feedforward filters, the P-TR equalizer will be less sensitive to channel estimation errors than either the L-MMSE or CE-DFE equalizers. The following section describes conditions under which $\mathbf{R}_{E_G}$ will meet this requirement.

Equation (16) is counterintuitive in that the excess error does not appear to depend upon either the feedback filter coefficients or the errors in these coefficients. This result can be explained as follows. Combining equations (4) and (7) results in

$$\hat{d}_s[n] = \mathbf{h}^h_{ff}[n] (\mathbf{g}_0 d[n] + \mathbf{G}_{fb} \mathbf{d}_{fb}[n] + \mathbf{v}[n] + \mathbf{G}_o \mathbf{d}_o[n]) + \mathbf{h}^h_{fb}[n] \hat{\mathbf{d}}_{fb}[n].$$

Substituting in (9) for $\mathbf{h}_{fb}[n]$, (15), and rearranging terms yields

$$\hat{d}_s[n] = \mathbf{h}^h_{ff}[n] (\hat{\mathbf{g}}_0 d[n] + \mathbf{v}[n] + \mathbf{G}_o \mathbf{d}_o[n]) + \mathbf{h}^h_{ff}[n] (\hat{\mathbf{G}}_{fb} \mathbf{d}_{fb}[n] - \hat{\mathbf{G}}_{fb} \hat{\mathbf{d}}_{fb}[n]) + \mathbf{h}^h_{ff}[n] \mathbf{E}_G \mathbf{d}[n].$$

The expected value magnitude squared of the difference between the first term and the actual data symbol, $d[n]$, is the minimum achievable error. The second term is the only term that depends on the feedback filter coefficients. Assuming that the past symbol decisions are accurate (i.e., $\hat{\mathbf{d}}_{fb}[n] = \mathbf{d}_{fb}[n]$), this term equals zero. The third term represents the excess error. The expected value of the magnitude squared of this term equals (16). Assuming that the channel estimation error is independent of the channel estimate, the transmitted data, and the observation noise, this equals the excess error, $\sigma^2_{\varepsilon_{dfe}}$.

CHANNEL ESTIMATION ERROR

To gain insight into the structure of $\mathbf{R}_{E_G}$ and the impact of channel structure and dynamics on the error in estimating the channel IDSF, consider the performance characteristics of the commonly used Exponentially Weighted Least Squares algorithm. With this algorithm, the estimate of the channel IDSF is given by

$$\hat{\tilde{\mathbf{g}}}[n] = \arg\min_{\mathbf{g}} \sum_{m=-\infty}^{n} \lambda^{(n-m)} \mid u[m] - \mathbf{g}^h \tilde{\mathbf{d}} \mid^2,$$

where λ is a constant "forgetting factor" between zero and one. Assume that the channel IDSF, $\tilde{\mathbf{g}}[n]$, is a zero mean, wide sense stationary random process with correlation matrix $\mathbf{R}_{\tilde{g}}[m] \triangleq E\left[\tilde{\mathbf{g}}[n]\tilde{\mathbf{g}}^h[n+m]\right]$. Then, the error correlation matrix $\mathbf{R}_{\varepsilon} \triangleq E\left[(\hat{\tilde{\mathbf{g}}}[n] - \tilde{\mathbf{g}}[n+1])(\hat{\tilde{\mathbf{g}}}[n] - \tilde{\mathbf{g}}[n+1])^h\right]$ is given by

$$\mathbf{R}_{\varepsilon} = \frac{2}{(1+\lambda)} \mathbf{R}_{\tilde{g}}[0] - \frac{1}{(1+\lambda)} \sum_{m=0}^{\infty} \left(\frac{\lambda^m}{W}\right) \left(\mathbf{R}_{\tilde{g}}[m+1] + \mathbf{R}_{\tilde{g}}^h[m+1]\right) + \frac{1}{(1+\lambda)} \left(\frac{\sigma_v^2}{W}\right) \mathbf{I},$$

where $W \triangleq \sum_{m=0}^{\infty} \lambda^m = (1-\lambda)^{-1}$. To relate this result to the matrix $\mathbf{R}_{E_G}$ used in (16), (17), and (18), note that $\mathbf{R}_{E_G}$ is well approximated by a Toeplitz matrix. Furthermore, the elements along the ith diagonal of $\mathbf{R}_{E_G}$ are equal to the sum of the elements along the ith diagonal of $\mathbf{R}_{\varepsilon}$. This will be used in the following section to develop intuition regarding the impact of channel dynamics on equalizer robustness.

For wide sense stationary, uncorrelated scattering (WSSUS) channels, $\mathbf{R}_{\varepsilon}$ is a diagonal matrix thus resulting in $\mathbf{R}_{E_G}$ equaling the trace of $\mathbf{R}_{\varepsilon}$ times the identity matrix. Therefore, evaluation of (16), (17), and (18) for this case shows that the excess error for each equalizer equals the trace of $\mathbf{R}_{\varepsilon}$ times the magnitude squared of the feedforward filter coefficient vector for each equalizer. This result is independent of the distribution of the IDSF estimation error among the taps of the IDSF vector. While these correlation matrices are not conditioned upon the channel estimate (or equivalently, the calculated feedforward filter weights) as required to properly evaluate (16) through (18), they do lend insights into the channel and equalizer characteristics that impact robustness with respect to channel estimation errors.

CONCLUSION

Expressions for the components of the soft decision error for linear MMSE, decision feedback, and passive time-reversal equalizers have been presented. The error is decomposed into a minimum achievable error and the excess error. Evaluation of the expressions shows that the ranking of the three equalizers from best to worst in terms of minimum achievable error is the CE-DFE, the L-MMSE equalizer, and the P-TR equalizer. However, with simplifying assumptions it is shown that the P-TR equalizer is the least sensitive of the three to channel estimation errors. Insights are presented into the channel characteristics that most significantly impact equalizer performance.

REFERENCES

1. P.A. Bello, "Characterization of randomly time-variant linear channels," *IEEE Trans. Commun. Syst.,* **CS-11**, pp. 360-393, (1963).
2. M. Stojanovic, J. Proakis, J. Catapovic, "Analysis of the Impact of Channel Estimation Errors on the Performance of a Decision-Feedback Equalizer in Fading Multipath Channels," *IEEE Trans. Comm.,* **COM-43**, pp. 877-886, (1995).
3. D. Rouseff, D.R. Jackson, W.L.J. Fox, C.D. Jones, J.A. Ritcey, D.R. Dowling, "Underwater acoustic communications by pasive phase conjugation: Theory and experimental results," *IEEE Journal of Oceanic Engineering,* **26**, pp. 821-831, (2001).

Spatio-Temporal Focusing for Elimination of Multipath Effects in High Rate Acoustic Communications

Milica Stojanovic

Massachusetts Institute of Technology, MIT E38-376, Cambridge, MA 02139

Abstract. High rate underwater communications have traditionally relied on equalization methods to overcome the intersymbol interference (ISI) caused by multipath propagation. An alternative technique has emerged in the form of time-reversal, which comes at virtually no cost in computational complexity, but sacrifices the data rate and relies on large arrays to reduce ISI. In this paperer, optimal multipath suppression using spatio-temporal processing is addressed analytically. A communication link between a single element and an array is considered in several scenarios: uplink and downlink transmission, with and without channel state information and varying implementation complexity. Transmit/receive techniques are designed which simultaneously maximize the data detection SNR and minimize the residual ISI, while maintaining maximal data rate in a given bandwidth and satisfying a constraint on transmitted energy. The performance of various techniques is compared on a shallow water channel operating in the 15 kHz band. Results demonstrate benefits of optimal focusing whose performance is not conditioned on the array size.

INTRODUCTION

High rate acoustic communications have traditionally relied on adaptive equalization methods to overcome the intersymbol interference (ISI) caused by multipath propagation. Excellent performance of these receivers comes at a price of high computational complexity [1]. In standard equalization, all of the signal processing is performed at the receiver, while the transmitter uses standard signaling waveforms. A different, and possibly better approach is to split the signal processing between the transmitter and receiver. Such an approach forms the basis of spatio-temporal focusing.

In its simplest form, focusing is achieved by transmitting a time-reversed (or equivalently, phase-conjugated in the frequency domain) replica of a probe signal received earlier from the source location. This technique has been used for medical imaging, therapy, and material testing, while recent research has hailed it as a method that can replace traditional equalization, and eliminate the associated computational burden (e.g., [2]-[5]). Several research groups have been involved in application of time-reversal (TR) arrays to undersea acoustic communications, addressing active phase-conjugation for two-way communication [2], as well as passive phase-conjugation for one-way communication from a point source to an array [3]. These groups have been engaged in experimental work, emphasizing low-complexity processing using TR only. In parallel, analytical work addressed the use of adaptive channel estimation and low complexity equalization in conjuction with TR [4]. The common goal of these efforts was to eliminate ISI by use of TR. Ensuring ISI-free transmission in a system that has multiple transmit/receive elements is a major asset in a channel whose bandwidth is severely limited. In particular, it lays ground for the exploitation of bandwidth-efficiency improvement available from multi-input multi-output (MIMO) signal processing [6].

A statement commonly encountered in the relevant literature is that TR "undoes" the effects of multipath. However, as the experimental results have shown, suppression of multipath effects by TR is achieved at the expense of reduced data throughput and/or the need for a large array, a consequence of relying on TR to eliminate ISI. TR recombines multipath energy in a manner of matched filtering, whose function is to maximize the SNR, and not to eliminate ISI. In fact, matched filtering increases temporal dispersion of the signal, and in a communication system where a sequence of pulses is transmitted at high rate, it must be followed by a sequence estimator or an equalizer. Increasing the number of elements in a TR array only helps to reduce residual ISI, but it does not eliminate it. The use of equalization in conjunction with TR may become necessary, but the advantage of this approach to standard equalization is not apparent. The use of spatio-temporal focusing for complete suppression of multipath effects thus remains an open question.

In this paper, a solution is proposed to the following problem: If the channel responses between a single element and an array are known, determine the optimal transmit/receive technique that the two can use to simultaneously (1) eliminate ISI *and* (2) maximize SNR, while maintaining *maximal* data rate in a given bandwidth and satisfying a constraint on transmitted energy. Note that because it allows for transmitter as well as receiver optimization, the solution differs from standard equalization. Also, because it explicitly requires minimization of ISI, it differs from TR. The resulting system does not depend on the number of array elements to minimize the multipath distortion, but instead provides an answer for a variety of applications that cannot afford large arrays. Optimal configurations are intended as a basis for adaptive system implementation in which channel estimates will replace the unknown, time-varying responses.

SYSTEM OPTIMIZATION

System optimization is addressed for uplink and downlink communication (to/from array), as shown in Fig.1. Performance is assessed using SNR as the figure of merit, and compared to that of TR, standard linear equalization, and TR in conjunction with

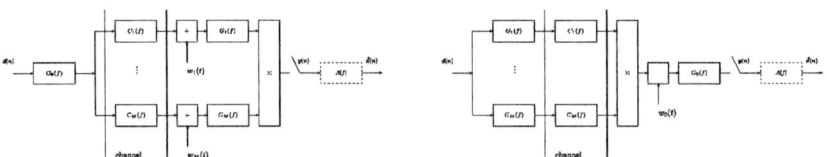

FIGURE 1. Uplink (left) and downlink (right) transmission. An equalizer may or may not be used.

Transmitter/receiver optimization for no ISI. The data sequence $d(n)$ is transmitted at symbol rate $1/T$. The problem is to find transmit/receive filters $G_0(f)$ and $G_1(f),\ldots G_M(f)$ such that the SNR at the receiver is maximized, subject to the constraint that there is no ISI in the decision variables $\hat{d}(n) = y(nT)$, and that finite transmitted energy per bit E is used. The channel responses $C_m(f), m = 1\ldots M$ and the power spectral density $S_w(f)$ of the uncorrelated noise processes $w_m(t), m = 0\ldots M$ are assumed to be known.

The composite channel response is denoted by $F(f) = G_0(f) \sum_{m=1}^{M} G_m(f) C_m(f)$. The received signal after filtering is then given by $y(t) = \sum_n d(n) f(t - nT) + z(t)$, where the noise $z(t)$ has power spectral density

$$S_z(f) = S_w(f) \sum_{m=1}^{M} |G_m^2(f)| \text{ for uplink, and } S_z(f) = S_w(f)|G_0^2(f)| \text{ for downlink.} \quad (1)$$

The requirement for no ISI is expressed as $F(f) = X(f)$, where $X(f)$ is Nyquist, i.e., bandlimited to $\pm 1/T$ and its waveform $x(t)$ satisfies $x(nT) = x_0 \delta_{n,0}$. $X(f)$ can be chosen as raised cosine, and without loss of generality we take that $X(f) = |X(f)|$. When there is no ISI, the sampled received signal is $y(nT) = d(n)x_0 + z(nT)$, and $SNR = \sigma_d^2 x_0^2 / \sigma_z^2$, where $\sigma_d^2 = E\{|d^2(n)|\}$ and $\sigma_z^2 = \int_{-\infty}^{+\infty} S_z(f) df$. The energy constraint is expressed as

$$E = \sigma_d^2 \int_{-\infty}^{+\infty} |G_0^2(f)| df \text{ for uplink, and } E = \sigma_d^2 \sum_{m=1}^{M} \int_{-\infty}^{+\infty} |G_m^2(f)| df \text{ for downlink.} \quad (2)$$

Let us first consider uplink transmission. Taking into account the no ISI requirement and the energy constraint, the SNR is expressed as

$$SNR = E x_0^2 \left[\int_{-\infty}^{+\infty} \frac{X^2(f)}{|\sum_m G_m(f) C_m(f)|^2} df \int_{-\infty}^{+\infty} S_w(f) \sum_{m=1}^{M} |G_m^2(f)| df \right]^{-1}. \quad (3)$$

This function is to be maximized with respect to the receive filters $G_m(f)$. To do so, we use a two-step procedure, each involving one Schwarz inequality. The first inequality states that

$$\left| \sum_{m=1}^{M} G_m(f) C_m(f) \right|^2 \le \gamma(f) \sum_{m=1}^{M} |G_m^2(f)|; \; \gamma(f) = \sum_{m=1}^{M} |C_m^2(f)| \quad (4)$$

where the equality holds for

$$G_m(f) = \alpha(f) C_m^*(f). \quad (5)$$

We note similarly with TR in that receiving filters should be proportional to the phase-conjugate of the channel transfer functions. However, there is room for additional improvement through optimization of the function $\alpha(f)$. Applying a second Schwarz inequality to the denominator of the SNR bound yields:

$$\int_{-\infty}^{+\infty} \frac{X^2(f)}{\gamma(f) \sum_m |G_m^2(f)|} df \int_{-\infty}^{+\infty} S_w(f) \sum_{m=1}^{M} |G_m^2(f)| df \ge \left[\int_{-\infty}^{+\infty} \frac{X(f)}{\sqrt{\gamma(f)}} \sqrt{S_w(f)} df \right]^2 \quad (6)$$

where the equality holds for

$$X(f)/\sqrt{\gamma(f)S_w(f)} = \beta \sum_{m=1}^{M} |G_m^2(f)| \tag{7}$$

and β is a constant. Combining the two conditions (5) and (7) we obtain the optimal value of $\alpha(f)$, which determines the the transmit filter $G_0(f) = X(f)/[\alpha(f)\gamma(f)]$, and the constant β then follows from the energy constraint (2). The desired solution is

$$G_0(f) = K(f)\sqrt{X(f)}\gamma^{-1/4}(f); \; G_m(f) = K^{-1}(f)\sqrt{X(f)}\gamma^{-3/4}(f)C_m^*(f), m = 1...M$$
$$\text{where } K(f) = \sqrt{E/\sigma_d^2}\left[\int_{-\infty}^{+\infty} S_w^{1/2}(f)X(f)\gamma^{-1/2}(f)df\right]^{-1/2} S_w^{1/4}(f). \tag{8}$$

This selection of filters achieves maximal SNR,

$$SNR_2 = Ex_0^2 \left[\int_{-\infty}^{+\infty} \sqrt{S_w(f)} \frac{X(f)}{\sqrt{\gamma(f)}} df\right]^{-2} \tag{9}$$

where index '2' indicates that both sides of the link adjust their filters in accordance with the channel. Optimization in the downlink case gives filters in identical form, except that the factor $K(f)$ is the reciprocal of that given in (8). The same maximal SNR is achieved.

We now turn to the situation in which one side of the link is constrained to have minimal complexity, such as when limited processing power is available at the single-element end. The no ISI condition still must hold, $F(f) = X(f)$, but the filter $G_0(f)$ may no longer be a function of the channel responses. A similar optimization procedure results in the following solution:

$$\text{uplink: } G_0(f) = K\sqrt{X(f)}; \; G_m(f) = K^{-1}\sqrt{X(f)}\gamma^{-1}(f)C_m^*(f), m = 1...M$$
$$\text{where } K = \sqrt{E/(\sigma_d^2 x_0)} \tag{10}$$

$$\text{downlink: } G_0(f) = K^{-1}(f)\sqrt{X(f)}; \; G_m(f) = K(f)\sqrt{X(f)}\gamma^{-1}(f)C_m^*(f), m = 1...M$$
$$\text{where } K(f) = \sqrt{E/\sigma_d^2}\left[\int_{-\infty}^{+\infty} S_w(f)X(f)\gamma^{-1}(f)df\right]^{-1/2} S_w^{1/2}(f). \tag{11}$$

This selection of filters achieves maximal SNR available with one-side adjustment,

$$SNR_1 = Ex_0 \left[\int_{-\infty}^{+\infty} S_w(f) \frac{X(f)}{\gamma(f)} df\right]^{-1}. \tag{12}$$

Comparing the SNR available with and without complexity constraint, we find that $SNR_1 \leq SNR_2$. The two SNRs are equal only when $\gamma(f)$ is proportional to $S_w(f)$. In what follows, we shall focus on the usual case of white noise, $S_w(f) = N_0$. Note that $K(f)$ then becomes a constant, and the same set of filters may be used for uplink and downlink transmission.

TR performance with residual ISI. When no care is taken to ensure focusing, samples of the received signal, $y(nT) = \sum_k f(kT)d(n-k) + z(nT)$, contain residual ISI. Assuming uncorrelated data symbols, the SNR is given by $SNR_0 = \sigma_d^2 |f^2(0)|/[\sigma_d^2 \sum_{k\neq 0} |f^2(kT)| + \sigma_z^2]$. We consider the following situation. On the uplink, the transmitter uses $G_0(f) = K_u\sqrt{X(f)}$, and the receiver uses $G_m(f) = G_0^*(f)C_m^*(f)$. This scenario is analogous to ideal (noiseless) passive phase-conjugation. On the downlink, the transmitter uses $G_m(f) = K_d\sqrt{X(f)}C_m^*(f)$. This scenario is analogous to active phase-conjugation. The receiver filter is $G_0(f) = \sqrt{X(f)}$. The constants K_u, K_d are determined from the energy constraint (2). The SNR in either case is obtained as

$$SNR_0 = \frac{E/N_0}{\rho E/N_0 + \frac{x_0}{\int_{-\infty}^{+\infty} X(f)\gamma(f)df}}; \rho = \frac{\sum_{k\neq 0}|f^2(kT)|}{|f^2(0)|} = \frac{\int_{-1/2T}^{+1/2T}|X\gamma[f]|^2 dfT}{\int_{-\infty}^{+\infty} X(f)\gamma(f)df} - 1 \quad (13)$$

where $X\gamma[f]$ denotes the folded spectrum of $X(f)\gamma(f)$, $X\gamma[f] = \frac{1}{T}\sum_k X(f+\frac{k}{T})\gamma(f+\frac{k}{T})$. We note that as the noise vanishes, i.e. $E/N_0 \to +\infty$, unlike with optimal focusing where $SNR_{1,2} \to +\infty$, the performance of TR saturates, $SNR_0 \to 1/\rho$. The value of ρ depends on the channel through the function $\gamma(f)$, and on the system bandwidth through $X(f)$.

TR performance with equalization. The performance of TR saturates because of residual ISI. To overcome this limitation, an equalizer may be used. An optimal (MMSE) linear processor used on the downlink consists of the matched filter, $G_0(f) = [\sum_{m=1}^M G_m(f)C_m(f)]^*$, and a symbol-spaced equalizer with transfer function $A[f] = \sigma_d^2 F^*[f]/(\sigma_d^2|F[f]|^2 + S_z[f])$, where $F[f]$ is the folded spectrum of the overall response $F(f) = |G_0^2(f)|$, and $S_z[f] = N_0 F[f]$ is the power spectral density of the discrete-time noise process $z(nT)$. The SNR at the equalizer output is

$$SNR = \left[\int_{-1/2T}^{+1/2T} \frac{dfT}{1+(\sigma_d^2/N_0)F[f]}\right]^{-1} - 1. \quad (14)$$

For the transmit filter selection as in active phase-conjugation, $G_m(f) = K_d\sqrt{X(f)}C_m^*(f)$, the SNR is

$$SNR_{3,tr} = \left[\int_{-1/2T}^{+1/2T} \frac{dfT}{1+\frac{E/N_0}{\int_{-\infty}^{+\infty} X(f)\gamma(f)df}X\gamma^2[f]}\right]^{-1} - 1 \quad (15)$$

where $X\gamma^2[f]$ is the folded spectrum of $X(f)\gamma^2(f)$.

Equalizer performance. A standard equalizer does not rely on TR at the transmitter, but instead uses pre-determined, channel-independent filters $G_m(f) = K_d\sqrt{X(f)}$. The SNR is computed from (14) as

$$SNR_{3,down} = \left[\int_{-1/2T}^{+1/2T} \frac{dfT}{1+\frac{E/N_0}{Mx_0}X\Sigma^2[f]}\right]^{-1} - 1 \quad (16)$$

where $X\Sigma^2[f]$ is the folded spectrum of $X(f)|\sum_m C_m(f)|^2$.

In the uplink scenario, the MMSE linear processor consists of a bank of matched filters, $G_m(f) = G_0^*(f)C_m^*(f)$ as in passive phase-conjugation, followed by the equalizer whose optimal transfer function now depends on $F(f) = |G_0^2(f)|\gamma(f)$. For the standard transmit filter selection, $G_0(f) = K_u \sqrt{X(f)}$, the SNR (14) becomes

$$SNR_{3,up} = \left[\int_{-1/2T}^{+1/2T} \frac{dfT}{1 + \frac{E/N_0}{x_0} X \gamma[f]} \right]^{-1} - 1. \tag{17}$$

Comparing uplink and downlink equalization, we have that $SNR_{3,up} \geq SNR_{3,down}$. The two are equal if the M channel transfer functions $C_m(f)$ are identical and constant within the signal bandwidth. It is not clear, however, how $SNR_{3,down}$ compares with $SNR_{3,tr}$, i.e. is there an advantage to using transmit TR in conjunction with equalization. This question gives rise to a broader one of optimal transmit filtering for equalization.

Equalizer performance with optimized transmit filter. If the requirement for no ISI is relaxed in the optimal system design, and an equalizer is used at the receiver, the question is what transmit/receive filtering should be used to maximize the SNR. Note that because this optimization criterion is less restrictive than that of optimal focusing (the no ISI constraint has been removed) improved performance may be expected.

Maximization of SNR (14) with respect to the transmit filter(s) is accomplished using the Lagrange method, yielding the following uplink/downlink solution when the system operates in minimal bandwidth $B = 1/T$ (see [7] for details):

$$SNR_4 = \left\{ 1 - \int_{B_L} dfT + [\int_{B_L} \frac{dfT}{\sqrt{\gamma(f)}}]^2 [\frac{E}{N_0} + \int_{B_L} \frac{dfT}{\gamma(f)}]^{-1} \right\}^{-1} - 1 \tag{18}$$

where $B_L = \{f : \gamma(f) \geq \gamma_L\}$, and γ_L is the smallest value of $\gamma(f)$ for which

$$\left[\frac{E}{N_0} + \int_{B_L} \frac{dfT}{\gamma(f)} \right] \left[\int_{B_L} \frac{dfT}{\sqrt{\gamma(f)}} \right]^{-1} \geq 1/\sqrt{\gamma_L}. \tag{19}$$

When $B_L = [-1/2T, 1/2T]$, it is easy to show that $SNR_4 = SNR_2 + SNR_2/SNR_1 - 1 \geq SNR_2$. Thus, this signaling scheme outperforms optimal focusing.

PERFORMANCE COMPARISON

To compare the performance of various techniques, a channel model based on the geometry of shallow water multipath is used. We look at repeated surface-bottom reflections and take into account P multipath arrivals, each characterized by a gain c_p, delay τ_p and angle of arrival θ_p. Nominal acoustic propagation loss that occurs for practical spreading at carrier frequency f_c=15 kHz is used to compute the gains. The channel transfer functions, observed at d-spaced elements $m = 1...M$ are computed as

$$C_m(f) = \sum_{p=1}^{P} c_{m,p} e^{-j2\pi f \tau_p}, \text{ where } c_{m,p} = c_p e^{-j(m-1)\varphi_p} \text{ and } \varphi_p = 2\pi \frac{f_c}{c} d \sin \theta_p. \tag{20}$$

As an example, we use a channel of depth 75 m, range 3 km, and the system mounted near the bottom. The resulting multipath profile for $P = 3$ is shown in Fig.2. The channel function $\gamma(f)$ is shown for $M = 4$ and 32, together with the desired system response $X(f)$, chosen to provide maximal bit rate for ISI-free transmission in bandwidth $B = 1/T$=5 kHz (10 kbps with 4-PSK, or 15 kbps with 8-PSK). The impulse response of the overall system obtained with TR is also shown, and is evidently far from ideal. As the number of array elements is increased, $\gamma(f)$ tends to flatten out, resulting in better, but not complete suppression of multipath through TR.

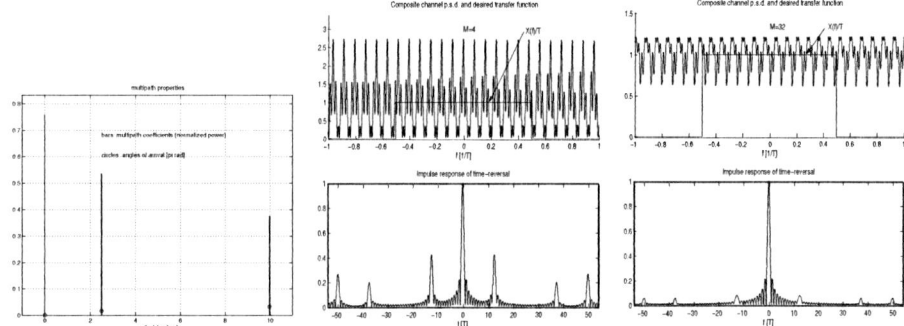

FIGURE 2. Multipath profile of the example channel (left). Composite channel power spectral density is $\gamma(f)$, and the impulse response of TR corresponding to $X(f)\gamma(f)$. Multipath coefficients are normalized such that $M \sum_p |c_p^2| = 1$, and half wavelength spacing between array elements is used.

Figure 3 summarizes performance results. Looking at $M = 4$ case, we first confirm that two-sided adjustment (9, solid '△') outperforms one-sided adjustment (12, dashed '△') in optimal focusing, but more interestingly, we observe that the difference in performance is small. This is an encouraging observation from the viewpoint of designing a practical system with restricted processing complexity. The performance of TR (13, dashed) is inferior to optimal focusing and to all other schemes at practical SNR, with loss becoming quite large even at a moderate E/N_0 of 10 dB - 15 dB. It saturates thereafter at a value $1/\rho$. Some of the loss is recovered by the use of equalizer in conjunction with TR (15,'+'); however, this system compares poorly with the standard equalizer (16, dashed 'o') as E/N_0 increases to more than a few dB. Finally, we confirm that equalization using optimized transmit filters (18, '*') provides an upper bound on the performance of all other schemes. More importantly, we observe that this scheme offers negligible improvement over focusing, which allows a much easier implementation.

With $M = 32$, the performance of TR is improved; nonetheless, saturation is still notable. Equalization in conjunction with TR now outperforms standard equalization, while the performance of both focusing methods tends to the same optimal curve.

CONCLUSION

Because it ignores residual ISI, TR exhibits performance saturation, and strongly depends on the use of a large array. When this can be afforded (e.g. in a network whose base station uses a large array to spatially isolate multiple users) TR offers a solution for

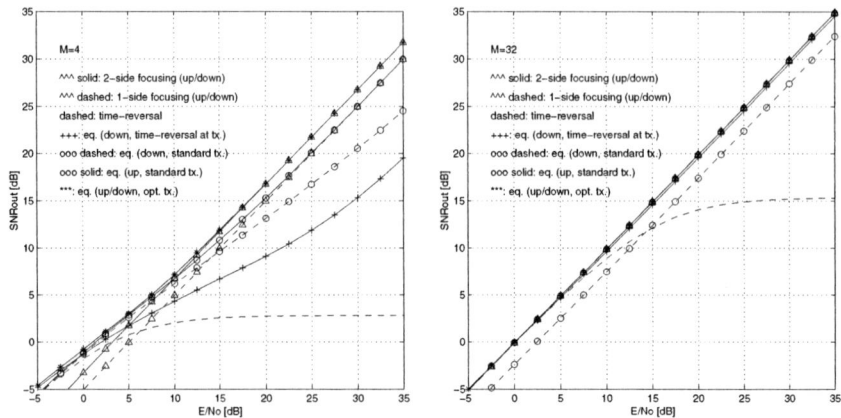

FIGURE 3. Performance of various techniques on the example channel.

minimal-complexity processing. With a smaller array, however, standard equalization outperforms TR, and the use of equalization in conjunction with TR does not guarantee performance improvement over standard equalization. Spatio-temporal focusing proposed in this paper guarantees maximal SNR and elimination of ISI for an arbitrary array size. It outperforms TR at the expense of additional filtering. If filter adjustment is constrained to the array side only, one-sided focusing offers an excellent trade-off between complexity and performance. It thus represents a solution for systems that cannot deploy large arrays and have limited processing power.

Future work will concentrate on experimental validation of spatio-temporal focusing aided by adaptive channel estimation. Two types of errors will guide the practical system performance: error due to noise and error due to time-variability of the channel. Analytical work will address system optimization with imperfect channel knowledge.

REFERENCES

1. M.Stojanovic, J.Catipovic and J.Proakis, "Reduced-complexity multichannel processing of underwater acoustic communication signals," *J. Acoust. Soc. Am.*, vol.98 (2), Pt.1, pp.961-972, Aug. 1995.
2. G.Edelmann, W.Hodgkiss, W.Kuperman and H.C.Song, "Underwater acoustic communication using time-reversal," in Proc. *IEEE Oceans'01 Conference*, Nov. 2001.
3. D.Rouseff et al., "Decision-directed passive phase-conjugation for underwater acoustic communication: experimental results," in Proc. *IEEE Oceans'02 Conference*, Oct. 2002.
4. J.Gomes and V.Barroso, "Time-reversed communication over Doppler-spread underwater channels," in Proc. *ICASSP'02 Conference*, pp.2849-2852(III), 2002.
5. T.C.Yang, "Temporal resolutions of time-reversal and passive phase-conjugation for underwater acoustic communications," *IEEE J. Oceanic Eng.*, vol.28, No.2, pp. 229-245, Apr. 2003.
6. D.Kilfoyle, J.Preisig and A.Baggeroer, "Spatial modulation over partially coherent multi-input / multi-output channels," *IEEE Trans. Sig. Proc.*, vol.51, No.3, pp.794-804, March 2003.
7. M.Stojanovic, "Retrofocusing techniques for high rate acoustic communications," MIT SG Internal Report, Jan. 2004. Available upon request (millitsa@mit.edu).

Synthetic Undersea Acoustic Transmission Channels

Dale Green*, Joseph Rice[¶]

*Benthos, Inc., 49 Edgerton Dr., North Falmouth, MA 02556
[¶]SPAWAR Systems Center, & Department of Physics, Naval Postgraduate School, Monterey, CA 93943

Abstract. Achieving effective through-water acoustic digital signaling (telesonar) requires an ability to adaptively accommodate a complex and possibly time-varying acoustic channel. Variable combinations of noise, interference, multipath, and motion impair real-world telesonar channels. When considering any one of these factors individually, performance degradation may be predicted from theory. But the combination of these factors can confound our theoretical predictive capabilities. A computer simulation of the acoustic channel is useful for developing telesonar waveforms and modems. The simulation directly drives the modem receiver with a virtually propagated analog signal, enabling us to test the performance of the synthetic end-to-end telesonar link. We have observed a close correlation between simulation-based performance and observed performance in at-sea channels exhibiting similar characteristics. The fact that telesonar performance is now quite predictable in a wide variety of channels is due, in large measure, to the use of channel simulation. The simulation presented here does not rely on physical modeling of the channel. Rather, it is based on the combination of theoretical multipath models (e.g., a Rician channel) with rapidly time-varying impulse response functions, where the statistics are derived from at-sea experiments or governed by values derived from independent physics-based models (e.g., PC-SWAT, Bellhop, etc). Noise and other additive interference are combinations of theoretical and stored data, and range rate-induced compression and dilation are incorporated.

INTRODUCTION

The achievement of practical underwater acoustic communications, or telesonar, owes much to the earlier development of terrestrial wireless communications. Indeed, the fundamental theoretical precepts of wireless communications apply equally well to telesonar, and the physical layer signaling techniques used for telesonar would be recognized in most communications textbooks. The primary difference between the two is the adjustments made to accommodate to the physical channel. In particular, telesonar deals with very limited signal bandwidths, with frequency-dependent fading caused by multipath, with long latencies, and with time-varying channel impulse response functions. These factors have forced the development of a relatively unique structure for the implementation of effective communications in the underwater acoustic channel.

A good example of the differences between telesonar and RF wireless communications is the use of power control in the RF world to enable multi-access communications. In the code division multiple access (CDMA) cell phone system used in the U.S., a base station maintains instantaneous links with many individual cell phones via a secondary channel unseen by the cell phone user. This link is used to control the transmitted power levels of all phones so that all primary signals are received at the base station with approximately equal power. This enables the use of a type of signal which has auto- and cross-correlation properties similar to those of uncorrelated, Gaussian, white noise. However, in the acoustic domain, several issues conspire to make such a network difficult to achieve:

mobile nodes at some distance from a base station may well move into quite different channels by the time a power control signal could be received from the base; furthermore, the very limited bandwidths and transducer systems currently available have to date prevented the development of full-duplex signaling.

Rapid temporal variations in the impulse response functions may severely constrain the use of phase coherent signaling but have markedly less impact on non-coherent signaling techniques. This fundamental issue with the physics of telesonar channels has been addressed with some success by the use of adaptive channel equalization, especially via the decision feedback equalizer (DFE) originally developed by Proakis [1]. However, there still are important, unresolved issues with channel equalizers which make their performance difficult to predict.

The tremendous variety of "real-world" telesonar channels, combined with the very real difficulty of obtaining sufficient physical information to fully characterize a given channel, has substantially constrained the ability of physics-based propagation models to predict the performance of telesonar systems. Even in those situations in which good physical characterization is available, the real-time computational burden required for physics-based modeling of high frequency, time-varying channels is considerable.

Our approach to telesonar simulation is to rely on statistics which describe: the (possibly) time-varying impulse response function of the channel; the character of the interference (both "noise" and discrete interferences); and certain properties of the communicating platform (e.g., range rate). These statistics may be obtained from physics-based propagation models (e.g., PC-SWAT, Bellhop, etc), or they may be obtained from pertinent at-sea experiments.

Our approach is to identify telesonar performance in statistically describable channels, and not to replicate any specific physical channel. Performance is based on metrics obtained from real telesonar modems with analog input provided by a simulator. In this way we test the end-to-end link as one system. We evaluate performance in well-understood, benign channels, then search for worst case channel conditions which "break" the modem.

SIGNAL DISTURBANCES

The phenomena we discuss which disturb the transmitted signal are those which have been observed empirically as detracting in a substantial way from modem performance. In this paper we discuss modeling of the signal received over a complex multipath channel, representation of noise, modeling of discrete interferences and jammers, and constant velocity motion (range rate). There is no particular importance of ordering in the following descriptions. Indeed, on a specific modem, the importance of a particular channel characteristic is determined by the effect it has on modem performance: rapid temporal variation in the impulse response function may severely constrain the use of phase coherent signaling, but have virtually no impact on non-coherent signaling techniques.

Modeling and Simulating The Acoustic Channel

Modeling

In the following, we first describe the receipt of a perfectly observable signal, but we quickly recognize that empirical estimation of channel characteristics will be constrained by the nature of the signal probes we use. In particular, we would employ a broadband (but band-limited) pulse processed with a replica correlator to examine the (band-limited) approximation to the channel impulse response. This specifically limits the observation of those components of the impulse response which are temporally closer than approximately 1/W, with W the pulse bandwidth.

The impulse response function describes the (possibly time-varying) temporal and spectral distribution of energy presented to a receiver by the channel. This function has been extensively studied [2,3], and it is not our purpose to replicate the detailed mathematical derivations that have been developed by others. We simply present our version of the received signal as follows. The received signal is ideally described as a collection of N amplitude-weighted, delayed versions of the transmitted signal s(t) with amplitude $a_n(t)$ and delay τ_n: For convenience, we assume that s(t) is analytic (1-sided spectrum).

$$r(t) = \sum_n a_n(t) s(t - \tau_n). \qquad (1)$$

In eqn (1), we assume the following:
1. $a_n(t)$ is a circular (complex) Gaussian random variable with zero mean and with power (variance) $b_n^2 = E_a |a_n(t;\tau)|^2$, where E_a is the expectation operator over the amplitudes for the nth path.
2. b_n^2 is constant for the duration of a single transmission, although the complex nature of $a_n(t)$ may change during that duration.
3. The delays τ_n reflect theoretical predictions or at-sea observations of time spread. They are constant for the duration of a single transmission.

Because the path power is constant, the time-varying nature of the <u>individual</u> path amplitude is described by the (assumed) wide sense stationary coherence function of the path, ρ_n:

$$\rho_n(\Delta t) = E_a |a_n^*(t) a_n(t + \Delta t)|^2 / b_n^2 \qquad (2)$$

We observe that ρ_n, being a coherence function, has a duality in the Fourier domain which may be interpreted as a power spectrum $P_n(f)$ with integrated power equal to b_n^2. The effective bandwidth of $P_n(f)$ is determined by the time difference at which $\rho_n(\Delta t)$ drops to an agreed-upon value. We define B_n to be the (3-dB) bandwidth of this path, or the inverse of Δt for which $\rho_n(\Delta t) = 0.75$. We also observe that the Fourier transform of $a_n(t)$ is $A_n(f)$, with the property that

$$P_n(f) = E_a |A_n(f)|^2. \qquad (3)$$

Now, we assumed that $a_n(t)$ was a Gaussian function, so $A_n(f)$ is likewise Gaussian at every frequency component.

As an example, consider a condition in which we had observed a channel with two dominant paths, with a 4 ms temporal separation between them, a relative power of 1 and 0.5, respectively, and coherence times of 50 ms and 20 ms, respectively. The upper trace of Figure 1 shows a realization of the path weight $|a_1(t)|$, while the lower trace shows the sample auto-coherence properties of this weight. The dashed line in the upper plot indicates the duration of the signal relative to this weight. Figure 2 shows corresponding sample results for the second path. Note the width of the coherence function for the two paths, which correspond well with the specified (*a priori*) coherence time.

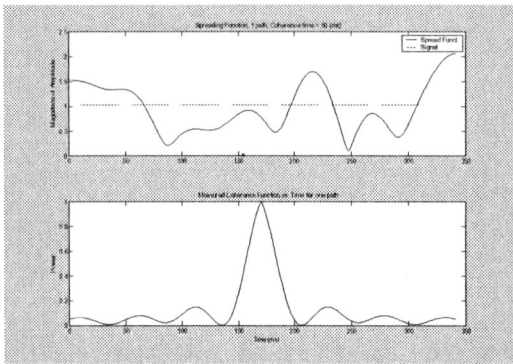

Figure 1. A sample temporal weighting function reflecting a single path in an acoustic channel. the coherence time (75% confidence level) for this path is specified to be 50 ms.

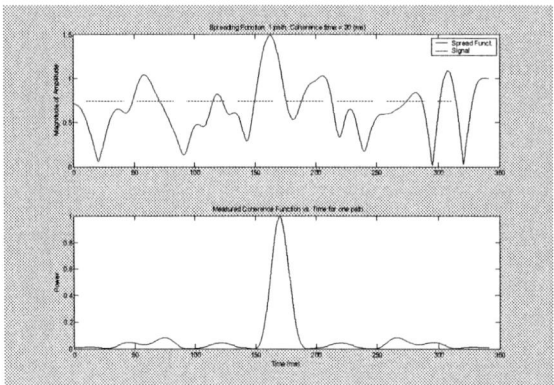

Figure 2. A sample temporal weighting function reflecting a single path in an acoustic channel. The coherence time (75% confidence level) for this path is specified to be 20 ms.

According to Eqn (1), the desired signal, s(t), is independently multiplied by each of the weights, and the products are delayed according to the respective path delays, and the results summed. The upper plot of Figure 3 shows a sample waveform received over our example 2-path channel. A standard signal processing technique is to process this signal with a matched filter, the filter being a copy of the transmitted waveform. The lower plot of Figure 3 shows the output of such a matched filter. It is seen that this is a useful tool in evaluating the gross characteristics of the channel: the delay between the two paths is accurate, and the relative powers are reasonably close to the specification.

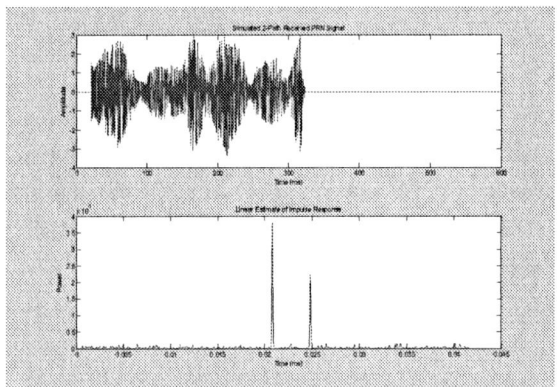

Figure 3. A PRN waveform received over a simulated 2-path channel (upper plot). The lower plot shows a classic matched filter output when the entire waveform is used as the filter.

We now consider the limitations we face in measuring and estimating these channel characteristics. In particular, the assumption implicit in eqn (1), that the received components of r(t) are observable cannot be justified or measured because the channel impulse response will be estimated via the correlation function of a band-limited waveform. The correlation function will have a temporal resolution approximately equal to the inverse of the waveform bandwidth, W. Thus, if several paths deliver the waveform with a temporal separation less than this resolution, the estimation will reflect a correlation among these arrivals. Furthermore, in a typical analysis system, the temporal sampling interval δt will be considerably less than $1/W$, so the information contained in each temporal sample of the correlation is itself correlated with adjacent samples. Thus we introduce yet another correlation function $\kappa_{n,m}$ which describes the similarity between the nth and m^{th} measured path components.

$$\kappa_{n,m} = E_\tau |a_n^* a_m|^2 \qquad (4)$$

Our fifth assumption is thus
   4. $\kappa_{n,m} = 0$, if $|n-m|\, \delta t > 1/W$

Simulation

At this point we consider only the sampled received signal, with a sample rate fs and a sample interval of $\delta t = 1/fs$. The transmitted signal is of duration T, so there are $K = T/\delta t$ samples. Thus, each path weight $a_n(p\delta t)$ is represented by K samples, with $1 \leq p \leq K$, and the Fourier transform $A_n(k\delta f)$ is represented by K frequency samples, with $1 \leq k \leq K$, with frequency spacing $\delta f = fs/K$. We choose to ignore the expectation operator, and work with individual realizations, obtaining thereby estimates $\tilde{a}_n(p\delta t)$ and $\tilde{A}_n(k\delta f)$. The intention is to fill the individual spectral bins of $\tilde{A}_n(k\delta f)$ with independent, zero-mean, complex Gaussian random numbers. We use unit variance Gaussian random numbers to fill the spectral bins lying under a window function, with the resulting spectrum normalized such that the integrated power is b_n^2. We use a Hanning window, with 3-dB power levels positioned at f_1 and f_2. We now compute a realization of $\tilde{a}_n(p\delta t)$ as the inverse Fourier transform of $\tilde{A}_n(k\delta f)$.

We have now computed all $\tilde{a}_n(p\delta t)$ as independent random realizations. That is, every possible delay τ_n reflects an independent path arrival. However, eqn (4) prescribes the requisite path-to-path temporal correlation. We approximate this correlation via application of a square root filter in the delay domain:

$$\tilde{a}_n(p\delta t) = (1-(\kappa_{n,n-1})^{1/2})\,\tilde{a}_n(p\delta t) + (\kappa_{n,n-1})^{1/2}\,\tilde{a}_{n-1}(p\delta t) \qquad (5)$$

The correlation function κ is empirically determined.

Comments and Observations

We obtain a path-dependent, time-varying temporal weighting function for every possible delay time via eqn (5). Each weight is of the same duration (or greater) as is the sampled signal, $s(p\delta t)$. Following eqn (1), therefore, we perform the multiplication $\tilde{a}_n(p\delta t)s(p\delta t)$. For all delay times τ_n we delay the product and add it to all previous products. The result is the received signal, $r(q\delta t)$, $1 \leq q \leq (K+\max(\tau_n/\delta t))$.

Estimation of Channel Characteristics

Although the process described above for simulating a received waveform applies to any band-limited transmitted waveform, for purposes of estimation of channel characteristics, we use a pseudo-random (PRN) waveform for $s(t)$. A properly constructed PRN signal admits to separation into K components, each of which is effectively independent of all others. Each component may thus be used as a "sub-correlator" which is substantially immune to the presence of other components in the received waveform. We can take advantage of this feature by performing a correlation (matched filter) of the entire received waveform with each of the components. Because the temporal relationships of all components are known, the output of the sub-correlators can be arranged in a waterfall display, as shown in Figure 4. Here we observe the variations across time of the power measured at any given lag time. It is clear that this presentation reveals much more of the time-varying nature of the example channel than does the classical matched filter. Indeed, although the presentation in Figure 4 shows the magnitude-squared (power) of the sub-correlator outputs, the underlying complex outputs contain all of the information required to construct our channel model.

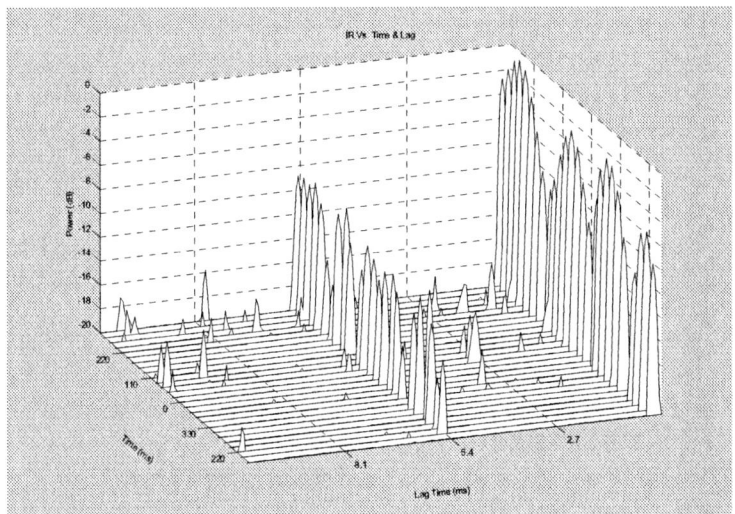

Figure 4. A waterfall display of the output of multiple sub-correlators, each a component of the transmitted waveform, each of which is used to process the entire received waveform.

Modeling and Simulating Range Rate

Range rate is simply constant relative speed between a source and a receiver. The effect of range rate on a waveform is to cause a dilation or contraction in the temporal duration of the waveform. With a broadband waveform, the upper frequencies will shift more than will the lower frequencies, hence causing a contraction or dilation in frequency. It is only with a tonal waveform that contraction or dilation is equivalent to the so-called Doppler shift.

We consider an arbitrary waveform $x(t;f_1;f_2;\varphi)$ which is a function of time (t), a start frequency (f_1), a stop frequency (f_2), and a trajectory φ which defines the time-frequency path by which the waveform migrates from f_1 to f_2. Time is described such that $0 \leq t \leq T$. A waveform subject to range rate may be modeled as:

$$x_\alpha = x(\alpha t; \beta f_1; \beta f_2; \varphi) \qquad (6)$$

where α is a constant multiplier, with $\alpha \approx 1$, and β is a function of α, both to be defined.

The Fourier Transform of $x(t)$ is $X(f)$, of $x_\alpha(t)$ it is $X_\alpha(f)$, and it can be shown that

$$X_\alpha(f) = X(f/\alpha) \qquad (7)$$

hence $\beta = 1/\alpha$. A useful description of α arises from considerations of constant velocity motion:

$$\alpha \approx (1 + v/c) \qquad (8)$$

where v is the relative velocity between a source and a receiver, and c is the propagation speed in the transmission channel. It is seen that $\beta \approx (1-v/c)$.

In the following we demonstrate a simple method for imposing compression/dilation on arbitrary (passband) waveforms. Let $r(p\delta t)$ be any passband waveform with a carrier frequency of f_0, and an envelope function $a(p\delta t)$. There are K samples taken over T seconds. For convenience, we assume that r(pδt) is analytic, with baseband representation:

$$\hat{r}(\alpha p\delta t) = a(\alpha p\delta t)\exp(i2\pi f_0 \alpha p\delta t), \quad 0 \leq p \leq (K-1)\alpha \qquad (9)$$

From eqn (8) we have our compression/dilation factor α. For every time value of (pδt) there is a corresponding value of (αpδt) which reflects the compressed/dilated time equivalent. Note that, in eqn (9), in the absence of range rate we would have an exact baseband replica of the transmitted signal.

We now investigate an interpolative approach to imposing range rate effects on the baseband formulation of the "transmitted" waveform to give it the requisite characteristics of the received, distorted signal. First, we must interpolate the waveform <u>envelope</u> a(pδt) by a suitable amount to obtain a(nδt'). Our experience has been that a factor of M = 8 is appropriate. That is, there are now $0 \leq n \leq MK-1$ samples, each sample taken at Mfs samples per second, with a new sample interval of δt' = δt/M. To compute the distorted envelope required by eqn (9), we compute an index n' for which

$$n' = \frac{\min}{n}[|n - \alpha p\delta t/\delta t'|], \qquad (10)$$

$$\hat{y}(\alpha p\delta t) = a(n'\delta t') \qquad (11)$$

Finally, recognizing the residual effect of speed on the carrier frequency, f_0, we multiply by the complex exponential to obtain the final "received" signal

$$\hat{r}(\alpha p\delta t) = \hat{y}(\alpha p\delta t)\exp(i2\pi f_0 p\delta t(\alpha - 1)) \qquad (12)$$

Figure 5 shows spectral comparisons between a transmitted LFM signal, an analytical representation of the signal received from a 20 kt source, and the signal received with 20 kts, as produced by the numerical process. Note the offsets in frequency caused by the range rate, which is the same offset for the model and for the numerical techniques.

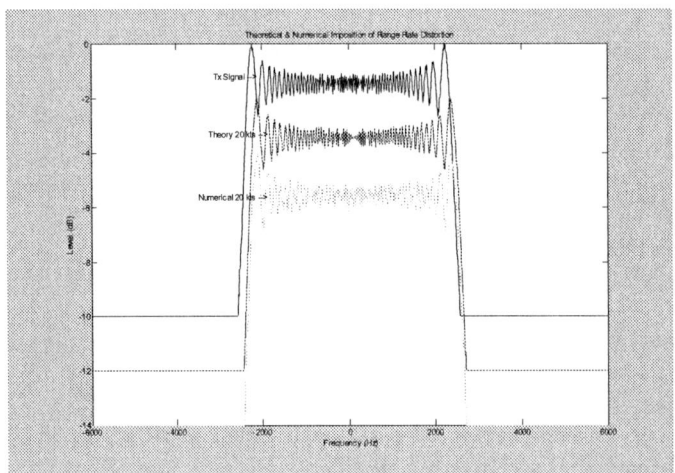

Figure 5. Comparison of the Power spectra of the original transmitted LFM (upper trace), and the "received" signal perturbed by 20 kts of range rate. The middle trace reflects a theoretical model, while the third trace reflects a numerical imposition. They are virtually identical. The vertical offsets in the plots are included only to clarify the presentation.

Modeling and Simulating Noise

We have little to add to the decades-long development and understanding of additive noise, other than to describe our use of it in modem development. Whether we use artificially-generated noise or experimentally recorded noise, whether it is represented analytically or as a "real" component, we always pass the noise through a bandpass filter prior to adding it to any signal. The filter has approximately the same bandwidth as does the signal of interest. Our received signals (see eqn (1)) are always normalized to have unit power (variance) and the filtered noise is then normalized to have a power (variance) such that the ratio of signal power to noise power meets user-specified criteria for signal-to-noise-ratio (SNR). We note a consequence of this definition of SNR: because we maintain unit signal power, the SNR available on any individual channel path is reduced as the number of paths increases. This is especially detrimental for phase-coherent signaling in which an adaptive equalizer attempts to remove multipath influence from a time-varying channel: a path which is occasionally strong may at times be quite weak, and the overall response of the equalizer will be to treat the path signal as noise bursts.

CHANNEL SIMULATION (CHANSIM)

Acomms performance is determined as much by acquisition, frequency alignment, and timing as it is by demodulation and decoding, and our experience is that these "preliminaries" are often the more demanding part of successful communications. We therefore have found it necessary to consider the effect of the channel (and platforms) on all aspects of the transmitted signal. In particular, we have found it necessary with our simulations to generate passband waveforms in an analog form and provide that directly into the output of the preamplifier of our modem receiver. We supply a few tens of seconds of "interference only" prior to the waveform to test the entire receiver, especially with regard to the automatic gain control (AGC) and false alarm performance.

Benthos has received funding from the US Navy via several modem-related programs dating back to 1998 for the purpose of developing a simulator adequate to meet the needs of modem development. This channel simulator (CHANSIM) is a MATLAB-based, GUI-driven, real-time emulation of real-world acoustic channels. It provides for most recognized effects of the channel, using statistical characteristics to control the realizations, which are converted via a digital-to-analog converter to a voltage signal which can drive the entire modem. Although the intended use of CHANSIM is for modem development, any band-limited waveform may be used as a "transmitted" signal. As such, the simulator is valid for sonar applications as well as telesonar applications.

CHANSIM has many features both to characterize the channel and to control the flow of the simulation. Table 1 lists the settable channel and control features of CHANSIM.

Table 1. CHANSIM Channel Characteristics & Simulation Control

Noise Types 1. band-limited AWGN 2. stored noise files from experiments, appropriately bandlimited & resampled to match the signal band 3. No noise	Signal-to-noise ratio number of realizations Waveform basebanding & decimation bandwidth control sample rate control
Interference Types 1. tonals, impulses 2. partial-band, short-term noise 3. recorded animal sounds (sea lions, whales, etc.) 4. External (stored files) waveforms (e.g., for multi-access interference)	External triggers Range rate to +/- 40 kts
Impulse response function characterization 1. manual entry (via cursor/mouse) 2. theoretical (Rayleigh, Rician,etc.) 3. from stored statistics, especially for time-varying channels	Analysis and plotting tools 1. spectrogram 2. magnitude 3. power spectrum 4. matched filter 5. plot individual impulse response realizations

CONCLUSIONS

There are two basic portions of a waveform used for acoustic communications: acquisition/alignment/timing components, and the modulated signal. In most practical situations the two cannot be separated when evaluating performance of a modem. We have developed a real-time simulator which tests a modem for most of the observed influences of the acoustic channel. Those influences are described by statistical parameters, which can be obtained from physics-based propagation models, or from at-sea experiments. Although the simulation incorporates many channel characteristics, the two addressed in this paper reflect the modeling of time-varying impulse response functions, and the simulation of platform range rate as it affects an arbitrary waveform.

REFERENCES

[1] M. Stojanovic, J.A. Catipovic, J.G. Proakis, "Adaptive multi-channel combining and equalization for underwater acoustic communications," J. Acoustic. Soc. Amer., vol 94, pp.1621-1631, 1993

[2] P. Bello, "Characterization of randomly time-variant linear channels," IEEE Trans. Commun. Syst., vil . CS-11, pp. 360-393, 1963

[3] R. Kennedy, *Fading Dispersive Channels*, New York: Wiley, 1969

Acoustic Communication Using Time-Reversal Signal Processing: Spatial and Frequency Diversity

Daniel Rouseff, John A. Flynn, James A. Ritcey and Warren L. J. Fox

Applied Physics Laboratory, College of Ocean and Fishery Sciences
University of Washington, 1013 NE 40th Street, Seattle, WA 98105

Abstract. Time-reversal signal processing can be viewed as a form of matched filtering that operates both in time and in space. Acoustic communication represents a promising potential application of the processing. In designing a communications system, constraints are imposed by the available bandwidth and by the geometry of the time-reversal array. In the present paper, the interplay between bandwidth and array geometry is examined. If the bandwidth is large relative to the symbol rate, time-reversal processing can be successful with sparse arrays. If the array is well populated, the required bandwidth can be reduced. Results from experiments and data-driven simulations are presented.

INTRODUCTION

The principle of acoustic time-reversal can be used to design both elegant physics experiments but also practical devices [1]. In active time-reversal, also called phase conjugation [2], a measured acoustic signal is rebroadcast but in a time-reversed fashion. Ideally, the backpropagated field will focus at the location of the original source. Invoking reciprocity, Dowling [3] showed how similar pulse compression could be achieved passively using a receive-only array. Acoustic communications represents a plausible application of time-reversal processing in the ocean. Both active[4-6] and passive [7-9] versions of the processing have been tested in experiments.

In the present paper, we consider three interrelated factors relevant in designing an acoustic communications scheme based on passive time-reversal signal processing. We first outline how a decision-directed technique can be used to update the matched filters. Updating the matched filters is necessary to compensate for a changing environment. The role of spatial diversity is then studied. Data from an experiment are processed using different subsets of a 14-element receiving array with the communications performance quantified in terms of the resulting bit-error rate. Finally, a form of frequency diversity is studied. Results from a broadband experiment are used to predict communications performance at reduced bandwidths and with different modulation schemes.

DECISION-DIRECTED PASSIVE PHASE CONJUGATION

As implemented in Rouseff et al. [7], passive phase conjugation processing begins by transmitting a single probe pulse. The response to this probe pulse is recorded at each element in the distant receiving array. The data stream is then transmitted. The measured probe responses serve as the matched filters; at each array element, the associated probe response is cross-correlated with the received data stream. The cross-correlation is done in parallel at each element with the outputs then combined across the array. The combined signal is then detected to infer the transmitted data.

At the high frequencies relevant to acoustic communication, the measured probe responses might accurately characterize the acoustic channel for only a fraction of a second. Small changes in the oceanographic environment from factors like internal waves, turbulence or surface waves can change the acoustic environment sufficiently to render the measured probe responses obsolete. Changes in the source or receiver positions can have a similar effect. One approach to compensating for these changes is to break a long data stream into small sections and intersperse additional probe pulses. While this approach has been applied successfully [7], the method is inefficient, as no data can be transmitted while the environment is being reprobed.

Flynn et al. [10,11] proposed an alternative method for inferring the matched filters. Rather than send an isolated probe, the procedure begins by sending an extended probing sequence that is known at the receiver. Combining this knowledge with the observed responses, an initial estimate for each channel's matched filter is generated. These matched filters are then applied to the subsequent data stream. After combining across the array, the demodulator output is quantized to give symbol estimates that are then fed back into a channel estimation algorithm. The estimation algorithm updates the matched filters that are then used to process the next block of data. In this way, past decisions for the symbols direct the form of the matched filters.

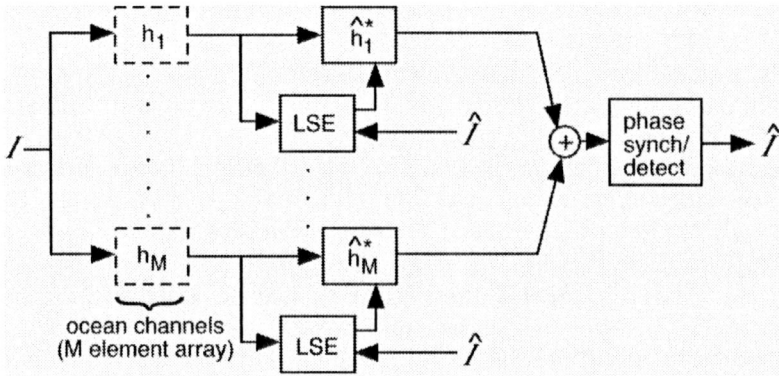

FIGURE 1. Baseband-equivalent of decision-directed passive phase conjugation processing. Data symbols represented by I and carets used for estimated quantites.

Figure 1 sketches the baseband-equivalent of decision-directed passive phase conjugation. The sequence of data is represented by I and h_m is the channel response for the mth element in the receiving array. Carets are used to denote estimated quantities. The LSE blocks represent the channel estimation step. Note that each channel is estimated independently from the other channels. This implies that the processing burden scales only linearly with the number of array elements M and suggests significant computational savings compared to joint equalization as M gets large. In practice, the LSE step can be efficiently implemented using a fast iterative method; see the references for the mathematical details [10,11].

As a byproduct, the algorithm produces an estimate for the time-varying channel response at each element in the array. Figure 2 is a sample result for the "drifting source" data set described previously [7]. The figure shows how the channel response evolves over a 5 s window for the deepest element in the receiving array. The bulk time shift is due to the increasing range as the source moves away from the array. A horizontal slice through the figure shows the channel response at a moment in time. Strong multipathing is evident with the later arriving paths typically showing the most variability. The data stream extends over the band from 5-18 kHz while the symbol rate is 2.17 kilosymbols/s. In this example, the channel response was modeled as being 35 symbols in duration corresponding to a delay spread of 16 ms. The channel was updated every 50 symbols, and 100 symbols were used to do the estimation. Communications performance for this case is discussed in the following section.

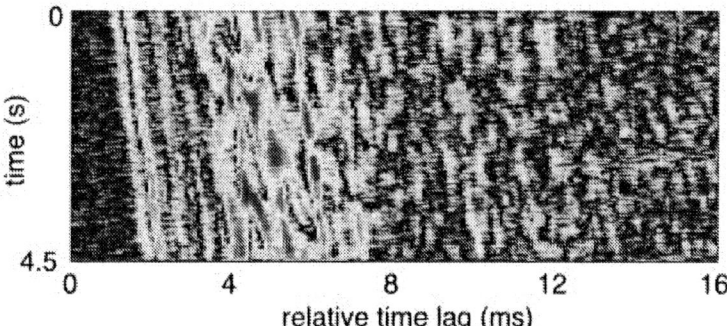

FIGURE 2. Evolving impulse response from Puget Sound experiment. Estimated as byproduct of decision-direct passive phase conjugation processing using full available bandwidth

SPATIAL DIVERSITY: EXPERIMENTAL RESULTS

Time-reversal signal processing exploits spatial diversity by using an array of receivers. The number of array elements and the spacing between elements are important considerations in designing an experiment. At moderate frequencies, it may be practical to assemble a vertical array that spans the water column with elements spaced every half-wavelength. Such a configuration is relatively easy to analyze because the orthogonality of the acoustic modes supported by the ocean waveguide can be exploited [12,13]. At the higher frequencies relevant to acoustic communications, however, a long array with densely spaced elements is unrealistic.

To quantify the effect of spatial diversity at communications frequencies, data from the May 2000 Puget Sound experiment [7] were reexamined. The experiment featured a 14-element receiving array with adjustable spacing between the elements. Five second long sections of Binary Phase Shift Keying (BPSK) data were sent over a 13 kHz bandwidth at 2.17 kilosymbols/s. Measurements were made at various ranges and water depths. In reprocessing the data, communications performance was assessed using subsets of the full 14-element array. The results were quantified in terms of the Bit Error Rate (BER).

Figure 3 shows a typical result plotting the BER versus time for various sized arrays. The range is 4.6 km and the array elements are spaced at 2 m in water 28 m deep. Results are for the same case that was considered in Fig. 2. When all 14 elements are used, the communication is error free. For a reduced number of channels, some errors are apparent. Only when the array is reduced to a single channel, however, is the tracking lost and does the method fail completely. Even with just three elements, the BER is less than 10^{-2} without any error-correction coding. It should be observed that the error rates are relatively stable; the method tolerates erroneous feedback symbols I (Fig. 1) up to error rates beyond the regime typically accepted for data links. The results shown in Fig. 3 represent averages over different combinations of array elements. For example, the seven-channel result is an average of using the top seven, the middle seven, and the bottom seven elements in the array. Interestingly, the BER is relatively insensitive to which array elements are used. The number of array elements is more important than their precise spatial distribution in depth.

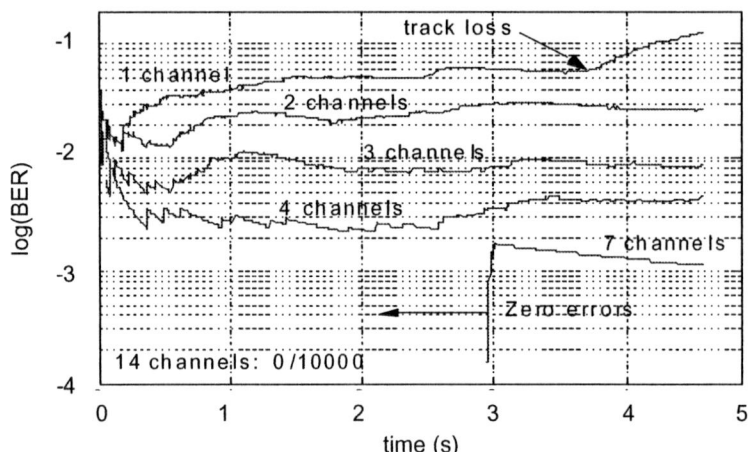

FIGURE 3. Communications performance for Puget Sound experiment. Bit error rate versus time using various subsets of full 14 element array.

FREQUENCY DIVERSITY: DATA-DRIVEN SIMULATIONS

Derode et al. [14] conducted laboratory demonstrations of active time-reversal processing in strongly multiple-scattering environments. If the signal has a wide bandwidth, strong refocusing of the backpropagated signal could be achieved using a single element without need for an array. In a subsequent paper [15], they discussed the implication of this result for communications. The results presented in Fig. 3 can be interpreted in a similar light; because the bandwidth (13 kHz) is large compared to the data rate (2.17 kilosymbols/s), decision-directed passive phase conjugation can be successful with a modest number of hydrophones. The drawback to such an approach for communications is that it is inefficient as a better use of the available bandwidth might allow the data rate to be increased.

Using a bandwidth that is large relative to the data rate represents a form of frequency diversity [16]. Using an array of hydrophones represents a form of spatial diversity. In the present section, we examine the interplay between these two forms of diversity on passive phase conjugation processing. Our approach is to use the results from broadband experiments to predict performance at reduced bandwidths. The mathematical details [17] are beyond the scope of the present short communication; here, we merely sketch how these data-driven simulations are performed and present numerical results.

As noted earlier, a byproduct of decision-directed passive phase conjugation is an estimate for the time-evolving channel response. Figure 2 is an example generated using the full 13 kHz bandwidth of the experiment. These estimates for the channel are used as input to the simulator. For the purposes of the simulator, the estimates generated using the full bandwidth data are treated as being the true time-evolving channel responses. The simulator is then driven using novel synthetic data streams having a bandwidth less than what was actually used in the experiment. Gaussian noise is added to produce a time series for each element in the array. The processor shown in Fig. 1 is applied yielding the synthetic demodulation output. Simulation parameters that can be varied include the bandwidth, the modulation scheme, the SNR and the number of array elements used in the processing. For a fixed set of parameters, the simulations are repeated many times for different realizations of the noise and the data with the results then averaged.

Figure 4 shows the predicted BER as a function of SNR for four combinations of bandwidth and modulation scheme. In Fig. 4(a) and 4(b), the bandwidth is 5.4 kHz while in Fig. 4(c) and 4(d) it has been reduced to 2.7 kHz. In Fig. 4(a) and 4(c), BPSK modulation has been simulated while Fig. 4(b) and 4(d) are QPSK. Because QPSK has two bits per symbol, it represents a doubling of the data rate as compared to BPSK. For each combination, the calculations are repeated using different subsets of the full 14 array elements.

Several observations can be made from Fig. 4. For the case in Fig. 4(a), a BER of 10^{-2} can be achieved at zero SNR if all 14 array elements are used. Similar performance can be achieved with fewer array elements at higher SNR. The results in Figs. 4(b) and 4(c) are similar to one another. This might be expected since the efficiency (defined as the ratio of the data rate to the bandwidth) is the same for the two cases. The case shown in Fig. 4(c) is four times as efficient as that shown in Fig.

4(a). The price paid for this improved efficiency is an error floor; increasing the SNR has little or no effect on the observed BER. The performance in this case is limited by the intrinsic intersymbol interference (ISI) produced by the processor, not by the noise.

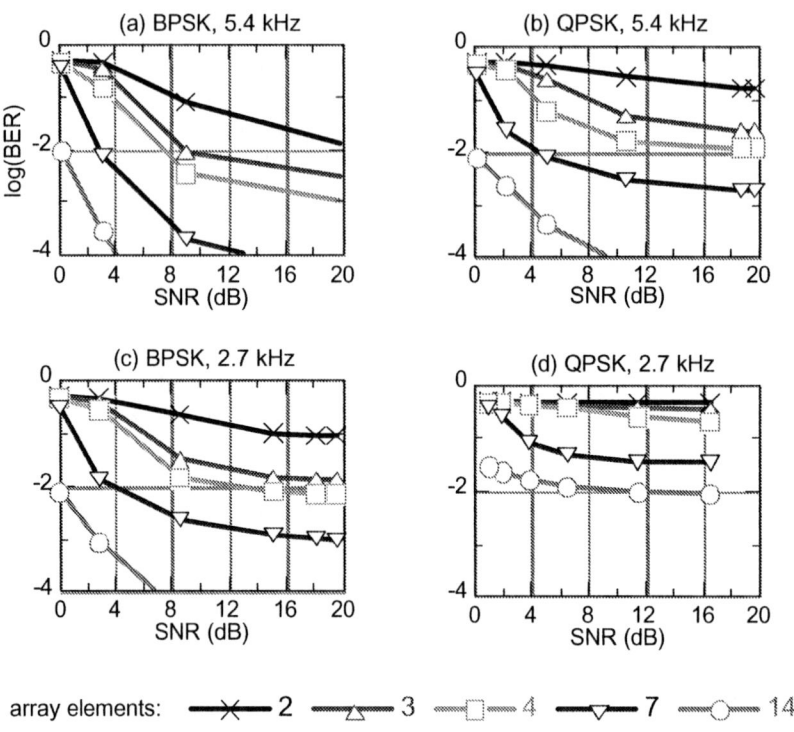

FIGURE 4. Effect of spatial and frequency diversity on communications performance. Bit error rate versus SNR for various modulation schemes (BPSK and QPSK) and bandwidths (5.4 and 2.7 kHz). The data rate in all cases is 2.17 kilosymbols/s. Results are from data-driven simulations.

SUMMARY

Passive phase conjugation is a form of time-reversal processing that uses a multi-element, receive-only array to do acoustic communication. At each array element, the received signal is matched filtered. The decision-directed version of passive phase conjugation outlined in this paper gives a method for updating the matched filters to compensate for the changing environment. A key point is that the computational burden scales only linearly with the number of elements in the array.

Results from field experiments and data-driven simulations demonstrate the interplay between spatial and frequency diversity in time-reversal processing. For our communications problem, an acceptable bit error rate can be achieved with a relatively small number of array elements provided that the bandwidth is large compared to the data rate. As the bandwidth is reduced, however, more array elements are necessary to achieve the same level of communications performance.

ACKNOWLEDGMENTS

The authors thank Dr. Darrell Jackson for his many valuable suggestions over the course of this project. The Office of Naval Research supported this work under the ARL Program.

REFERENCES

1. Fink, M., Cassereau, D., Derode, A., Prada, C., Roux, P., Tanter, M., Thomas, J-L, Wu, F., "Time reversed acoustics," *Reports on Prog. in Physics*, **63**, 1933-1995 (2000).
2. Jackson, D. R. and Dowling, D. R., "Phase conjugation in underwater acoustics," *J. Acoust. Soc. Am.*, **89**, 171-181 (1991).
3. Dowling, D. R., "Acoustic pulse-compression using passive phase-conjugate processing," *J. Acoust. Soc. Am.* **95**, 1450-1458 (1994).
4. Akal, T., Edelmann, G., Kim, S., Hodgkiss, W.S., Kuperman, W.A., and Song, H-C., in Proc. 5th European Conf. on Underwater Acoustics, 2000, pp. 989-994.
5. Edelmann, G. F., Akal, T., Hodgkiss, W. S., Kim, S., Kuperman, W. A., and Song, H. C., "An initial demonstration of underwater acoustic communication using time reversal," *IEEE J. Oceanic Eng.* **27**, 602-609 (2002).
6. Heinemann, M., Larraza, A., Smith, K.B., "Experimental studies of applications of time-reversal acoustics to noncoherent underwater communications," *J. Acoust. Soc. Am.* **113**, 3111-3116 (2003).
7. Rouseff, D., Jackson, D. R., Fox, W. L. J., Ritcey, J. A. and Dowling, D. R., "Underwater acoustic communication by passive-phase conjugation: theory and experimental results," *IEEE J. Oceanic Eng.* **26**, 821-831 (2001).
8. Hursky, P., Porter, M.B., Rice, J.A., and McDonald, V.K., "Passive phase-conjugate signaling using pulse-position modulation," in IEEE Oceans Conf. Record **4**, 2001, pp. 2244-2249.
9. Silva, A.J. and Jesus, S. M., "Underwater communications using virtual time reversal in a variable geometry channel," in IEEE Oceans Conf. Record **4**, 2002, pp. 2416-2421.
10. Flynn, J. A., Ritcey, J. A., Fox, W. L. J., Jackson, D. R. and Rouseff, D., "Decision-directed passive phase conjugation: equalisation performance in shallow water," *Elect. Let.* **37**, 1551-1553 (2001).
11. Flynn, J. A., Ritcey, J. A., Fox, W. L. J., Jackson, D. R. and Rouseff, D., "Decision-directed passive phase conjugation for underwater acoustic communications with results from a shallow-water trial," Conference Record of 35th Asilomar Conf. on Sig., Sys. and Comp., vol 2, 2002, pp.1420-1427.
12. Siderius, M., Jackson, D.R., Rouseff, D., and Porter, R.P., "Multipath compensation in shallow water environments using a virtual receiver," *J. Acoust. Soc. Am.* **102**, 3439-3449 (1997).
13. Kuperman, W. A., Hodgkiss, W. S., Song, T. Akal, H. C., Ferla, C. and Jackson, D. R., "Phase conjugation in the ocean: Experimental demonstration of an acoustic time-reversal mirror," *J. Acoust. Soc. Am.*, **103**, 25-40 (1998).
14. Derode, A., Tourin, A., and Fink, M., "Time reversal versus phase conjugation in a multiple scattering environment," *Ultrasonics* **40**, 275-280 (2002).
15. Derode, A., Tourin, A., de Rosny, J., Tanter, M., Yon, S., and Fink, M., "Taking advantage of multiple scattering to communicate with time-reversal antennas," *Phys. Rev. Let.* **90**, 014301 (2003).
16. Proakis, J. G., Digital Communications, New York: McGraw Hill, 1995, pp.777-806.
17. Flynn, J.A., Ritcey, J. A., Fox, W.L.J., and Rouseff, D., *Technical Report UWEETR-2004-0003*, Dept. of Elect. Eng., Univ. of Washington, Box 32500, Seattle, WA 98195-2500.

Environmental Effects On Phase Coherent Underwater Acoustic Communications: A Perspective From Several Experimental Measurements

T. C. Yang, Naval Research Lab., Washington DC 20375

Abstract. This paper presents a summary of results from several experiments using phase coherent underwater acoustic communications in shallow waters with weak to strong internal waves. Measurements of temporal coherence time of the underwater acoustic channels are presented and related to the equalizer performance.

INTRODUCTION

Phase coherent communications provide a high data rate for a given bandwidth (compared with phase incoherent communications) and are particularly useful for wireless communications in a band-limited underwater acoustic channel. Successes of phase coherent underwater acoustic communications (ACOMMS) have been recently demonstrated after the introduction of adaptive channel decision feedback equalizer (DFE) in combination with the phase locked loop (PLL) [1]. The problem, however, is the reliability of the equalizer in terms of bit error rate (BER) over a large number of packets. Typically, about a fraction (e.g., one third) of the packets are error-free before error decoding. The unreliability is largely attributed to the "catastrophic" error propagation problem of the DFE. As symbol errors occur when the tap coefficients of the equalizer are incorrectly estimated, the errors propagate to later symbols through the use of incorrect tap coefficients (before they are updated/corrected). It is generally noted that the performance of DFE is sensitive to the channel condition, particularly the temporal variation of the channel. The question is then how is BER affected by the channel condition.

To track the channel variation and counter the error propagation, the equalizer is adaptively updated using the training or decision symbols (during the training and message data phases). One finds empirically that the DFE is updated for more than 99% of the time (or symbols) even for successfully transmitted data packages. This has led to the suggestion that the acoustic communication channel is rapidly varying at the time scale of symbols. Is the BER related to the rate of temporal variation?

The purpose of this paper is to investigate the rate of temporal variation of the underwater acoustic channel and how it affects ACOMMS. If the channel is indeed changing at the symbol rate, then the physical mechanism of what causes such rapid channel changes needs to be understood. To investigate the channel fluctuation rate, we conducted several experiments in shallow water environments with weak to strong internal wave activities. We use consecutive m-sequences to measure the channel impulse response functions at the scale of tens to hundreds of symbols. The motivation was that if the channel were rapidly changing at the scale of symbols, we would definitely observe changes in the channel impulse response functions at the scale of tens to hundreds of symbols.

The result of this investigation is reported here based on measurements from six experiments at different frequencies and different locations. We find that in general, the channel impulse response functions change rapidly from packet to packets (usually separated by a few minutes). But what is relevant for ACOMMS is not the inter-packet coherence but the intra-packet coherence. We find that at mid-frequencies, most of the data packets have intra-packet coherence > 0.9 indicating that the channel is very stable within a packet. However, the symbols do incur an extra phase, which changes rapidly at the scale of a few symbols. We find that the decision feedback equalizer (DFE/PLL) in these cases still shows a high (>99%) update rate, the reason being that both the DFE and PLL try to compensate for the rapid symbol phase change. We determine that the high update rate is an artifact of the processor and does not reflect the true condition of the channel. The channel can be equalized with almost no update using a different approach.

TIME SCALE OF THE OCEANOGRAPHIC PROCESSES

Shallow (coastal) water is known for its complexities. The effect of the random medium on acoustic signal propagation can be analyzed by the time scale of the various oceanographic processes. The intrusion of fresh water and eddies turns the coastal shallow water into a range dependent environment for signal propagation. These processes are slowly varying and are generally stable over a time scale of tens of minutes. The internal waves have a time scale of ~1-10 minutes. The turbulence, rough sea surfaces etc. can have a much shorter time scale (seconds). How the internal waves affect signal propagation has been a subject of great interest recently. Several experiments have been conducted in the last 10 years to study the effect of the internal waves on signal propagation at low frequencies, as reported in the open literature. At low (<1 kHz) frequencies, it has been reported that the signal characteristics change drastically at the scale of ~2 minutes. The most noticeable changes include the depth where the signal intensity peaks, the multipath arrival (time) pattern, and the signal amplitude and phase as a function of depth. The changes are more pronounced when solitary internal waves are present, which cause significant mode coupling and consequently the change of the mode amplitudes (of the signals) with respect to range. These changes are important for low frequency (< 1kHz) ACOMMS since the packet length is of the order of 1-2 minutes for a data packet of 5-10 kbits.

At mid (2-5 kHz) to high (>15 kHz) frequencies, the internal waves are not expected to have any significant effect on ACOMMS since the packet length is < 5-10 sec. At this time scale the internal waves can be treated as a deterministic (range dependent) environment. However, the internal waves can cause severe signal fading, which is a signal-to-noise ratio problem. At mid and high frequencies, the turbulence and sea surface waves are believed to have a (more) significant effect on signal propagation. The effects of these processes on signals are not yet completely characterized (quantitatively). For a downward refractive sound speed profile, the sea surface effect is less significant at lower frequencies, particularly if the source and receiver are in the duct.

What physical processes can cause channel variation at the symbol rate? The answer depends on the carrier frequency. Let us consider, for example, ACOMMS with a bandwidth of 500 Hz. Each symbol is 2 msec long. At a (carrier) frequency of 20 kHz, the micro and fine structure of the ocean medium has a spatial scale comparable to the acoustic wavelength, and therefore could easily affect the acoustic signals and produce temporal variations at the scale of ~10 msec (5 symbols). At mid frequencies, the ocean inhomogeneities that have a scale comparable to the acoustic wavelength include the turbulence, and high wavenumber spectral components of the internal waves. Most of these processes have a temporal scale of seconds. At low frequencies, the dominant oceanic process is likely to be the internal waves, which have a temporal scale of minutes. Other processes such as ocean fronts, eddies, and intrusion of fresh water have a temporal scale of hours. Hence, at low and mid frequencies, it seems unlikely that any of the above oceanographic processes would cause fluctuation at the scale of ~10 msec (5 symbols) or less.

The above issues prompted us to conduct a systematic measurement of the channel coherence as a function of time at different frequencies.

EXPERIMENTAL MEASUREMENTS

To measure the environmental impact on ACOMMS, data were collected in dynamic shallow water using a fixed source and fixed receivers. To measure the effect of platform motion, data were collected in a relatively stable environment using a moving source and fixed receivers. The rate of the channel fluctuation can be characterized by the signal Doppler spread or the temporal coherence time.

The results of six experiments are reported here. Advent99 was conducted by NURC (previously SACLANTCEN) on the Adventure bank west of Sicily in the Mediterranean [2,3]. Water depth was ~80 m. The source-receiver range is 10 km. Conductivity, temperature and Depth (CTD) data indicated the presence of weak internal waves. The ASCOT01 experiment was also conducted by NURC in the Massachusetts Bay east of Boston along the contour of 100m [4]. The same source-receiver geometry was used. This area is known to have non-negligible internal wave activities. The SWARM95 and RAGS03 experiments were conducted off the New Jersey coast in water of 40-80 m depth using a fixed source and VLA receivers [5]. The source to NRL array range was 40 km in SWARM 95 and 3 km in RAGS03. Strong internal wave activities were reported [5]. The RDS3 experiment was conducted in the Adriatic Sea off the coast of Bali, Italy. The water depth was ~110m. CTD data indicated a stable sound speed profile over a period of hours. The MREA03 experiment was conducted north of Elba in the Mediterranean. The water depth was ~88 m. The sound speed profile was stable over a period of hours. Both RDS3 and MREA03 experiments used a towed source and a fixed VLA.

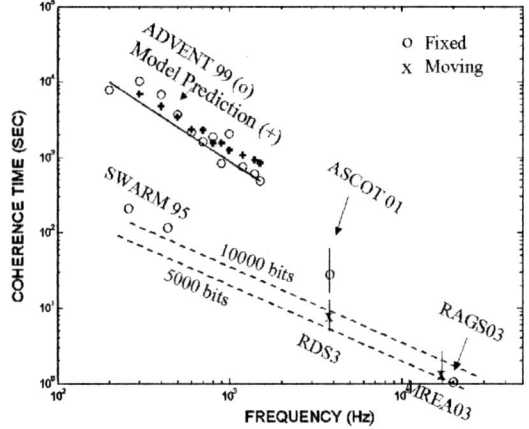

FIGURE 1. Temporal coherence time as a function of frequency from six experiments. The dashed lines indicate the packet duration as a function of the carrier frequency for 5 and 10 kbits of data respectively (see text). The vertical bar indicates the variation of the coherence time over the data samples.

For the temporal coherence measurements, multitone data in the band of 200-1,500 Hz were used for the Advent99 exp. The rest of the experiments used consecutive m-sequences, which led to measurements of channel impulse response functions at the scale of hundreds of msec. Temporal coherence is defined as the cross correlations of an initial signal vector with a later signal vector normalized by the norm of the two vectors. Temporal coherence time is defined as the time when the temporal coherence between the two signals drops to 0.8. We use 0.8 instead of the standard 0.5 because (1) temporal coherence usually does not follow an exponential dependence as normally assumed, and (2) acoustic communication normally requires high (>0.8) coherence.

Figure 1 shows the measured temporal coherence time from the various experiments. The coherence time measurements for the Advent 99, ASCOT01, RDS3 and SWARM99 experiments have been reported previously [2,3,6,7].

Also shown in Fig. 1 are two dashed lines which show the communication packet duration (length) as a function of the carrier frequency for data packets of 5 and 10 kbits respectively. We assume binary phase-shifted keying signal with a bandwidth one quarter of the carrier frequency. In general, we want the packet length to be less than or equal to the channel coherence time, beyond which the signal becomes less coherent. We note that for the ASCOT01 environment, the channel can support packet transmission of up to 20 kbits data (lasting 20 sec), but for RDS3 experiment, the channel can only support data packets of 5-10 kbits. As shown in Fig. 1, the packet length at low (<1 kHz) frequencies is rather long due to the limited bandwidth. When the packet length approaches the channel coherence time as is the case for the SWARM 95 environment, appreciable variation in the impulse response function will impede the equalizer performance. One finds that the data are successfully equalized (BER is <1%) only for those (portions of the) packets with high (> 0.7-0.8) intra-packet coherence. Figure 1 also shows that at high (>15 kHz) frequencies, the channel coherence time is rather short allowing only short packets to be transmitted successfully.

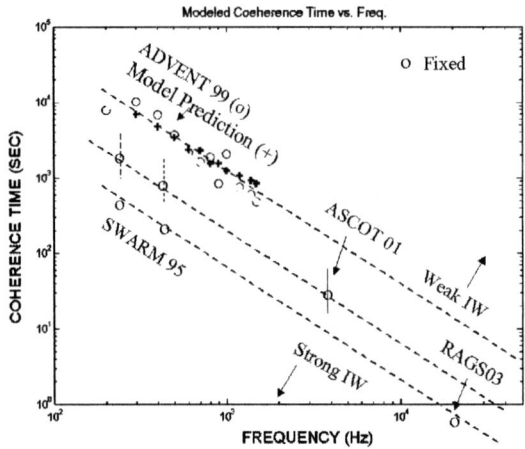

FIGURE 2. Temporal coherence time scaled to a range of 10 km. The dashed lines indicate −1.5 power of frequency dependence. The data are separated into weak, intermediate and strong internal wave regions.

We note that the temporal coherence time is shorter for a moving source than a stationary source. The temporal coherence time for a moving source is given by the horizontal coherence length divided by the platform speed, which is often smaller than the (intrinsic) temporal coherence time associated with the medium. Compare RDS3 data with the ASCOT01 data.

Internal waves seem to also reduce the channel coherence time. Figure 2 shows the channel coherence time using only fixed source-receiver data. The coherence time in Fig. 2 is scaled to a fixed range of 10 km assuming the coherence time is inversely proportional to the square-root of range. Figure 2 indicates that the coherence time τ can be divided into regimes of weak, intermediate and strong internal waves with coherence time given by

$$\tau \sim \alpha\, R^{1/2}\, f^{-1.5}$$

with $\alpha(10km) = (0.25 - 4)\cdot 10^5$ $(meter\ sec)^{1/2}$.

PERFORMANCE ANALYSIS

For performance analysis, we shall concentrate on the ASCOT01 data and illustrate the environmental influence [6]. We plot in Fig. 3 the channel impulse response functions at the scale of minutes. The channel impulse response functions were measured from the LFM signals in each packet which were separated by ~ 2 min. We see that the intensity and arrival time of the multipath arrivals vary significantly from packet to packets. This is confirmed by the variation of the arrival time of the peak intensities as a function of transmission time. The temporal variation is apparently caused by the presence of internal waves.

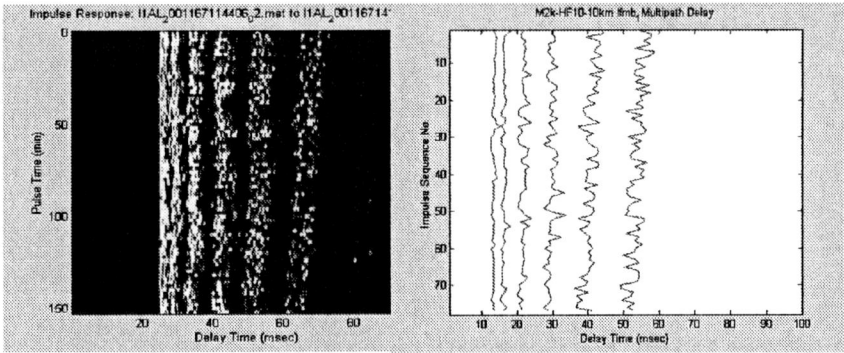

Figure 3. Channel impulse responses functions at the scale of minutes from the ASCOT01 experiment (left). The arrival time of the peak intensities versus the transmission time (right)

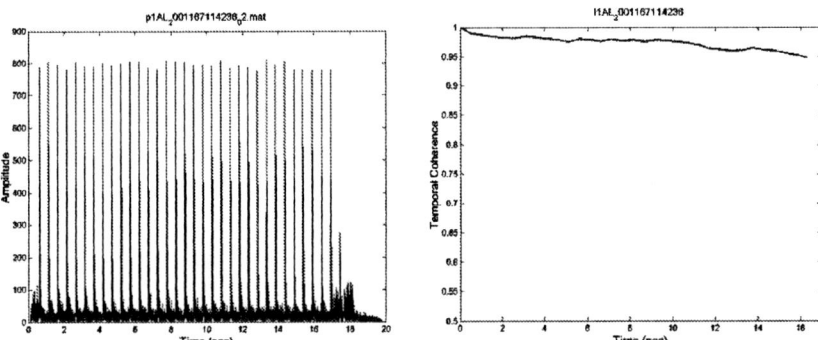

FIGURE 4. Measured channel impulse responses within a packet (left). Temporal coherence of the impulse response functions with respect to the first one (right).

Figure 4 shows the channel impulse response functions measured with consecutive m-sequences within the first packet. It shows little variation of the intensity level as a function of time within the packet. The temporal coherence remains very high (>0.9) as shown in Fig. 4 (right figure). In this case, the DFE works very well – one finds zero BER as shown in Fig. 5. Figure 6 plots the BER and percentage of update for 78 packets transmitted during the two hours time. We see that the BER is < 0.2% except for one packet. The DFE processor is updated more than 99% (Fig. 6) even for the packets with zero bit error and high coherence > 0.9 (e.g. the first packet). If the channel is changing >99% of time, the channel could not have such a high coherence value as shown in Fig. 4. It turns out that symbol phase is changing rapidly at the symbol rate (See Ref. 8). The >99% update results from the fact that both DFE and PLL attempt to remove the rapid symbol phase change [9]. The conclusion is that at the channel is stable at the scale of tens of seconds (therefore no mystery about the rapid channel change). A different equalizer by applying first passive-phase-conjugation to the data showed that the channel can be equalized with < 5% update.

Figure 7 plots the coherence time of the channel as a function of the time. It also plots the output signal-to-noise ratio (SNR) as a function of time. We find a high correlation (~0.9 coherence value) between the channel coherence time and output SNR as shown by the figure on the right. Note that high temporal coherence improves the processor performance in removing ISI and hence a higher output SNR.

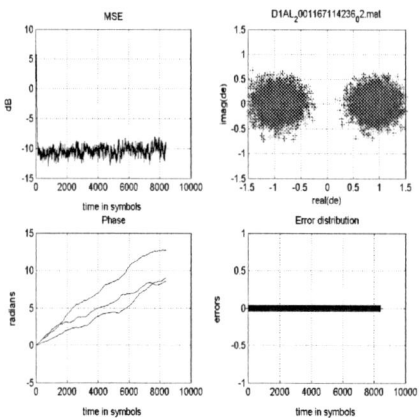

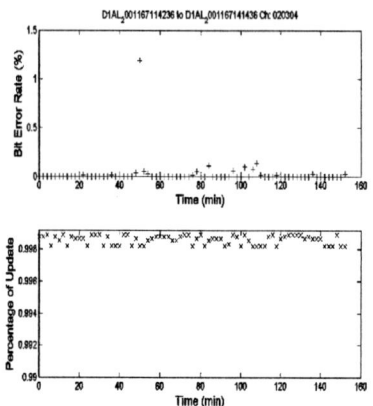

FIGURE 5. Output of the DFE showing the MSE (upper left), symbol phase (lower left), symbol constellations (upper right) and symbol error as a function of symbol number.

FIGURE 6. BER and percentage of update as a function of transmitted packets displayed by the transmission time.

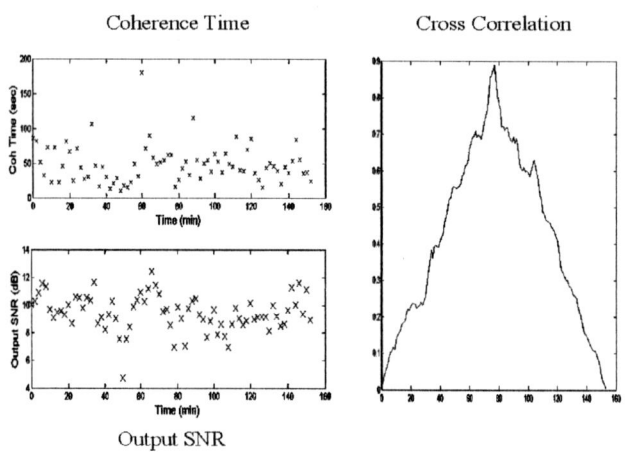

FIGURE 7. The channel coherence time and average output SNR as a function of packet time (left). The normalized cross correlation between coherence time and output SNR (right).

SUMMARY AND DISCUSSIONS

Temporal coherence of the channel impulse responses plays a vital role for underwater acoustic communications. At mid (2-5 kHz) frequencies, the length of a communication packet (5-10 sec for 5-10 kbits) is often shorter than the signal coherence time. The acoustic environment, while randomly changing from packets to packets, presents an instantaneous "deterministic" environment from the point of acoustic communications. In this case, the environmental impact can be mitigated by adequately sampling the channel impulse response function. We use m-sequences to independently verify the temporal variations of the channel within a packet and corroborate with the DFE results.

At low (< 1 kHz) frequencies, the packet length for the same amount of data will be long (due to the limited bandwidth) and comparable to the average signal coherence time. The environmental impact is appreciably noticeable by the temporal variations of the impulse responses within a packet (intra-packet). We show in a separate publication [10] that the channel variation can be successfully compensated by a DFE for those (portions of the) packets with high intra-packet coherence.

At high (>15 kHz) frequencies, we note that the average signal coherence time is relatively short. This implies that short packets need to be used to minimize BER.

ACKNOWLEDGEMENT

This work is supported by the Office of Naval Research. We thank NURC for the opportunities to participate in the Advent99 and ASCOT01 experiments.

REFERENCES

1. M. Stojanovic, "Recent advances in high-speed underwater acoustic communications," IEEE J. Oceanic Eng. Vol. 21, pp. 125-136, 1996 and references therein.
2. T.C. Yang and M. Siderius, "Temporal coherence and fluctuation of acoustic signals in shallow water," Proc. of the Fifth European Conference on Underwater Acoustics, pp. 63-68. Lyon, France, July 11-14, 2000.
3. T.C. Yang, K. Yoo, and M. Siderius "Internal waves and its effect on signal propagation in the Adventure Bank," Proc. of 8th International Congress On Sound And Vibration, pp.3001-3008, Hong Kong (2 - 6 July, 2001).
4. J. Sellschopp, P. Nielsen and M. Siderius, "Combination of acoustics with high-resolution oceanography," *Impact of littoral environmental variability on acoustic prediction and sonar performance*, edited by N.G. Pace and F. B. Jensen, Kluwer Academic publishers, pp 19-26 (2002)
5. J. R. Apel et al, "An Overview of the 1995 SWARM Shallow Water Internal Wave Acoustic Scattering Experiment," IEEE J. Oceanic Eng. 22,465-500 (1997).
6. T. C. Yang, "Temporal fluctuations of broadband channel impulse functions and underwater acoustic communications at 2-5 kHz," Proc. of MTS OCEANS 2002, Vol. 4, pp. 2395-2400, Biloxi (2002).
7. D. Rouseff et al, "Coherence of acoustic modes propagating through shallow water internal waves," J. Acoust. Soc. Am. 111, 1655-1666 (2002).
8. T. C. Yang, "Temporal resolution of time-reversal and passive-phase conjugation for underwater acoustic communications," IEEE J Oceanic Eng. **28**, 229-245 (2003).
9. T. C. Yang, "Differences between passive-phase conjugation and decision-feedback equalizer for underwater acoustic communication," IEEE J Oceanic Eng. (2004).
10. T.C. Yang "The effect of internal waves on low frequency underwater acoustic communications," Proc. of International Congress of Acoustics, (2004)

Environmental and Motion Effects on Orthogonal Frequency Division Multiplexed On-Off Keying

Paul J. Gendron and T.C. Yang

Naval Research Laboratory, 4555 Overlook Avenue SW, Washington DC 20375

Abstract. Orthogonal frequency division multiplexing with on-off keying (OFDM-OOK) offers a means to near BPSK signaling rates with the simplicity of non-coherent processing. At high frequencies source receiver acceleration induces time varying signal dilations that adversely effect both frequency alignment and multicarrier orthogonality. Dilation process estimation is coupled with a decision directed frequency domain channel magnitude response estimator for non-coherent equalization of OFDM-OOK. Joint co-channel estimation is presented for near optimal decisions under loss of orthogonality due to Doppler spreading. Data from a moving source experiment at ranges from .8 km to 2.0 km, conducted in the shallow water off the coast of Elba Italy were used to test the feasibility of OFDM-OOK at 18kHz center frequency with 4 kHz bandwidth. The effects of source receiver relative motion, frequency selectivity, and Doppler spreading on bit error rates were assessed. Mollification of frequency selective fading by diversity combining is demonstrated.

INTRODUCTION

Modulation strategies for underwater communication are typically categorized broadly as coherent and non-coherent. Coherent strategies require the fairly accurate estimation of the channel impulse response to attain relatively high rates (> 1 bps/Hz). Diverse phase coherent approaches have been proposed and demonstrated, from phase shift keying (PSK) schemes [1] to multi carrier orthogonal frequency division multiplexing (OFDM) [2]. Single carrier PSK schemes are well suited to computationally fast in-time recursions for channel and Doppler tracking; the LMS and RLS algorithms for channel response estimation and the phase locked loop for Doppler correction are naturally employed in the PSK framework. The equalization of coherent OFDM signals against frequency selective fading is usually aided with training/pilot symbols [3][2] similar to the PSK approach with Doppler estimation critical just as in PSK signaling [4][2].

Low rate non-coherent approaches, such as frequency shift keying (FSK) on the other hand, rely simply on energy detection requiring little or no knowledge of the channel and have demonstrated success in severely Doppler spread channels. In the power-limited and bandwidth-unconstrained case these simple energy detection approaches attain channel capacity for both time invariant and fading channels [6]. Indeed for this limiting case the attainable rate is $(1 - 2B_d \tau_{max})P/N_o [bps]$ where B_d and τ_{max} are the Doppler spread and delay spread of the channel respectively, P is the signal power available and N_o the power of the noise process per Hz.

The goal of this paper is to consider OFDM with on-off keying (OFDM-OOK) on

TABLE 1. MREA03 OFDM-OOK data packet parameters

center freq.	bandwidth	T_s	duration	τ_{max}	bps	bps/Hz
18kHz	4 kHz	.5 sec	8 sec	.125 sec	3.2 kbps	.8

each tightly spaced (i.e. $\Delta f = 1/T_s$) subcarrier as a noncoherent scheme for higher rates than are typically expected from energy detection strategies. OFDM is ideally suited for time-dispersive channels as typically encountered in RF communications, however for frequency dispersive underwater accoustic channels with accelerating sources significant interference between subcarriers leads to bit errors as decision variables become dependent across neighboring subcarriers [5]. Frequency selective fading poses constraints on OFDM systems requiring the use of time-frequency interleaving with error correcting coding to attain frequency and time diversity. It is demonstrated with at sea data that uncoded bit error rates are largely associated with channel frequency bands exhibiting low spectral levels.

A sub-optimal receiver processor is used to test the feasibility of OFDM-OOK. Doppler and symbol timing estimation is performed via a spline model of the time varying dilation process. Channel magnitude response measurement is by decision directed estimation without pilot symbols. Symbol decision rules that incorporate Doppler induced dependencies across neighboring sub-carriers is presented to minimize the error rate due to intercarrier interference. Since OFDM is particularly susceptible to motion induced Doppler the scheme is tested from an accelerating source to demonstrate the feasibility of the approach.

MREA03 Experiment

An OFDM acoustic communication experiment was conducted during MREA03 on June 20 2003 off the north coast of Elba Italy in 87m of water. The sound speed profile is downward refracting and repeated interaction with the soft bottom result in a multipath delay spread of $\tau_{max} = .125\ sec$ and an assumed coherence time of $T_c > T_s = .5\ sec$. From these constraints OFDM-OOK communication packets of 8 seconds duration consisting of ≈ 24 kbits per packet as shown in Table 1 were tested. With the receiver array moored at a mean depth of 31 m, the source was towed from behind a ship at an approximate depth of 25m with speed increasing to 4 knots.

Figure 1a) depicts the Doppler shift measured over the duration of the 8 OFDM-OOK signals at ranges from .8 km to 2.0 km. The source track includes both cross range, azimuthally accelerating (first packet), outbound accelerating (second and third packets) as well as constant range rate packets. All of the packets have associated intra-packet time varying Doppler as depicted in Figure 1b) for the fourth packet.

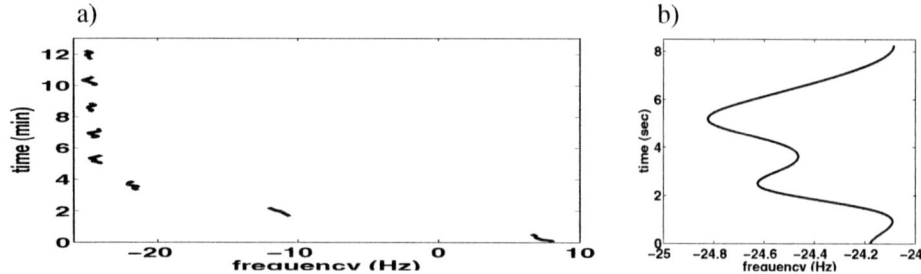

FIGURE 1. a) Doppler associated with the 18kHz MREA03 OFDM-OOK experiment on June 20 2003. Experiment starts at .8km with negative range rate accelerating to $2m/s$ for the remainder of experiment (time scale of packets are extended for display). b) To the right, intra-packet Doppler at 5 minutes.

OFDM-OOK SIGNALING MODEL

Under the assumption that τ_{max} is known the source signal consists of a sequence of L symboling intervals of duration T_s seperated by τ_{max}. Let $\alpha_l(t)$ represent the dilation process at the source over the lth signaling interval. The lth dilated source symbol $s_{\alpha,l}(t)$ is a superpostion of $T'_s = \int_{(l-1)T_o}^{lT_o - \tau_{max}} \alpha'(t)dt$ duration dilated tones spaced $\approx 1/T'_s$ Hz apart.

$$s_{\alpha,l}(t) = A\sqrt{\frac{d\alpha}{dt}} \sum_{k=-K}^{K-1} c_{k,l} g(\alpha(t-lT_o))e^{j2\pi f_k \alpha(t) + \phi_k} \qquad c_{k,l} \in \{0,1\}$$

where $f_k = f_c + k/T_s$, is the kth subcarrier and $T_o = T_s + \tau_{max}$. With bandwidth B there are $2K = BT_s$ such tones per symboling interval. The windowing function $g(t-lT_o) = 1$ for $(l-1)(T_o) < t < lT_o$ and 0 otherwise. The phases ϕ_k are uncorrelated and serve to minimize the peak to average signal level. The channel symbols, $\mathbf{c}$, encode the information bits $\mathbf{b}$, b_n $n = 1,\ldots 2RKL$ via the $R = 1/2$ rate convolutional code with a time-frequency interleaver as $\mathbf{c} = C\mathbf{b}$ where C represents both the encoding and interleaving operator.

At depth d the contribution from $s_{\alpha,l}(t)$ at the receiver is $r_{d,l}(t) = \int h_{d,l}(\tau)s_{\alpha,l}(t-\tau)d\tau + e(d,t)$ where $e(d,t)$ is the ambient noise at depth d with spatial coherence Σ and is uncorrelated in time due to a prewhitening filter of insignificant duration relative to $h_{d,l}(\tau)$. Figure 2 depicts the noise process spatial coherence at 30 m depth for a 1.8 m aperature over a 12 minute interval.

Doppler spread associated with source acceleration. To isolate the effect of the time-varying source dilation on the received OFDM signal let the dilation process for the moving source be described by a quadratc spline of sufficient knots. For the lth symbol interval $t' = \alpha_l(t) = \alpha_{0,l} + \alpha_{1,l}t + \alpha_{2,l}t^2$ where $\alpha_{1,l} \approx 1$ v/c and $\alpha_{2,l} \approx a/c$ is proportional to the acceleration of the source divided by the average sound speed at the source such that acceleration towards (away) the receiver implies $\alpha_{2,l} > 0$ ($\alpha_{2,l} < 0$). From two pilot tones the dilation process is estimated by fitting a spline to the

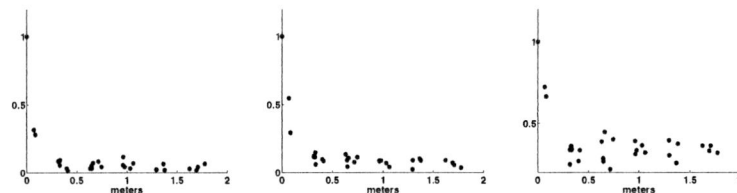

FIGURE 2. Spatial coherence of noise at vertical receiver array on June 20, 2003, at start of experiment (left) at 6 minutes (center) and at 12 minutes (right).

instantaneous frequency $d\alpha(t)/dt = f(t)/f_c$. A high resolution Fourier method is used for this purpose. The method is made computationally efficient with the chirp transform algorithm. The resulting expectation is weighted least squares with variances at each sample of the frequency spectrum approximated by the inverse Hessian [7]. Denote these spline parameters by $\hat{\alpha}_l(t)$. Figure 3 displays the source motion induced time varying Doppler for a single tone at one phone and compares the residual Doppler spread averaged over two tones and all 6 phones associated with the spline model of Doppler and a constant Doppler model. The spline model effectively demodulates the received signal. To assess the effect of source motion on the modulation strategy at high signal-

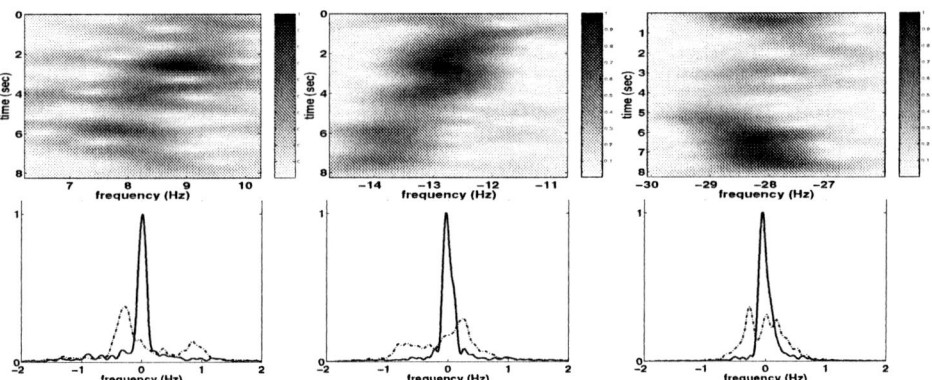

FIGURE 3. Time varying Doppler of 20.429 kHz tone at phone # 1 and averaged spectra of over all 6 phones with Doppler mean compensation (dashed) and spline based Doppler compensation (solid).

to-noise ratio assume perfect estimation of the dilation process $\hat{\alpha}_l(t) = \alpha_l(t)$. Consider the receiver contribution at depth d, letting source dilated time $t' = \alpha_l(t)$ with inverse $t = \zeta_l(t')$ it follows that

$$r(d, \zeta_l(t'), l) \approx A\sqrt{\frac{d\alpha_l}{dt}} \sum_k c_{k,l} e^{j2\pi f_k t'} \int_0^{\tau_{max}} \check{h}(\tau, l) e^{j2\pi f_k \alpha_{1,l} \tau} g(\alpha(t - lT_o - \tau)) \times$$
$$e^{j2\pi f_k \alpha_{2,l} \zeta_{1,l} t' \tau} d\tau + e(t') \quad (1)$$

where $\check{h}(d,\tau,l) = h(d,\tau,l)e^{j\pi f_k(\alpha_{2,l}\zeta_{0,l}+\alpha_{2,l}\tau^2)}$. Assume that $\alpha_{2,l}\zeta_{2,l} << 1$ and it follows that the dilated received signal $r_\zeta(d,t',l) = \sqrt{d\zeta_l/dt'}\, r(d,\zeta_l(t'),l)$ yields the finite duration Fourier transform

$$\tilde{r}(d,f,l) = \int_{(l-1)T_o+\tau_{max}}^{lT_o} r_\zeta(d,t',l)e^{-j2\pi ft'}dt' =$$

$$AT_s\sum_k c_{k,l}\int_0^{\tau_{max}} \check{h}(d,\tau,l)e^{-j2\pi f_k\alpha_{1,l}\tau}\text{sinc}\,(T_s(f_k(1-\alpha_{2,l}\zeta_{1,l}\tau)-f))\,d\tau + \tilde{e}(f). \quad (2)$$

Equation 2 is a superpostion (in τ) of *sinc* kernals (each frequency shifted by $\approx \alpha_{2,l}\tau$) and weighted by the chirp modulated impulse response. For constant velocity motion $\alpha_{2,l} = 0$ and the time invariant assumption over the symboling interval T_s, it follows that $r(f,l) = T_s\sum_k c_{k,l}\tilde{h}(f_k\alpha_{1,l},l)\text{sinc}\,(T_s(f_k-f))$; no ICI at the receiver. For sources radially accelerating towards the receiver, energy from $c_{k,l}$ in the received demodulated signal is roughly spread to lower frequencies $f_k(1-\alpha_{2,l}\tau_{max}) < f < f_k$. Likewise for a source accelerating away from the receiver, $\alpha_{2,l} < 0$ and spreading of the source tone at f_k is skewed to the frequency band $f_k < f < f_k(1-\alpha_{2,l}\tau_{max})$. It is implied then that with conventional processing (i.e. demodulate, dilate and integrate) of OFDM signals the channel is effectively time variant even over durations as short as the symbol interval T_s. Expand $\text{sinc}\,(T_s(f_k(1-\alpha_{2,l}\zeta_{1,l}\tau)-f)) = \sum_m \gamma_{m,k}(\tau)\text{sinc}\,(T_s(f_k-f-f_m))$ where $|\gamma_m|^2 \propto 1/m^2$ to explicitly display the loss of orthogonality. In short, radial acceleration imparts a spread that is roughly proportional to the product of acceleration and channel delay spread. Each channel coefficient $c_{k,l}$ associated with frequency f_k and time lT_o contributes to the frequency slots of its neighbors even under time invariant channel conditions. In addition differential Doppler associated with different dilations along individual ray paths lead to further Doppler spreading for source environment scenarios where eigenrays exhibit signficant launching angles.

Detection. With on-off keying, non-coherent detection is employed at the receiver with the ICI model truncated to neighboring frequencies. The resulting model of the received signal at the D phones is $\tilde{\mathbf{r}}(k,l)|H(k,l),\mathbf{c}_k^l,\Sigma \sim \mathcal{N}(H(k,l)\mathbf{c}_k^l,\Sigma)$ where $H(k,l)$ represents the C^{Dx3} channel transfer function from source frequencies $f_{k-1}..f_{k+1}$ to received frequency f_k at symboling instant l, $\mathbf{c}_k^l = [c_{k-1,l},c_{k,l},c_{k+1,l}]'$. Σ, assumed known, is the spatial covariance of the noise. This assumption is reasonable for high SNR in cases where the receiver has time-bandwidth sufficient to accurately estimate Σ. With a view towards on-off keying (i.e. $c_k^2 = c_k$) this truncated ICI approximation implies that $\rho_{k,l} = \tilde{\mathbf{r}}'(k,l)\Sigma^{-1}\tilde{\mathbf{r}}(k,l)$ is distributed as [8] [9] $p(\rho_{k,l}|H(k,l),\mathbf{c}_k^l,\Sigma) = \chi^2_{2D}(\mathbf{c}_k^{l'}H(k,l)'\Sigma^{-1}H(k,l)\mathbf{c}_k^l)$ where $\chi^2_{2D}(\lambda^2)$ is a non-central χ^2 distribution with non-centrality parameter λ^2. Approximate the noncentrality parameter, ignoring cross terms as $\mathbf{c}_{k-1,k,k+1}^{l'}H(k,l)'\Sigma^{-1}H(k,l)\mathbf{c}_{k-1,k,k+1} \approx \sum_d \lambda^2_{k-1,k,l}c_{k-1,l} + \lambda^2_{k,k,l}c_{k,l} + \lambda^2_{k+1,k,l}c_{k+1,l}$ where $\lambda^2_{k-1,k} = \sum_d |H_{d,k-1}(k,l)|^2/\sigma_{d,l}$. For scenarios where the channel is time invariant over the symboling interval and the source is not accelerating there is no ICI and only $\lambda^2_{k,k,l}$ persists. The channel gains $\lambda^2_{k\pm 1,k,l}$ are approximated with $\lambda^2_{k\pm 1,k,l} = \delta_\pm \lambda^2_{k,k,l}$ where the factors $\delta_\pm$ are empirical Doppler spread estimates from the out of band tones depicted in Figure 3.

To estimate data in the presense of ICI, iterate over the conditional expectations of each of the unknown parameters; data, c_l, and channel gains, Λ_l, without consideration of dependencies associated with the convolutional code. The approach is warranted for its computational efficiency; with interleaving over time it is necessary to estimate the source data $c_{k,l}$ without Viterbi decoding so that estimation of the channel transfer function can be accomplished. Final estimation of data and channel can be made via the Viterbi decoder once the entire channel response is initally estimated. With this in mind start with the conditional expectation of the channel bits given adjacent bits and the channel magnitude response

$$\hat{c}_{k,l} = E[c_{k,l}|c_{k-1,l}, c_{k+1,l}, \Lambda_{k,l}, \Sigma, \rho(k,l)] \approx$$

$$\frac{\chi^2_{2D}(\lambda^2_{k-1,k,l}\hat{c}_{k-1,l} + \lambda^2_{k,k,l} + \lambda^2_{k+1,k,l}\hat{c}_{k+1,l})}{\chi^2_{2D}(\lambda^2_{k-1,k,l}\hat{c}_{k-1,l} + \lambda^2_{k+1,k,l}\hat{c}_{k+1,l}) + \chi^2_{2D}(\lambda^2_{k-1,k,l}\hat{c}_{k-1,l} + \lambda^2_{k,k,l} + \lambda^2_{k+1,k,l}\hat{c}_{k+1,l})}. \quad (3)$$

Initialing with $\hat{c}_l = 0$, two iterations are sufficient for convergence.

Channel Estimation. Estimation of the channel magnitude $E[|H|^2_l|\mathbf{r}_{l,\ldots,1},\hat{\mathbf{c}}_{l,\ldots,1}]$ is made by use of an RLS-like recursion from estimates $E[|H|^2_{l-1}|\mathbf{r}_{l-1,\ldots,1},\hat{\mathbf{c}}_{l-1,\ldots,1}]$ and $E[\mathbf{c}_l|\mathbf{r},E[|H|^2_{l-1}]]$. Channel estimates in time-frequency slots k,l for which $c_{k,l}=0$ have zero precision implying that the channel estimate relies solely on the previous estimate based on $\hat{c}_{k,l-1,\ldots,1}$. The algorithm uses an empirical Bayes approach twice iterating between estimates of channel and data. This has proved sufficient for convergence. Figure 4 depicts the channel transfer function magnitude, estimated via the proposed method, for ranges from .8 km to 2.0 km demonstrating the frequency selective nature of the channel and the changing striation pattern with range. At closer ranges, Figures 4a), b) and c) there is greater striation and an implied decrease in coherence time. The vertical coherence length is summarized in Figure 5 and gives a measure of the diverisy present for this configuration. Typical vertical coherence lengths for these channels is between 5 and 10 wavelengths.

RESULTS

This signal model and receiver structure were tested for uncoded bit error rate (BER) performance. Figure 6 lists BERs for 6 different ranges and source motion scenarios with a symetric ICI model ($\delta_\pm = \delta$). Since SNR varies over range the received SNR is augmented by adding Gaussian white noise to the packet to control for SNR and compare more effectively performance over different ranges and Doppler motion scenarios. It is evident that the two accelerating source packets yield BERs that are typical of the sets while packets at ranges 1 km, 1.6 km, and 2.0 km which experience the greatest degree of nulling in the channel transfer function have relatively large BERs. This suggests that channel frequency selectivity is a dominant factor for OFDM-OOK performance.

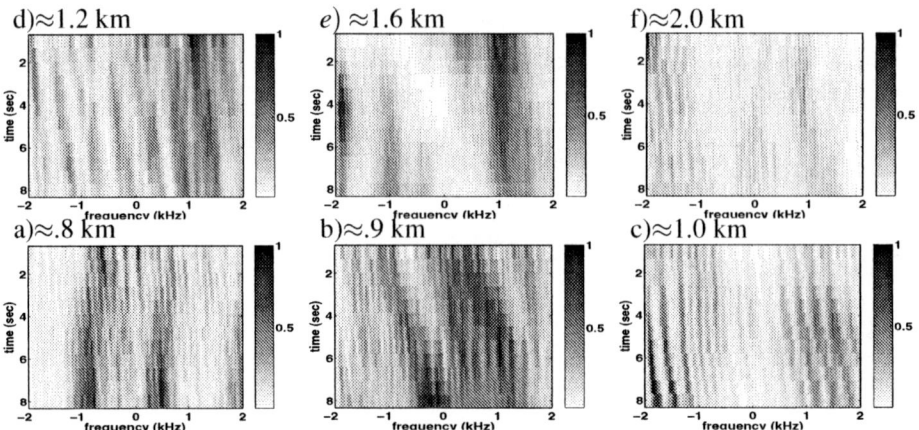

FIGURE 4. Channel transfer function magnitude estimated via decision directed RLS algorithm at ranges from .8km to 2.0km showing frequency selective fading.

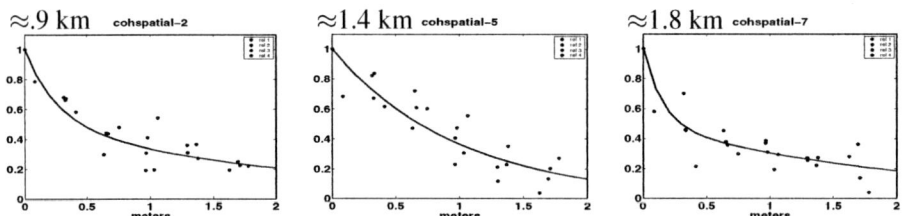

FIGURE 5. Vertical coherence of channel at various ranges over 1.8 m aperature at receiver depth of 30 m.

SUMMARY AND DISCUSSION

OFDM with on-off keying is considered as a means of attaining near BPSK rates with non-coherent processing. This paper evaluates OFDM-OOK bit error rates under time varying Doppler conditions (towed source speed at $\approx$ 4 knots and non-negligible accelerations) contrasted with previous data limited to small Doppler shifts. Doppler spread estimation is made empirically from out of band tones and gives a measure of the ICI used to remove bias in the estimation of channel bits at each subcarrier. The algorithm employs a decision directed RLS-like channel magnitude estimator to yield stable performance at SNRs below 12 dB. Lastly the utility of modeling skew in Doppler spread is a current topic of interest and its causes such as source radial velocity, acceleration and wind induced surface waves require further study.

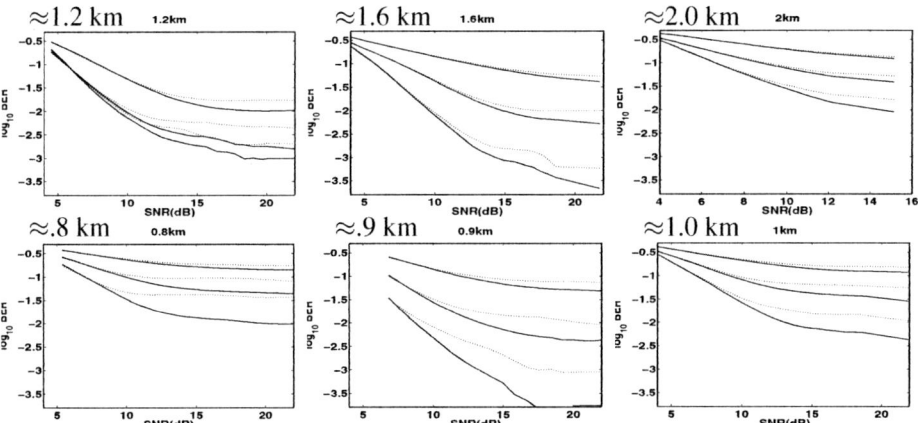

FIGURE 6. Uncoded bit error rates for diversity levels of 1, 3 and 6 at moderate SNRs and ranges from .8 km to 2.0 km. Dashed line shows performance without ICI model and with channel assumed perfectly coherent across symboling frames.

ACKNOWLEDGMENTS

This work supported by the Office of Naval Research. Thanks to NURC for the opportunity to participate in the MREA03 experiment; Jeff Schindall, Mike McCord and Wen-Bin Yang for experiment execution; Bruce Pasewark and Thomas Hayward for helpful discussions.

REFERENCES

1. Stojanovic, M., *Journal of Oceanic Engineering*, **21**, 125–136 (1996).
2. Kim, B., and Lu, I., *Journal of Acoustical Society of America*, **109**, 2477 (2001).
3. Li, Y., Cimini, L., and Sollenberger, N., *IEEE Transactions on Communications*, **46**, 902–915 (1998).
4. Luise, M., and Reggiannini, R., *IEEE Transaction on Communications*, **44**, 1590–1598 (1996).
5. Davies, J. J., and Pointer, S. A., *Oceans '98 Coference Proceedings*, **2**, 1022–1027 (1998).
6. Telatar, I., and Tse, D., *IEEE Transactions on Information Theory*, **46**, 1384–1400 (2000).
7. Rife, D., and Boorstyn, R., *IEEE Transactions on Information Theory*, **20**, 591–598 (1974).
8. Mäkaläinen, T., *Commentationes Physico-Mathematicae, Societas Scientarium Fennica*, **31**, 1–6 (1966).
9. Johnson, N., and Kotz, S., *Continuous Univariate Distributions-2*, John Wiley & Sons, Inc., 605 Third Avenue, New York, NY 10158, 1972, pp. 189–194.

High-Frequency FH-FSK Underwater Acoustic Communications: The Environmental Effect and Signal Processing

Wen-Bin Yang and T. C. Yang

Naval Research Laboratory
Washington DC 20375, USA

Abstract. This paper analyzes the environmental effect and signal processing approaches for frequency-hopped frequency-shift-keyed (FH FSK) underwater acoustic communications based upon data collected in two experiments: the RDS4 experiment near Halifax, Canada and MREA03 experiment near Elba, Italy. The FH FSK signals have a bandwidth of 4 kHz centered at 17 and 20 kHz. The source was towed at ~4 knots. The signals were received on a vertical line array anchored to the bottom. The acoustic environments at both sites have the same downward refractive sound speed profiles but very different bottom properties. The multipath spread last ~20 msec in the MREA03 experiment and is > 1 sec in the RDS4 experiment. The different lengths of multipath delays have a significant effect on the bit error rate (BER) and the appropriate signal processing needed to reduce the BER. We present the data analysis results, the signal processing approaches, including multi-channel beamforming and spatial diversity combining, and discuss the implications for the use of FH-FSK for multi-user communications.

INTRODUCTION

Underwater acoustic communications (ACOMM) are required for command, control and networking of autonomous underwater vehicles (AUVs). For multiple users, networking between users requires multi-access communications due to latency in signal transmission, particularly when AUVs are spread out in different ranges. Two signaling approaches using spread-spectrum are the frequency-hopped and direct sequence methods, known as the frequency-hopped frequency-shift-keying (FH FSK) and Direct Sequence-Code Division Multiple Access (DS-CDMA) modulation schemes [1]. DS-CDMA spreads the information bits in different codes and is subject to near-far problem as the signal can be degraded by nearby interference sources. It has the advantages of higher processing gain through code compression. FH FSK spreads the information bits in different frequency bins. To minimize the number of times that different users transmit in the same bands (frequency collision), a common approach is to use a prime number of frequency bins and assign a hopping pattern, which increments through the frequency set by n where n is the user number [2]. In this manner, any pair of users will occupy the same frequency bin only once per hopping pattern. To reduce the bit errors due to collisions between different users, one resorts to the use of an appropriate error-correction code. We concentrate on FH FSK in this paper.

The performance (e.g., bit error rate) of the FH FSK signaling approach depends on the signal design and the acoustic channel characteristics in which it operates. To maximize the number of users, the number of frequency bins should be large. In principle, the tone duration should be long with respect to the channel clearing time (the channel multipath spread) so that there is little signal beyond the symbol period. This results in many narrow bands, which are subject to frequency selective fading, and are vulnerable to uncertainty in the Doppler shift. To minimize the sensitivity to Doppler shift, a wider frequency bin (larger than the expected uncertainty in the Doppler shift estimation) can be used. This minimizes the frequency selective fading but the drawback is the signal energy exceeding the symbol duration (inter-symbol interference) unless a sufficient guard time is built into the signal design. Also bit error rate increases due to frequency collision, as a small number of frequency bins are available. The performance of FH FSK clearly depends on the multipath delay time and the ability to estimate the Doppler shift.

For a single user, the transmitter can select any hopping pattern, which hops through the (selected) frequency bins only once per hopping pattern. The total duration of the hopping sequence needs to exceed the (maximum) multipath delay.

In this paper, we study the FH FSK signaling performance for a single user under two very different shallow water environments: one with a multipath delay of ~20 msec and the other with a multipath delay >1 sec at 18-20 kHz. The source was towed at a speed of 4 knots in both cases. The uncoded bit-error-rates (BERs) are determined as a function of the signal-to-noise ratio using the conventional approach, which detects the signal within one-symbol duration. As expected, BER is significantly higher for the second case than the first case. To reduce the BER a modified algorithm is presented. Implications for multi-users are discussed.

ENVIRONMENTAL EFFECT AND SIGNAL DESIGN

Environmental impact on the FH FSK signaling performance is studied using two experimental data sets: one from the MREA03 (Military Rapid Environment Assessment) experiment which was conducted in June of 2003, in water of ~103m depth north of Elba, Italy, and the other from the RDS4 (Rapidly Deployable Systems) experiment which was conducted off the coast of Halifax, Canada in water of ~80m depth. A vertical line array of 8 receivers was used in both experiments. The source was towed behind a ship at 4 knots. The FH FSK signals covered a range of 18-22 kHz and 16-20 kHz in MREA03 and RDS4 experiments respectively.

Environmental Effect

The signal characteristics are very different between the MREA03 and RDS4 environments. In the MREA03 environment, the multipath delay spread is around 20 msec as shown in Fig. 1a. In contrast, one finds a multipath delay lasting more than 1 second in the RDS4 experiment area as shown in Fig. 1b. The differences between the multipath delay time are due to the different bottoms' absorption coefficients. The

RDS4 area has very thin sediment overlay a hard bottom resulting in a low loss environment for acoustic propagation.

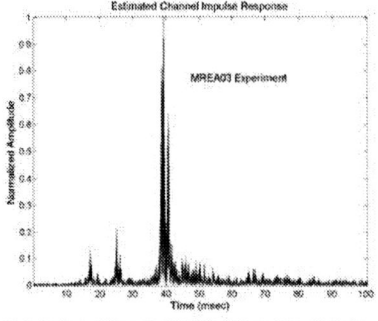
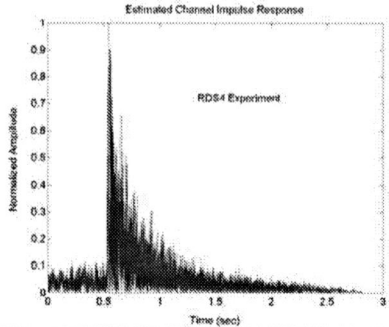

FIGURE 1. Estimated Channel Impulse Response functions. 1a: MREA03 (left figure) and 1b: RDS4 (right figure).

Fig. 2a shows the Doppler shift of the signal measured using the pilot tones, which are located outside the frequency band of the FH FSK signal. Fig. 2b shows the variance of the Doppler shift over 6 minutes, where the Doppler shift is 40.5 ± 0.5 Hz. The Doppler shift estimate is used to adjust and realign the frequency bins of the FH FSK symbols. Precise Doppler shift estimation is needed for symbol synchronization (e.g., determine the beginning of the signal).

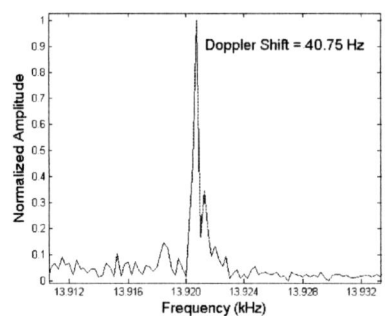
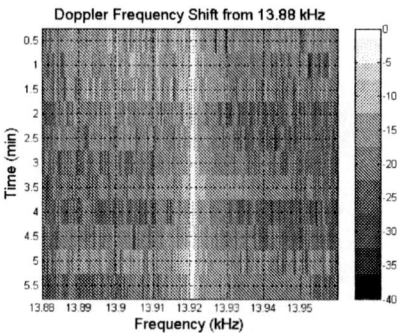

FIGURE 2. Doppler Frequency Shift Estimate. 2a: at a time instant (left figure) and 2b: vs. time (right figure).

FH FSK Signal Structure

The FH FSK packet is structured as shown in Fig. 3, which includes a Linear Frequency Modulation (LFM) signal, followed by the synchronization bits (also used for user identification) and then data messages with gaps between them. Two pilot (narrowband) tones outside of the FH FSK frequency bands are added to the signal to estimate the Doppler shift. An LFM signal is appended to the end of the packet. The LFM signals are used to estimate the channel impulse response functions and also the

Doppler shift, by way of the signal dilation. Either LFM or synchronization bits can be used for symbol synchronization/acquisition. The synchronization bits and data messages are modulated using FH-FSK signals with a pre-determined hopping frequency pattern.

Gap 0	LFM	Gap 1	Sync	Gap 2	Data	Gap 3	LFM	Gap 4

FIGURE 3. FH FSK Frame Structure

For both experiments, the FH FSK signals are patterned after the WHOI design [3] as opposed to the Benthos design [4]. The signal structure is shown in Table 1. Also shown is the Benthos signaling parameters for comparison. Note that both signaling schemes use the same bandwidth. Benthos scheme allows more frequency blocks and hence more users. On the other hand, WHOI's scheme yields a bit rate >4 times higher than the Benthos scheme (for a single user). Since its symbol duration is one twelfth of that of the Benthos signal, it is more prone to multipath induced symbol interference (or frequency collision) when the multipath delay is much longer than the symbol duration. In other words, the multipath effect is expected to be stronger on the WHOI signals than the Benthos signal. (Table 1 shows the trade off in the signal design.) The effects will be studied below.

TABLE 1. Comparison of WHOI vs. Benthos FH-FSK Signal Parameters

	WHOI	Benthos
M-ary FSK	M=2	M=8
Bandwidth	4 kHz	4 kHz
Frequency Resolution	160 Hz	11.3 Hz
# of Frequency Blocks	13	43
Symbol Duration	12.5 msec	150 msec
Guard Time	0	150 msec
Information Bits	800	102
Packet Length	29.425 sec	16.2 sec
Information Rate	27.2 bps	6.3 bps

It is worth noting that the frequency bin size (for both schemes) is twice the inverse of the symbol duration. This is a feature commonly adapted for FH FSK signals so that the frequency leakage from one frequency bin to its neighboring frequency bins is minimized. The feature is built in to minimize the vulnerability to the Doppler estimation error so that the frequency orthogonality is not exactly required.

INCOHERENT RECEIVER

FH-MFSK uses a sequence of M tones per hopping pattern. Thus a simple incoherent FH-MFSK energy detector [2] shown in Fig. 4 can be used efficiently to detect the signal as long as there is no interference from the multipath delay or a different user. The detector contains a bandpass filter of bandwidth (W) around the center frequency, a signal acquisition (or synchronization) processor, followed by a T-

second energy integrator (reset in the beginning of each symbol) and a comparator. The integrator is applied to the hopping frequency band. By design, T should be of the order of the symbol duration. This is the conventional processor. But in some oceans, the multipath delay may be longer than the symbol duration. In that case, the integration time T may be increased to capture the multipath arrivals. The benefit is improved BER performance. The drawback is more vulnerability to interference from other users and more noise. The comparator weighs the integrator output to determine the transmitted symbol. We shall study the BER performance by varying the integration time.

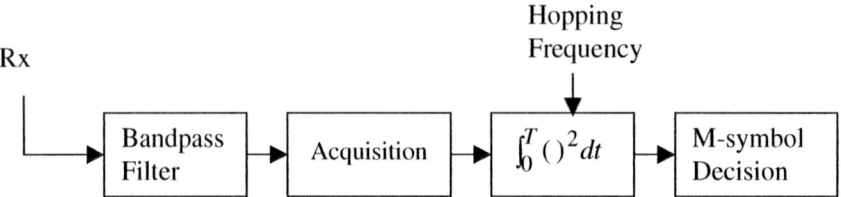

FIGURE 4. FH MFSK Energy Detector

Multiple channel signal combining methods such as beamforming and spatial diversity can also be used to improve BER performance. We use a conventional delay-and-sum (DS) beamforming technique to boost signal-to-noise ratio. We also use a spatial diversity technique of selective diversity combining [5] to lower BER value. The technique of selective diversity combining is based upon the principle of selecting the best signal among all of the signals received from different channels. The best signal is defined as the maximum signal-to-noise ratio at each pair of hopping frequencies. A combination of long integration time, beamforming and spatial diversity techniques is studied.

RESULTS AND DISCUSSIONS

The hopping pattern used in the experiments consists of 13 blocks of hopping frequencies and two frequencies within each block to represent 0 and 1. To determine the environmental impact, the uncoded BER is deduced from the experimental data. The coded BER is much smaller than the coded BER after error correction.

MREA03 Experiment

The channel impulse response function for the MREA03 environment (as shown in Fig. 1a) lasts about 20 msec. This multipath delay spread is less than two-symbol periods. Although part of the signal spreads into the next time frame, the conventional energy detector seems to work well as shown below. Eleven packets are analyzed using 8 single receivers. The source is at a range of 3.8 - 4.5 km from the receivers. Fig. 5 shows the uncoded BER of the MREA03 data. Also shown is the theoretical BER curve with only additive white Gaussian noise (AWGN). The observed BER is higher than the AWGN curve. The difference is presumably caused by the multipath induced interference. We model the (uncoded) BER data using the following formula, $p_e = \frac{1}{2}\exp\left(-\frac{\alpha}{2}\frac{E_b}{N_0}\right)$, where $\alpha = 1/(1 + \beta E_b/N_o)$, and β represents the effective interference-to-signal ratio. For MREA03, we find $\beta = 0.06$ (-12 dB). Note that the uncoded BER over the packets is less than 9%; most of them are about 1%. When the (observed) uncoded BER is < 9%, zero BER is obtained after error decoding and correction. In this case, a 1/2 convolutional code with a constraint length of 9 is used. We conclude that the conventional energy detector (which is implemented in sea-going acoustic modems) with a single receiver yields a satisfactory result (at a range of 3-5 km) for the MREA03 environment. Consequently, there is no need to explore beamforming or spatial diversity.

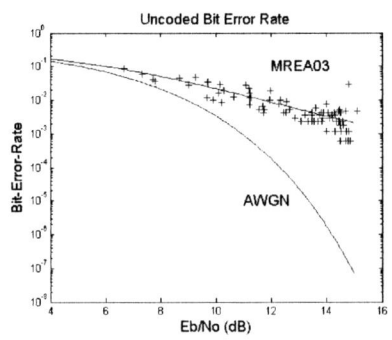

FIGURE 5. Uncoded Bit-Error-Rate: MREA03 Channel vs. AWGN Channel.

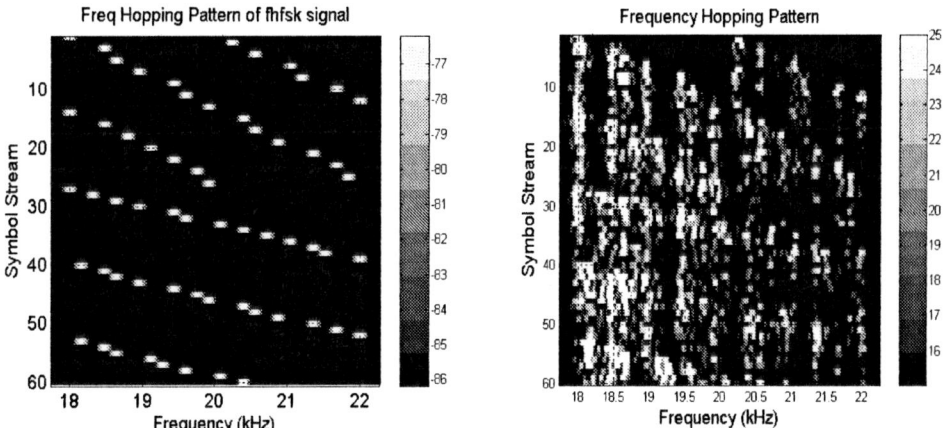

FIGURE 6. RDS4 Hopping Pattern (a) Transmit Signal (left figure) and (b) Received Signal (right figure)

RDS4 EXPERIMENT

From the RDS4 environment, the channel impulse response function (Fig. 1b) lasts more than one second, which is ~80 symbols long. Figs. 6a and 6b show the frequency-hopping pattern of transmitted and received signals at the beginning of the signal. The first 26 symbols are synchronization bits, while the others are data message bits. The synchronization bits and the data message bits were modulated using different hopping patterns in the experiment. One sees clearly in Fig. 6b a multipath delay extending over 13 symbol periods. Note that Fig. 6b displays a dynamic range of only 10 dB. The multipath spread is significantly more than 13 symbol periods when a higher dynamic range is used.

We first use a conventional energy detector to analyze the experimental data. In this case, the conventional energy detector picks up predominantly the signal energy associated with the first (main) path since the integration time is only one symbol long. Due to the severe multipaths, a much higher uncoded BER is found for the RDS4 environment (compared with MREA03) using the conventional receiver. Table 2 shows the uncoded BER values over 3 received packets. The uncoded BER is greater than 14%. For these cases, error correction does not work very well. Substantial coded BERs remain. Hence, it is necessary to modify the conventional receiver.

TABLE 2. Uncoded BER Using Conventional Energy Detector

Received Packet #	310	328	346
Distance from TX source	0.5 km	1.2 km	2.0 km
SNR (dB) @ Ch. #1	3.87	3.18	3.02
Uncoded BER @ Ch. #1	14.35 %	21.25%	22.02 %

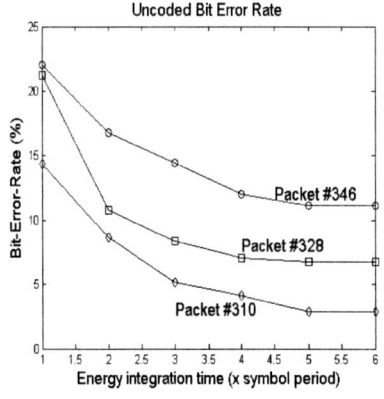

FIGURE 7. Uncoded BER vs. Integration Time

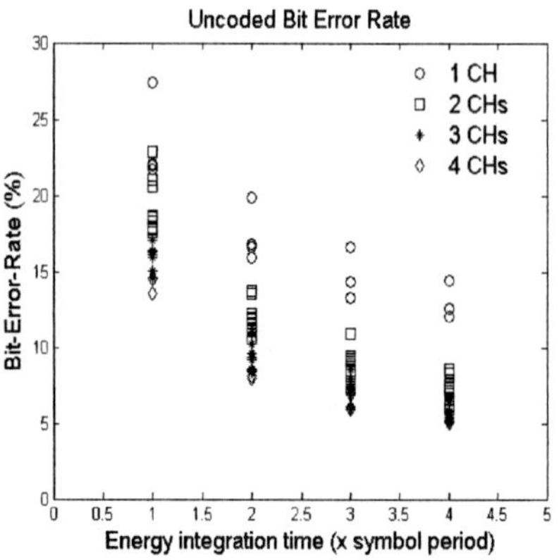

FIGURE 8. Multi-channel Combining Uncoded BER vs. Integration Time

The energy detector is modified to include a longer integration time for multipath energy combination in the same spirit as the RAKE receiver. The SNR is increased and the BER is lowered. The relationship between the uncoded BER and integration time is shown in Fig. 7. We note that the BER values are reduced dramatically as the integration time increases up to 4-symbol period. Increasing the integration time further (e.g., to include the 5th and 6th symbol periods) did not significantly improve the BER due to the decreasing signal energy in the later arrivals.

In addition to increasing integration time, we use spatial diversity and beamforming to improve BER performance. The multi-channel combining techniques are based on an example of 4-channel receivers, which are spaced at 4, 4 and 1 wavelengths. The BER results are shown in Fig. 8. One notes that the uncoded BER is uniformly reduced by ~10% using 4-channel spatial diversity or beamforming versus a single channel. For this example, we do not see any significant difference between spatial diversity and beamforming or a combination of them. We find that the original data message can be recovered perfectly (i.e., coded BER=0) when the uncoded BER is lower than 13% (using ½ convolutional code with a constraint length of 9 and interleaving schemes).

SUMMARY AND DISCUSSIONS

The BER performance of FH FSK signals is critically dependent on the signal design, namely, the symbol duration relative to the multipath delay. For environments in which the multipath delay is much longer than the symbol duration, the uncoded BER error rate can be high (>20%) using the conventional processor, which detects signal energy with a symbol period. BER can be reduced significantly by extending the signal integration to several symbol periods and using spatial diversity/beamforming, but this precludes the use of multi-users due to signal collisions. As a result, the use of FH FSK is limited to environments with short multipath delay. To remove this constraint, a method to remove the multipath spread either by a channel equalizer or passive-phase conjugation will be needed and will be presented in a separate paper.

ACKNOWLEDGMENTS

This work is supported by the Office of Naval Research. The authors thank J. Schindall, M. McCord and P. Gendron for their experimental effort.

REFERENCES

1. Freitag, L., Stojanovic, M., Singh, S. and Johnson, M, "Analysis of Channel Effects on Direct-Sequence and Frequency-Hopped Spread-Spectrum Acoustic Communication", *IEEE Journal of Oceanic Engineering,* Vol. 26, No. 4, October 2001, pp. 586-593.
2. Simon, M.K., Omura, J.K., Scholtz, R.A. and Levitt, B.K., *Spread Spectrum Communications Handbook*, New York: McGraw-Hill, 1994.
3. "Multi-user Frequency-Hopping Underwater Acoustic Communication Protocol", Ver. 1.02, Woods Hole oceanographic Institution (May, 2000), unpublished.
4. Green, M.D. and Rice, J.A., "Channel-Tolerant FH-MFSK Acoustic Signaling for Undersea Communications and Networks", *IEEE Journal of Oceanic Engineering,* Vol. 25, No. 1, January 2000, pp. 28-39.
5. Lee, W.C.Y., *Mobile Communications Engineering*, New York: McGraw-Hill, 1993.

Underwater Acoustic Communication Channel Capacity: A Simulation Study

Thomas J. Hayward and T. C. Yang

Naval Research Laboratory, Washington, DC 20375

Abstract. Acoustic communication channel capacity determines the maximum data rate that can be supported (theoretically) by an acoustic channel for a given source power and source/receiver configuration. In this paper, broadband acoustic propagation modeling is applied to estimate the channel capacity of a shallow water waveguide for a single source-receiver pair, both with and without source bandwidth constraints. Initial channel capacity estimates are obtained for a range-independent environment defined by the mean (time-averaged) sound speed profile measured at a site in the 1995 SWARM experiment. Without bandwidth constraints, estimated channel capacities approach 10 megabits per second at 1 km range, but after 2 km range they decay at a rate consistent with that of estimates by Peloquin and Leinhos [1], which were based on a sonar equation analysis for a generic underwater channel. Channel capacities subject to source bandwidth constraints are approximately 30-90% lower than the upper bounds predicted by the sonar equation analysis, and exhibit a significant wind speed dependence. Simulations of internal wave effects on channel capacity show minimal effects at low frequencies but, at 2500 Hz, show a significant increase in the channel capacity at longer ranges. Implications for underwater acoustic communication systems are discussed.

INTRODUCTION

Acoustic communication channel capacity determines the maximum data rate that can be supported (theoretically) by an acoustic channel for a given source power and source/receiver configuration. In practice, additional constraints are imposed on the source spectrum by the bandwidth limitations of physically realizable transducers. In this paper, broadband acoustic propagation modeling is applied to estimate the acoustic communication channel capacity of a shallow water waveguide as a function of range for a single source-receiver pair, both for unrestricted source bandwidth and for the limited bandwidths afforded by state-of-the-art transducers. We also present a preliminary examination, confined to a limited frequency band, of the effects of internal wave induced channel fluctuations on the channel capacity.

In early work in communication theory, the problem of determining the capacity of a linear communication channel with Gaussian source and perfectly known channel impulse response function was formulated as an extremal problem, leading to the explicit "waterfill" solution of [2]. Recent work has extended the determination of channel capacities to include simple channel models with, e.g., Rayleigh signal fading statistics. Significant work remains before these analyses can be extended to realistic representations of time-varying underwater acoustic channels.

The question of the maximum data rate has important practical implications. One would like to know what is theoretically possible and what causes degradation in the data rate. To obtain an upper bound on the channel capacity, we will assume that both the transmitter and receiver know the channel transfer function exactly for each transmission, and that the channel function does not vary during each transmission. These assumptions idealize the actual underwater acoustic communication scenario, in which there is a finite time, determined by the temporal coherence function, during which the channel transfer function varies minimally and can be learned, to considerable accuracy, through probe signals. The data rates of very short packets are experimentally measurable and can be compared with the theoretical channel capacity. The difference between the experimental and theoretical values can then provide a measure of the communication system performance.

CHANNEL CAPACITY ESTIMATION

Reference 2 derives the capacity of a known time-invariant linear channel with transfer function $H(f)$ for a Gaussian source having power spectral density $X(f)$ and average power P, and additive Gaussian noise having power spectral density $N(f)$. In that case, the channel capacity is obtained as the maximum of the integral

$$I = \int_0^\infty \log_2\left(1 + \frac{H(f)X(f)}{N(f)}\right) df, \tag{1}$$

subject to the source power constraint

$$\int_{-\infty}^\infty X(f)\, df = P. \tag{2}$$

By a straightforward application of the method of Lagrange multipliers, the maximum is obtained when the source spectrum satisfies

$$X(f) = \max\left\{L - \frac{N(f)}{H(f)},\ 0\right\}, \tag{3}$$

where the value L is chosen so that Equation (2) is satisfied. In acoustic communication, the channel capacity is a function of the source-receiver range through the (implicit) dependence of the acoustic transfer function $H(f)$ on range.

Note that the optimal source spectrum defined by Equations (1) and (3) has the maximum spectral content supported by the channel, i.e., the set of all frequencies f for which $L - N(f)/H(f) > 0$. The bandwidth of this optimal source may, in fact, exceed that which is physically realizable by a single source transducer. In practice, the bandwidth is often limited to a fraction of the carrier frequency, and the carrier frequency is often set to be inversely proportional to the square root of the receiver range. Channel capacity subject to these constraints can be found by restricting the frequencies in Equations (1)–(3) to the specified frequency bands.

COMPUTATIONS FOR A RANGE-INDEPENDENT MEDIUM

The first simulations reported here are for a range-independent shallow-water environment derived from yo-yo CTD measurements taken during the 1995 SWARM experiment. [3] Figure 1 shows the mean water-column and bottom sound speed profiles, based on two hours of yo-yo CTD data and geoacoustic measurements reported in [3].

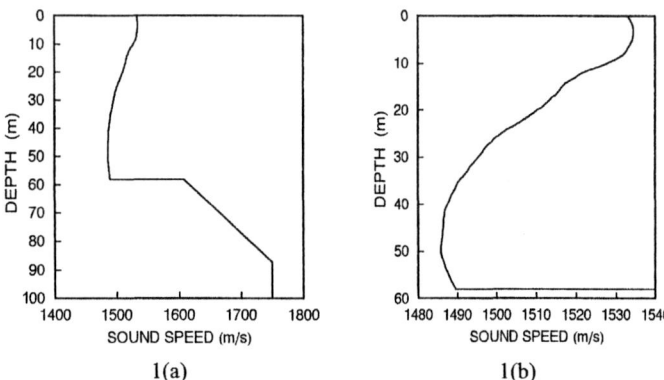

FIGURE 1. Mean sound speed profiles for the SWARM yo-yo CTD site: (a) water column, sediment and sub-bottom; (b) detail of water column.

Acoustic propagation losses were computed using a Gaussian-beam propagation code (Bellhop) [4] for a source at 10 m depth and a receiver at 30 m depth. Effects of bottom sound speed and attenuation were represented by reflection coefficients derived from the geoacoustic measurements. Surface scattering losses were computed using the SRFLOS component of the Oceanographic and Atmospheric Master Library (OAML) [5] for wind speeds of 3 kts, representing the upper end of Sea State 0, and 20 kts, representing Sea State 4. Volume attenuation was represented by Thorp's Law. [6] The ambient noise spectrum was constructed based on typical shallow-water noise spectra given in [7]. The acoustic fields were computed for frequencies ranging from 100 Hz to slightly less than 1 MHz in increments of 1/14 octave. This upper frequency limit was found to be sufficiently high to achieve the channel capacity in the simulations for ranges greater than 1.2 km.

Results for Unconstrained Source Bandwidth

Figure 2 shows the noise-to-channel function ratio $N(f)/H(f)$ and the optimum source spectral density $X(f)$ for a source power of 193 dB μPa^2 and receiver ranges of 2, 5, 10 and 20 km. The upper frequency limits of the optimal spectra are approximately 90, 40, 23, and 20 kHz, respectively. Because of the high signal to noise ratio for the 193 dB source, the optimal spectrum is nearly flat, eliminating the need for a "shaped" source spectrum.

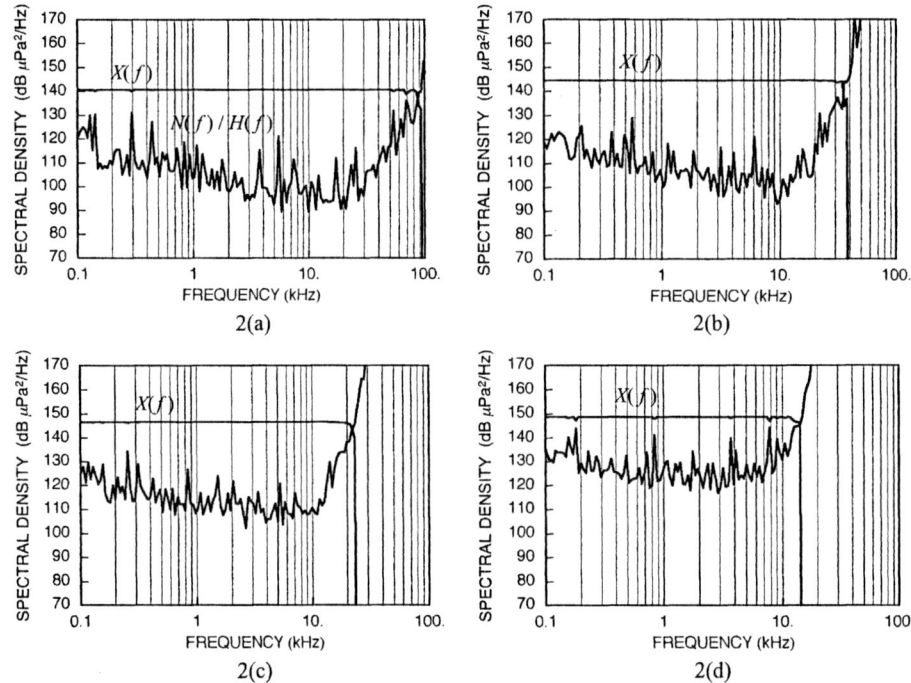

FIGURE 2. Noise-to-channel function ratio $N(f)/H(f)$ and optimal signal power spectral density $X(f)$ for ranges of (a) 2 km, (b) 5 km, (c) 10 km, and (d) 20 km. The source power is 193 dB μPa^2, the source depth is 10 m, and the receiver depth is 30 m.

Note that the optimal source spectra most likely contain energy at frequencies below the lower frequency limit of 100 Hz used in the computations. Inclusion of these frequencies would increase the estimated channel capacity very slightly (by less than 1%). In practice, however, high-bandwidth systems would not likely include low frequencies in the transmitted spectrum.

Figures 3(a),(b) show the computed channel capacity for the mean sound speed profile as a function of range for wind speeds of 3 kts and 20 kts, respectively. The effect of the higher wind speed is to decrease the channel capacity by approximately 30–60%, except for certain ranges, at which waterborne paths most likely dominate the acoustic field. Note again that these estimates are based on the assumption of a perfectly known channel function, and thus represent upper bounds rather than practically achievable data rates.

The channel capacity estimates in Figs. 3(a),(b) assume full utilization of the available bandwidth in constructing the optimal source spectrum $X(f)$ given by Equations (1)-(3). In practice, the bandwidth of a single transducer is limited to some fraction of the carrier frequency. (Note, however, that the theoretical capacity could be approached by using multiple transducers.)

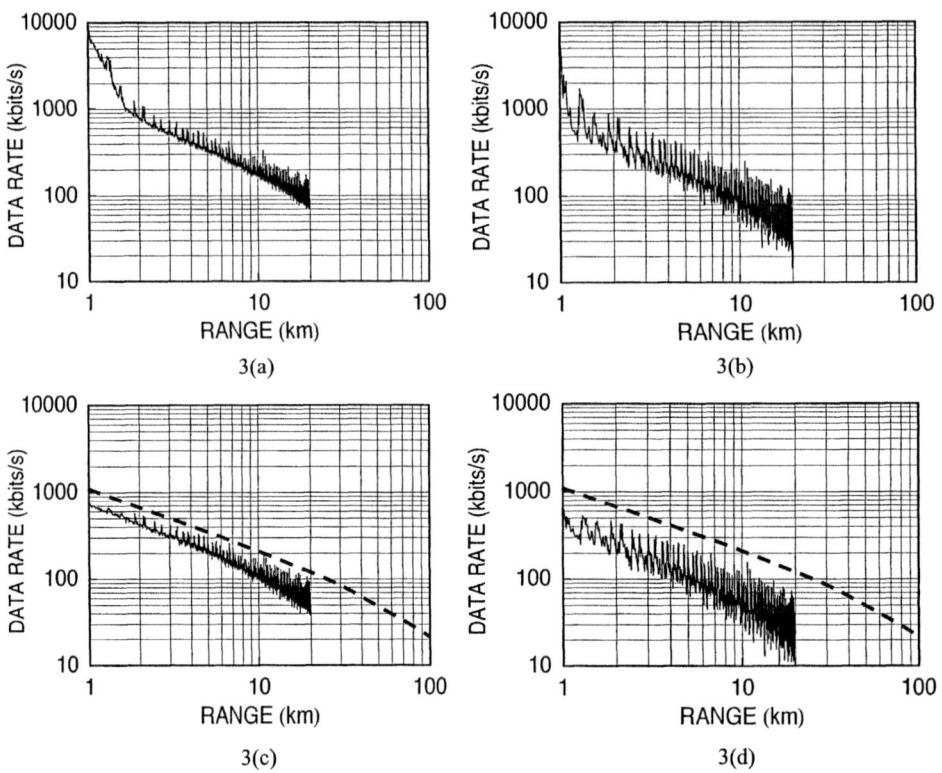

FIGURE 3. Solid curves: Channel capacity estimates for the mean sound speed profile at the SWARM yo-yo CTD site for source power 193 dB μPa^2, source depth 10 m and receiver depth 30 m. (a) Unconstrained bandwidth, 3 kt wind speed. (b) Unconstrained bandwidth, 20 kt wind speed. (c) Constrained bandwidth, 3 kt wind speed. (d) Constrained bandwidth, 20 kt wind speed. Dashed curves: Channel capacity upper bounds for Sea State 0, subject to bandwidth equal to carrier frequency, obtained by Peloquin and Leinhos [1] for a generic underwater channel based on sonar-equation analysis.

Results for Constrained Source Bandwidth

Channel capacities were computed assuming that the source bandwidth W equals the carrier frequency, and that the carrier frequency is given by $F_c = a/\sqrt{r}$, where a is determined by $F_c = 20$ kHz at range $r = 5$ km. Figs. 3(c) and 3(d) show the channel capacities calculated for the mean sound speed profile at the SWARM site, subject to the frequency band constraints, for wind speeds of 3 kts and 20 kts, respectively. Also shown is the upper bound obtained by Peloquin and Leinhos [1] for Sea State 0, based on a sonar-equation analysis for a generic underwater channel, with bandwidth constrained to be equal to carrier frequency. The computed channel capacities for the

SWARM environment are approximately 30–90% below the sonar equation based upper bounds, depending on wind speed and range.

CAPACITY VARIATION IN AN INTERNAL WAVE FIELD

This section examines the effect on channel capacity of random sound speed variations induced by (diffuse) internal gravity waves. Again, it is assumed that, for each transmission, both the source and receiver know the channel function. This corresponds in practice to transmissions with very short packet lengths. When longer-length packets are used, the achievable data rates will decrease unless the receiver processor is able to track the channel variation. Ideally, it would be desirable to be able to calculate the channel capacity in the case of an uncertain channel. However, the problem is extremely difficult, and explicit results are known only for a few simple statistical models of the channel variation that fall short of a realistic representation of a time-varying, multipath underwater acoustic channel.

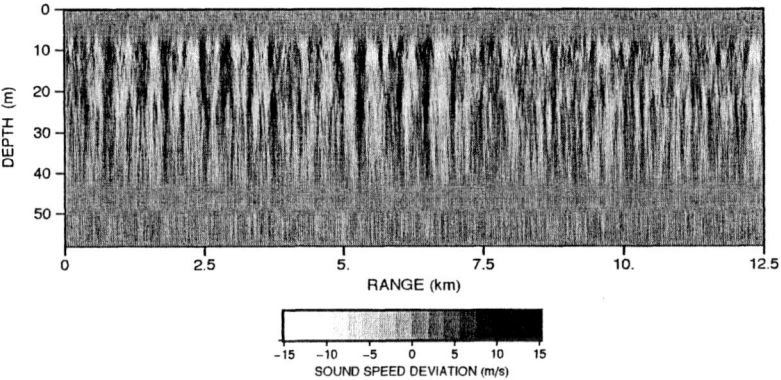

FIGURE 4. Sound speed deviation from the mean as a function of range and depth for a typical internal wave field realization.

Modeling acoustic propagation in a strongly range dependent medium requires extensive numerical computation. Because of the computation times required, this initial study examines the variations in channel capacity over limited frequency bands.

Realizations of linear internal wave fields were generated for the SWARM site, based on the spectrum of the internal waves derived from yo-yo CTD measurements taken at a time during which nonlinear internal waves were absent or negligible [3,8]. Figure 4 shows the sound speed deviation from the mean as a function of range and depth for a typical internal wave field realization. Geoacoustic properties were represented as described in the previous section.

Acoustic fields were computed using a parabolic equation code (RAM) [9] for frequencies of 300-400 Hz in increments of 1 Hz, for each of 200 realizations, and for frequencies of 900-1000 Hz for 100 realizations. Acoustic field computations for the 2500-2600 Hz band were based on a coupled normal mode model (C-SNAP) [10].

Figure 5 shows the mean values, over all of the realizations, of the ratios of the channel capacities of the internal-wave perturbed environments to the channel capacities of the mean environment, plotted as a function of range for frequency bands of 300-400 Hz and 900-1000 Hz. Assuming perfect knowledge of the channel, the effects of the internal waves on the channel capacity are minimal.

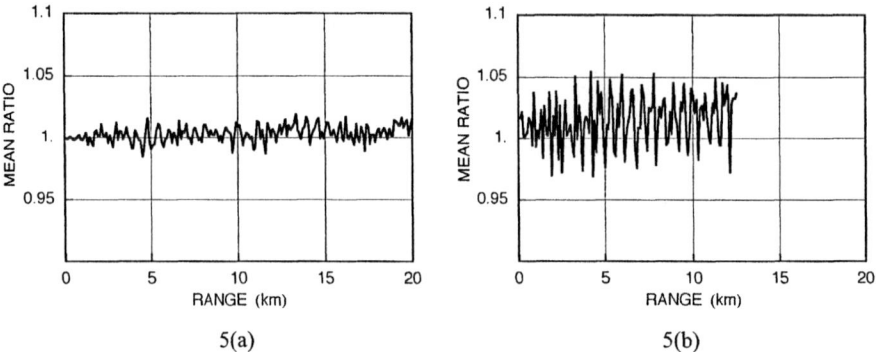

FIGURE 5. Mean values of the ratios of the channel capacities of the internal wave perturbed environments to the channel capacities of the mean environment, plotted as a function of range for frequency bands of (a) 300-400 Hz, and (b) 900-1000 Hz.

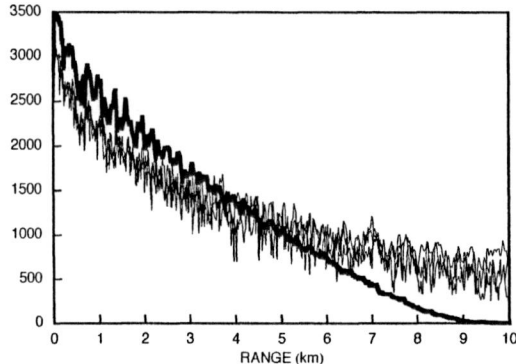

FIGURE 6. Heavy curve: Channel capacity for the mean environment for source power 193 dB μPa^2 and frequency constrained to 2500–2600 Hz. Light curves: Channel capacities for three realizations of the internal wave perturbed environment for the same frequency band.

Figure 6 shows the computed channel capacity for the mean environment for source power 193 dB μPa^2 and frequency constrained to 2500–2600 Hz (heavy curve). Also shown are the computed channel capacities for the same source power and frequency band for three realizations of the internal wave field. The effect of the internal waves is to enhance the acoustic field (hence the channel capacity) at longer ranges, but to decrease it at shorter ranges. These effects presumably result from the scattering of acoustic energy into lower propagation angles; this low-angle energy then propagates to longer ranges.

SUMMARY AND DISCUSSION

Acoustic communication channel capacities were computed for a single source-receiver pair for a shallow water waveguide derived from the 1995 SWARM experiment, both for the range-independent mean environment and for the same environment perturbed by simulated linear internal wave fields. Without bandwidth constraints, estimated channel capacities for the mean environment approach 10 megabits per second at 1 km range for a source power of 193 dB μPa^2. Channel capacities subject to commonly imposed source center frequency and bandwidth constraints are up to 90% lower at ranges less than 2 km. This suggests that further study of the optimal source frequency band as a function of range is warranted.

Effects of internal waves on theoretical channel capacity were examined in an initial study for limited (100-Hz) frequency bands centered at 350, 950 and 2550 Hz. Assuming that the channel is known, the internal waves had minimal effect on the channel capacity at the lower frequencies, but, for the 2500-2600 Hz band, the internal waves significantly increased the capacity at longer ranges, presumably due to scattering of acoustic energy into lower propagation angles.

The foregoing computations assume perfect knowledge of the channel function. Further study is needed to determine the extent to which the practically achievable data rates are limited by channel uncertainty, and to determine the optimal design of source spectra, coding, and signal processing to maximize the data rate.

ACKNOWLEDGMENTS

This work was supported by the Office of Naval Research and by a grant of computer time from the Department of Defense High Performance Computing Modernization Program at the Army Research Laboratory Major Shared Resource Center. The authors thank Paul Gendron of NRL for helpful discussions.

REFERENCES

1. Peloquin, R. F., and Leinhos, H., unpublished (1997).
2. Shannon, C. E., *Proc. IRE* **37**, 10-21 (1949).
3. Apel, J. H. et al., *IEEE J. Oceanic Eng.* **22**, 465-500 (1997).
4. Porter, M. B., and Bucker, H. P., *J. Acoust. Soc. Am.* **82**, 1349-1359 (1987).
5. Naval Oceanographic Office, OAML-SRS-37, Stennis Space Center, Mississippi, 1992.
6. Thorp, W. H., *J. Acoust. Soc. Am.* **42**, 270 (1967).
7. Urick, R. J., *Principles of Underwater Sound*, 3rd ed., McGraw-Hill, New York, 1983, p. 213.
8. Yang, T. C., and Yoo, K. B., *IEEE J. Oceanic Eng.* **24**, 333-345 (1999).
9. Collins, M. D., *J. Acoust. Soc. Am.* **93**, 1736-1742 (1993).
10. Ferla, C. M., Porter, M. B. and Jensen, F. B., SACLANTCEN Memorandum SM-274, LaSpezia, Italy, 1993.

BOUNDARY SCATTERING AND VOLUME FLUCTUATIONS

Progress and Research Issues in High-Frequency Seafloor Scattering

Darrell R. Jackson

Applied Physics Laboratory, University of Washington, Seattle, WA 98105

Abstract. The status of model-data comparisons for seafloor scattering is reviewed. The small-roughness perturbation and small-slope approximations have proven accurate and useful in treating scattering by interface roughness, and have been extended from the fluid case to the elastic and poroelastic cases. The problem of scattering by sediment heterogeneity is more difficult, but the method of small perturbations has been tested successfully in several experiments. Further tests are needed, particularly with respect to scattering by discrete inclusions. Stratification has been modeled in both roughness and volume scattering, but only a few experimental tests have been attempted.

INTRODUCTION

Substantial progress has been made in recent years toward understanding and modeling the physical processes responsible for acoustic scattering by the seafloor. This has been possible due to the increasing effort devoted to characterization of the sediment physical parameters that control scattering. Seafloor scattering is most often described in terms of the scattering cross section per unit area per unit solid angle or its decibel equivalent, scattering strength. In most modeling approaches, the cross section is assumed to be a sum of two terms, one representing scattering due to interface roughness and the other representing scattering due to volume heterogeneity of the sediment. This idealization is avoided in some recent work [1].

SCATTERING BY INTERFACE ROUGHNESS

The two most commonly used approximations for scattering by seafloor roughness are the small-roughness perturbation method (sometimes known as Rayleigh-Rice perturbation theory) and the Kirchhoff approximation (also known as the tangent-plane approximation). Each has its own separate domain of validity, with perturbation theory tending to be most accurate for scattering at wide angles relative to the specular (flat-interface reflection) direction and the Kirchhoff approximation being better for scattering near the specular direction. The small-slope approximation [2, 3] provides some of the best properties of the two in a logical expansion scheme.

The small-perturbation, Kirchhoff, and small-slope approximations yield equations whose general form does not depend upon the particular wave theory employed. That is, these forms are the same whether the sediment is modeled as a fluid, a viscoelastic solid or a poroelastic medium. The general forms for the bistatic scattering cross-section in these three scattering approximations are, in the small-roughness perturbation approximation:

$$\sigma = k_w^4 |A|^2 W_2(\Delta \mathbf{K}) \,, \tag{1}$$

in the Kirchhoff approximation:

$$\sigma = \frac{|V(\theta_{is})|^2}{8\pi} [\frac{\Delta k^2}{\Delta K \Delta k_z}]^2 I_K \,, \tag{2}$$

and, in the small-slope approximation:

$$\sigma = \frac{k_w^4 |A|^2}{2\pi \Delta K^2 \Delta k_z^2} I_K \,. \tag{3}$$

The wavenumber in water is denoted k_w, and the argument, Δk, appearing in Equation (2), is the magnitude of the difference of the scattered and incident wave vectors. Similarly, $\Delta \mathbf{K}$, is the horizontal component of this difference, the so-called "Bragg wave vector". The magnitude of this difference is denoted ΔK and appears in Equations (2) and (3) along with the vertical component of the difference, Δk_z. These variables are related by $\Delta k^2 = \Delta K^2 + \Delta k_z^2$. The "Kirchhoff integral"

$$I_K = \frac{\Delta K^2}{2\pi} \int e^{-i\Delta \mathbf{K} \cdot \mathbf{R}} [e^{-\frac{1}{2}\Delta k_z^2 S(\mathbf{R})} - e^{-\Delta k_z^2 h^2}] d^2 R \tag{4}$$

does not depend upon the choice of wave theory and is common to both the Kirchhoff and small-slope approximations. The spectrum, $W_2(\Delta \mathbf{K})$, appearing in the small-roughness cross-section and the structure function, $S(\mathbf{R})$, appearing in the Kirchhoff integral are related by a transform, so that knowledge of one is sufficient to determine the other. The factor A depends upon choice of wave theory, but is common to both the perturbation and small-slope approximations. The reflection coefficient, V, appearing in the Kirchhoff expression depends upon the choice of wave theory, and is evaluated at the grazing angle

$$\theta_{is} = \sin^{-1}\left(\frac{\Delta k}{2k_w}\right) \tag{5}$$

that corresponds to specular reflection from the source to the receiver with the rough surface tilted in such a way as to provide such a reflection.

In summary, a particular wave theory must be applied to obtain A for small-roughness perturbation and small-slope approximations or, correspondingly, V for the Kirchhoff approximation. Once these theoretical tasks are complete, only numerical computation remains, including evaluation of Equation (4) [4].

Attention will be focused on the small-roughness perturbation approach, as it is applicable over a wide range of angles, excluding near-specular directions. In addition, its predictions are essentially the same as those of the small-slope approximation in this angular range for typical seafloor parameters. The simplest and most widely used model treats the sediment as a fluid. In this case, the factor A in (1) and (3) is given by [11]

$$A = \frac{1}{2}[1 + V(\theta_i)][1 + V(\theta_s)]G \,, \tag{6}$$

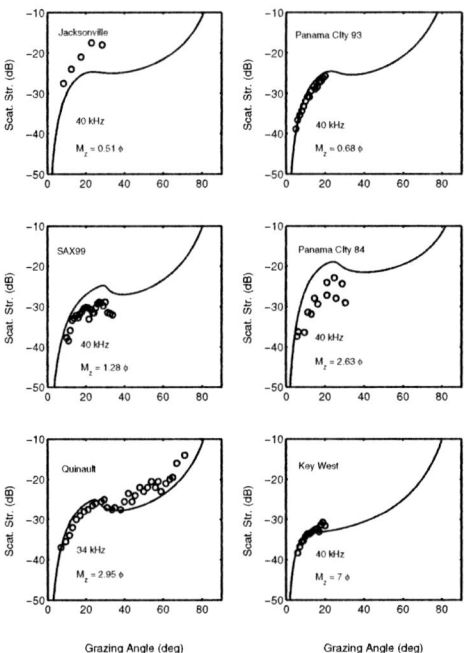

FIGURE 1. Comparison of fluid small-roughness perturbation model with data from six experiments conducted at sandy sites. The examples are given in order of increasing grain size, M_z, and the primary acoustic and geoacoustic data sources are: Jacksonville [5], Panama City 93 [6], SAX99 [7], Panama City 84 [8], Quinault [9], and Key West [10].

where

$$G = (1/a_\rho - 1)[\cos\theta_i \cos\theta_s \cos\phi_s - \frac{\sin\theta_{pi} \sin\theta_{ps}}{a_p^2 a_\rho}] + 1 - \frac{1}{a_p^2 a_\rho}, \quad (7)$$

and

$$\sin\theta_{pi} = \sqrt{1 - a_p^2 \cos^2\theta_i}, \quad (8)$$

$$\sin\theta_{ps} = \sqrt{1 - a_p^2 \cos^2\theta_s}. \quad (9)$$

In these expressions, θ_i and θ_s are the incident and scattered grazing angles, ϕ_s is the "bistatic angle," equal to zero in the specular direction and equal to π in the backscattering direction. The ratio of sediment mass density to water mass density is denoted a_ρ, and a_p is the corresponding (complex) ratio for compressional wave speeds.

Figure 1 compares the small-roughness, fluid-sediment model with data from six experiments conducted at sandy sites. The fit between the model and data appears to be reasonably good, with the greatest model-data difference occurring for the two intermediate grain size cases, SAX99 and Panama City 84. A detailed error analysis has been performed for the SAX99 data [7], with the conclusion that the model-data

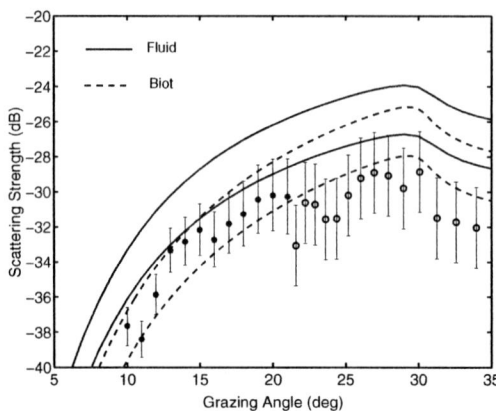

FIGURE 2. Comparison of measured bottom backscattering strength at 40 kHz with small-roughness perturbation fluid model (solid curves) and Biot model (dashed curves). The upper and lower curves in each case give the 95% uncertainty bounds for the models based on the uncertainty bounds for the roughness data. The vertical error bars give the corresponding 95% uncertainty bounds for the acoustic measurements. The data denoted by filled circles were taken at the same site as the roughness measurements, whereas the data denoted by open circles were obtained using a different apparatus at a site removed by a few hundred meters. The slight offset between these data sets may be due to either acoustic calibration error, difference in roughness at the two sites, or a combination of these two. (Figure adapted from [7])

difference is significant and likely due to neglect of poroelastic (Biot) effects. The similar model-data differences seen at the Panama City 84 site [8] may be due to the same cause. One significant point of agreement between model and data is the scattering strength maximum occurring near the critical angle, visible in the data from SAX99, Panama City 84, and Quinault. No maximum is predicted or seen for the finer grain sediment of the Key West site.

The Biot model has been compared with backscattering strength data in [7].
Figure 2 shows that, within the uncertainties of acoustic measurement and the uncertainties in model inputs, the Biot model provides a better fit to the data than the fluid model. A nearly identical fit would be provided by the "effective density" approximation [12] in which the sediment is modeled as a fluid having a reduced value of density, this value being computed in terms of a few of the Biot parameters.

Roughness scattering by viscoelastic seafloors has been treated in the perturbation [13, 14] and small-slope [15, 16, 17] approximations. Shear effects are negligible for sandy seafloors [18], but significant for rock seafloors. Testing of available models at well-characterized rocky sites presents a difficult challenge.

Gradients in sediment physical properties can influence scattering owing to the incidence of energy from below the sediment-water interface. Ivakin has developed a general formalism [1] that lends itself to layered seafloors, including cases where the distinction between roughness and volume scattering is not clear. Less generally, one may simply include the upward reflected wave in the perturbation treatment [10]. This leads to the relatively simple result presented in Equation (6) in which the effect of the energy inci-

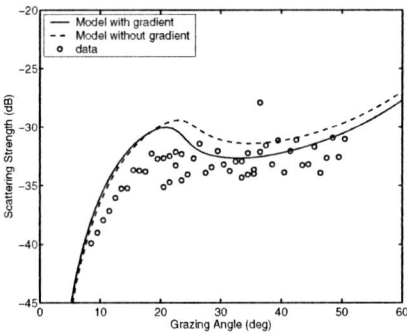

FIGURE 3. Effect of gradients in roughness scattering model using acoustic and geoacoustic data from [19]. The frequency is 140 kHz and the seafloor is a silty sand. Gradients are treated using Equation (6).

dent from below is contained in the reflection coefficient, V, which must be computed including the effects of stratification. In the measurements reported by Pouliquen and Lyons [19], gradients had a significant effect, and the inclusion of gradient effects via a numerically computed reflection coefficient improves the model-data fit, as shown in Fig. 3.

SCATTERING BY VOLUME HETEROGENEITY

As with roughness scattering, attention will be focused on the method of small perturbations as applied to sediment volume scattering. It is in this arena that the most rigorous model-data comparisons have been made. If the sediment is treated as a fluid, an effective interface bistatic cross section can be defined as follows:

$$\sigma = \frac{|[1+V(\theta_i)][1+V(\theta_s)]|^2 \sigma_v}{2a_\rho^2 \Im[k_p(\sin\theta_{pi} + \sin\theta_{ps})]}, \quad (10)$$

where the volume scattering cross section per unit solid angle per unit volume is

$$\sigma_v = \frac{\pi |k_p|^4}{2}\left[W_{\kappa\kappa}(\Delta\mathbf{k}_p) + 2\Re\{\frac{\mathbf{k}_{ps}\cdot\mathbf{k}_{pi}\,k_p^{*2}}{|k_p|^4}W_{\rho\kappa}(\Delta\mathbf{k}_p)\} + \frac{|\mathbf{k}_{ps}\cdot\mathbf{k}_{pi}|^2}{|k_p|^4}W_{\rho\rho}(\Delta\mathbf{k}_p)\right]. \quad (11)$$

The wavenumber for compressional waves in the sediment is $k_p = k_w/a_p$, and the function $W_{\kappa\kappa}$ is the three-dimensional spectrum for the fluctuations in normalized sediment compressibility. The normalization consists in dividing the fluctuating part of the compressibility (which is the inverse of the bulk modulus) by the mean compressibility. Similarly, $W_{\rho\rho}$ is the spectrum for normalized density fluctuations, and $W_{\rho\kappa}$ is the cross-spectrum that expresses the correlation between these two random variables.

The wave vectors, $\mathbf{k}_{pi}$ and $\mathbf{k}_{ps}$, for the incident and scattered compressional waves in the sediment have the same horizontal components as the corresponding wave vectors in

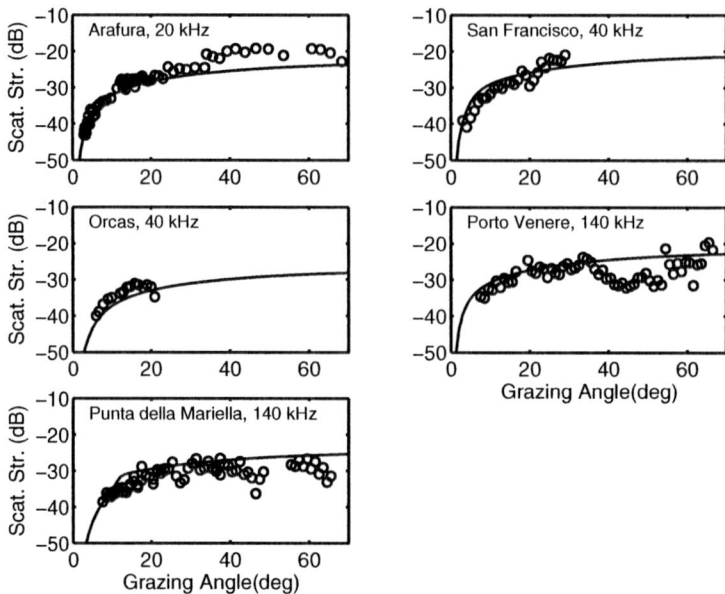

FIGURE 4. Comparison of the fluid volume scattering perturbation model with backscattering strength measured at five sites having soft sediments. The primary acoustic and geoacoustic data sources are: Arafura [9], San Francisco [9], Orcas [20], Porto Venere [19], Punta della Mariella [19]. The volume heterogeneity parameters for the Arafura and San Francisco sites were taken from [21].

water and complex z-components determined by Equations (8) and (9). The spectra in Equation (11) are evaluated at the Bragg wave vector, $\Delta \mathbf{k}_p$:

$$\Delta \mathbf{k}_p = \Re\{\mathbf{k}_{ps} - \mathbf{k}_{pi}\} . \qquad (12)$$

The Bragg wave vector is the real part of the wave vector difference. If the complex difference were required instead, one would be faced with the issue of analytic continuation of the spectrum to complex arguments. This issue cannot be avoided when treating volume scattering in elastic or layered media.

Figure 4 compares higher-frequency backscattering data obtained by several different investigators with the perturbation model. These model-data comparisons are the same, except for details, as those given in the primary references. The model-data agreement is satisfactory, but some of the comparisons are not as rigorous as one might wish. In particular, the volume heterogeneity data for the Arafura, San Francisco, and Orcas sites were obtained from small-diameter, short core samples and were analyzed in the vertical coordinate only. The three-dimensional spectra were assumed to be isotropic. The Arafura site had a large concentration of buried shell fragments, calling into question the use of the perturbation approach. In all cases, the density–compressibility cross spectrum was not determined by measurement, but assigned a default value.

Further effort is required to obtain accurate estimates of sediment heterogeneity statistics. Nevertheless, the results of these model-data comparisons are encouraging and indicate that the perturbation model for sediment volume scattering is a reasonable approximation in many cases of interest.

Corresponding perturbation expressions are available for volume scattering in viscoelastic material, [22, 14, 18], but have not been tested experimentally. The poroelastic volume scattering problem has been treated in the perturbation approximation [23], but results are not yet available in a form appropriate to the high-frequency seafloor problem.

REFERENCES

1. Ivakin, A., *J. Acoust. Soc. Am.*, **103**, 827–837 (1998).
2. Voronovich, A., *Soc. Phys. J. ETP*, **62**, 65–70 (1985).
3. Thorsos, E., and Broschat, S., *J. Acoust. Soc. Am.*, **97**, 2080–2093 (1995).
4. Drumheller, D., and Gragg, R., *J. Acoust. Soc. Am.*, **110**, 2270–2275 (2001).
5. Stanic, S., Briggs, K., Fleischer, P., Sawyer, W. B., and Ray, R., *J. Acoust. Soc. Am.*, **85**, 125–136 (1989).
6. Jackson, D., Briggs, K., Williams, K., and Richardson, M., *IEEE J. Oceanic Engr.*, **21**, 458–470 (1996).
7. Williams, K., Jackson, D., Thorsos, E., Tang, D., and Briggs, K., *IEEE J. Oceanic Eng.*, **27**, 376–387 (2002).
8. Stanic, S., Briggs, K., Fleischer, P., Ray, R., and Sawyer, W., *J. Acoust. Soc. Am.*, **83**, 2134–2144 (1988).
9. Jackson, D., and Briggs, K., *J. Acoust. Soc. Am.*, **92**, 962–977 (1992).
10. Briggs, K., Williams, K., Jackson, D., Jones, C., Ivakin, A., and Orsi, T., *Marine Geology*, **182**, 141–159 (2002).
11. Moe, J., and Jackson, D., *J. Acoust. Soc. Am.*, **96**, 1748–1754 (1994).
12. Williams, K., *J. Acoust. Soc. Am.*, **110**, 2956–2963 (2001).
13. Essen, H., *J. Acoust. Soc. Am.*, **95**, 1299–1301 (1994).
14. Jackson, D., and Ivakin, A., *J. Acoust. Soc. Am.*, **103**, 336–345 (1998).
15. Yang, T., and Broschat, S., *J. Acoust. Soc. Am.*, **96**, 1796–1803 (1994).
16. Gragg, R., Wurmser, D., and Gauss, R., *J. Acoust. Soc. Am.*, **110**, 2878–2901 (2001).
17. Soukoup, R., and Gragg, R., *J. Acoust. Soc. Am.*, **113**, 2501–2514 (2003).
18. Ivakin, A., and Jackson, D., *J. Acoust. Soc. Am.*, **103**, 346–354 (1998).
19. Pouliquen, E., and Lyons, A., *IEEE J. Oceanic Eng.*, **27**, 388–402 (2002).
20. Jones, C., and Jackson, D., "Temporal Fluctuations of Backscattered Field Due to Bioturbation in Marine Sediments," in *High Frequency Acoustics in Shallow Water*, edited by N. Pace, E. Pouliquen, O. Bergem, and A. Lyons, NATO SACLANT Undersea Res. Ctr., La Spezia, Italy, 1997, pp. 275–282.
21. Yamamoto, T., *J. Acoust. Soc. Am.*, **99**, 866–879 (1996).
22. Ivakin, A., *Soviet Physics–Acoust.*, **36**, 377–380 (1990).
23. Gurevich, B., Sadovnichaja, A., Lopatnikov, S., and Shapiro, S. A., *Geophys. J. Int.*, **133**, 91–103 (1998).

Modeling Shallow Water Propagation With Scattering From Rough Boundaries

Eric I. Thorsos, Frank S. Henyey, W. T. Elam, Steve A. Reynolds, and Kevin L. Williams

Applied Physics Laboratory, University of Washington, 1013 NE 40th Street, Seattle, WA 98105-6698

Abstract. Results of PE simulations at 3 kHz will be described for propagation in a shallow water waveguide with a rough sea surface and a flat water-sediment interface overlying absorbing sediments. Pressure fields are simulated in two space dimensions and have been obtained using a wide-angle PE code that accounts for scattering from a rough sea surface. The PE simulation approach is a Monte Carlo method, requiring solutions for many surface realizations in order to obtain results for field moments. To obtain a fast, yet accurate model for average field quantities, a transport theory method based on coupled modes has been developed for both the first and second moments of the field. Estimates for the fourth moment of the field have also been obtained based on the transport theory results for the first and second moments. Transport theory results for the coherent and total intensity, and for the scintillation index, will be presented and compared with Monte Carlo simulations.

INTRODUCTION

Modeling of shallow water propagation can be challenging because of the need to treat 3-D spatial and temporal variations of the sound speed field, spatial bathymetric variations, and multiple boundary interactions for the propagating field. In practice, the problem is typically made even more difficult by an incomplete knowledge of the environmental conditions. Here we focus on the effects of rough surface scattering at sea surface boundary interactions, and for simplicity, take the sound speed to be constant within the water (1,500 m/s) and within the seafloor sediment (1,600 m/s), and assume the water column has a constant average depth (50 m) to a flat water-sediment interface. In addition, the field in the homogeneous sediment is attenuated at 0.5 dB/λ, and the sediment-to-water density ratio is 2.0.

For the problem described, and for typical surface roughness conditions, the modeling task is simpler at low frequencies (< 1 kHz), because the propagating field is dominated by the coherent field, and the surface scattered incoherent component is relatively small by comparison. (The coherent field, or the first moment of the field, is given by the average of the complex field over an ensemble of rough surface realizations.) In this case, one could proceed iteratively by first obtaining in a self-consistent manner the coherent field as a function of range and depth, and then using the coherent field to obtain the scattered field [1-3]. As the acoustic frequency increases, however, the scattered incoherent field becomes larger relative to the coherent field, and it becomes more important to treat the effects of multiple scattering in forward propagation. We consider a frequency of 3 kHz where simulations to be described shortly show that the incoherent and coherent fields are of comparable magnitude for ranges up to about 5 km, while at longer ranges the coherent field dominates. To obtain "ground truth" for the effects of surface scattering on shallow water propagation at 3 kHz, we have used numerical simulations based on rough surface PE. In addition, we have developed a fast, yet accurate method based on transport theory for modeling both the coherent field and the average total intensity of the field that applies for all relative levels of the incoherent field. Results for the scintillation index have also been obtained.

RESULTS

The propagation simulations were done using a wide-angle PE method developed by Rosenberg [4] that we believe accurately accounts for forward scattering from a rough sea surface (in two space dimensions). The Rosenberg propagation model is an extension to a wide-angle PE propagation model developed by Collins [5]. Rough sea surface realizations generated for use in the simulations are consistent with a one-dimensional cut through a two-dimensional isotropic spectrum of a Pierson-Moskowitz form [6]. For examples shown, the surface waves are produced by a wind speed of 7.7 m/s (15 knots). A point source is located at the mid depth of 25 m, and a vertical beam pattern has been applied with a full width of 20° and with the beam center aimed up at a 10° grazing angle. The simulations have been done for a CW source.

PE simulation results for the coherent intensity (first moment of the field) and the average total intensity (second moment of the field) based on 50 surface realizations are shown in Fig. 1. The simulation is done in two space dimensions, and cylindrical divergence is not included. The color bar denotes the field intensity in dB. At short range the total intensity is noticeably greater than the coherent intensity due to the presence of the incoherent field scattered from the rough sea surface. At longer range the total intensity approaches the coherent intensity, except in regions where the coherent intensity has nulls. These trends can be understood qualitatively from both ray and normal mode perspectives [7]. Monte Carlo simulations of this type give reliable results but are computer intensive and time consuming. It would, therefore, be advantageous to have a more practical method that yields these average results without depending on a Monte Carlo approach.

We have developed a transport theory method for computing directly the first, second, and fourth moments of the propagating field. Space does not permit a detailed exposition of the method here, and these details will be presented elsewhere. Instead, we briefly indicate the basic ideas, and then show how well the transport theory results agree with those obtained with rough surface PE. The starting point is to expand the field in unperturbed normal modes and then obtain the evolution equations for the mode amplitudes, accounting for mode coupling due to scattering from a particular realization of the rough surface. Small surface height perturbation theory is used to evaluate the mode coupling terms. Stopping at this stage would yield an alternative Monte Carlo simulation method. Transport theory gives evolution equations for moments of the mode amplitudes at the cost of some approximations. The evolution equations for the first moment of the mode amplitudes are obtained by formally averaging the set of mode amplitude equations and using transport theory approximations as given, for example, by Van Kampen [8]. The final result is a first-order evolution equation of form

$$\frac{dA}{dx} = RA \tag{1}$$

for the N-component vector A of average mode amplitudes with R an $N \times N$ matrix. The solution to (1) is given by

$$A(x) = \exp(Rx)A(0). \tag{2}$$

For the second moment, one first obtains the evolution equation for all products of two mode amplitudes with one amplitude complex conjugated. Formally averaging and making transport theory approximations again leads to equations of the form of (1) and (2), but now A is an N^2-component vector and R is an $N^2 \times N^2$ matrix.

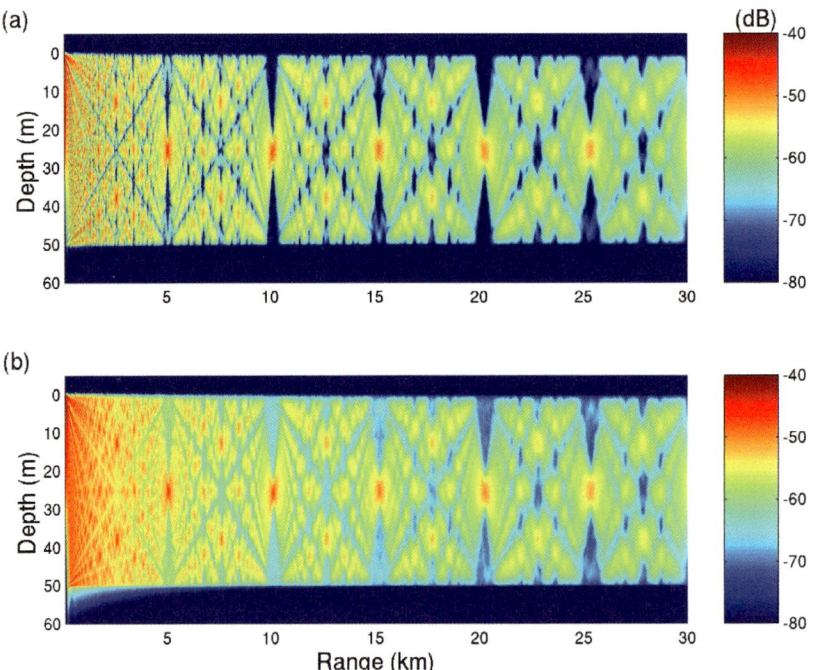

FIGURE 1. PE simulation for coherent intensity (a) and total intensity (b) obtained by averaging over the results for 50 rough surface realizations.

For our problem there would be 74 discrete modes (or "propagating modes") plus continuum modes if the sediment were modeled as a lossless, infinite half-space. In practice, we have taken the computational region to consist of a 50 m water layer and a 150 m sediment layer; this converts the continuum modes into a set of closely spaced discrete modes. (Because sediment attenuation is also included, some of the continuum modes are promoted to discrete modes; this will be described elsewhere.) We have investigated in detail the number of modes that must be retained in transport theory to obtain convergence with PE results. This was done by comparing transport theory and PE results for mode amplitude decay with range, using both the average mode amplitudes (the first moments) and the average of the absolute squares of the mode amplitudes (a subset of the second moments). The PE mode amplitudes were obtained by projecting the PE fields onto the mode functions as a function of range for each rough surface realization and then performing averages over realizations of the complex mode amplitudes and their absolute squares. We found that true convergence between transport theory and PE results could be obtained by taking $N = 74$, the total number of propagating modes, each of which corresponds to rays that reflect from the bottom at grazing angles below the critical angle (about 20°). However, to obtain this convergence, we also found it necessary to extend sums over modes that occur in the matrix elements of R up to 200. This means that coupling terms are included that couple energy from below to above the critical angle, where it would be rapidly lost into the bottom; this energy loss needs to be included. Since we do not transport modes that correspond to rays that are above the critical angle, this energy is effectively removed from the problem at the point the coupling occurs. However, this has almost the same effect as transporting these higher modes, since they would be rapidly attenuated by absorption in the bottom. Thus, restricting the transported modes to the propagating modes turns out to be a good approximation when internal matrix element sums are extended to sufficiently high values (200 in our case).

Obtaining the solution for the second moment with Eq. (2) would be entirely impractical for an N as large as 74. Fortunately, a major simplification can be made with essentially no loss in accuracy: we assume the effect of cross-mode coherences on the mode intensities can be neglected in Eq. (1). Our rationale is that these effects should largely average out over relatively short ranges due to range dependence of the phases on the right-hand side. We refer to this as the Dozier-Tappert approximation, since the same approximation was made in their transport theory treatment of acoustic propagation through internal waves [9, 10]. However, we do not neglect cross-mode coherence in constructing the intensity field; rather, we assume only that the incoherent contributions to the cross-mode coherence vanish. In the Dozier-Tappert approximation for the second moment, A becomes an N-component vector and R becomes an $N \times N$ matrix. We have implemented transport theory with and without the Dozier-Tappert approximation, and found that for our problem the Dozier-Tappert approximation is perfectly adequate and yields excellent agreement with rough surface PE results. We make a related approximation for the first moment, which reduces R to diagonal form, and which we have also found to be highly accurate by comparing results with and without the approximation. Though R becomes diagonal, coherent mode amplitude decay due to rough surface scattering is still accurately taken into account through internal sums over modes in the diagonal elements that we extend up to 200. In what follows, the Dozier-Tappert approximation will be taken to imply the approximations described for both the first and second moments of the field; this approximation has been used for all transport results shown.

Transport theory results for the coherent intensity and the total intensity are shown in Fig. 2. A comparison of the intensities from PE (Fig. 1) with the corresponding transport theory predictions shows remarkable agreement. The coherent intensities in part (a) of these two figures are difficult to distinguish. (The difference above the mean surface occurs simply because the domain for the PE simulations extends well above the mean surface though the coherent field vanishes there, while the domain for the transport method terminates at the mean surface and the plot has not been modified to show a vanishing field above the mean surface.) The average total intensities (coherent plus incoherent) in part (b) of the figures are also in very good agreement in the water column.

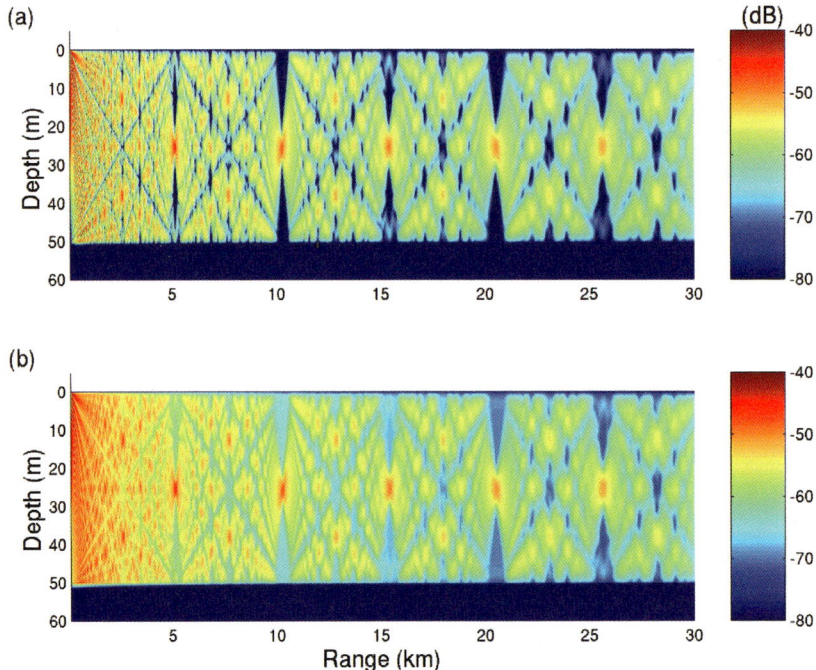

FIGURE 2. Transport theory results for coherent intensity (a) and total intensity (b) obtained in the Dozier-Tappert approximation.

Note that the PE simulation shows scattered intensity in the sediment at short range that is not present in the transport theory plot. This difference follows from our restriction to propagating modes with transport theory and has two elements. First, we begin by projecting the starting field onto the set of 74 propagating modes. However, the starting field employed has some initial energy propagating at angles relative to the horizontal that are greater than the critical angle. This energy is rapidly lost into the bottom in the PE simulation, but is not included in the transport model; differences downrange due to this should be negligible. Second, rough surface scattering continually promotes energy from discrete propagating modes to modes that are rapidly attenuated in the bottom (the set of closely spaced discrete modes that represent the continuum modes plus the "promoted" discrete modes). Or, equivalently, scattering continually transfers energy from below to above the critical angle, and this energy shows up in the PE simulation as it is being lost into the bottom. As discussed previously, in our transport method this energy effectively vanishes in the water column at the point it scatters to angles above the critical angle, and therefore does not appear as intensity in the sediment. Since this energy loss is being properly taken into account, differences downrange should again be negligible. If the field in the sediment were of primary interest, the number of modes retained in the transport method could be increased.

In addition to being accurate, the transport method is very fast in the Dozier-Tappert approximation. The PE results in Fig. 1 took on the order of a day of computer time to obtain, while the time required for the transport method was on the order of a minute. Therefore, the transport theory approach appears to be an attractive alternative to brute force Monte Carlo methods.

Figure 3 shows the comparison at short range between PE and transport theory for the total intensity. The high quality of the agreement in the water column is again evident, as are the differences in the sediment as discussed in reference to Fig. 2. Other approaches, such as ray tracing where the effect of boundary roughness is incorporated as a loss into a boundary reflection coefficient, may be capable of accurately modeling the coherent intensity [11]. However, we believe the ability to accurately and rapidly model the coherent plus incoherent intensity in the mid frequency region is new.

In addition to the average intensity, the fluctuations in the average intensity are of interest as the realization of surface roughness changes with time. One characterization of these fluctuations is given by the scintillation index, a fourth moment of the field. In principle, transport theory can be extended to obtain evolution equations for the fourth moments of the mode amplitudes. We have taken a simpler and more approximate approach. The total pressure is the sum of the coherent and incoherent pressure. If one assumes that the incoherent complex pressure obeys Gaussian statistics, it is straightforward to derive an expression for the scintillation index in terms of the coherent and total intensities. The result is

$$SI = \frac{\langle I \rangle^2 - I_c^2}{\langle I \rangle^2} = \frac{\langle |p|^2 \rangle^2 - |\langle p \rangle|^4}{\langle |p|^2 \rangle^2}, \qquad (3)$$

where $\langle p \rangle$ and $\langle |p|^2 \rangle$ can be given by transport theory. Thus, we obtain a transport result for the scintillation index that can be compared with the PE result for the same quantity where no approximations (other than rough surface PE) are made. The comparison is shown in Fig. 4. To obtain good convergence to the fourth moments with PE simulations, the number of rough surface realizations was increased to 1,000 for this example. The agreement in the water column between transport theory and PE results for the scintillation index is very good, in spite of the approximations made.

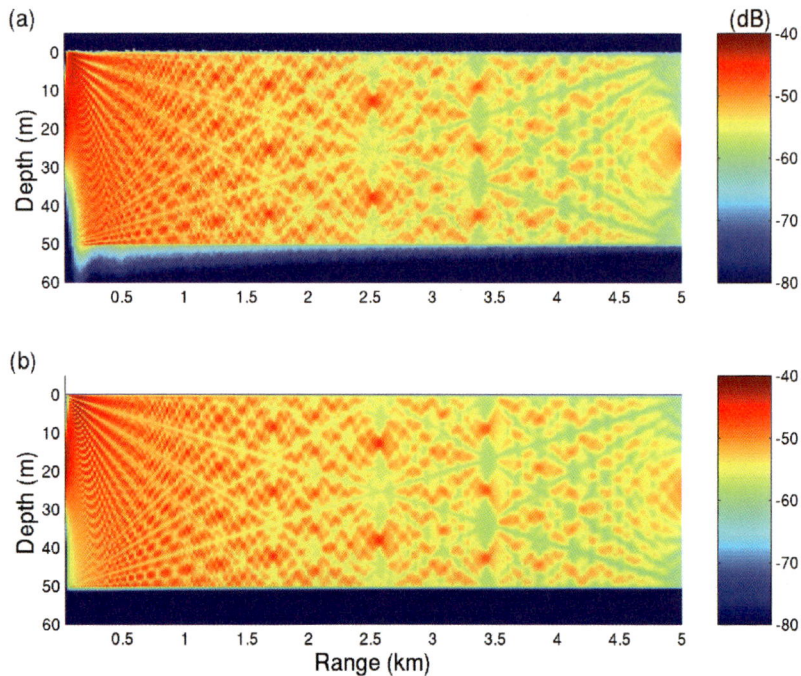

FIGURE 3. Comparison of (a) PE simulation for total intensity obtained by averaging over the results for 50 rough surface realizations with (b) transport theory result.

DISCUSSION

The results presented indicate that transport theory in the Dozier-Tappert approximation is a promising approach for modeling shallow water propagation at mid frequencies where boundary scattering can play an important role. The method as outlined is accurate and computationally fast for the problem studied. In addition to providing the coherent field and the total intensity, the scintillation index can be obtained with good accuracy at essentially no extra effort for the case treated.

The problem considered includes several significant simplifications. The geometry is two-dimensional, roughness is confined to the surface, sound speed variations (e.g., internal waves) and bathymetric variations are ignored. We believe that some, and perhaps all, of these restrictions can be relaxed in future work.

ACKNOWLEDGMENT

This work was supported by the Office of Naval Research.

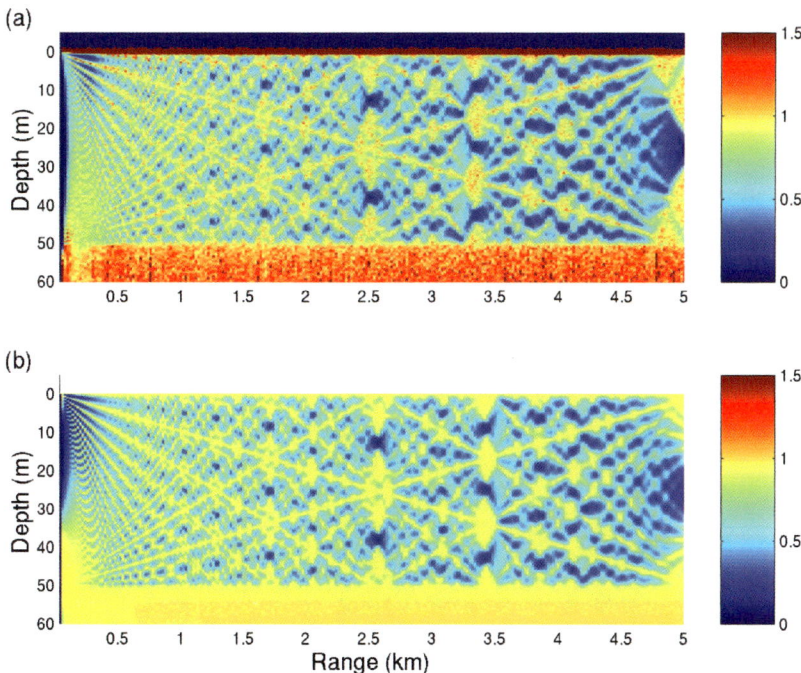

FIGURE 4. Comparison of (a) PE simulation for the scintillation index obtained by averaging over the results for 1000 rough surface realizations with (b) transport theory result.

REFERENCES

1. W. A. Kuperman and H. Schmidt, "Self-consistent perturbation approach to rough surface scattering in stratified elastic media," *J. Acoust. Soc. Am.* **86**, 1511-1522 (1989).
2. H. Schmidt and W. A. Kuperman, "Spectral representations of rough interface reverberation in stratified ocean waveguides," *J. Acoust. Soc. Am.* **97**, 2199-2209 (1995).
3. B. H. Tracy and H. Schmidt, "Seismo-acoustic field statistics in shallow water," *IEEE J. Oceanic Eng.* **22**, 317-331 (1997).
4. A. P. Rosenberg, "A new rough surface parabolic equation program for computing low-frequency acoustic forward scattering from the ocean surface," *J. Acoust. Soc. Am.* **105**, 144-153 (1999).
5. M. D. Collins, "A split-step Padé solution for the parabolic equation method," *J. Acoust. Soc. Am.* **93**, 1736-1742 (1993).
6. E. I. Thorsos, "Acoustic scattering from a 'Pierson-Moskowitz' sea surface," *J. Acoust. Soc. Am.* **88**, 335-349 (1990).
7. E. I. Thorsos, F. S. Henyey, K. L. Williams, W. T. Elam, and S. A. Reynolds, "Simulations of Temporal and Spatial Variability in Shallow Water Propagation," in *Impact of Environmental Variability on Acoustic Predictions and Sonar Performance*, edited by N. G. Pace and F. B. Jensen, Kluwer, Dordrecht, The Netherlands, 2002, pp. 337-344.
8. N. G. Van Kampen, *Stochastic Processes in Physics and Chemistry*, North-Holland, Amsterdam, 1997, chapter XVI.

9. L. B. Dozier and F. D. Tappert, "Statistics of normal mode amplitudes in a random ocean. I. Theory," *J. Acoust. Soc. Am.* **63**, 353-365 (1978).
10. L. B. Dozier and F. D. Tappert, "Statistics of normal mode amplitudes in a random ocean. II. Computations," *J. Acoust. Soc. Am.* **64**, 533-547 (1978).
11. K. L. Williams, E. I. Thorsos, and W. T. Elam, "Examination of Coherent Surface Reflection Coefficient (CSRC) approximations in shallow water propagation," submitted to *J. Acoust. Soc. Am.*

Mid-Frequency Sonar Backscatter Measurements from a Rippled Bottom

Joseph L. Lopes, Raymond Lim, and Kerry W. Commander

Naval Surface Warfare Center – Panama City, 110 Vernon Ave, Panama City, FL 32407

Abstract. Seafloor ripple reverberation is associated with a peak in the scattering frequency spectrum at a frequency around $c/(2\lambda_r \cos\theta)$, where c is the sound speed in water, λ_r is the ripple wavelength, and θ is the incident grazing angle. In the vicinity of this peak, perturbation theory predicts the reverberation level to be high enough to be a concern for detection of targets buried under ripple. In order to validate such predictions, an experiment was conducted in the Naval Surface Warfare Center Panama City (NSWC-PC) Facility 383, which is a 13.7-m deep, 110-m long, 80-m wide test pool that has 1.5 m of sand covering the bottom. Backscatter reverberation levels from two bottom configurations were measured using a parametric source that was operated in the 1 to 10 kHz frequency range. One bottom configuration corresponded to a non-rippled, near-flat bottom. The second was a rippled bottom with a Gaussian spectrum centered on a wavelength of 20 cm. The rippled bottom was artificially formed with the aid of a sand scraper. Results showed the reverberation levels were significantly higher in the 3 to 5 kHz frequency range for the rippled bottom than for the non-rippled bottom. The maximum reverberation level for the rippled bottom occurred at 4 kHz, which is consistent with perturbation theory predictions.

INTRODUCTION

There is an interest in using sonar systems that operate in the Mid Frequency (MF) regime, taken here to be 1 to 10 kHz, for seafloor reconnaissance in littoral areas. In this frequency range, longer detection ranges are possible when compared to sonar systems that operate at higher frequencies. In addition, in the MF regime, penetration into a sandy bottom is possible and may permit the detection of buried targets. A critical component impacting sonar performance against bottom targets is bottom reverberation. In this frequency range, bottom reverberation will consist of scattering from the interface, within the sediment volume, and any layers within the sediment volume. The dominant mechanism is dependent upon sonar frequency, grazing angle, and the water-sediment interface roughness. If ripples are present, scattering from the interface may be a significant source of reverberation. The frequency of the dominant scattering peak associated with rippled interface roughness readily follows from first-order perturbation theory to be

$$f = c/(2\lambda_r \cos\theta) . \tag{1}$$

Here c is the sound speed in water, λ_r is the ripple wavelength, and θ is the incident grazing angle. Thus, for a 20° grazing angle and ripple wavelengths of 20 cm and 100 cm, the scattering peak will be near 4 kHz and 0.8 kHz, respectively. This may impact MF sonar's capability against bottom targets, whether these targets are proud, partially buried, or completely buried.

This point is clearly illustrated in Fig. 1, which shows model predictions of backscatter from a buried target insonified at a subcritical grazing angle.[1] In this model the roughness on the interface over the buried target is represented as ripple with a shifted Gaussian spectral distribution in a given direction, combined with an isotropic small-scale roughness with a power law spectral distribution. The model uses perturbation theory to calculate penetration and reverberation. The figure shows predictions associated with a ripple orientation of 0° (acoustic propagation direction perpendicular to the ripple crest) and with mean ripple wavelengths of 25, 50, and 75 cm. In each case the projected beam is incident on bottom sediment at a 10° grazing angle, the reverberating area is assumed to be 40 cm long by 15 cm wide, and the target has a target strength of about -11 dB. The ripple has a 2-cm root-mean-square (RMS) height, and the top of the target is buried under 6.4 cm of sand. The solid lines designate the backscatter from the target, while the dashed lines correspond to reverberation levels. The salient point to note is that, in part of the MF region, the model predicts a negative signal-to-noise ratio due to high reverberation levels from these ripples. In addition, the frequency in which these high reverberation levels appear are dependent upon ripple spacing as described above. Thus, a sonar operating at a higher frequency than an MF system may detect buried targets while an MF system might not detect the same target.

The scientific community lacks specific information on MF bottom scattering strength that is necessary for careful evaluation of sonar systems. The objective of this work is to measure the reverberation levels from a rippled bottom under controlled conditions in the MF frequency range and compare these levels to predictions of the model. This paper documents the progress to date of this effort. In particular, the backscatter levels obtained from a rippled bottom are compared to those collected from a non-rippled bottom as well as to predictions of a model that uses first-order perturbation theory.

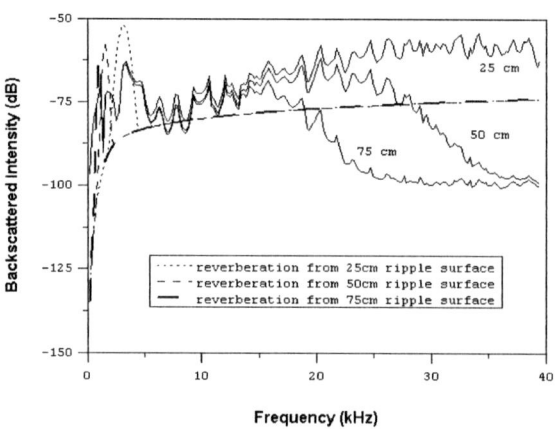

FIGURE 1. Backscatter predictions for various ripple wavelengths.

MEASUREMENT SETUP

The measurement was conducted in the Naval Surface Warfare Center - Panama City (NSWC-PC) Facility 383 test pool, which is shown in Fig. 2. This is a freshwater pool that is 13.7 m deep, 110 m long and 80 m wide with approximately 1.5 m of sand covering the bottom. A filtration system provided approximately 12 m (~ 40 ft) water visibility and mixed the water column. The sound speed in the water was measured to be 1495 m/s with no velocity gradients.

The scattering geometry is depicted in Fig. 3, which corresponds to a view from directly overhead. The scattering region consisted of a bottom area that had rippled and non-rippled regions, and the experimental equipment included a rail system with a sonar tower, a parametric sonar, a transducer located next to the parametric sonar, and a free-field transducer. In addition, a sand-scraping apparatus was used to create the ripple profile on the bottom sediment. Each of these components is described below.

FIGURE 2. Aerial view of Facility 383.

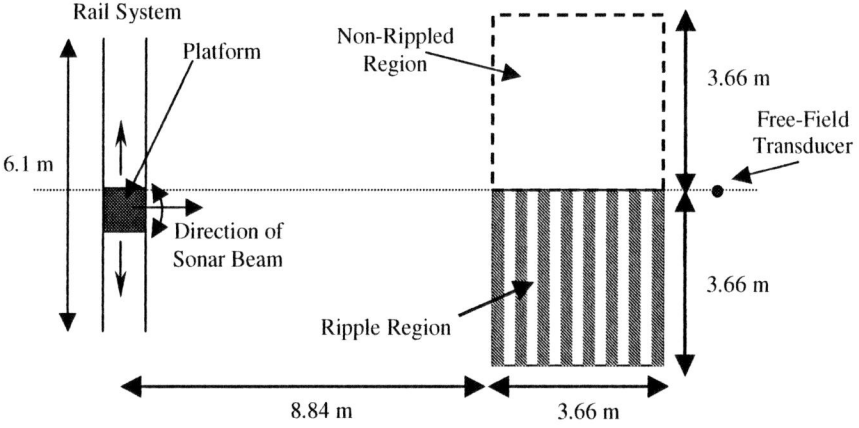

FIGURE 3. Scattering geometry.

The bottom area, from which the backscattered signals were recorded, was approximately 7.32 m in width by 3.66 m in length. This area started about 8.84 m from the rail system (see Fig. 3) and consisted of two adjacent bottom regions. One region corresponded to a non-rippled bottom while the second was a rippled bottom. Both of these regions were about 3.66 m in length by 3.66 m in width. The non-rippled bottom was somewhat flat and was created by divers dragging a weighted bar over its surface. The contour of the rippled bottom was artificially formed with the aid of a sand scraper, which consisted of a frame and a "rake" that glided along the frame. The rake was pulled across the sand using two winches, one located on each side of the test pool. The ripple profile was determined by an insert placed on the rake. Previous measurements using the sand scraper have shown good agreement between the intended ripples (wavelength and RMS height) and those formed.[2,3] In this measurement, the one ripple profile formed had a 0.57-cm RMS height and was consistent with a Gaussian spectrum having a 20 cm center wavelength and 0.000987cm^{-2} wavenumber variance.

A rail system that sat on the bottom sediment was also used in the measurement. A photo of the rail system is shown in Fig 4. The system included a 6.1-m (20-ft) long rail, a platform on wheels that was translated along the rail using a translation motor, and a sonar tower. A 2.1-m long extender attached the sonar tower to the platform and permitted the sonar tower to stand 3.89 m above the bottom sediment. This geometry provided insonification at the center of the two bottom regions at a 20° grazing angle.

The sonar tower supported a parametric sonar, a receiver, and scanning (horizontal pan and vertical tilt) motors such that both transducers had an almost 360° (180°) rotational (tilt) capability. The parametric sonar was developed for NSWC-PC by the Naval Underwater Warfare Center – Newport (NUWC/NPT) and was employed as the projector. This sonar produced a conical beam with a 3-dB beam width of about 5° with side lobes that are down by approximately 50 dB across its entire operational frequency band of 1 to 20 kHz. This sonar was oriented such that the direction of its main response axis (MRA) was perpendicular to the rail. An International Transducer Corporation (ITC) 1001 transducer was located next to the parametric sonar as shown in Fig. 5. This transducer has an omni-directional response and was used to record the backscattered signals from the bottom. The parametric sonar and ITC 1001 transducer could be translated linearly to any position along the rail, enabling the acoustic beam to be projected to, and received from, either the rippled or non-rippled bottom region. An encoder that employed a wire cable attached to the platform was used to verify the position of the platform as it translated along the rail, and a pendulum tilt sensor was used to monitor the inclination angle of the parametric sonar's MRA.

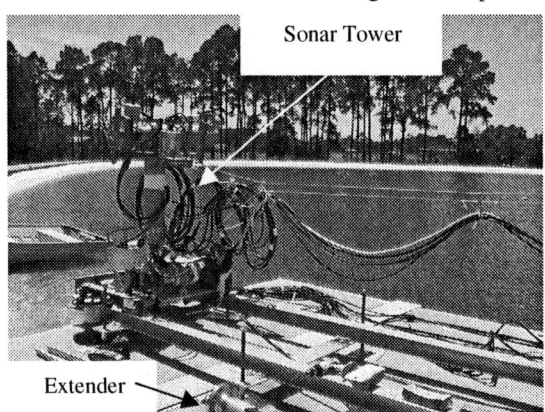

FIGURE 4. Photo of rail system.

FIGURE 5. Photo of parametric sonar (bottom) and ITC 1001 transducer (top).

The free-field transducer indicated in Fig. 3 was an ITC 1001 transducer. This transducer was placed on the bottom about 12.9 m from the rail system and was used to record the waveforms and levels transmitted by the parametric sonar.

Transmitted signals were generated using a National Instruments DAQ Card-6062E digital-to-analog board at a sample frequency of 500 kHz. These signals were amplified and then sent to the parametric sonar. Transmit signals were sinusoidal waveforms with a pulse length of 1 ms. The received signals were amplified and filtered, then digitized by a GageScope analog-to-digital card. All receive signals were sampled at a frequency of 1 MHz.

The procedure for preparing the scattering regions was as follows. First, divers deployed the rail system to the desired location in the facility. Next, the non-rippled, bottom region was flattened by divers. Third, the sand scraper was carefully positioned at the appropriate location from the rail system. Fourth, ripples were formed with the sand scraper, which took several iterations to ensure a well-formed bottom contour. Next, divers carefully removed the sand scraper, and then they visually re-inspected the formed bottoms by swimming either well above, or around the perimeter of, the two bottom regions. After the divers confirmed that the two bottom contours were reasonably well formed, acoustic data were acquired.

Data were obtained in the frequency range of 2 to 10 kHz at an average grazing angle of $20°$ by translating the rail platform and taking data in about 2.5-cm (1-inch) increments. The waveforms recorded at each rail location were the resultant of an 8-ping average.

RESULTS AND DISCUSSION

Data Reduction

MATLAB code was written to read and analyze the collected data. The data were first processed and displayed in a backscattered intensity image. This image is a plot of the backscatter intensity (in dB) in range versus sonar location along the rail system. The processed data were further analyzed to determine the calibrated bottom backscatter level from the rippled bottom region. An estimate of the reverberation level was determined by taking an average of the reverberation intensities in a patch of about 1.3 m wide in cross-range by 1 m long in range. The calibrated backscatter level, EL in dB, was calculated using,

$$EL = EL_{MEASURED} - SPL_{INTERFACE}. \qquad (2)$$

Here $EL_{MEASURED}$ is the measured backscatter level in dB and $SPL_{INTERFACE}$ is the sound pressure level in dB incident on the rippled bottom. $SPL_{INTERFACE}$ was obtained by using the level measured with the free-field transducer and accounting for the difference in propagation loss to the location where the beam is incident at the rippled bottom with the parametric sonar code CONVOL.[4] Both $EL_{MEASURED}$ and

SPL$_{INTERFACE}$ were obtained after correcting for the particular receiver's system response (transducer with pre-amplifier).

Backscattered Intensity Images

Figures 6, 7, 8, and 9 illustrate bottom backscatter intensity images corresponding to frequencies of 3, 4, 5, and 6 kHz, respectively. In each figure there are two images. The image on the left refers to the non-rippled bottom while the image on the right is associated with the rippled bottom. To facilitate comparison of the non-rippled and rippled bottom backscatter intensities within the figures, the same dB-level gray scale is used for all images. The images in the figures clearly show increased reverberation levels in the 3 to 5 kHz frequency range for the rippled bottom when compared to the corresponding non-rippled bottom. In addition, the maximum reverberation level for the rippled bottom occurred at 4 kHz, corresponding to that predicted using Eq. (1).

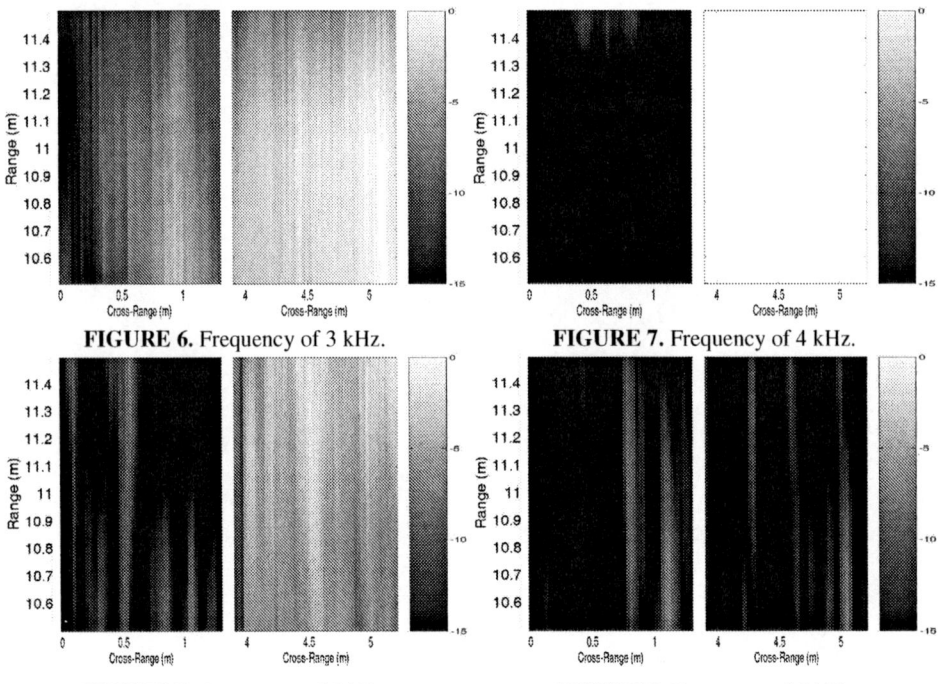

FIGURE 6. Frequency of 3 kHz. **FIGURE 7.** Frequency of 4 kHz.

FIGURE 8. Frequency of 5 kHz. **FIGURE 9.** Frequency of 6 kHz.

Calibrated Backscattered Levels

Predicted (solid lines) and measured (filled circles) calibrated backscatter intensity levels from the rippled bottom are compared in Fig. 10. The model assumes a unit amplitude, monochromatic plane wave incident on the bottom. Intensity predictions are calculated using a steady-state Rayleigh-Rice perturbation theory to account for ensemble-averaged scattering from interface roughness. Since the statistical roughness parameters were not measured, the small-scale roughness superimposed on the scraped Gaussian ripple profile was assumed to be similar to the small-scale roughness observed in past measurements with sinusoidal ripple profiles.[3] Therefore, four curves are plotted, corresponding to four different assumptions for the statistical parameters of the superimposed roughness. In regions exhibiting significant differences in these curves, the measured reverberation level is expected to be spanned by the reverberation range of these curves. An average of the steady-state predictions over a moving 2 kHz window weighted by the spectrum of the pulse employed in the experiments was performed in order to compare with the measured results. Also, the overall level of the reverberation must be scaled to the area of the effective bottom patch contributing to the detected reverberation. This area is the product of the range and cross-range resolutions in the measurements. For a 1 ms source pulse incident on the bottom at a $20°$ grazing angle, the range resolution is about 0.79 m. For a $5°$ beam width, the cross-range resolution at the rippled bottom is approximately 0.93 m. The measured calibrated level at each frequency represents the mean reverberation level from patches observed between 10.5 and 11.5 m in range by 1.3 m in cross-range of the ripple region depicted in Fig. 3. This patch size corresponds to almost two sonar resolution cells.

The error range for the measured data points is indicated by vertical bars and is based on the sum of (a) the statistical uncertainty in our estimate of the mean reverberation intensity, (b) an estimate for the uncertainty in transducer calibration, and (c) the CONVOL predicted variation of the incident field at the interface where the reverberation is calculated. The statistical uncertainty is taken to be $\pm \sigma/\sqrt{N}$, where σ is the standard deviation of reverberation intensity, and N is the number of independent resolution cells in the 1 m by 1.3 m region used to obtain the mean background noise level. The uncertainty in transducer calibration was estimated to be about ± 0.7 dB.

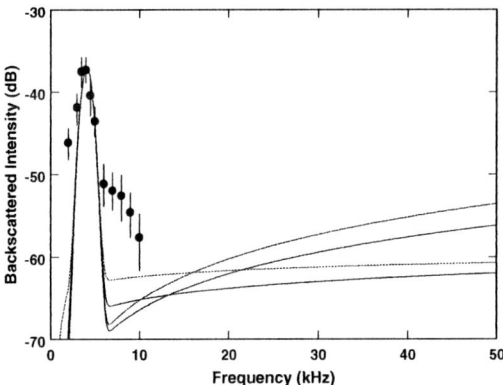

FIGURE 10. Measured (filled circles) and predicted backscatter intensity levels (solid lines).

The measured backscatter levels in Fig. 10 appear in good agreement with the model predictions around the spectral peak, with the maximum level occurring at 4 kHz as expected. However, even when taking into account the range in errors of the data points, there is some discrepancy in the data-model comparison at frequencies of 7 kHz and higher. The cause for this discrepancy is unknown, but it could be due to either inadequate assumptions used in the model and/or data analysis, or deviations from the scraped ripple profile. The assumption that the statistics of these deviations can be described by the same small-scale roughness statistics observed superimposed on roughness profiles created in previous measurements may not be accurate.

SUMMARY

A laboratory-type measurement was conducted to investigate reverberation levels from a rippled bottom in the MF range. Measured backscatter levels obtained from a rippled bottom with a 20 cm average ripple wavelength and a RMS height of 0.57 cm were compared to those collected from a non-rippled bottom. The measured levels from the rippled bottom were further compared to predictions of a model based on first-order perturbation theory.

The results showed increased reverberation levels in the 3 to 5 kHz frequency range when compared to the corresponding non-rippled bottom. The maximum reverberation level for the rippled bottom occurred at 4 kHz, which corresponds to that predicted using Eq. (1). In addition, the measured calibrated scattering levels were in good agreement with model predictions in the vicinity of the spectral peak. Data-model agreement compared very well for frequencies less than or equal to 6 kHz. The comparison at 7 kHz and above is not as good, and remains to be resolved.

Future work includes: (a) measuring and verifying the parametric sonar projected sound pressure levels as functions of range, (b) conducting additional measurements using ripple profiles centered on different ripple wavelengths to validate the trend associated with Eq. (1), and (c) analyzing the data to determine the calibrated levels for each ripple configuration and comparing the results to model predictions.

ACKNOWLEDGMENTS

The authors gratefully acknowledge support from the NSWC-PC ILIR Program, SERDP, and ONR Code 32OA. The authors also wish to thank Carrie Nesbitt and Edmund Kloess of NSWC-PC for their participation and help in this effort, as well as Eric Thorsos and Kevin Williams of the Applied Physics Laboratory at the University of Washington for their helpful discussions with this work.

REFERENCES

1. R. Lim, K. L. Williams, and E. I. Thorsos, "Acoustic scattering by a three-dimensional elastic object near a rough surface," J. Acoust. Soc. Am. **107**, 1246-1262 (2000).
2. J. L. Lopes, C. L. Nesbitt, R. Lim, D. Tang, K. L. Williams, and E. I. Thorsos, "Shallow Grazing Angle Sonar Detection of Targets Buried Under a Rippled Sand Interface," Proceedings of Oceans 2002 MTS/IEEE, pp. 461-467.
3. J. L. Lopes, C. L. Nesbitt, R. Lim, K. L. Williams, E. I. Thorsos, and D. Tang, "Subcritical Detection of Targets Buried Under a Rippled Interface: Calibrated Levels and Effects of Large Roughness" Proceedings of Oceans 2003 MTS/IEEE, pp. 485-493.
4. M. B. Moffett and R. H. Mellon, "CONVOL: A Computer Program for Parametric Source Nearfield and Farfield Beam Patterns," NUSC Technical Memorandum, No. 791132, Naval Underwater Systems Center, New London, CT, July 1979.

The Dependence of Long-Range Reverberation on Bottom Roughness

Roger Gauss, David Fromm, Kevin LePage, and Robert Gragg

Acoustics Division, Naval Research Laboratory, Washington, DC 20375-5350, USA

Abstract. At long-range, shallow-water reverberation can be driven by sub-critical-angle scattering, i.e. by rough interface scattering. The Naval Research Laboratory has recently developed a small-slope model for elastic seafloors that provides physics-based estimates of the dependence of scattering on the incident and scattered angles, and physical descriptors of the environment. In this paper, this incoherent model is used as kernels in reverberation models, which in turn are used to assess the sensitivity at 3.5 kHz of long-range monostatic reverberation to the roughness of the water-sediment interface. It is shown that when sub-critical-angle scattering dominates, the acoustic field could be quite sensitive to the parameter values of the roughness, thus arguing for the need for regional in-situ methods for its estimation.

INTRODUCTION

Bottom reverberation is a major source of interference for active sonar systems in shallow water that is caused by the interaction of acoustic energy with environmental features at or in the seafloor. The rough water-sediment (and sediment-sediment) interfaces and the sediment volume contribute to the acoustic reverberation. Often at long ranges, low scattering angles prevail. Under these conditions, rough interface scattering can be a dominant scattering mechanism, particularly for non-soft bottoms.

Recently, the Naval Research Laboratory (NRL) has developed broadband, bistatic physics-based formulas that predict the dependence of scattering strength on the incident and scattered angles, the acoustic frequency, and environmental variables [1]. It has been demonstrated that the acoustic Scattering Strength (SS) of the bottom interface can depend quite strongly on the environmental features [1,2]. In this paper, we use the bottom-interface formula as kernels in two reverberation models, the ray-based BiRASP [3] and mode-based R-SNAP [4] to explore the sensitivity of long-range reverberation in shallow water at 3.5 kHz to the values of the interface-roughness parameters.

A range-independent waveguide of 150 m depth was assumed. Two bottom half spaces were used in the study: very fine sand and rock (basalt). The assumed geoacoustic values come from Hamilton [5-6], for sand and rock respectively: density ratios of 1.85 and 2.7, compressional speeds of 1708 and 5185 m/s, compressional attenuations of 0.12 and 0.02 dB/m/kHz, shear speeds of 100 and 2745 m/s, and shear attenuations of 25.0 and 0.07 dB/m/kHz. Two sound speed profiles were considered (Fig. 1a), a near-isospeed "winter" profile and a downward-refracting "summer" profile. For these profiles, the p-wave (compressional) critical angles are at ~27 and

~73 deg for sand and rock respectively, while the s-wave (shear) critical angle for the rock is at ~56 deg. The assumed bottom losses are also shown in Fig. 1b. For the sand, it is seen that the loss is very low at angles below critical. For the rock, the maximum loss is between its two critical angles, while below ~45 deg, the losses are very low.

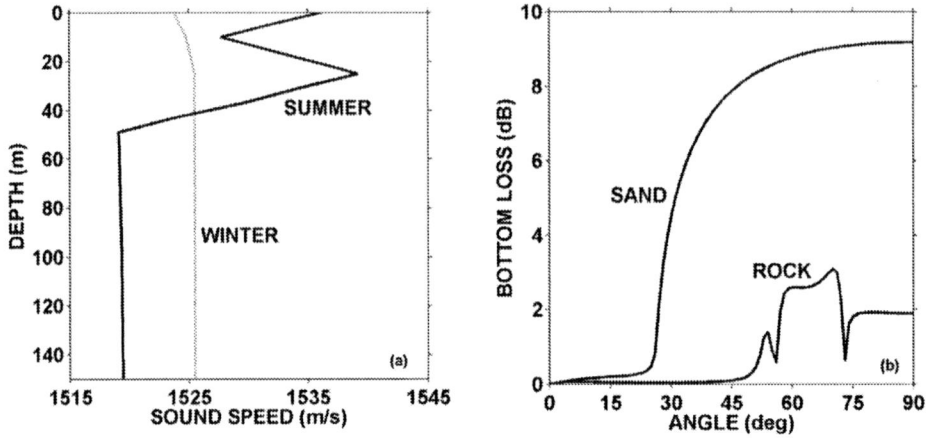

FIGURE 1. (a) Sound-speed profiles assumed in this study. (b) Bottom loss vs. angle at 3.5 kHz for rock and sand using REFLECT.

SCATTERING STRENGTH

The bottom scattering strength formula relies on lowest-order small-slope theory for scattering from the rough water-sediment interface [2] and a stochastic volume theory for scattering from the subbottom [7]. Figure 2 shows predicted monostatic bottom backscattering strengths vs. grazing angle at 3.5 kHz for very fine sand (left) and basalt (right) bottoms for three values of the bottom-interface roughness spectral exponent γ_2. In these plots, the other roughness parameter, the spectral strength w_2 was fixed at 0.001 m^4. In our reverberation model studies (next Section), we will also consider a w_2 value of 0.01 m^4, for a total of six cases.

Below the critical angle, the rough water-sediment interface is the dominant SS mechanism. At higher angles, the sediment volume contributes as well, especially when the bottom roughness is small; however, the particular volume model used predicts no sediment-volume scattering contribution below the critical angle. (This volume model ignores shear effects.)

Overlain in Fig. 2 are curves corresponding to the commonly-used (frequency-independent) Mackenzie's rule, i.e. Lambert's Law with $\mu = -27$ dB. It shows significant differences, especially at low grazing angles (even if translated vertically, i.e. varying μ).

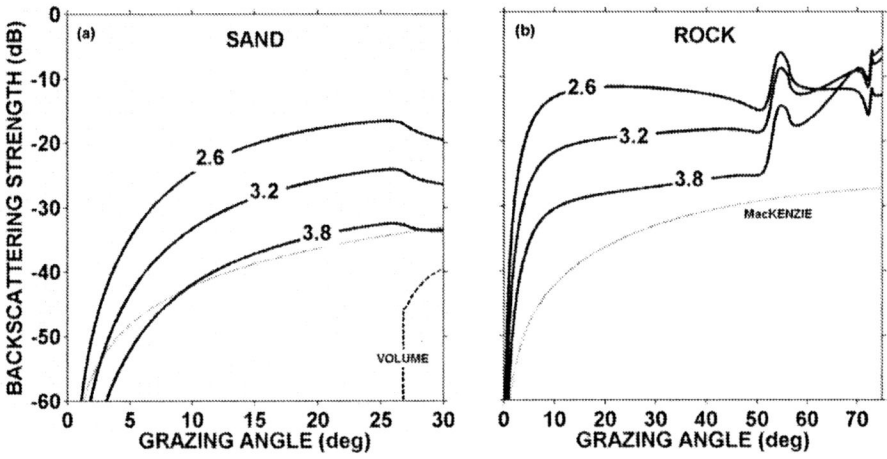

FIGURE 2. Monostatic bottom backscattering strength vs. grazing angle θ at 3.5 kHz for three values of the bottom roughness spectral exponent γ_2 for (a) sand to 30 deg and (b) rock to 75 deg. In each case, the bottom roughness spectral strength w_2 = 0.001 m^4. In (a), the solid curves represent the total (interface + volume) backscattering strength—the dashed curve shows the volume contribution. In (b), no volume contribution is included. For reference, the Mackenzie curve $\mu \sin^2\theta$ with $\mu = -27$ dB (dotted) is included in each plot.

LONG-RANGE REVERBERATION

To examine the sensitivity of long-range reverberation to environmental variables, the above scattering strength model was used as kernels in BiRASP and R-SNAP. The monostatic reverberation calculations assumed the co-located source and receiver (S-R) were at the same depths, with two depths considered: 10 and 75 m. A 0-dB source level and a 1-s (1-Hz-bandwidth) CW at 3.5 kHz were also assumed. (To obtain calibrated reverberation levels for a given source level, say, 200 dB re µPa at 1 m, simply add that number to the reverberation y-axis values.) Propagation included Thorp-based attenuation [8]. Noise was not included in the runs.

The reverberation predictions using the two models agreed very well, so for consistency this paper (and its companion, Ref. [9]) will present predictions from only those from one of the models, BiRASP. These runs also used monostatic scattering strengths as inputs. (We also did sets of bottom, surface and fish runs using bistatic scattering strengths as inputs in BiStaR, a bistatic version of R-SNAP. As expected for these simple monostatic scenarios, the results differed only slightly from the purely monostatic runs.)

Sensitivity Studies

Seasonal and geometry dependence are examined in Fig. 3a, where it can be seen the differences are very small (even at 70 s). A comparison of the rock and sand reverberation levels in Fig. 3a shows the significantly longer decay rates for the rock, e.g., over a 20-dB-higher Reverberation Level (RL) than for sand at most ranges. Such differences are not surprising given rock's higher scattering strengths (Fig. 2) and lower bottom losses (Fig. 1b) over more angles.

Sensitivity to roughness values is examined in Fig. 3b, using rock for the summer/S-R at 75 m case as a representative example. Significant differences of up to 20 dB or more can be seen at all ranges. As after a few seconds sub-critical-angle scattering dominates, this argues that for such cases in the field, one needs to either acquire accurate estimations of the local roughness or measure in-situ the local backscattering strength over a range of grazing angles.

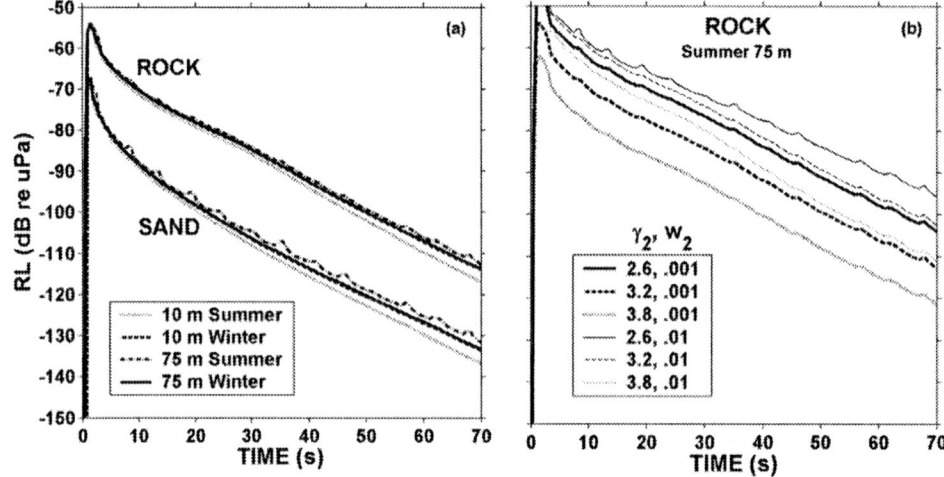

FIGURE 3. At 3.5 kHz: (a) RL for rock and sand for four scenarios for $\gamma_2 = 3.2$ and $w_2 = 0.001$ m^4; and (b) RL for rock for six pairs of roughness values for one scenario.

BiRASP's ability to deconstruct the reverberation by average grazing angle is illustrated for the summer profile/S-R at 75 m case in Fig. 4 for sand (a) and rock (b). (The corresponding RL curves in Fig. 3a are the sum of these curves.) It is seen that at 20 km (~26 s), angles up to 50 deg are still contributing in the rock case, but only angles up to 30 deg are in the sand case. This reflects both rock's higher critical angle(s) and lower bottom loss.

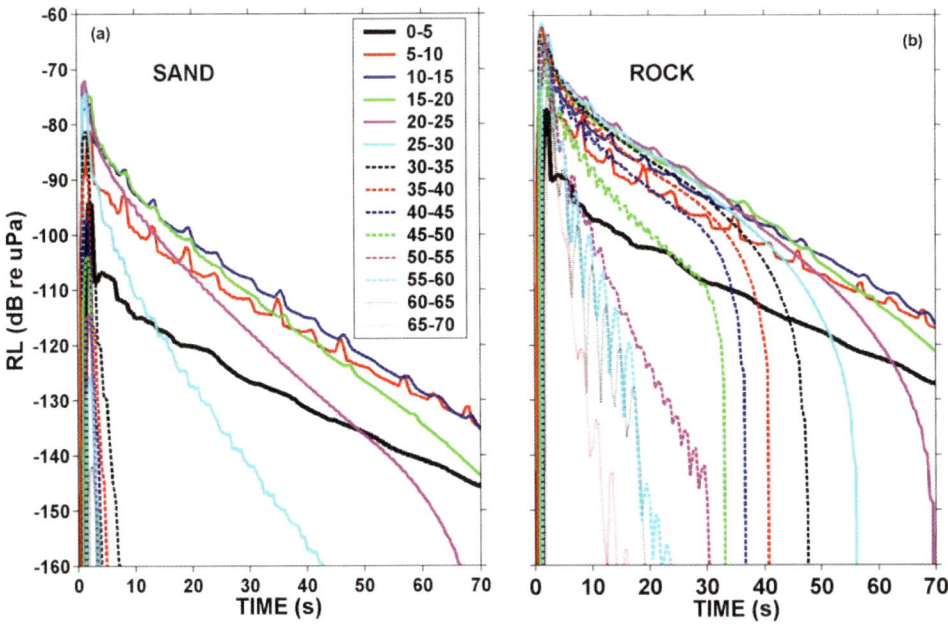

FIGURE 4. At 3.5 kHz for (a) sand and (b) rock: RL contributions by mean grazing angle in 5-deg bins for the $\gamma_2 = 3.2$ and $w_2 = 0.001$ m^4 case. Summer profile and S-R at 75 m.

To highlight the sensitivity of reverberation to the two bottom roughness parameters, Fig. 5 presents scattering strengths (top) and reverberation levels (bottom) for sand (left) and rock (right) *relative* to their values for the $\gamma_2 = 3.2$ and $w_2 = 0.001$ m^4 case.

For the scattering strengths, over the angles controlling the long-range reverberation, i.e. those below the p-wave critical angle for the sand and below the s-wave critical angle for the rock, the differences are generally fairly flat with angle (except above ~20 deg for the $w_2 = 0.01$ m^4 rock case), with generally increased backscatter in these cases the larger the w_2 (for a given γ_2) or the smaller the γ_2 (for a given w_2). (In general, increasing w_2 for a fixed γ_2, or decreasing γ_2 for a fixed w_2, will not always lead to stronger backscatter [2]—e.g., above the shear critical angle in Fig. 2b.) A non-obvious result is that the relative differences of the six cases are the same for rock and sand—note the different x-axis scales. This is because the scattering strength can be expressed as a product of two factors, one that depends on the bottom's material properties, but is independent of interface roughness (and frequency), and one that depends on the roughness (and frequency), but not the bottom properties [2]. So, when dividing by a reference case, the first factor is divided out, leaving a dependence only on the roughness parameters.

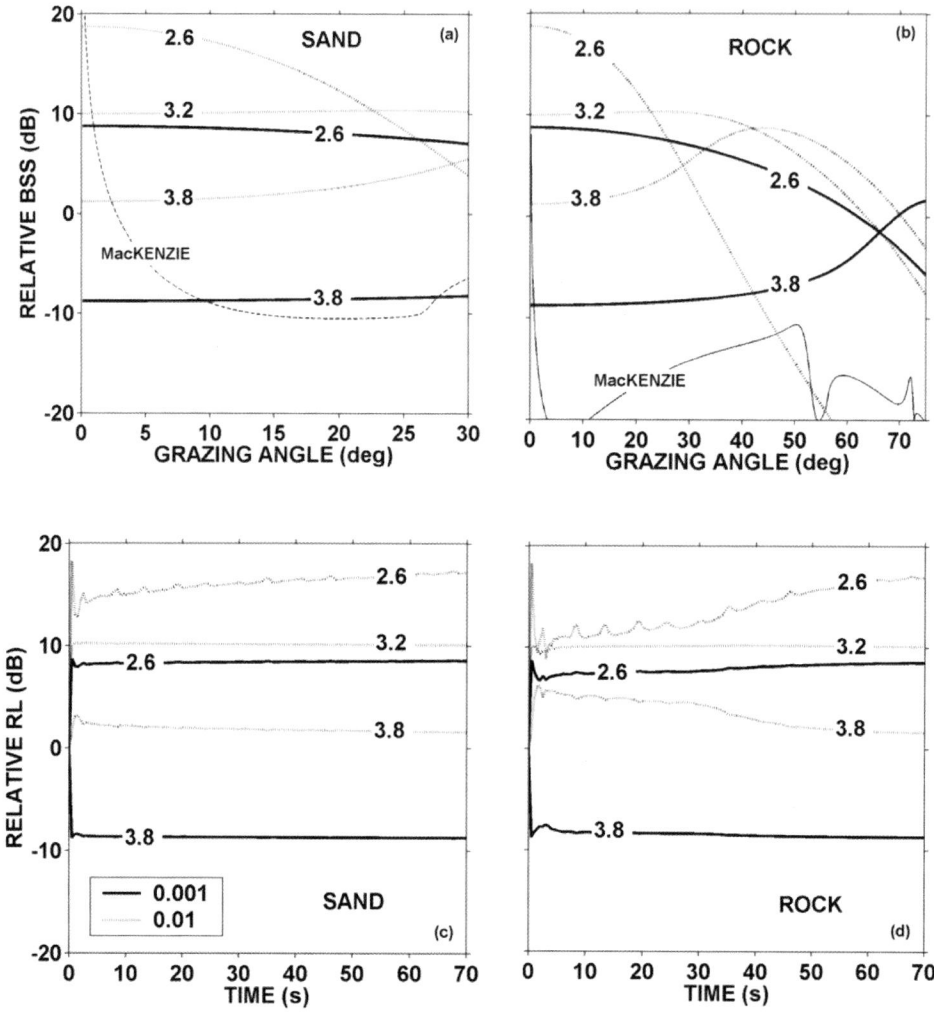

FIGURE 5. At 3.5 kHz for sand (left) and rock (right): scattering strengths (top) and reverberation levels (bottom) *relative* to the $\gamma_2 = 3.2$ and $w_2 = 0.001$ m^4 case. Summer profile, S-R at 75 m. The solid and dotted curves correspond to w_2's of 0.001 and 0.01 m^4, respectively, with the γ_2's as shown.

For the reverberation, similar trends emerge. One difference that can be seen is the smaller spreads in the $w_2 = 0.01$ m^4 rock values for times less than ~30s, a reflection of the complex γ_2 dependence at the higher angles (coupled with the range of contributing grazing angles for rock—cf. Fig. 4b).

SUMMARY AND DISCUSSION

Sub-critical-angle scattering, i.e. rough interface scattering, can drive shallow water reverberation. Application of an interface-scattering model for rough elastic seafloors in reverberation models suggests that the long-range acoustic field could be quite sensitive (up to 25 dB for the cases considered) to the parameter values of the roughness. As bottom roughness is a difficult quantity to accurately measure in the field even with state-of-the-art instrumentation, especially on regional scales, alternative methods for its in-situ estimation are needed. One method would be to use acoustic inversion with an elastic surface roughness model (as in [10]). A key feature of these scattering models is that the only frequency dependence is through the roughness spectral exponent γ_2 [2], arguing for multiple-frequency direct-path measurements to nail down its value. Given parameter values for the bottom, physics-based scattering strength models could then be used as kernels in reverberation models to predict the acoustic response at other frequencies and angles (especially bistatically).

This study also underscores the importance of the knowledge of the spatial and frequency dependence of the critical angle for predicting long-range reverberation. When considering sub-critical-angle scattering, a potential competing mechanism is near-bottom fish. As fish backscatter has a fairly flat grazing angle response, depending on their densities, sizes and depths, when present they can be significant low-angle scattering mechanism. See Ref. [9] for more details.

Finally, we note that for range-dependent environments, the presence of bathymetric features can excite higher angles at long range thus potentially enhancing reverberation variability at these ranges (especially if the feature is rock, i.e. of harder composition than the surrounding seabed).

ACKNOWLEDGMENTS

Work supported by the Office of Naval Research.

REFERENCES

1. Gauss, R. C., Gragg, R. F., Nero, R. W., Wurmser, D., and Fialkowski, J. M., "Broadband Models for Predicting Bottom, Surface, and Volume Scattering Strengths," NRL/FR/7100—02-10,042, Washington, DC: Naval Research Laboratory, September 30, 2002.
2. Gragg, R. F., Wurmser, D., and Gauss, R. C., "Small-slope scattering from rough elastic ocean floors: General theory and computational algorithm," *J. Acoust. Soc. Am.* **110**, 2878-2901 (2001).
3. Fromm, D. M., Crockett, J. P., and Palmer, L. B., "BiRASP – The Bistatic Range-dependent Active System Performance Model," NRL/FR/7140—95-9723, Washington, DC: Naval Research Laboratory, September 30, 1996.
4. LePage, K. D., "Monostatic Reverberation in Range Dependent Waveguides: The R-SNAP Model," SACLANT Undersea Research Centre SR-363, La Spezia, Italy, 2002.
5. Hamilton, E. L., "Geoacoustic Modeling of the Sea Floor," *J. Acoust. Soc. Am.* **68**, 1313-1319 (1980).
6. Essen, H.-H., "Scattering from a rough sedimentary seafloor containing shear and layering," *J. Acoust. Soc. Am.* **95**, 1299-1310 (1994).

7. Turgut, A., "Inversion of bottom/subbottom statistical parameters from acoustic backscatter data," *J. Acoust. Soc. Am.* **102**, 833-852 (1997).
8. Jensen, F. B., Kuperman, W. A., Porter, M. B., and Schmidt, H., *Computational Ocean Acoustics*, AIP Press, 1994, p. 38.
9. Gauss, R. C., Fromm, D. M., LePage, K. D., Fialkowski, J. M., and Nero, R. W., "The Influence of the Sea Surface and Fish on Long-Range Reverberation," in *High Frequency Ocean Acoustics Conference*, edited by M. B. Porter, T. M. Siderius, and W. A. Kuperman, San Diego, CA, March 2004.
10. Soukup, R. J., and Gragg, R. F., "Backscatter from a limestone seafloor at 2—3.5 kHz: Measurements and modeling," *J. Acoust. Soc. Am.* **113**, 2501-2514 (2003).

The Influence of the Sea Surface and Fish on Long-Range Reverberation

Roger Gauss, David Fromm, Kevin LePage, Joseph Fialkowski, and Redwood Nero*

Acoustics Division, Naval Research Laboratory, Washington, DC 20375-5350, USA
* Acoustics Division, Naval Research Laboratory, Stennis Space Center, MS 39529-5004, USA

Abstract. Acoustic detection for active sonars involves identifying target signatures in the presence of environmental effects, such as acoustic scattering from the ocean boundaries and fish. The Naval Research Laboratory has recently developed 3D broadband models that provide physics-based estimates of the dependence of scattering from the sea surface, bubble clouds and near-boundary fish (including boundary-interference effects) on the incident and scattered angles, and physical/biological descriptors of the environment. In this paper, these models and a surface-loss model are used as kernels in reverberation models, which in turn are used to assess the sensitivity at 3.5 kHz of long-range reverberation to environmental variables. It is shown that the acoustic field in shallow water waveguides could be quite sensitive to the values of sea surface (wind speed) and fish (density, size, depth) parameters, and that physics-based models are needed for accurate field characterization.

INTRODUCTION

Reverberation is a major source of interference for active sonar systems that is caused by the interaction of acoustic energy with environmental features at or near the ocean boundaries. In low-to-moderate sea states, the seafloor, the rough air-sea interface, and fish contribute to the acoustic reverberation. When wave breaking is significant, air becomes entrained in the near-surface zone in the form of subsurface bubbles. Under these conditions, bubble clouds also contribute to the acoustic reverberation.

Recently, the Naval Research Laboratory (NRL) has developed broadband, bistatic physics-based formulas that predict the dependence of surface and near-boundary fish scattering strengths on the incident and scattered angles, the acoustic frequency, and environmental variables [1]. It has been demonstrated that the acoustic Scattering Strength (SS) can depend quite strongly on the environmental features [1,2]. In this paper, we use these formulae as kernels in two reverberation models, the ray-based BiRASP [3] and mode-based R-SNAP [4] to explore the sensitivity of long-range reverberation in shallow water at 3.5 kHz to the values of both sea surface (wind speed) and fish (density, size, depth) parameters.

A range-independent waveguide of 150 m depth was assumed. Two bottom half spaces were used in the study: very fine sand and rock (basalt). The assumed geoacoustic values come from Hamilton [5-6]; for sand and rock respectively: density ratios of 1.85 and 2.7, compressional speeds of 1708 and 5185 m/s, compressional attenuations of 0.12 and 0.02 dB/m/kHz, shear speeds of 100 and 2745 m/s, and shear attenuations of 25.0 and 0.07 dB/m/kHz. The resulting bottom losses are shown in Fig. 1b of [7]. Two sound speed profiles were considered (Fig. 1a of [7]), a near-isospeed "winter" profile and a downward-refracting "summer" profile. For these profiles, the p-wave (compressional) critical angles are at ~27 and ~73 deg for sand and rock respectively, while the s-wave (shear) critical angle for the rock is at ~56 deg. For this paper, the assumed bottom roughness values [7] are $\gamma_2 = 3.2$ and $w_2 = 0.001$ m^4.

The monostatic reverberation calculations assumed the co-located source and receiver (S-R) were at the same depths, with two depths considered: 10 and 75 m. A 0-dB source level and a 1-s (1-Hz-bandwidth) CW at 3.5 kHz were also assumed. (To obtain calibrated reverberation levels for a given source level, say, 200 dB re µPa at 1 m, simply add that number to the reverberation y-axis values.) Propagation included Thorp-based attenuation [8]. Noise was not included in the runs.

SURFACE REVEBERATION

The surface scattering strength formula relies on lowest-order small-slope theory for scattering from the rough air-sea interface and a stochastic volume theory for scattering from the bubble clouds [1]. Model fits to open ocean data yielded a formula that environmentally depends only on the wind speed (measured at an elevation of 10 m). By its semi-empirical nature, the discrete nature of bubble clouds and attenuation effects are embedded in the effective surface-scattering formula. Figure 1a shows predicted sea surface backscattering strengths vs. grazing angle at 3.5 kHz for a set of wind speeds. At low scattering angles, bubble cloud backscattering is the dominant SS mechanism (when wave breaking is significant), and the rough air-sea interface at high scattering angles.

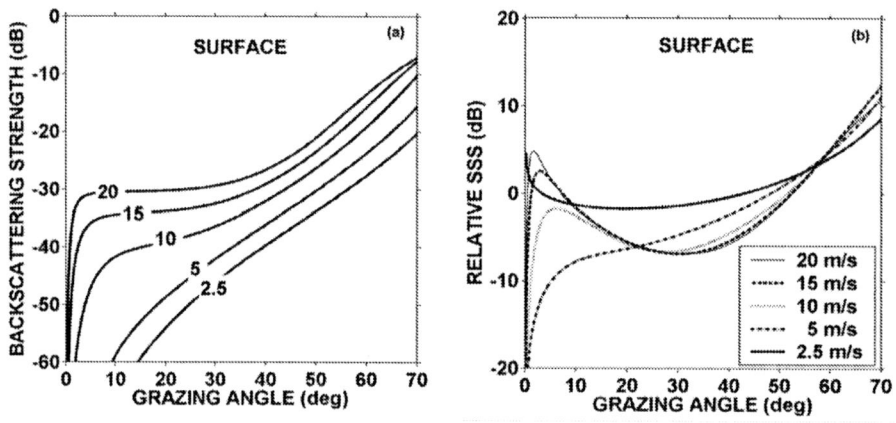

FIGURE 1. Predictions of monostatic surface backscattering strength at 3.5 kHz as parameterized by wind speed at 10 m (m/s): (a) NRL model and (b) the difference of two model predictions, NRL's minus Chapman-Harris's.

Figure 2 illustrates how long-range surface reverberation can depend not only on wind speed, but also on the waveguide characteristics (bottom type, sound speed profile) and geometry (S-R depth). Figure 2a shows significant differences in Reverberation Level (RL)—15 to 25 dB at 20 km (~26 s)—depending on whether the seafloor is sand or rock. The significantly longer decay times for a rock seafloor compared to a very fine sand seafloor are a function of how much energy propagates to a given range, a result of the lower bottom loss and higher critical angle for rock [7]. As illustrated in Fig. 2b, RL exhibited more sensitivity to the sound-speed profile and S-R depth when the seafloor was sand.

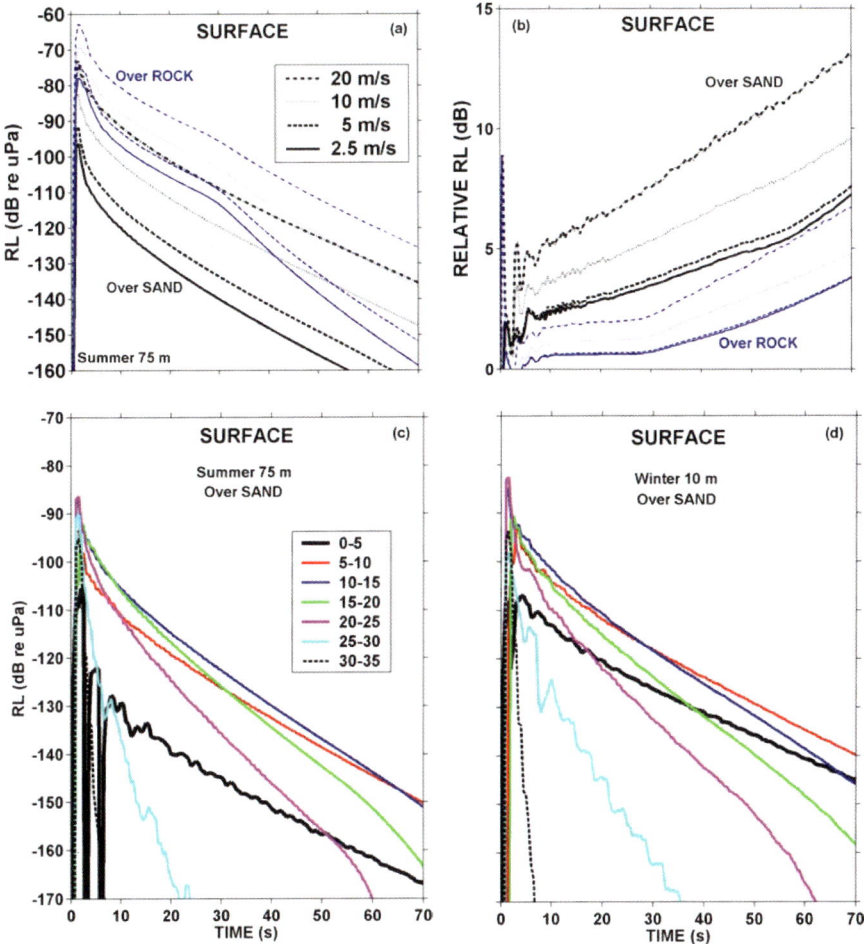

FIGURE 2. Predictions of monostatic surface reverberation at 3.5 kHz. (top) As parameterized by wind speed U for both sand and rock seafloors, (a) RL vs. time for the summer profile/75 m S-R case, and (b) relative enhancement in RL in going from the summer profile/75 m S-R case to the winter profile/10 m S-R case. (c-d) RL contributions by mean grazing angle in 5-deg bins when U = 10 m/s for two scenarios.

The latter effect is explored in Fig. 2c-d. Using BiRASP's ability to deconstruct the reverberation by average grazing angle, it is seen that in the downward-refracting conditions of the summer profile/mid-water S-R case (left), the higher-angle energy is more quickly stripped away and lower-angle interactions more shielded compared to the mild ducting conditions of the winter-profile/shallow S-R case (right). The relative absence of this effect for a rock seafloor over this time window is due to rock's higher critical angle, coupled with lower bottom loss (cf. Fig. 4 in Ref. [7]).

Figure 2a suggests that RL grows increasingly sensitive to wind speed U with increasing range (time), i.e. as lower and lower grazing angle backscattering dominates the reverberation. This is made more apparent in Fig. 3, which presents SS (left) and RL (right) *relative* to their values at U = 10 m/s. It is seen that the biggest RL differences occur at the lower wind speeds.

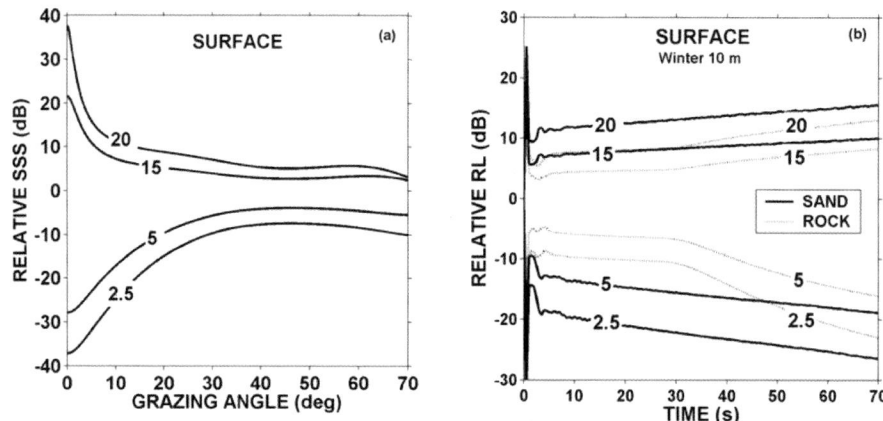

FIGURE 3. At 3.5 kHz for four wind speeds U *relative* to the U = 10 m/s case: (a) surface scattering strengths and (b) surface RL over sand and rock for the winter profile, S-R at 10 m case.

SURFACE LOSS

The above RL predictions (and all those in Ref. [7]) ignored the effects of surface loss on the propagation. In this section, we incorporate the incoherent high-frequency Surface Loss (SL) model of APL/UW [9-10] (extrapolated down to 3.5 kHz) into the propagation calculations of BiRASP. Figure 4a shows predictions of SL vs. grazing angle at 3.5 kHz for four wind speeds. It seen that the losses per bounce are largest at low grazing angles, especially at high wind speeds. (Bubble attenuation is the primary surface loss mechanism [10].) Incorporating this SL model in the BiRASP runs significantly reduced the predicted RL at long range as demonstrated by comparing the curves in Fig. 4b,c with the corresponding curves for sand in Fig. 2a,c. To make the impact clearer, Fig. 4d displays the characteristic reduction in RL when SL is included as a function of wind speed. Only for the 2.5- and 5-m/s wind-speed cases are the reductions not large.

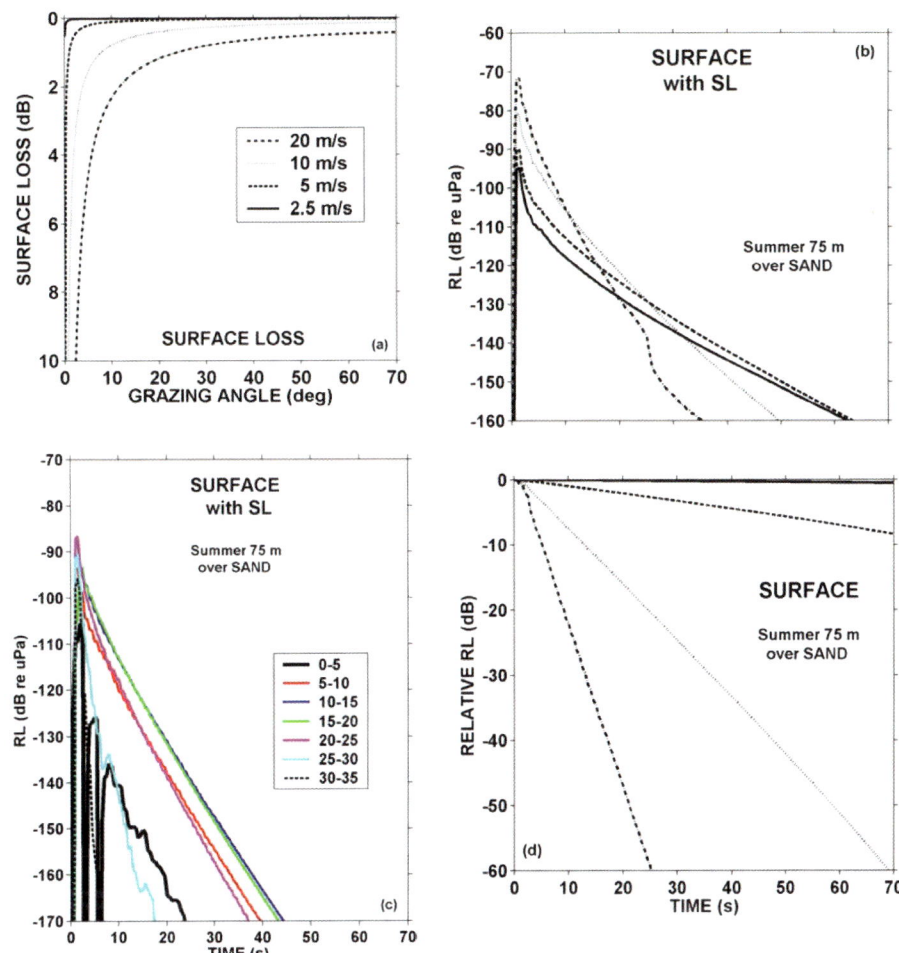

FIGURE 4. 3.5-kHz predictions at four wind speeds U: (a) SL vs. grazing angle. For the summer profile / S-R at 75-m over a sandy seafloor case: (b) surface RL including SL; (c) reverberation contributions to the U = 10 m/s case of (b) by mean grazing angle in 5-deg bins; and (d) the reduction in surface reverberation when SL is included in the propagation calculations.

Figure 5 includes SL in predictions of bottom reverberation for two scenarios. The assumed bottom scattering strengths are described in Ref. [7] ($\gamma_2 = 3.2$, $w_2 = 0.001$ m^4 case). Marked differences between the scenarios are apparent, with SL having relatively little impact at long range in the summer profile/75 m S-R case, but a significant impact in the winter profile/10 m S-R case where there are significantly more low-angle surface interactions (cf. Fig. 2c-d).

These runs stress the importance of surface loss on reverberation, whether it be from the bottom, sea surface, or fish and, so, the importance of having accurate SL models. (Small changes in dB-per-bounce values can have major cumulative effects on long-range reverberation levels at mid-frequencies.)

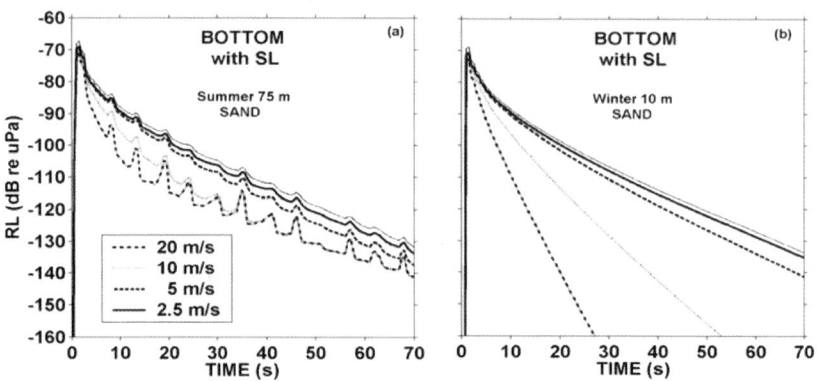

FIGURE 5. Effects on bottom reverberation of including surface loss in the propagation modeling for a sand bottom for two scenarios: (a) summer profile with S-R at 75 m and (b) winter profile with S-R at 10 m. The topmost curve (thin solid line) in each plot corresponds to a zero wind speed [7].

REVERBERATION FROM FISH

The primary physical drivers of the acoustic response of fish are their density, and size and depth distributions [11]. When fish are near the ocean surface or bottom, boundary-interference effects must also be accounted for. In Refs. [1-2], the fish target strength model of Love [11] was extended by convolving a Lloyd-mirror model with fish density and target strength over their depths to yield a formula effectively equivalent to surface scattering algorithm (allowing a simple implementation in reverberation models). For fish near the bottom, bottom properties are also important (but not the roughness parameters) [1]. This paper assumes the fish are uniformly distributed throughout the layer and ignores fish-attenuation effects. The RL calculations in this section assume flat boundaries and ignore SL effects.

Figure 6a shows predictions of fish backscattering strength vs. grazing angle at 3.5 kHz for mean fish lengths of 0.3 and 0.1 m, and for layer depths of 0.5-2 and 2-10 m below the sea surface. (The density was fixed at 0.01 fish per m^3. Changing fish density by a factor of k raises or lowers such curves by $10\log_{10}(k)$ dB.) It is seen that at these depths, the scattering response is basically flat with grazing angle.

Figure 6b shows the sensitivity of near-surface fish reverberation to fish depth and scenario. A noticeable dependence on scenario is seen with levels up to 10 dB higher at 20 km (~26 s) in the winter profile/10 m S-R case. As expected given the flat dependence of scattering strength on grazing angle, the reverberation differences for different layer depths for a given scenario are basically range independent.

A comparison of Fig. 6a with Fig. 1a suggests that whether the sea surface or fish dominates near-surface reverberation can depend on grazing angle and a number of environmental factors: wind speed and fish density, sizes, and depths. For example, comparing Fig. 6b with Fig. 2a shows that for the chosen fish parameters, near-surface fish reverberation is dominant over surface reverberation at low wind speeds and comparable to surface reverberation at high wind speeds. Recall changes in fish density can raise or lower the RL curves. (Frequency is another consideration.)

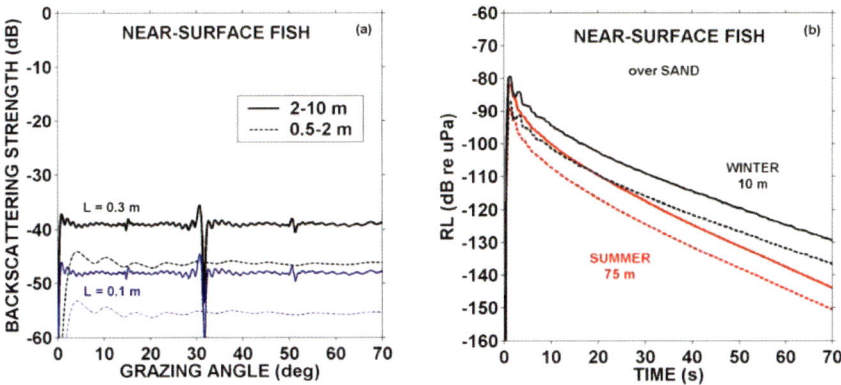

FIGURE 6. At 3.5 kHz for fish in layers 0.5-2 and 2-10 m below a flat sea surface at a density of 0.01 m^{-3}: (a) monostatic backscattering strength vs. grazing angle for two fish sizes, 0.3 and 0.1 m, and (b) near-surface fish reverberation for two scenarios over sand.

Similarly, for fish near the bottom, their scattering strength response with angle is basically flat (Fig. 7a), and whether the bottom or fish dominates the bottom-zone reverberation depends on the grazing angle, and fish sizes, depths and densities. This is illustrated in Fig. 7b, where for the particular parameter values chosen, near-bottom fish can easily dominate bottom reverberation from a sandy seafloor, but can be easily dominated by bottom reverberation from a basalt seafloor. Near-bottom fish reverberation was relatively insensitive to scenario, with ~5 dB differences at 20 km.

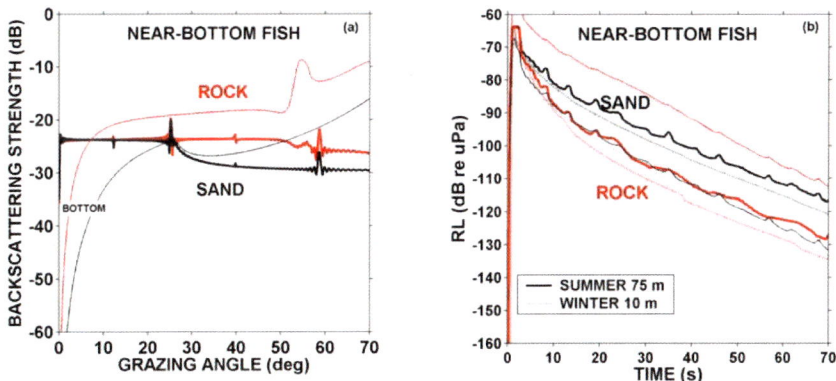

FIGURE 7. At 3.5 kHz for 0.3-m-long fish 0-25 m above flat rock and sand seabeds at a density of 0.1 m^{-3}: (a) monostatic backscattering strength vs. grazing angle for fish over rock and sand (thick curves). Also shown are corresponding bottom backscattering strengths at 3.5 kHz (thin curves). (b) Comparison of near-bottom fish (thick curves) and bottom (thin curves) RL for rock and sand for two scenarios.

In real shallow-water environments, fish are not uniformly distributed in space. Hence, RL curves would exhibit range and azimuth dependence beyond such as that shown in Figs. 6 and 7 (e.g., fish echoes may only be seen at particular ranges in a limited set of beams). An additional complication is that fish exhibit a variety of temporal (short-term, day/night and seasonal) behavior.

DISCUSSION

The model studies in this paper suggest that the sea surface and near-boundary fish could have a significant impact on reverberation levels, and moreover those levels could be quite sensitive to geometry, and oceanographic (wind speed, sound speed profile), geoacoustic, and biological (fish depth, size and density) variables. As the latter are generally unknown for a given environment, there is a particular need to assess the local fish populations (e.g., perhaps via echosounders) coupled with models such as these to estimate their contribution to the reverberation. Knowledge of fish behavior (e.g., typical day/night depths) can then help optimize sonar settings to minimize their impact.

ACKNOWLEDGMENTS

Work supported by the Office of Naval Research.

REFERENCES

1. Gauss, R. C., Gragg, R. F., Nero, R.W., Wurmser, D., and Fialkowski, J. M., "Broadband Models for Predicting Bottom, Surface, and Volume Scattering Strengths," NRL/FR/7100—02-10,042, Washington, DC: Naval Research Laboratory, September 30, 2002.
2. Gauss, R. C., Fialkowski, J. M., and Wurmser, D., "Assessing the Variability of Near-Boundary Surface and Volume Reverberation Using Physics-Based Scattering Models," in *Impact of Littoral Environmental Variability on Acoustic Predictions and Sonar Performance*, edited by N. G. Pace and F. B. Jensen, Dordrecht: Kluwer Academic, 2002, pp. 345-352.
3. Fromm, D. M., Crockett, J. P., and Palmer, L. B., "BiRASP – The Bistatic Range-dependent Active System Performance Model," NRL/FR/7140—95-9723, Washington, DC: Naval Research Laboratory, September 30, 1996.
4. LePage, K. D., "Monostatic Reverberation in Range Dependent Waveguides: The R-SNAP Model," SACLANT Undersea Research Centre SR-363, La Spezia, Italy, 2002.
5. Hamilton, E. L., "Geoacoustic Modeling of the Sea Floor," *J. Acoust. Soc. Am.* **68**, 1313-1319 (1980).
6. Essen, H.-H., "Scattering from a rough sedimental seafloor containing shear and layering," *J. Acoust. Soc. Am.* **95**, 1299-1310 (1994).
7. Gauss, R. C., Fromm, D. M., LePage, K. D., and Gragg, R. F., "The Dependence of Long-Range Reverberation on Bottom Roughness," in *High Frequency Ocean Acoustics Conference*, edited by M. B. Porter, T. M. Siderius, and W. A. Kuperman, San Diego, CA, March 2004.
8. Jensen, F. B., Kuperman, W. A., Porter, M. B., and Schmidt, H., *Computational Ocean Acoustics*, AIP Press, 1994, p. 38.
9. Thorsos, E. I., "Surface Forward Scattering and Reflection," APL-UW 7-83, Applied Physics Laboratory, University of Washington, May 1984.
10. Dahl, P. H., "Revisions and Notes on a Model for Bubble Attenuation in Near-Surface Acoustic Propagation," APL-UW TR 9411, Appl. Physics Laboratory, University of Washington, July 1994.
11. Love, R. H., "Resonant Acoustic Scattering by Swimbladder-Bearing Fish," *J. Acoust. Soc. Am.* **64**, 571-580 (1978).

Environmental Effects of Waveguide Uncertainty on Coherent Aspects of Propagation, Scattering and Reverberation

Kevin D. LePage and B. Edward McDonald

Naval Research Laboratory, 4555 Overlook Ave. SW, Washington, DC 20375

Abstract. The robustness of the coherence of waveguide propagation to environmental uncertainty becomes an important consideration for systems that seek to exploit coherence for gain. Examples include matched field processing for passive localization and Time Reversal Mirrors (TRMs) for active systems. Here efficient normal mode representations of mid-frequency time domain propagation using the narrowband and adiabatic approximations are used to explore the deterioration of active system predictability and performance in the presence of environmental fluctuations (i.e. sound speed perturbations in the water column and/or bottom, or bathymetry fluctuations). Results show that for TRMs the reverberation level at the focal range is increased, and the scattering from an illuminated object is reduced for ensembles over environmental uncertainty. Results are obtained analytically as formal averages and are believed to represent a lower limit on the deterioration of TRM performance in the presence of environmental uncertainty for actual waveguides.

INTRODUCTION

Uncertainty in waveguide properties causes commensurate uncertainty in acoustic propagation, scattering and reverberation in shallow water waveguides. In cases where knowledge of the spatial scales of waveguide variability and the associated variances are known, it is possible to quantify the uncertainty of the acoustic quantities of interest under the statistical hypothesis of homogeneity to various levels of fidelity (travel time perturbations to rays or modes, vs. full wave modeling). Representations based on the assumption that environmental perturbations cause only phase and travel-time perturbations offer value for estimating the lower bound of acoustic uncertainty. Here the adiabatic approach developed by Krolik [1] for modeling the predictability of the co-intensity of coherent propagation of normal modes through internal wave fields at a single frequency is extended to the time domain, and consistent expressions are obtained for modeling the predictability of boundary reverberation and target scattering. Such expressions are believed to represent a lower bound for the propagation of environmental uncertainty into acoustic uncertainty.

THEORY

The adiabatic theory for time domain propagation in normal modes has been derived previously [2,3]. However, for the sake of completeness, we briefly review. Complex envelope theory may be used to integrate adiabatic mode solutions to the Helmholtz equation [4] over a Gaussian weighted frequency to obtain

$$p(t,r,z_s,z|\omega,\Delta\omega) = \frac{2}{\rho(z_s)\sqrt{r}} \operatorname{Re}\left\{ e^{-i(\omega t + \pi/4)} \sum_{n=1}^{N} \frac{\exp\left(i\langle k_n\rangle_r r - \frac{(t-\langle S_n\rangle_r r)^2/2}{(\Delta\omega^{-2} - i\langle D_n\rangle_r)}\right)}{\sqrt{\langle k_n\rangle_r (\Delta\omega^{-2} - i\langle D_n\rangle_r)}} \phi_n(z_s,0)\phi_n(z,r) \right\}, \quad (1)$$

where $\Delta\omega$ is the bandwidth, $\langle\ \rangle_r$ indicates range average and S_n and D_n are the first and second derivatives w.r.t. frequency of the wavenumber k_n. It is assumed that k_n and S_n deviate from their mean value by their first perturbations w.r.t. the environmental sound speed defect $\Delta c(r,z)$ [5]

$$k_n = k_n^o - \frac{\omega}{k_n^o} \int_{-\infty}^{0} \frac{\Delta c(r,z)}{c^3(z)\rho(z)} \phi_n^2(z)\,dz, \quad (2)$$

and

$$S_n = S_n^o - \left(2 - S_n^o \frac{\omega}{k_n^o}\right)\frac{\omega}{k_n^o}\int_{-\infty}^{0}\frac{\Delta c(r,z)}{c^3(z)\rho(z)}\phi_n^2(z)\,dz - 2\frac{\omega^2}{k_n^o}\int_{-\infty}^{0}\frac{\Delta c(r,z)}{c^3(z)\rho(z)}\frac{\partial\phi_n(z)}{\partial\omega}\phi_n(z)\,dz \quad (3)$$

It is also assumed that the modal dispersions D_n are independent of the environmental variability for reasons of analytic tractability.

An EOF decomposition of the sound speed perturbations is adopted [1]

$$\Delta c(r,z) = \sum_{e=1}^{E} g_e(r)\varphi_e(z). \quad (4)$$

Equation (4) assumes that the perturbations are separable into orthogonal depth functions φ_e and uncorrelated random amplitudes g_e. These latter are distributed Gaussian in amplitude and are characterized spatially by a Gaussian correlation function with length scale l_e. Under this model the range integrated wavenumbers and slownesses are

$$\langle k_n\rangle_r r = k_n^o r - \frac{\omega}{k_n^o}\sum_{e=1}^{E} N(0,\sigma_g^2 r l_e)\int_{-\infty}^{0}\frac{\varphi_e(z)}{c^3(z)\rho(z)}\phi_n^2(z)\,dz, \quad (5)$$

and

$$\langle S_n\rangle_r r = S_n^o r - \frac{\omega}{k_n^o}\sum_{e=1}^{E} N(0,\sigma_g^2 r l_e)\left(\left(2-\frac{\omega}{k_n^o}\right)\int_{-\infty}^{0}\frac{\varphi_e(z)}{c^3(z)\rho(z)}\phi_n^2(z)\,dz + \omega\int_{-\infty}^{0}\frac{\varphi_e(z)}{c^3(z)\rho(z)}\frac{\partial\phi_n(z)}{\partial\omega}\phi_n(z)\,dz\right), \quad (6)$$

where $N(0,\sigma^2)$ is a zero mean Gaussian random variable with variance σ^2.

Propagation Uncertainty

Equations (5) and (6) indicate that the uncertainties in the total accumulated phase and travel time have variances proportional to the horizontal correlation length scale of the environmental sound speed defects multiplied by the range. The short-time average of the acoustic co-intensity from a TRM may be written as [2]

$$\langle p_{TRM}(t_1,r,z_1) p_{TRM}(t_2,r,z_2) \rangle = 2\text{Re}\left\{ \frac{4\pi^2}{\Delta z^2 \rho^2(z_p) r R_p} \sum_{n=1}^{N}\sum_{m=1}^{M} \frac{e^{i(k_n-k_m^*)r - i(k_n-k_m)R_p}}{|k_n||k_m|} \phi_n(z_p)\phi_m(z_p)\phi_n(z)\phi_m(z) \right.$$

$$\times \frac{1}{4\pi^2}\int_{-\infty}^{\infty} d\omega_1 \int_{-\infty}^{\infty} d\omega_2 \, e^{-i\omega_1(t_1 - S_n r + S_n R_p)} e^{i\omega_2(t_2 - S_m r + S_m R_p)}$$

$$\left. \times \prod_{e=1}^{E} e^{-(\Delta k_{en} - \Delta k_{em}^* + \omega_1 \Delta S_{en} - \omega_2 \Delta S_{em})^2 r l_e \sigma_{ge}^2 / 2} \right\} \quad , (7)$$

where

$$\Delta k_{en} = \frac{\omega}{k_n^o} \int_{-\infty}^{o} \frac{\varphi_e(z)}{c^3(z)\rho(z)} \phi_n^2(z) dz, \quad (8)$$

and

$$\Delta S_{en} = -\left(2 - \frac{\omega}{k_n^o}\right)\frac{\omega}{k_n^o}\int_{-\infty}^{o}\frac{\varphi_e(z)}{c^3(z)\rho(z)}\phi_n^2(z)dz - \frac{\omega^2}{k_n^o}\int_{-\infty}^{o}\frac{\varphi_e(z)}{c^3(z)\rho(z)}\frac{\partial \phi_n}{\partial \omega}\phi_n(z)dz. \quad (9)$$

Equation (7) may be integrated w.r.t. ω_1 and ω_2 giving

$$\langle p_{TRM}(t_1, \Delta r + R_p, z_1) p_{TRM}(t_2, \Delta r + R_p, z_2) \rangle = \frac{2\pi}{\Delta z^2 \rho^2(z_p) r R_p} \sum_{n=1}^{N}\sum_{m=1}^{M} \frac{e^{i(\text{Re}(k_n - k_m)\Delta r - \text{Im}(k_n + k_m)(\Delta r + 2R_p))}}{|k_n||k_m|} \phi_n(z_p)\phi_m(z_p)\phi_n(z)\phi_m(z)$$

$$\times \exp(-\Theta_{nm}^2)\left(\Theta_{22}^2 + (\Delta\omega^2 + iD_m\Delta r)/2\right)^{-1/2}\left(\Theta_{11}^2 + (\Delta\omega^2 - iD_m\Delta r)/2 - \frac{\Theta_{12}^2}{(4\Theta_{22}^2 + 2(\Delta\omega^2 + iD_m\Delta r))}\right)^{-1/2} \quad (10)$$

$$\times \exp\left(\frac{\Theta_2^2 - i2\Theta_2^2(t_2 - S_m\Delta r) - (t_2 - S_m\Delta r)^2}{4\Theta_{22}^2 + 2(\Delta\omega^2 + iD_m\Delta r)}\right) \exp\left(\frac{\left(\Theta_1^2 + i(t_1 - S_n\Delta r) - \frac{2\Theta_2^2(t_2 - S_m\Delta r) - 2\Theta_2^2 \Theta_2^2}{4\Theta_{22}^2 + 2(\Delta\omega^2 + iD_m\Delta r)}\right)^2}{4\Theta_{11}^2 + 2(\Delta\omega^2 - iD_m\Delta r) - \frac{\Theta_{12}^2}{\left(\Theta_{22}^2 + (\Delta\omega^2 + iD_m\Delta r)/2\right)}}\right),$$

where

$$\Theta_{nm}^2 = \sum_{e=1}^{E}\left(\Delta k_{en}^2 - 2\Delta k_{en}\Delta k_{em} + \Delta k_{em}^2\right)\sigma_{ge}^2 l_e r / 2 \quad \Theta_2^2 = \sum_{e=1}^{E}(\Delta k_{em} - \Delta k_{en})\Delta S_{em}\sigma_{ge}^2 l_e r$$

$$\Theta_1^2 = \sum_{e=1}^{E}(\Delta k_{en} - \Delta k_{em})\Delta S_{en}\sigma_{ge}^2 l_e r \quad \Theta_{12}^2 = \sum_{e=1}^{E}\Delta S_{en}\Delta S_{em}\sigma_{ge}^2 l_e r / 2 \quad , (11)$$

$$\Theta_{11}^2 = \sum_{e=1}^{E}\Delta S_{en}^2 \sigma_{ge}^2 l_e r / 2 \quad \Theta_{22}^2 = \sum_{e=1}^{E}\Delta S_{em}^2 \sigma_{ge}^2 l_e r / 2$$

Δz is the inter-element spacing of the TRM, and the probe source is at $[R_p, z_p]$.

Reverberation Uncertainty

An expression equivalent to Equation (10) has been obtained for boundary reverberation [4]. The reverberation intensity from a TRM may be written as

$$\langle p_{rev_{TRM}}(t_1,z_1) p_{rev_{TRM}}(t_2,z_2) \rangle = \frac{8\pi^3}{\Delta z^2 \rho^2(z_p)\rho^2(z_{rev})} \int_0^\infty dr_1 \int_0^\infty dr_2 \sum_{n=1}^N \sum_{m=1}^M \frac{e^{i\left(k_m r_1 - k_n^* R_p + k_{n'm'}^* r_2 + k_n R_p\right)}}{R_p |k_n||k_{n'}|\sqrt{k_m k_{m'}}} \frac{\exp\left(-(r_1-r_2)^2/2l^2\right)}{l\sqrt{2\pi r_1 r_2}}$$

$$\times \frac{\exp\left(-\Theta_{nn'mm'}^2\right) \phi_n(z_p) \phi_n(z_1) \phi_n^- \phi_{n'}^- ss_{nm} ss_{n'm'} \phi_m^+ \phi_{m'}^+ \phi_m(z_1) \phi_{m'}(z_2)}{\sqrt{\left(\Theta_{22}^2 + (\Delta\omega^{-2} + iD_{n'm'}r_2)/2\right)}\sqrt{\Theta_{11}^2 + (\Delta\omega^{-2} - iD_{nm}r_1)/2 - \frac{\Theta_{12}^4}{\left(4\Theta_{22}^2 + 2(\Delta\omega^{-2} + iD_{n'm'}r_2)\right)}}}$$

$$\times \exp\left(\frac{\Theta_2^4 - i2\Theta_2^2(t_2 - S_{n'm'}r_2) - (t_2 - S_{n'm'}r_2)^2}{4\Theta_{22}^2 + 2(\Delta\omega^{-2} + iD_{n'm'}r_2)}\right) \exp\left(\frac{\left(\Theta_1^2 + i(t_1 - S_{nm}r_1) - \frac{2i\Theta_{12}^2(t_2 - S_{n'm'}r_2) - 2\Theta_{12}^2}{4\Theta_{22}^2 + 2(\Delta\omega^{-2} + iD_{n'm'}r_2)}\right)^2}{4\Theta_{11}^2 + 2(\Delta\omega^{-2} - iD_{nm}r_1) - \frac{\Theta_{12}^4}{\left(\Theta_{22}^2 + (\Delta\omega^{-2} + iD_{n'm'}r_2)/2\right)}}\right), \quad (12)$$

where $k_{nm} = k_n + k_m$, $S_{nm} = S_n + S_m$, $D_{nm} = D_n + D_m$, l is the correlation length scale of the scatterers, ϕ_n^- and ϕ_m^+ are the downgoing and upgoing planewave decomposition of the mth mode, both obtained at the depth of the reverberating surface z_{rev}, and the angular dependence of scattering from the surface is described by the functions ss_{nm} and $ss_{n'm'}$. Note in Equation (12) that the definitions of the Θ_{ij} are taken from Equations (11) with Δk_n replaced with Δk_{nm}, Δk_m replaced with $\Delta k_{n'm'}$, etc. When the two integrals over range are evaluated an expression for the expected value of the reverberation intensity integrated over the ensemble of possible environments is obtained

$$\langle p_{rev_{TRM}}(t_1,z_1) p_{rev_{TRM}}(t_2,z_2) \rangle = \frac{8\pi^4}{\Delta z^2 \rho^2(z_p)\rho^2(z_{rev})R_p} \sum_{n=1}^N \sum_{m=1}^M \frac{e^{i\left(-k_n^* R_p + k_n \cdot R_p\right)} \exp\left(-\Theta_{nnhm'}^2\right)}{l\sqrt{t_1 t_2 \frac{2\pi}{S_{nm} S_{n'm'}}} |k_n||k_{n'}|\sqrt{k_m k_{m'}}}$$

$$\times \phi_n(z_p) \phi_n(z_1) \phi_n^- \phi_{n'}^- ss_{nm} ss_{n'm'} \phi_m^+ \phi_{m'}^+ \phi_m(z_1) \phi_{m'}(z_2) \left(\Theta_{22}^2 + \left(\Delta\omega^{-2} + i\left(\frac{D_{n'm'}}{S_{n'm'}}\hat{t}_2 - D_n R_p\right)\right)/2\right)^{-1/2}$$

$$\times \left(\Theta_{11}^2 + \left(\Delta\omega^{-2} - i\left(\frac{D_{nm}}{S_{nm}}\hat{t}_1 - D_n R_p\right)\right)/2 - \Theta_{12}^4 \left(4\Theta_{22}^2 + 2\left(\Delta\omega^{-2} + i\left(\frac{D_{n'm'}}{S_{n'm'}}\hat{t}_2 - D_n R_p\right)\right)\right)^{-1}\right)^{-1/2}, \quad (13)$$

$$\times \left(\frac{1}{2l^2} + \left(\frac{1}{D_1} + \frac{C_4^2}{D_2}\right) S_{n'm'}^2\right)^{-1/2} \left(-2\frac{C_4}{D_2} S_{nm} S_{n'm'} + \left(\frac{1}{D_1} + \frac{C_4^2}{D_2}\right) S_{n'm'}^2 + \frac{S_{nm}^2}{D_2} - \frac{G_2^2}{4A}\right)^{-1/2}$$

$$\times \exp\left(\frac{BB^2}{4AA} - \frac{G_1^2}{4A} + \frac{C_1}{D_1} + \frac{C_3^2}{D_2} - \left(2i\frac{C_4 C_3}{D_2} + i\frac{C_2}{D_1}\right)\hat{t}_2 + 2i\frac{C_3}{D_2}\hat{t}_1 + 2\frac{C_4}{D_2}\hat{t}_1\hat{t}_2 - \left(\frac{1}{D_1} + \frac{C_4^2}{D_2}\right)\hat{t}_2^2 - \frac{1}{D_2}\hat{t}_1^2\right)$$

where $\hat{t}_1 = t_1 + S_n R_p - \min(S_n) R_p$ and $\hat{t}_2 = t_2 + S_{n'} R_p - \min(S_{n'}) R_p$ and where

$$D_1 = 4\Theta_{22}^2 + 2/\Delta\omega^2 + 2i\hat{t}_2 D_{n'm'}/S_{n'm'} - 2iD_{n'}R_p$$

$$C_3 = \Theta_1^2 + 2\Theta_2^2 \Theta_{12}^2 / D_1$$

$$C_4 = 2\Theta_{12}^2 / D_1$$

$$D_2 = 4\Theta_{11}^2 + 2/\Delta\omega^2 - 2i\hat{t}_1 D_{nm}/S_{nm} + 2iD_n R_p - 4\Theta_{12}^4 / D_1$$

$$G_1 = ik_{n'm'} - (2iC_3C_4/D_2 + 2i\Theta_2^2/D_1)S_{n'm'} + 2C_4 S_{n'm'}\hat{t}_1/D_2 - 2(1/D_1 + C_4^2/D_2)S_{n'm'}\hat{t}_2$$

$$G_2 = -2C_4 S_{nm} S_{n'm'} + 2(1/D_1 + C_4^2/D_2)S_{n'm'}^2$$

$$A = 1/2l^2 + (1/D_1 + C_4^2/D_2)S_{n'm'}^2$$

$$AA = -2C_4/D_2 S_{nm} S_{n'm'} + (1/D_1 + C_4^2/D_2)S_{n'm'}^2 + S_{nm}^2/D_2 - G_2^2/4A$$

$$BB \square -i(k_{nm} - k_{n'm'}) - (2iC_3C_4/D_2 + 2i\Theta_2^2/D_1)S_{n'm'} + 2iC_3 S_{nm}/D_2 + 2C_4(S_{nm}\hat{t}_2 + S_{n'm'}\hat{t}_1)$$

$$-2(1/D_1 + C_4^2/D_2)S_{n'm'}\hat{t}_2 - 2S_{nm}\hat{t}_1/D_2 - 2G_1G_2/4A$$

(14)

Scattering Uncertainty

The backscattering from an object ensonified by a TRM has a very similar form to Equation (13) with the exception that objects may scatter any of four ways in amplitude, yielding sixteen cross terms for the scattered intensity

$$\langle p_{scat_{TRM}}(t_1,z_1) p_{scat_{TRM}}(t_2,z_2) \rangle = \frac{4\pi^2}{\Delta^2 \rho^2(z_p)\rho^2(z_{scat})R_{scat}^2 R_p} \sum_{n=1}^{N}\sum_{m=1}^{N}\sum_{n'=1}^{N}\sum_{m'=1}^{N} \frac{e^{i\left((k_{nm}-k_{n'm'}^*)R_o - (k_n^* - k_n)R_p\right)}}{|k_n||k_{n'}|\sqrt{k_m k_{m'}}}$$

$$\times \frac{\phi_n(z_s)\phi_{n'}(z_s)T_{nmn'm'}\phi_m(z_1)\phi_{m'}(z_2)\exp(-\Theta_{nmn'}^2)}{R_o\sqrt{(\Theta_{22}^2 + (\Delta\omega^{-2} + iD_{n'm'}R_o)/2)}\sqrt{\Theta_{11}^2 + (\Delta\omega^{-2} - iD_{nm}R_o)/2 - \frac{\Theta_{12}^4}{(4\Theta_{22}^2 + 2(\Delta\omega^{-2} + iD_{n'm'}R_o))}}}$$

$$\times \exp\left(\frac{\Theta_2^2 - i2\Theta_2^2(\hat{t}_2 - S_{n'm'}R_o) - (\hat{t}_2 - S_{n'm'}R_o)^2}{4\Theta_{22}^2 + 2(\Delta\omega^{-2} + iD_{n'm'}R_o)}\right) \exp\left(\frac{\left(\Theta_1^2 + i(\hat{t}_1 - S_{nm}R_o) - \frac{2i\Theta_{12}^2(\hat{t}_2 - S_{n'm'}R_o) - 2\Theta_2^2\Theta_{12}^2}{4\Theta_{22}^2 + 2(\Delta\omega^{-2} + iD_{n'm'}R_o)}\right)^2}{4\Theta_{11}^2 + 2(\Delta\omega^{-2} - iD_{nm}R_o) - \frac{\Theta_{12}^4}{(\Theta_{22}^2 + (\Delta\omega^{-2} + iD_{n'm'}R_o)/2)}}\right)$$ (15)

where $\hat{t}_1 = t_1 + S_n R_p - \min(S_n) R_p$, $\hat{t}_2 = t_2 + S_{n'}R_p - \min(S_{n'}) R_p$ and

$$T_{nmn'm'} = \begin{Bmatrix} \phi_n^-\phi_{n'}^-ss_{nm}^+ss_{n'm'}^+\phi_m^+\phi_{m'}^+ & +\phi_n^-\phi_{n'}^-ss_{nm}^-ss_{n'm'}^+\phi_m^+\phi_{m'}^- & +\phi_n^-\phi_{n'}^+ss_{nm}^-ss_{n'm'}^+\phi_m^+\phi_{m'}^- & +\phi_n^-\phi_{n'}^+ss_{nm}^-ss_{n'm'}^-\phi_m^+\phi_{m'}^+ \\ +\phi_n^-\phi_{n'}^-ss_{nm}^-ss_{n'm'}^+\phi_m^-\phi_{m'}^+ & +\phi_n^-\phi_{n'}^-ss_{nm}^-ss_{n'm'}^-\phi_m^-\phi_{m'}^- & +\phi_n^-\phi_{n'}^+ss_{nm}^-ss_{n'm'}^-\phi_m^-\phi_{m'}^+ & +\phi_n^-\phi_{n'}^+ss_{nm}^-ss_{n'm'}^+\phi_m^-\phi_{m'}^+ \\ +\phi_n^+\phi_{n'}^-ss_{nm}^+ss_{n'm'}^+\phi_m^+\phi_{m'}^- & +\phi_n^+\phi_{n'}^-ss_{nm}^+ss_{n'm'}^-\phi_m^+\phi_{m'}^- & +\phi_n^+\phi_{n'}^+ss_{nm}^+ss_{n'm'}^-\phi_m^+\phi_{m'}^- & +\phi_n^+\phi_{n'}^+ss_{nm}^-ss_{n'm'}^+\phi_m^+\phi_{m'}^+ \\ +\phi_n^+\phi_{n'}^-ss_{nm}^{++}ss_{n'm'}^-\phi_m^-\phi_{m'}^+ & +\phi_n^+\phi_{n'}^-ss_{nm}^{++}ss_{n'm'}^-\phi_m^-\phi_{m'}^- & +\phi_n^+\phi_{n'}^+ss_{nm}^{++}ss_{n'm'}^-\phi_m^-\phi_{m'}^- & +\phi_n^+\phi_{n'}^+ss_{nm}^+ss_{n'm'}^{++}\phi_m^-\phi_{m'}^+ \end{Bmatrix}.$$

(16)

In Equation (17) the up and downgoing mode decompositions $\phi_n^{\pm}$ are evaluated at the depth of the scatterer z_{scat}. For a sphere, the scattering functions $s_{nm}^{\pm\pm}$ are given by Ingenito [6].

RESULTS

We show an example of the deterioration of TRM performance due to uncertainty in the sound speed profile. The environment is shown in the left panel of Fig. 1. A downward refracting sound speed profile in 140 m of water lies over a 5 m thick slow sediment layer with a density of 1 g/cm^3, a sound speed of 1,482 m/s and a bulk attenuation of 0.06 dB/λ. The basement has a sound speed of 1,562 m/s, a density of 1.8 g/cm^3 and a bulk attenuation of 0.1 dB/λ. An internal wave field generated by PROSIM [7,8] superimposes sound speed perturbations on the water column. A realization of these perturbations is shown in the right panel of Fig. 1. A full depth-spanning monostatic TRM sonar is deployed at the origin and ensonifies the waveguide at 2kHz with a 0 dB gain time reversed version of signals it receives from a 0 dB probe source deployed at a range of 10 km and a depth of 20 m. The TRM has a source spacing of 1 m. The TRM sonar receives reverberation from the sediment-water interface and scattering from a 10 m radius vacuum spherical target deployed at the probe source location. The sediment water interface has a correlation length scale of 8 cm (eliminating Bragg scattering effects) and a scattering strength conforming to Lambert's law

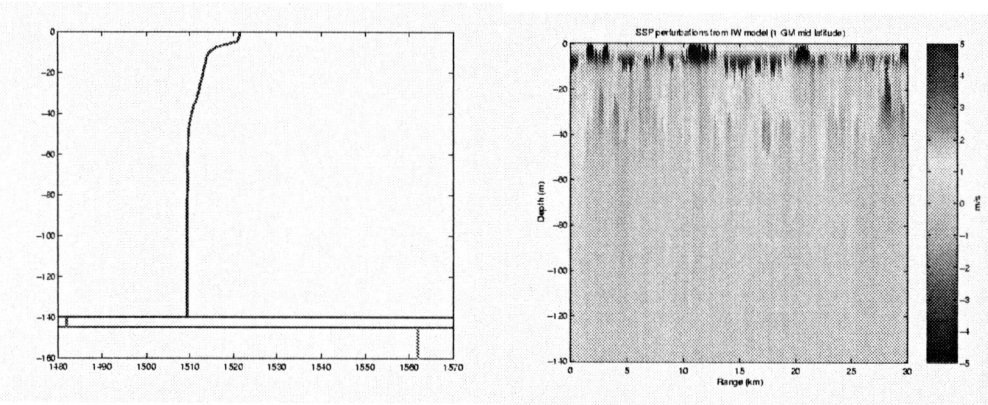

FIGURE 1. Sound speed profile of a typical shallow water environment (left). Superimposed sound speed perturbations caused by internal wave activity at the level of 1 GM (right).

$$SS_{nm} = -27 + 10\log 10\left\{\sin\left[a\cos\left(k_n/k_o\right)\right]\sin\left[a\cos\left(k_m/k_o\right)\right]\right\}, \quad (17)$$

leading to the following expression for the surface scattering amplitude required in Equation 17

$$ss_{nm} = 10^{-27/20}\sqrt{\sin\left[a\cos\left(k_n/k_o\right)\right]\sin\left[a\cos\left(k_m/k_o\right)\right]}. \quad (18)$$

In the absence of the internal wave activity, the sphere is very strongly ensonified by the focused field at the probe source location, causing a strong echo. At the same time, the reverberation from the bottom is reduced in the vicinity of the target echo on the order of 40 dB. The presence of the strong focus makes possible the detection of the object by both increasing the incident pressure on the object and reducing the field incident on the scatterers beneath it [9][10].

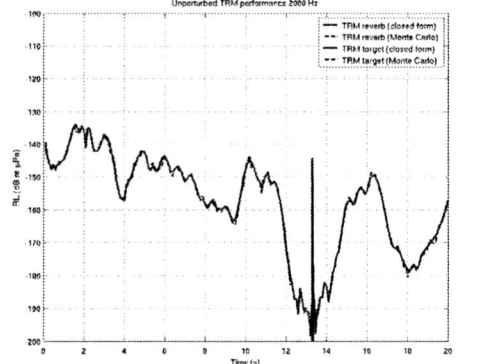 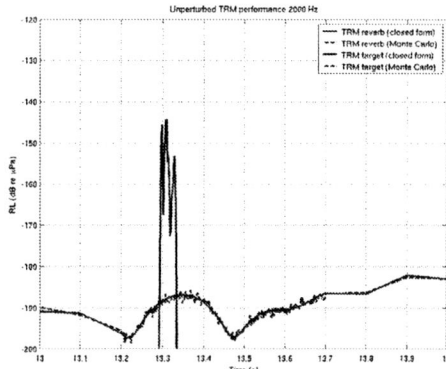

FIGURE 2. Reverberation and target echos in the environment of Figure 1 without internal wave activity. Left panel shows entire reverberation and echo time series, right panel is a close-up.

The uncertainty introduced by the presence of shallow water internal waves increases the expected value of the reverberation at the target range and reduces the expected value of the target echo, as shown in Figure 3.

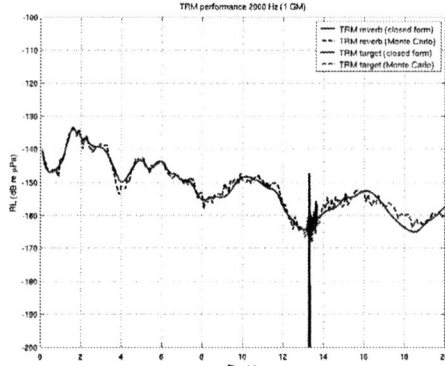 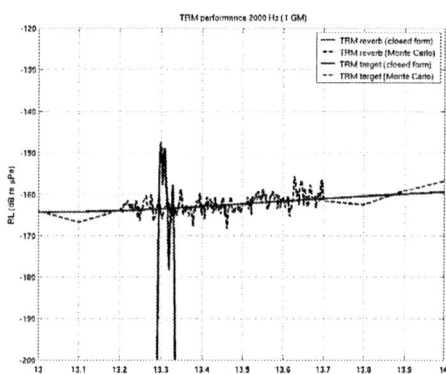

FIGURE 3. The expected value of bottom reverberation and target echo in the presence of internal waves. Left panel shows entire reverberation and echo time series, right panel is a close-up.

CONCLUSIONS

Expressions for the expected value of the second moment of acoustic pressure in the presence of environmental uncertainty have been obtained for propagation, reverberation and target echo caused by both point sources and TRMs. The expressions have been obtained for adiabatic propagation of normal modes through environmental variability. Results show that environmental uncertainty in the form of internal wave activity at the 1 GM level reduces the expected value of the coherent gain against bottom reverberation from 40 dB to 10 dB at 15 km for a 20 m probe source depth. The corresponding expected value of the backscattered intensity from a sphere at the probe source position is also reduced, by approximately 3 dB. The overall deterioration of SNR is, therefore, 33 dB. This prediction is believed to represent a lower bound on deterioration caused by environmental realizations. Inclusion of mode coupling effects and inhomogeneity of the environmental perturbations could be expected to seriously degrade the performance over and above the degradation predicted for adiabatic effects only.

ACKNOWLEDGMENTS

This work was funded under ONR's Capturing Uncertainty in the Common Tactical/Environmental Picture DRI.

REFERENCES

1. Krolik, J. L., *J. Acoust. Soc. Am.* **92**, 1408-1419 (1992).
2. LePage, K. D., *J. Comp. Acoust.* **9**, 1455-1474 (2001).
3. LePage, K. D., *Modeling Propagation and Reverberation Sensitivity to Oceanographic and Seabed Variability, SM-398*, La Spezia, Italy: SACLANT Undersea Research Centre, 2002.
4. Jensen, F.B., Kuperman, W.A., Porter, M.B., and Schmidt, H., *Computational Ocean Acoustics*, New York: AIP, 1993, pp. 274.
5. Sperry, B. J., McDonald, B.E., and Baggeroer, A.B., *J. Acoust. Soc. Am.* **114**, 1851-1860 (2003).
6. Ingenito, F., *J. Acoust. Soc. Am.* **82**, 2051-2059 (1987).
7. PROSIM (PROpagation channel SIMulator) First year technical report, TMS SAS 97/S/EGS/NC/084 AP (1998).
8. Elliot, A.J., and Jackson, J.F.E., "Internal Waves and Acoustic Variability," in *Proc. of the IEEE Oceans'98 Conference,* 1998, pp. 10-14.
9. Holland, C.W., and McDonald, B.E., *Shallow Water Reverberation from a Time Reversed Mirror, SR-326*, La Spezia, Italy: SACLANT Undersea Research Centre, 2000.
10. Lingevitch, J.F., Song, H.C., and Kuperman, W.A., *J. Acoust. Soc. Am.* **111**, 2609-2614 (2002).

Towards a Deterministic High-Frequency Shallow Water Ray Propagation Model

L. Pautet* and E. Pouliquen*

*NATO Undersea Research Centre, viale San Bartolomeo 400, 19038 La Spezia, Italy

Abstract. High frequency acoustic scattering from the ocean bottom and the ocean surface has been the subject of continuing interest for many years. Data and models have shown that as the signal frequency increases, the scattering pattern from such rough surfaces evolves from specular to quasi omni-directional. Multipath structures observed in shallow waters exhibit angular spreads that cannot be explained by assuming purely specular reflection. Most ray propagation models treat surface bounces as being essentially specular which poorly accounts for the multipath structures and in turn introduce errors in the estimation of the propagated and reverberated fields. For several applications such as high frequency Synthetic Aperture Sonar (SAS), minehunting and communications there is a need for reliable and fast propagation models able to correctly treat scattering at the rough waveguide boundaries. The Bellhop ray tracing model has been modified to introduce deterministic properties of the sea surface and sea bottom. Scattering from these surfaces is treated using solutions from the integral equation. Simulations display a significant effect of the angular spreading of the multipath structure increasing with roughness that is closer to the one observed on in-situ data.

INTRODUCTION

Applications such as communications, minehunting, high-frequency SAS have brought the focus on ocean acoustics modeling to shallow waters from tens to hundreds of metres and high-frequencies in the tens of kHz. To be accurate in these conditions, sound propagation modeling needs to take into account scattering occuring at the multiple interactions with the rough boundaries.

High frequency scattering from rough surfaces has been the object of many studies. Scattering models span from the bidimensional "exact" solution obtained with the integral equation [1] to bi- and tri-dimensional Kirchhoff models [1], small perturbation methods [2], composite-roughness models [3], small slope approximations [4, 5, 6], etc. The limits and the domain of validity of the different models and approximations have been investigated through extensive numerical work. However, only a few propagation models have included the effect of scattering by rough boundaries. The most basic approach has been to associate an angle-dependent surface loss to the specular direction discarding the fact that energy is also scattered in non-specular directions [7]. Some sophisticated parabolic equation models are able to take into account the whole scattered field by discretizing the surface in order to capture the large scale scattering of the rough boundaries [8, 9]. However, at high frequency, this approach becomes computationally challenging due to the required spatial discretization. There are also some questions about the accuracy of a stair-case discretization of a surface with a multi-scale roughness spectrum when the size of the wavelength is of the same order as the size of the RMS-roughness. Normal modes have also been used to calculate scattered field by combining a normal mode decomposition with a module describing the intermodal transfer of energy due to scattering [10]. Again, at high frequencies, the increasing number of modes makes this approach non suitable. In order to find a compromise between computational time and accuracy, a ray tracing approach is presented here.

High frequency scattering by a rough surface observed experimentally in a very shallow water environment shows that a significant portion of the energy is scattered in directions other than the specular direction [11, 12]. This calls into question the use of ray theory based on Snell's law to model propagation. Similar problems are encountered in the EM field. In atmospheric science, it can be solved by using a Monte Carlo propagation model to simulate the photons paths within the ocean-atmosphere system as they are scattered by particules and the air/sea interface [13, 14]. This paper presents an approach similar to the one used in the EM field applied to a commonly used ray propagation model, Bellhop [15]. A statistical scattering kernel based on the integral equation is combined with this ray tracer hereafter referred to as the "Bellodds" model. The implementation of the scattering kernel is described before presenting results from a few simulations.

SCATTERING KERNEL

The Bellhop ray tracing propagation model is modified by adding an incoherent scattering component. The interface patch on which the ray impinges can be considered as a small antenna, which creates a scattering beam pattern as illustrated in Fig. 1. Each ray impinging on a rough boundary with an incident angle θ_i is scattered in a direction θ_s, determined statistically as a function of the local scattering strength. The idea is that if enough rays are launched, the resulting acoustic field becomes statistically meaningful. Interface bistatic scattering coefficients are calculated for a series of generated rough profiles with chosen spatial statistics (e.g. Gaussian, power-law, Pierson-Moskowitz spectra) using the integral equation [1]. The scattering coefficient is treated as $P_{\theta_i}(\theta_s)$, the probability for a ray with an incident angle θ_i to be "reflected" in the direction given by the scattered angle θ_s:

$$P_{\theta_i}(\theta_s) = \frac{m(\theta_i, \theta_s)}{\int_{\alpha=-\pi/2}^{\pi/2} m(\theta_i, \alpha) \, d\alpha}, \tag{1}$$

where $\sigma(\theta_i, \theta_s) = 10\log(m(\theta_i, \theta_s))$ is the scattering strength. The principle of energy conservation is satisfied by an appropriate normalization but a larger number of rays need to be sent in order to obtain a statiscally meaningful acoustic field in the waveguide. Fig. 2 and Fig. 3 show how the averaged scattering strength and corresponding probabilities of scattering evolve with increasing surface roughness at a given frequency. The probability for the energy to be *non-specular* (i.e. incoherent) is significant and increases with roughness and/or frequency.

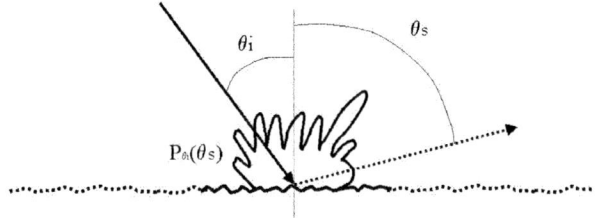

FIGURE 1. Schematic of the ray scattering process showing the scattering beam pattern in terms of probability and a ray being scattered at a non-specular angle.

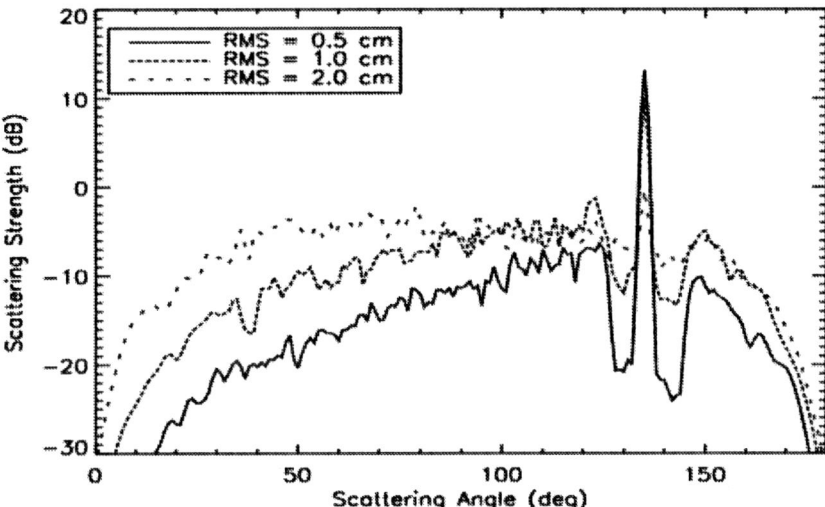

FIGURE 2. Scattering strength at 20 kHz from a surface having a saturated power-law spectrum ($\gamma = 3.6$, $k_s = 30$ rad/m) computed using the integral equation [1]. Results are averaged over 50 surfaces and correspond to a 45° incident angle.

RESULTS

With the use of a pre-calculated lookup table containing a series of scattering probabilities depending on the incident and scattered angles and various surface roughness

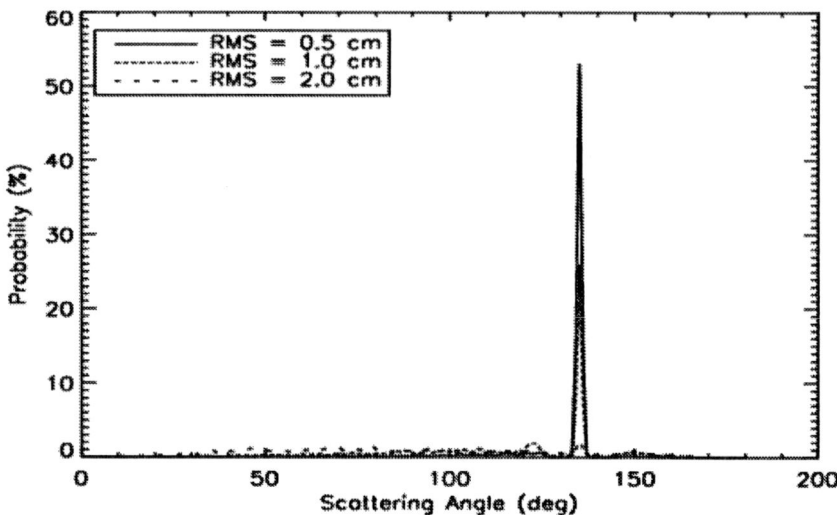

FIGURE 3. Scattering probabilities extracted from Fig. 2. The probability for the energy to be scattered in the exact specular direction is 55% for the smoothest surface, 27% for the intermediate surface, and less than 5% for the roughest surface.

properties, Bellodds can be used to simulate high frequency propagation and reverberation. Fig. 4 shows the results obtained in a monostatic case. A 100kHz source is placed in the middle of a 20m deep channel and transmits like a sidescan sonar towards the seabed. The bottom interface is generated using a saturated power-law spectrum ($\gamma = 3.6$, $k_s = 30$ rad/m) with a RMS-roughness of 1cm. The sea surface is considered to be quasi-flat with a RMS-roughness of 1mm. The clearly visible paths are the bottom path (B path), the bottom/surface path (BS path) and the bottom/surface/bottom path (BSB path). Each of them displays an angular spreading caused by the seabed roughness. There are two competing effects: an increase of the angular spread with the number of surface bounces and the classical attenuation loss. This explains why the bottom/surface/bottom shows the strongest angular spread in this case. Similar behavior was observed in data acquired in situ [11].

As the roughness increases, the scattering probability $P_{\theta_i}(\theta_s)$ becomes almost uniform (Fig. 3). In fact, the probability could be schematized as being uniform for all scattered angles except for the specular angle. Equation 1 can be simplified to the following expression.

$$P_{\theta_i}(\theta_s) = 1 - P_{spec} + \delta_{\theta_{spec}}(\theta_s)\left[2P_{spec} - 1\right], \qquad (2)$$

where P_{spec} is the probability for the scattered ray to be a specular ray, θ_{spec} is the specular angle, and $\delta_{\theta_{spec}}(\theta_s)$ is the Kronecker delta symbol.

An estimate of waveguide properties can then be calculated with only the specular probability p_{spec} as roughness input. Fig. 5 and Fig. 6 illustrate transmission loss results obtained using this approach. The channel used is a isovelocity channel, 20 m deep with a source placed at 10 m depth over a silt bottom. The calculations are made first without

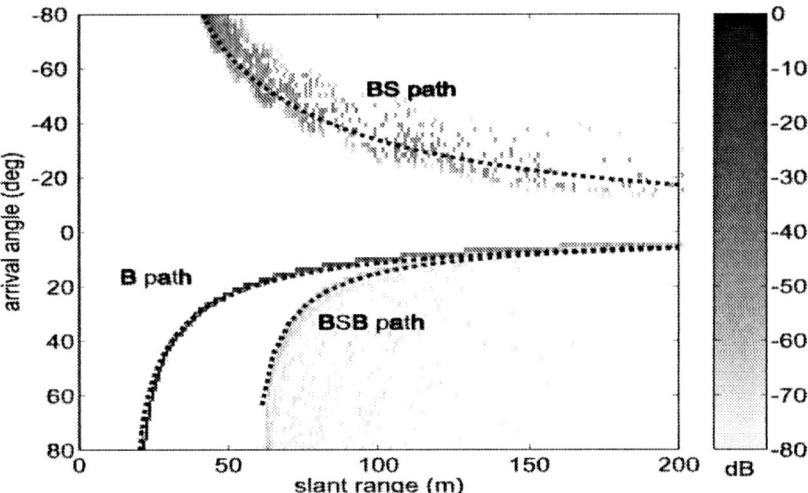

FIGURE 4. Estimate of reverberation versus vertical arrival angle and slant range for a flat sea surface and a bottom RMS roughness of 1 cm at 100 kHz. The lines represent the theoretical specular paths.

any roughness, and then with a 90 percent probability of having a specular reflection at each boundary, the rest of the rays being uniformly scattered in the other directions. As the roughness increases more energy is dissipated in the first bounces: rays launched at angles that would be above the critical angle may be scattered at angles below the critical angle and be absorbed by the bottom [16]. As expected, the multipath structure gets more and more disturbed as range increases. The refocusing of the fiel observed is coherent with a isovelocity scenario with source at mid depth. Any changes in sound speed profile or source depth would eliminate this regular focusing. The interest of using such a particular scenario is to illustrate the decay of the structure as the roughness increases, i.e. as scattering inhibits the perfect refocusing.

CONCLUSION

Through a statistical description of the scattering process, a deterministic ray-based model can be used to simulate the propagation or reverberation in a rough waveguide. As observed in data, simulations show an angular spreading of the multipath structure in reverberation and an increased loss in the propagation. In situ, water column instabilities and small source/receivers movements would have a similar impact on multipath structure and transmission loss. Unfortunately, it is impossible to separate these effects especially at these high frequencies. However boundary roughness appears to be a significant factor in channel coherence at high frequency. One advantage of Bellodds is that because of its statistical nature, each run is different from the next one, so the effects of even these small fluctuations can be simulated using a similar approach.

Future work will address volume scattering within the water column and the sediment.

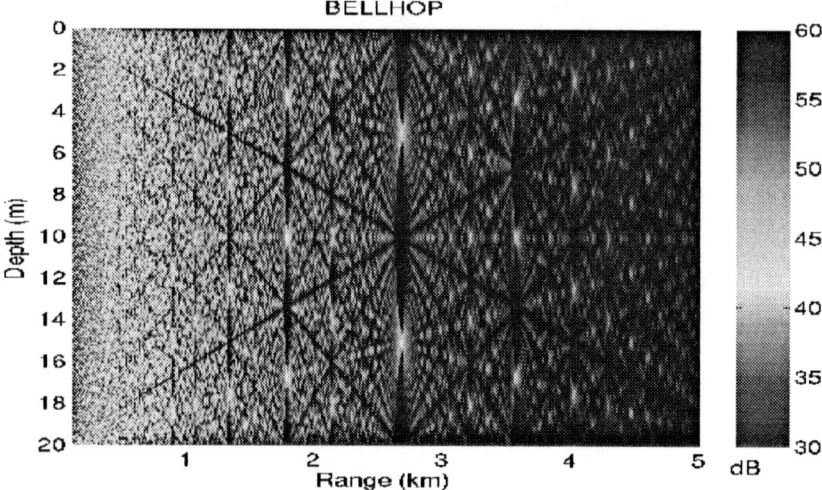

FIGURE 5. Transmission loss for a 20kHz source placed at 10 m depth in a 20 m wave guide with a flat silty bottom and sea surface.

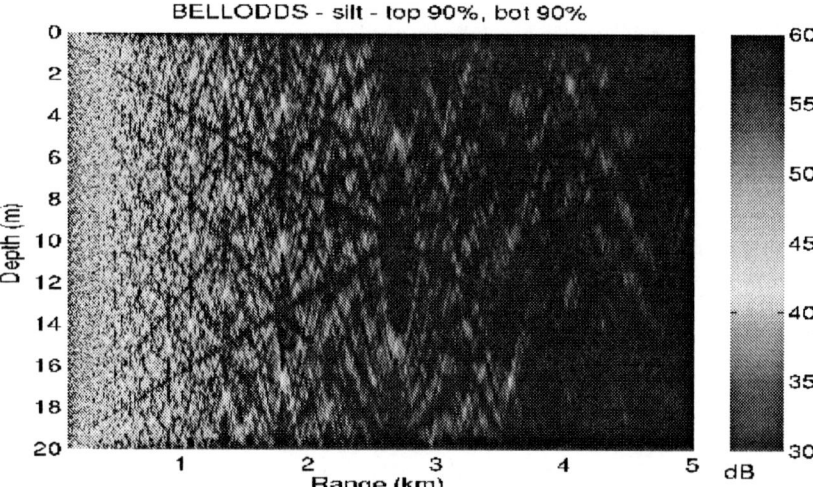

FIGURE 6. Transmission loss for a 20kHz source placed at 10 m depth in a 20 m wave guide with a slightly rough silty bottom and sea surface (90 percent of the energy is reflected in the specular direction).

REFERENCES

1. Thorsos, E., *Journal of the Acoustical Society of America*, **83**, 78–92 (1988).
2. Thorsos, E., and Jackson, D., *Journal of the Acoustical Society of America*, **86**, 261–277 (1989).
3. Daniel, S. M., and Gorman, A., *Journal of the Acoustical Society of America*, **73**, 1476–1486 (1983).
4. Thorsos, E., and Broschat, S., *Journal of the Acoustical Society of America*, **97**, 2082–2093 (1995).

5. Broschat, S., and Thorsos, E., *Journal of the Acoustical Society of America*, **101**, 2615–2625 (1997).
6. Yang, T., and Broschat, S., *IEEE Transactions on Antennas and Propagation*, **40**, 1505–1512 (1992).
7. Moore-Head, M., and Jobst, W., *Journal of the Acoustical Society of America*, **86**, 247–251 (1989).
8. Purrington, R., *Journal of Computational Acoustics*, **2**, 147–160 (1994).
9. Rosenberg, A., *Journal of the Acoustical Society of America*, **105**, 144–153 (1999).
10. Ellis, D., *Journal of the Acoustical Society of America*, **97**, 2804–2814 (1995).
11. Wang, L., Davies, G., Bellettini, A., and Pinto, M., "Multipath effect on DPCA micronavigation of a synthetic aperture sonar," in *Impact of Littoral Envionmental Variability on Acoustic Predictions and Sonar Performance*, 2002, pp. 465–472.
12. Dahl, P., *IEEE Journal of Oceanic Engineering*, **26**, 141–151 (2001).
13. Breon, F., *Journal of Atmospheric Science*, **49**, 1221–1232 (1992).
14. Frouin, R., Pouliquen, E., and Breon, F., "Ocean color remote sensing using polarization properties of reflected sunlight," in *Proceedings of the 6th international conference on Physical Measurements and Signatures in Remote Sensing*, 1994.
15. Porter, M., and Liu, Y., *Theoretical and Computational Acoustics*, **2**, 947–956 (1994).
16. Thorsos, E., Henyey, F., Williams, K., Elam, W., and Reynolds, S., "Simulation of temporal and spatial variability in shallow water propagation," in *Impact of Littoral Envionmental Variability on Acoustic Predictions and Sonar Performance*, 2002, pp. 337–344.

Nonlinear Bubble Dynamics And The Effects On Propagation Through Near-Surface Bubble Layers

Timothy G. Leighton

Institute of Sound and Vibration Research, University of Southampton, Highfield, Southampton UK

Abstract. Nonlinear bubble dynamics are often viewed as the unfortunate consequence of having to use high acoustic pressure amplitudes when the void fraction in the near-surface oceanic bubble layer is great enough to cause severe attenuation (e.g. >50 dB/m). This is seen as unfortunate since existing models for acoustic propagation in bubbly liquids are based on linear bubble dynamics. However, the development of nonlinear models does more than just allow quantification of the errors associated with the use of linear models. It also offers the possibility of propagation modeling and acoustic inversions which appropriately incorporate the bubble nonlinearity. Furthermore, it allows exploration and quantification of possible nonlinear effects which may be exploited. As a result, high acoustic pressure amplitudes may be desirable even in low void fractions, because they offer opportunities to gain information about the bubble cloud from the nonlinearities, and options to exploit the nonlinearities to enhance communication and sonar in bubbly waters. This paper presents a method for calculating the nonlinear acoustic cross-sections, scatter, attenuations and sound speeds from bubble clouds which may be inhomogeneous. The method allows prediction of the time dependency of these quantities, both because the cloud may vary and because the incident acoustic pulse may have finite and arbitrary time history. The method can be readily adapted for bubbles in other environments (e.g. clouds of interacting bubbles, sediments, structures, *in vivo*, reverberant conditions etc.). The possible exploitation of bubble acoustics by marine mammals, and for sonar enhancement, is explored.

NONLINEAR THEORY

Acoustic propagation through bubbly water has been modeled only with the introduction of the assumption of bubble linearity, or linearisation of the bubble dynamics, at an early stage. Probably the most notable example is the pioneering work of Commander and Prosperetti [1], which has been cited over 100 times since publication and used in many more acoustic investigations. If this linearisation is not done, not only do the formulations become inherently more complicated, but several useful mathematical techniques are not valid. These include complex representation of oscillations, small amplitude expansions, Green's function, Fourier transforms, superposition and addition of solutions. This paper describes a nonlinear approach.

Consider a cloud of bubbly water (having volume V_c and sound speed c_c and bulk modulus B_c). It is made up of a volume V_w of bubble-free water (having sound speed c_w and bulk modulus B_w) and a volume V_g of free gas (having sound speed c_g and bulk modulus B_g) distributed in a population of bubbles. Hence

$$V_c = V_w + V_g \tag{1}$$

Mass conservation is simply expressed by multiplication of the volumes with the respective densities (of cloud, ρ_c; bubble free water, ρ_w; and gas, ρ_g), i.e.

$$\rho_c V_c = \rho_w V_w + \rho_g V_g \tag{2}$$

When an acoustic wave passes through the bubbly liquid, the oscillatory pressures applied to the bubbles cause them to undergo pulsation. Under the assumption that each of the three media separately conserve mass, the differential of (2) with respect to the applied pressure P is, of course, zero. In an infinite body of either water or gas that contains no dissipation (an assumption which will shortly be examined further), sound speeds (c_w and c_g respectively) may be defined:

$$c_\varsigma^2 = \frac{B_\varsigma}{\rho_\varsigma} = \left(\frac{\partial P(\rho, S)}{\partial \rho}\right)_\varsigma \qquad (\varsigma = w, g) \tag{3}$$

where S is the entropy and the subscript ς refers to application to bubble-free water (w) or gas (g) throughout (3). Similarly, differentiation of (1) with respect to the applied pressure gives, with (3), the relationship between the bulk moduli

$$\frac{1}{B_c} = \frac{V_w}{V_c}\frac{1}{B_w} + \frac{V_g}{V_c}\frac{1}{B_g} \tag{4}$$

Let us define a function ξ_c (which is not an inherent property of the bubble cloud in the thermodynamic sense), equal to the root of the ratio of the bulk modulus of the bubbly water to its density, which with (4) gives:

$$\xi_c = \sqrt{\frac{B_c}{\rho_c}} = \sqrt{\left(\frac{V_c}{\rho_w V_w + \rho_g V_g}\right) \Big/ \left(\frac{V_w}{V_c B_w} + \frac{V_g}{V_c B_g}\right)} \approx c_w \left(1 + \frac{B_w V_g(t)}{V_c B_g}\right)^{-1/2} \tag{5}$$

where the final approximate form is valid under low void fraction conditions [2]. If there were no dissipation, the quantity ξ_c could be identified with the sound speed in bubbly water. However such an identity is not rigorous for lossy bubble clouds [2].

Evaluation of (5) requires calculation of the bulk modulus of the gas, as it is distributed through a (presumably) numerous population of bubbles pulsating with a broad range of amplitudes, phases, frequency content, damping and start times. The inhomogeneous bubbly water must be divided into volume elements which are sufficiently small to ensure that all the bubbles in that element are subjected to the same pressure change $dP(t)$ simultaneously (the use of d indicating an intention to use or calculate the quantity numerically). This would allow calculation of a value ξ_c for each volume element, since from (3) the bulk modulus B_{gl} of the gas within the lth volume element is related to the volume changes dV_i of the I bubbles in that volume element:

$$\frac{1}{B_{gl}} = -\frac{1}{V_{gl}} \sum_{i=1}^{I} \frac{dV_i}{dP_l} \tag{6}$$

where P_l denotes the pressure in the lth element. Consider one such volume element V_{c_l} of a cloud which has total volume $V_c = \sum_{l=1}^{L} V_{c_l}$. Substituting (6) into (5) gives ξ_{c_l}, the time history of ξ_c within the volume element V_{c_l}:

$$\xi_{c_l} \approx c_w \left(1 - \frac{\rho_w c_w^2}{V_{c_l}} \sum_{i=1}^{I} \frac{dV_i}{dP_l}\right)^{-1/2} \quad (7)$$

To understand the meaning of this quantity, consider that, if the system were linear, monodisperse and lossless, dV_i/dP_l would be a constant throughout the steady-state oscillatory cycle: In pressure-volume space, as one progressed throughout the oscillatory cycle one would move back and forth along a locus of points mapping out a straight line. The constant gradient of that line could be related to the sound speed in the cloud through (7), which would equal the constant ξ_{c_l}. If the monodisperse system were nonlinear and lossless, dV_i/dP_l would vary through the cycle, and the single line mapped out by the locus of points in pressure-volume space would not be straight. In this case the sound speed would vary through the oscillatory cycle, and could again be identified with ξ_{c_l} through (7). This could then be related to a sound speed for nonlinear propagation. If however dissipation occurs, the locus of points in the pressure-volume plane would, during a single oscillatory cycle, map out a finite area. In such circumstances ξ_{c_l} cannot strictly be identified with any sound speed. If dissipation is very small then one might identify a characteristic value of dV_i/dP_l which is not much different from the true value for most of the acoustic cycle; for the linearised case, this is in effect what Commander and Prosperetti do [1]. This will be discussed further in relation to Figure 4.

To evaluate (7), the bubble population of the volume element is classified into j discrete bins according to bubble size. Every individual bubble in the jth bin is replaced by another bubble which oscillates with radius $R_j(t)$ and volume $V_j(t)$ (about equilibrium values of R_{0_j} and V_{0_j}), such that the total number of bubbles N_j and total volume of gas $N_j V_j(t)$ in the bin remain unchanged by the replacement. If the bin width increment is sufficiently small (1 μm is normally chosen), the time history of every bubble in that bin should closely resemble $V_j(t) = V(R_{0_j}, t)$ (the sensitivity being greatest around resonance). Hence the total volume of gas in the lth volume element of bubbly water is $V_{g_l}(t) = \sum_{j=1}^{J} N_j(R_{0_j}, t) V_j(t) = V_{c_l} \sum_{j=1}^{J} n_j(R_{0_j}, t) V_j(t)$, where $n_j(R_{0_j}, t)$ is the number of bubbles per unit volume of bubbly water within the jth bin. Expressing (7) in terms of this bin scheme gives:

$$\xi_{c_l} \approx c_w \left(1 - \rho_w c_w^2 \sum_{j=1}^{J} n_j(R_{0_j}) \frac{dV_j}{dP_l}\right)^{-1/2} \quad (8)$$

Crucially (8) provides a *generic framework into which any bubbly dynamics model may be inserted* (giving $dV_j(t)/dP_l(t)$ appropriate to bubbles in free field or reverberation, *in vivo*, in structures or sediments, or in clouds of interacting bubbles, etc. as the chosen model dictates). This will be discussed later. First, the following section will use a speculative application to illustrate the general characteristics in the linear monochromatic steady state limit.

REDUCTION OF THEORY TO LINEAR LIMIT

Equation (8) contains low-void fraction limitations identical to those discussed by Commander and Prosperetti [1], to be used in an appropriate propagation model [1-4]. However so far no assumptions of small amplitude, steady state, monochromatic or linear bubble pulsations have been included, nor have the bubbles and their wall motions been assumed to be spherically symmetric. If these assumptions are introduced, the linear formulation of Commander and Prosperetti [1] is readily obtained from (8) [2]. The qualitative effect on the sound speed, predicted by such linear formulations when bubbles are added to previously bubble-free water, is well-known. In quasi-static conditions, the addition to gas bubbles to liquid will reduce its bulk modulus (since in (3) a given positive ∂P will cause a much larger compression $\partial \rho$ because of the reduction in volume of the bubbles). It will also reduce the density, but to a lesser amount, and therefore the net result is a reduction in the sound speed (see (3); alternatively, note that in a binomial expansion of (5), all the terms in $B_w V_g/(V_c B_g)$ are positive under quasi-static conditions). Consider now a monodisperse population of bubbles. As the driving frequency increases towards resonance, the amplitude of oscillation increases, and therefore so will the reduction in sound speed engendered by the presence of these bubbles. However, when the driving frequency exceeds the resonance of the bubbles, they are inertia-controlled: The compressive half-cycle of the driving field dP coincides with a bubble expansion, because of the change in the phase relation between them (which changes the sign of dV_i/dP_l in (7)). Hence at frequencies greater than resonance, the presence of bubbles causes an increase in sound speed. The magnitude of the effect becomes smaller as the frequency increases to values further from the bubble resonance (and the amplitude of pulsation decreases), until at very high frequencies the sound speed is identical to the bubble-free condition.

This linear scheme has been put to valuable use in modeling the propagation of sound through the near-surface bubble layer, where for frequencies sufficiently low to drive the bubbles in stiffness-mode, the zone can be upwardly refracting, generating a waveguide [5-7]. However a more fanciful application might explain the mystery of the mechanism by which humpback whales (*Megaptera novaeangliae*) exploit bubble nets to catch fish [8]. It has been known for decades that up to 30 whales might dive deep and then release bubbles to form the walls of a cylinder, the interior of which is relatively bubble-free (Fig. 1*a,b*) [9]. The prey are trapped within this cylinder, for reasons previously unknown, before the whales lunge feed on them from below (Fig. 1*c*). When the whales form such nets, they emit very loud, 'trumpeting feeding calls', the available recordings containing energy up to at least 4 kHz. A suitable void fraction profile would cause the wall to act as a waveguide.

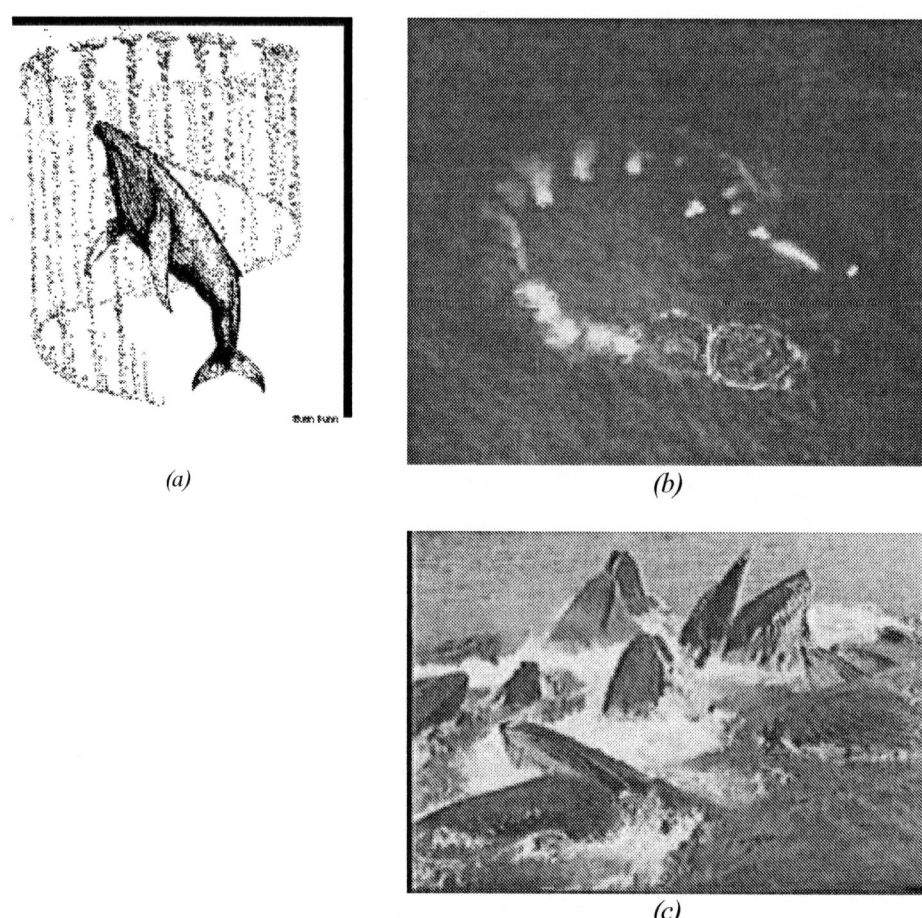

FIGURE 1. (*a*) Schematic of a humpback whale creating a bubble net. A whale dives beneath a shoal of prey and slowly begins to spiral upwards, blowing bubbles as it does so, creating a hollow-cored cylindrical bubble net. The prey tend to congregate in the center of the cylinder, which is relatively free of bubbles. Then the whale dives beneath the shoal, and swims up through the bubble-net with its mouth open to consume the prey ('lunge feeding'). Groups of whales may do this co-operatively (Image courtesy of Cetacea.org). (*b*) Aerial view of a humpback bubble net (photograph by A. Brayton, reproduced from [10]). (*c*) Humpback whales lunge feeding (Image courtesy of L. Walker, http://www.groovedwhale.com).

Assume the scales permit the use of ray representation. Figure 2 shows how, with tangential insonification,[i] the mammals could generate a 'wall of sound' around the net, and a quiet region within it. As fish approach the wall, swim bladder resonances may be excited. The natural schooling response of fish to startling by the intense sound as they approach the walls would, in the bubble net, be transformed from a survival response into one that aids the predator in feeding [8].

Figure 2*b* plots the raypaths (calculated using standard techniques [11]) from four whales for the stated launch conditions, for a bubble net in which the void fraction increases linearly from zero at the inner and outer walls, to 0.01% at the mid-line of the wall.

(a)   (b)

FIGURE 2. (*a*) Schematic of a whale insonifying a bubble-net (plan view), illustrating the sound speed profile in the cloud and, by Huygen's construction, sample ray paths. The sound speed profile assumes void fractions are greatest in the mid-line of the net wall, and assumes that the bubbles pulsate in stiffness mode. Hence the closer a Huygens wavelet is to the mid-line, the smaller the radius of the semicircle it forward-plots in a given time. Rays tend to refract towards the mid-line. (*b*) Four whales insonify an annular bubble net described in the text. The inner circle represents the inner boundary of the net wall. The outer boundary is obscured by the rays. Computed ray paths, where each whale launches 281 rays with an angular extent of 10°, refract as in (*a*). The rays gradually leak out, although some rays can propagate around the entire circumference. Plotting of a raypath is terminated when it is in isovelocity water and on a straight-line course which will not intersect the cloud. This refers to rays whose launch angles are such that they never intersect the net; and to rays which, having entered the net and undertaken two or more traverses of the mid-line, leave it. (Figure by T. G. Leighton, S. D. Richards and P. R. White [8]).

The actual acoustics of the cloud will, of course, be complicated by 3D effects and the possibility of collective oscillations; and even, speculatively, bubble-enhanced parametric sonar effects [12] which might be utilized by whales, for example to reduce beamwidth or generate harmonics, sum- and difference-frequencies etc.

The effect follows the frequency dependency described above. At frequencies sufficiently high to drive the bubble cloud in an inertia-controlled fashion, the bubbles produce in an increase in sound speed. The wall is outwardly-refracting, and rays are no longer trapped within the cloud. The refractive effect of these bubbles on sound speed becomes negligible at even higher frequencies, although of course acoustic attenuation and scatter may be great. A variety of ray behavior is possible, from reflecting straight off the net to traversing it and the interior with barely any refraction [8]. Such frequencies would not be effective in trapping prey, even if the prey could perceive them. However, if scattering losses permit (and it is by no means certain they would), is it possible that, given these refracted paths, such frequencies could be used for echolocation of the contents of the net?

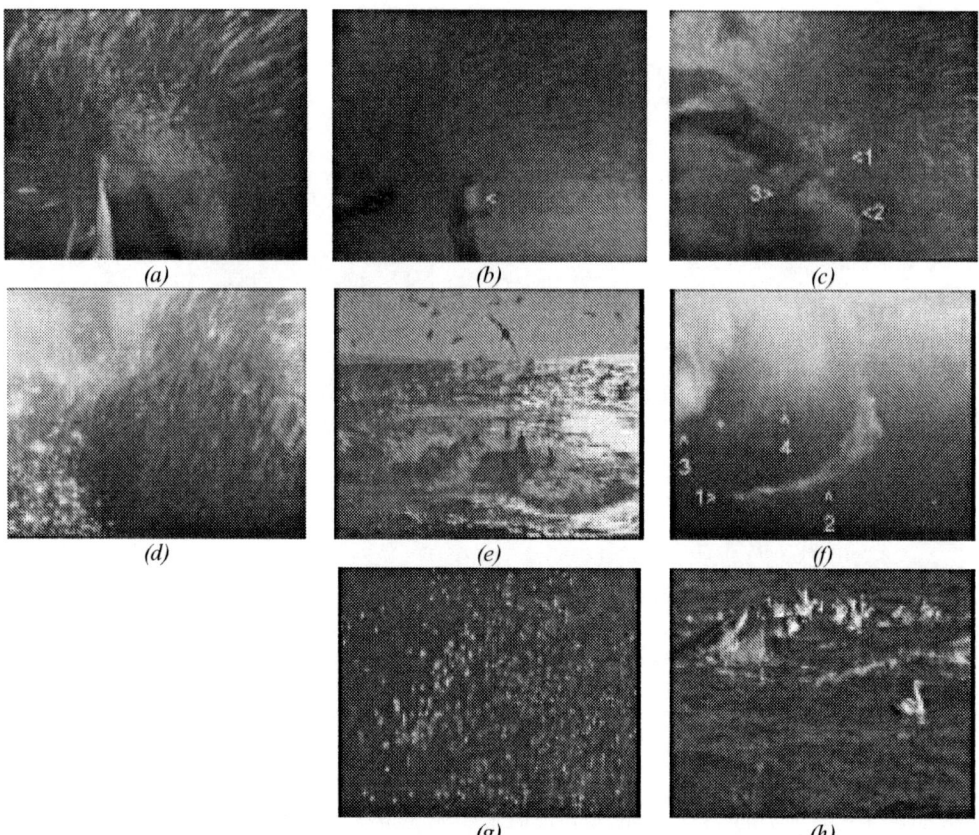

FIGURE 3. (*a*) Common dolphins herd sardines with bubble nets. (*b*) A dolphin starts to release a cloud of bubbles (arrowed) from its blowhole. A moment later (*c*) the dolphin (1) swims on, leaving behind the expanding cloud (2). Other dolphins (including the individual labeled '3') enter the frame. (*d*) The sardines school within a wall of bubbles that they are reluctant to cross, whilst (*e*) gannets dive into the sardine shoal to feed, folding their wings just before entry (arrowed). (*f*) On diving, a gannet (1) entrains a bubble plume (2). Plumes a few seconds old (3, with an older 4) have spread. (*g*) An aerial view shows hundreds of tight bubble plumes beneath airborne gannets. (*h*) A Bryde's Whale joins the feed. It surfaces with open mouth, which it then closes, sardines spilling from it. Images copyright of the The Blue Planet (BBC) and reproduced with permission. The accompanying book to the series is Byatt et *al.* [18].

With humpbacks the probability appears to be low. Echolocation is normally associated with *odontoceti*, and although there are suggestions that humpbacks may exploit it [13,14], there is to date no evidence that they have used it to locate schools of prey. Although there is evidence of directionality in the songs of humpbacks [15,16], Fig. 2*b* should not be interpreted as implying they can generate a 10° beam – we do not know one way or the other[i]. Similarly, the highest reported frequencies

---

[i] Even if the whales do not create beams that are sufficiently directional, and do not insonify tangentially, the bubble net might still function through its acoustical effects. The 'wall of sound' effect in Figure 2*b* is generated from those rays which impact the wall at low grazing angles. Those rays which never impact the wall do not contribute to the 'wall of sound'. If rays of higher grazing angle impact the net, they may cross into the net interior, though their amplitudes would be reduced by the bubble scattering, and attenuation alone would generate a quieter region in the center of the net.

generated by humpbacks correspond to harmonics in recordings in excess of 15 kHz [17] and 24 kHz [16], close to the bandwidth of the recording equipment. Exploitation of the inertia-controlled regime, as described above, would probably require higher frequencies. However, dolphins have also been observed to feed using bubble clouds [18] (Fig. 3a-d), and some can generate up to 170 kHz. It would be perhaps asking a lot for dolphins to identify fish among the strong bubble scatterers, although the environments which they naturally might encounter are similarly complex [19]. Either dolphins are accepting the loss of their echolocating abilities when they generate bubble nets to catch prey, or they have developed techniques for echolocating in bubbly water. Possible nonlinear ways to do this, which would suit the high amplitude signals and short ranges they would be working with, are discussed in relation to Fig. 5. The prospect of trapping low frequency sound in a bubble cloud to herd prey, while simultaneously echolocating with higher frequencies, is attractive but perhaps unlikely.

It may, however, be that exploitation of the schooling of fish in response to startling using bubble acoustics is more widespread, if perhaps less elegant, than the scheme of Figure 2b. The filming associated with Byatt et al. [18] shows bubble plumes generated by gannets (Fig. 3e-g) diving into a shoal of sardines which dolphins have herded to the sea surface. These plumes will no doubt complicate an underwater sound field already populated by the calls and bubble nets of dolphins, and the entrainment noise of the gannet bubble plumes, and could further stimulate the sardines to school [7]. Gannets, dolphins, sharks and whales etc. (Fig. 3h) all benefit from this, although to what extent this is intentional is unknown.

ATTENUATION AND SOUND SPEED IN NONLINEAR THEORY

Key to interpretation of (8) is the understanding that it is a generic framework. Depending on how $dV_j(t)/dP_i(t)$ is calculated for the bubbles within the population, it can be made applicable to linear or nonlinear bubble oscillation [2]; to bubbles in free field or reverberant conditions [20]; or to bubbles constrained by structures [21, 22], or surrounded by media other than pure water (such as tissue, sediment, or interacting bubbles) [23-26]. It is also through $dV_j(t)/dP_i(t)$ that a rigorous time-dependent attenuation can be calculated for bubbles undergoing nonlinear propagation [2].

Acoustic attenuation through bubble clouds has previously been predicted using the concept of acoustic cross-section. Predominantly this has been calculated for bubbles undergoing linear steady state pulsations [27]. However, in 1986 Akulichev et al. [28] produced a version which described the ring-up to steady state of the cross-section for semi-infinite monochromatic forcing, although this was strictly limited to resonant bubbles only. Two later studies [29, 30] attempted to extend Akulichev's formulation to off-resonant bubbles, but the method and results are unphysical. The cross-sections were constrained to ring up from an initial value (for which the quantitative basis was not strong) to the steady-state value, with the $\left(1-e^{-\beta_{tot}t}\right)H(t)$ time-dependency that Akulichev et al. had found for resonant bubbles. However, the pulsations of off-resonant bubbles do not display such a time-dependency [7]. A physics-based time-dependent cross-section (for both single bubbles and clouds) was formulated by Clarke and Leighton [31], which properly accounted for the bubble time-dependency and initial conditions. However, while the model of bubble dynamics was nonlinear, the damping upon which the energy loss was calculated was based upon the losses associated with linear monochromatic bubble pulsations. Therefore, a fully nonlinear time-dependent cross-section, taking account of damping correctly, was later developed [2].

Much of this interest, in the time-dependency of scatter or attenuation from bubble clouds, historically arose to address a specific problem. This was, whether or not it could be exploited to extract the signal returned from a solid body (such as a mine) from that scattered by a bubble cloud whose own echoes are hiding the solid body. The ability to include nonlinearity in addition to time dependency, as this paper provides, greatly enhances this possibility. An example will be given in Fig. 5, after a demonstration of how nonlinearity can be included into the calculation of sound speed and attenuation.

As stated above, the method derives its description of the bubble environment (free field, in sediment etc.) from the model used to calculate $dV_j(t)/dP_l(t)$. For the calculations of this paper, the Keller-Miksis model was used [2], with thermal damping calculated after the manner of Prosperetti *et al.* [32, 33] and Nigmatulin *et al.* [34]. The results are explained in Figure 4, with comparison to the result if the linear steady-state formulation of Commander and Prosperetti [1] is used to calculate $dV_j(t)/dP_l(t)$.

Recall the earlier discussion of the plot of gas volume against applied pressure, applied to a single bubble subjected to a semi-infinite driving pulse. The locus consists of a single point until the onset of insonification. From this moment on, the locus describes orbits until reaching steady-state, after which it repeatedly maps out a given orbit. Assume the gas is perfect. Its internal energy U is a state function, such that whenever an orbit crosses its previous path, at both moments represented by the intersection the value of U is the same. More specifically, consider that:

$$dU = \bar{d}Q + \bar{d}W = \bar{d}Q - PdV \tag{9}$$

where the notation indicates that both the incremental heat supplied to the bubble ($\bar{d}Q$) and the work done on the bubble ($\bar{d}W$) are not exact differentials, whilst dU is.

Because Fig. 4 uses the applied acoustic pressure $P(t)$, the area mapped out by any loop represents the energy subtracted from the acoustic wave by the bubble in the time interval corresponding to the perimeter of the loop. This is because the bubble dynamics (such as the Keller-Miksis with thermal losses used here) may be interpreted simply as a statement of the equality between that pressure difference (Δp) which is uniform across the entire bubble wall, and a summation of other terms. These terms relate to the pressure within the gas and vapour inside the bubble (p_i), surface tension pressures (p_σ), and the dynamic terms resulting from the motion of the liquid required when the bubble wall is displaced [12], which will here be termed p_{dyn}:

$$\Delta p = p_i - p_{dyn} - p_\sigma \tag{10}$$

The energy subtracted from the sound field by the pulsating bubble in each circuit of a loop is given by:

$$E_{loop} = -\oint p_i dV + \oint p_{dyn} dV + \oint p_\sigma dV \tag{11}$$

(noting that the details of the chemistry on the bubble wall may make the final integral non-zero). However Δp equals the spatial average over the bubble wall of the blocked pressure $\langle p_{blocked} \rangle$, which in the long-wavelength limit equals the applied acoustic pressure $P(t)$ that would be present at the bubble centre were the bubble not present. Substituting (10) into (11) therefore shows that the area mapped in a loop in the pressure-volume plane is the energy subtracted from the acoustic wave in the time interval corresponding to that loop:

$$E_{loop} = -\oint \Delta p dV = -\oint \langle p_{blocked} \rangle dV \approx -\oint P dV \quad (kR \ll 1) \tag{12}$$

Therefore, the rate at which the acoustic field does work on the bubble can be found by integrating the area in the pressure-volume plane enclosed by the loops formed by the intersections described above, and dividing energy so obtained by the time interval taken to map out that loop. In this way the rate at which the bubble subtracts energy from the driving acoustic field can be calculated as a function of time, for example during bubble ring-up; and whilst steady state is strictly only achieved as $t \to \infty$, loops approximating to it can readily be identified (Fig. 4, middle row).

Of particular interest is the bottom row of Fig. 4, which superimposes the steady-state nonlinear loops of the middle row (thin line) with the corresponding linear solution using the formulation of Commander and Prosperetti (which is of course steady-state [1]; thick line). At frequencies much greater than or less than resonance (not shown), both models predict loci indistinguishable from straight lines (dissipation and nonlinearities being negligible at such extremes). The gradients of these lines have opposite signs, in keeping with the π phase change which takes place between the stiffness- and inertia-controlled regimes; and that sound speed is reduced in the first and increased in the second (through the sign of the gradient of dV_i/dP_l, after (7)). In such cases a sound speed can be readily calculated from (7) or (8). Closer to resonance, increasing dissipation imparts finite areas to the loops, and the sound speed must be inferred from the 'spine' of the loop.

While in some cases the nonlinear model would impart a similar spine to its loop as would that of Commander and Prosperetti (Fig. 4a, bottom row), closer to resonance identification of the optimum spine becomes more difficult (Fig. 4b, bottom row; note that the conditions for resonance in the nonlinear and linear models are slightly different). The increasing dissipation and ill-defined nature of the spine may lead to inaccuracies, and indeed Commander and Prosperetti note that 'In the presence of resonance effects, the accuracy of the model is severely impaired'. This may not simply be due to the rapidity with which sound speed changes around a resonance, but also because errors associated with the free field assumption are greatest there [20].

In Fig. 4c, the nonlinear model displays a second harmonic (which is of course not apparent in the linear result). The 'spine' of this double loop would be curved, and its identification would allow calculation of nonlinear propagation through bubble clouds, waveform distortion, parametric signal generation etc.

With $dV_j(t)/dP_i(t)$ for the individual bubbles (and the cloud) varying through the oscillatory cycle, so too will the speed at which different regions of the acoustic wave will propagate. Hence, by combining nonlinear time-dependent data such as is presented in Fig. 4 for a single bubble, to generate the response of a given volume element of the bubble cloud, the sound speed for nonlinear propagation through the bubble cloud can be found in the usual manner [35].

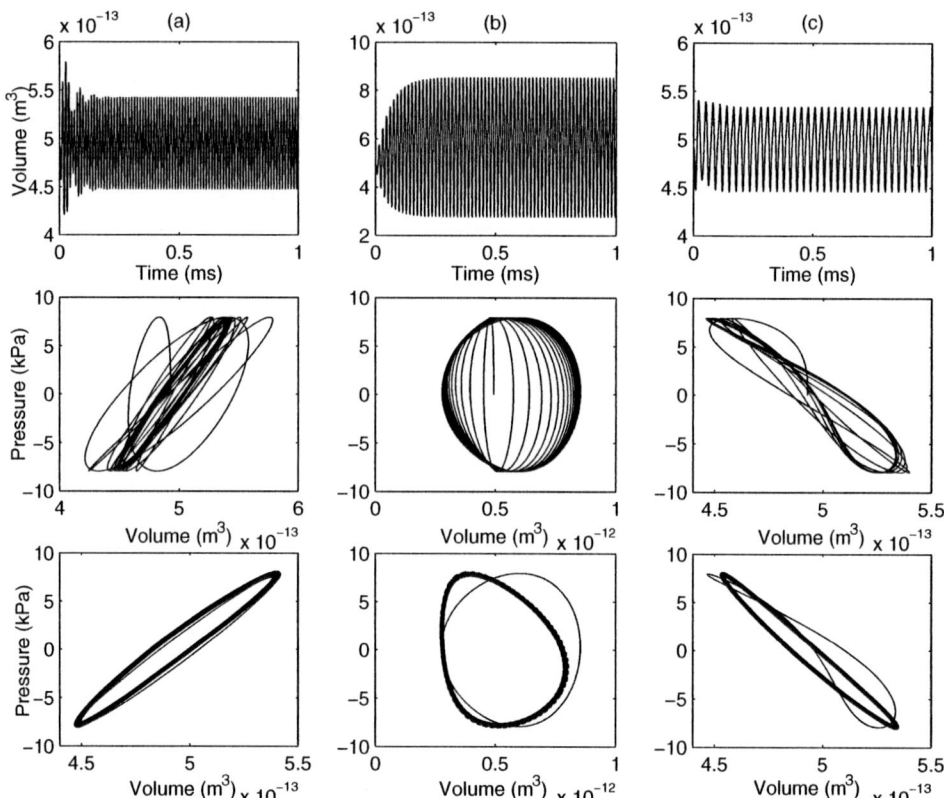

FIGURE 4. Bubble responses for a 49 μm bubble insonified by a semi-infinite pulse starting at $t=0$ with an amplitude of 7.95 kPa at (a) 84.2 kHz (b) 65.7 kHz and (c) 31.5 kHz. The top graph in each case shows the volume time history calculated using the Keller-Miksis equation (with damping after [32]). The middle graph in each case shows the corresponding pressure-volume curve. The darker area in each PV curve shows the steady state regime, where the successive loci overlap each other. Nonlinear components will cause crossovers in a loop (as in Fig. 4c where a second harmonic arises from driving the bubble close to half resonance frequency), such that the integration of (12) causes the areas of the clockwise loops to be subtracted from those of the anticlockwise. The bottom row superimposes the steady-state loops of the middle row (thin line) with the corresponding linear solution using the steady-state formulation of Commander and Prosperetti [1] (thick line). Figure by T. G. Leighton, S. D. Meers, and P. R. White [2].

APPLICATIONS FOR NONLINEAR PROPAGATION THROUGH BUBBLE CLOUDS

The above methodology has been used to:

(1) Invert measured acoustic propagation characteristics in the surf zone to determine the bubble size distribution [2, 36], and compare the results with inversions undertaken using the linear technique of Commander and McDonald [37], which exploits the linear propagation of Commander and Prosperetti [1];
(2) Predict the amplitude dependency of attenuation in oceanic bubble clouds [2];
(3) Compare the errors which might accrue through neglect of the nonlinearity of bubble pulsations in high amplitude fields, with those which occur through neglect of bubble-bubble interaction [2];
(4) Model the nonlinear response of biomedical contrast agents [38].

As stated at the outset, the ability to incorporate nonlinear bubble dynamics into models of acoustic propagation is not restricted to their use in systems where the void fraction is so high as to make high amplitude insonification an unavoidable necessity. With any bubble cloud, nonlinear pulsations can be generated and the results exploited as an additional diagnostic tool. Consider for example the problem scenario described earlier: Sonar fails to detect a linearly-scattering target (e.g. a solid body such as a mine, or a swim bladder insonified at frequencies much greater than its resonance), because the returned sonar signal is dominated by the scatter from bubble clouds in the vicinity of the target. If the scatter from the bubbles were linear, all that could be done to suppress their overwhelming contribution from the sonar return would be to try to exploit the time-dependence of the signal, as discussed earlier. However if the insonifying field were sufficiently high-amplitude to generate nonlinear response, it might be possible to enhance scatter from the target whilst simultaneously suppressing it from the bubbles. Consider if the insonifying field consisted of two high amplitude pulses, one having reverse polarity with respect to the other (Fig. 5, top line). Linear reflection from the target is shown in Fig. 5*b(i)*. The bubble generates nonlinear radial excursions (Fig. 5*a(i)*) and emits a corresponding pressure field (Fig. 5*a(ii)*). Normal sonar would not be able to detect the signal from the target (Fig. 5*b(i)*), as it is swamped by the return from the bubbles (Fig. 5*a(ii)*). If however the returned time histories are split in the middle and combined to make a time history half as long, enhancement and suppression occurs. If the two halves of the returned signals are added, the scattering from the bubble is enhanced (Figs. 5*a(iii)* and *a(iv)*), whilst the scatter from linear scatterers (such as the target) is suppressed (Fig. 5*b(ii)*). This can be used to enhance the scatter from biomedical contrast agents [7]. If however the two halves of each returned signal are subtracted from one another, the scattering from the bubbles is suppressed (Figs. 5*a(v)* and *a(vi)*) whilst the reflections from the linear target are enhanced (with the usual constraints imposed by increased signal-to-noise ratio) (Fig. 5*b(iii)*) [7].

CONCLUSIONS

This paper describes the method by which the sound speed and attenuation can be calculated for inhomogeneous bubble clouds subjected to pulses of arbitrary time history. The method provides a generic framework, such that the bubble cloud in question could be in a range of environments (such as in free field, in reverberation, in clouds of interacting bubbles, in sediments, in structures or *in vivo*) depending on the model used to calculate the dependence of the bubble volume on the insonifying field. Some applications are outlined.

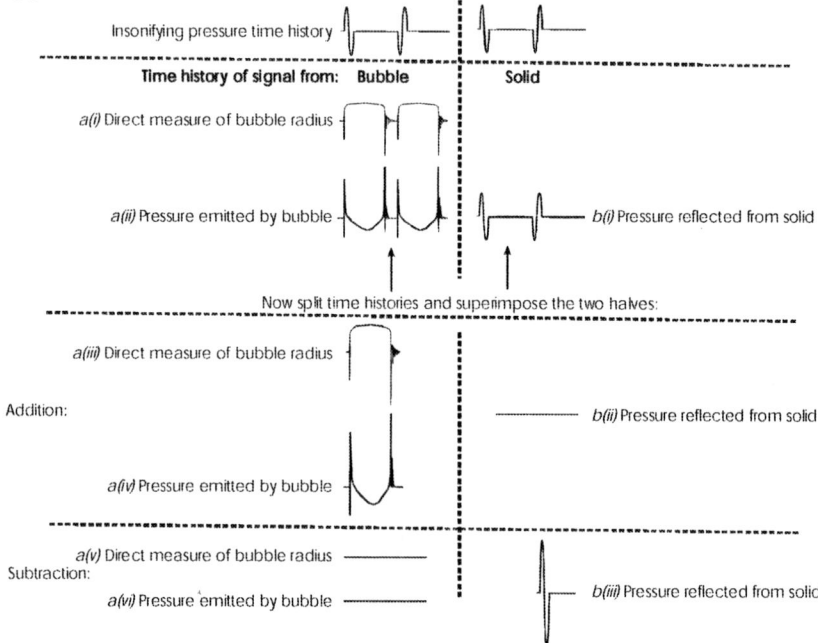

FIGURE 5. Schematic of a proposed 'Twin Inverted Pulse Sonar', whereby the scattering from a linear scatterer (such as a solid, a mine, or a swim bladder insonified at frequencies much greater than its resonance), and scattering from nonlinear scatterers (such as bubbles) can be enhanced and suppressed relative to one another (see text). The schematic bubble radius and time histories are justified in [7].

ACKNOWLEDGMENTS

The author would like to thank for the following for useful discussions: P. R. White, G. J. Heald, S. D. Richards, H. A. Dumbrell, C. L. Morfey. The author has made reasonable attempts to obtain permission to use the image in Figures 1*b*.

REFERENCES

1. Commander, K. W., and Prosperetti, A., *J. Acoust. Soc. Am.* **85**, 732-746 (1989).
2. Leighton, T. G., Meers, S. D., and White, P. R., Propagation through nonlinear time-dependent bubble clouds, and the estimation of bubble populations from measured acoustic characteristics, *Proceedings of the Royal Society* (in press; published Royal Society FirstCite website May), (2004).

3. van Wijngaarden, L., *J. Fluid Mech.* **33**, 465-474 (1968).
4. Caflisch, R. E., Miksis, M. J., Papanicolaou, G. C., and Ting, L., *J. Fluid Mech.* **153**, 259-273 (1985).
5. Farmer, D. M., and Vagle, S., *J. Acoust. Soc. Am.* **86**, 1897-1908 (1989).
6. Buckingham, M. J., *Phil. Trans. R. Soc. Lond. A* **335**, 513-555 (1991).
7. Leighton, T. G., From seas to surgeries, from babbling brooks to baby scans: The pressure fields produced by non-interacting spherical bubbles at low and medium amplitudes of pulsation, *International Journal of Modern Physics B* (in press) (2004).
8. Leighton, T. G., Richards, S. D., and White, P. R., *Acoustics Bulletin* **29**, 24-29 (2004).
9. Sharpe, F. A., and Dill, L. M., *Canadian Journal of Zoology-Revue Canadienne de Zoologie* **75**, 725-730 (1997).
10. Williams, H., *Whale Nation*, London: Jonathan Cape (now part of Random House), 1988.
11. Jensen, F. B., Kuperman, W. A., Porter, M. B. & Schmidt, H. *Computational Ocean Acoustics.* New York: Springer-Verlag, 2000.
12. Leighton, T. G., The Acoustic Bubble, London: Academic Press, 1994, pp. 58-59, 288-301, 302-308.
13. Frazer L. N., and Mercado, E., *IEEE J. Oceanic Engineering,* **25**(1), 160-182 (2000).
14. Mercado, E. and Frazer L. N., *IEEE J. Oceanic Engineering,* **26**(3), 406-415 (2001).
15. Levenson, C. "Characteristics of sound produced by humpback whales (Megaptera novaeangliae)," in *NAVOCEANO Technical Note 7700-6-72.* Washington, DC: Naval Oceanographic Office, 1972.
16. Au, W. W. L., Pack, A. A., Lammers, M. O., Herman, L., Andrews, K., and Deakos, M., *J. Acoust. Soc. Am.*, **113**, 2277 (2003).
17. Au, W., James, D., and Andrews, K., *J. Acoust. Soc. Am.*, **110**, 2770 (2001).
18. Byatt, A., Fothergill, A., Holmes, M., and Attenborough, Sir David, The Blue Planet, BBC Consumer Publishing (2001).
19. Leighton, T. G., and Heald, G. J., Chapter 21: Very high frequency coastal acoustics. In: *Acoustical Oceanography: Sound in the Sea*, edited by H. Medwin, Cambridge: Cambridge University Press, 2004 (in press).
20. Leighton, T. G., White, P. R., Morfey, C. L., Clarke, J. W. L., Heald, G. J., Dumbrell, H. A., and Holland, K. R., *J. Acoust. Soc. Am.* **112**, 1366-1376 (2002).
21. Leighton, T. G., White, P. R., and Marsden, M. A., *Acta Acustica* **3**, 517-529 (1995).
22. Leighton, T. G., Cox, B. T., and Phelps, A. D., *J. Acoust. Soc. Am.* **107**, 130-142 (2000).
23. Anderson, A. L., and Hampton, L. D., *J. Acoust. Soc. Am.*, **67**, 1865-1889 (1980).
24. Anderson, A. L., Abegg, F., Hawkins, J. A., Duncan, M. E., and Lyons, A. P., *Continental Shelf Research*, **18**, 1807-1838 (1998).
25. Kargl, S. G., Williams K. L., and Lim, R., *J. Acoust. Soc. Am.* **103**, 265-274 (1998).
26. Kargl, S.G., *J. Acoust. Soc. Am.* **111**, 168-173 (2002).
27. Medwin, H., and Clay, C. S., *Fundamentals of Acoustical Oceanography*, San Diego: Academic Press, 1998.
28. Akulichev, V. A., Bulanov, V. A., and Klenin, S. A., *Sov. Phys. Acoust.* **32**, 177-180 (1986).
29. Suiter, H. R., *J. Acoust. Soc. Am.* **91**, 1383-1387, (1992).
30. Pace, N. G., Cowley, A., and Campbell, A. M., *J. Acoust. Soc. Am.* **102**, 1474-1479 (1997).
31. Clarke, J. W. L., and Leighton, T. G., *J. Acoust. Soc. Am.* **107**, 1922-1929 (2000).
32. Prosperetti, A., Crum, L. A., and Commander, K. W., *J. Acoust. Soc. Am.*, **83**, 502-514 (1988).
33. Prosperetti, A., and Hao, Y., *Phil. Trans. R. Soc. Lond. A* **357**, 203-223 (1999).
34. Nigmatulin, R. I., Khabeev, N. S., and Nagiev, F. B. *Int. J. Heat Mass Transfer*, **24**, 1033-1044 (1981).
35. Morse, P. M., and Ingard, K. U., *Theoretical Acoustics*, Princeton: Princeton University Press, 1986 pp. 874-882.
36. Leighton, T. G., *J. Acoust. Am.*, **110**, 2694 (2001).
37. Commander, K. W., and McDonald, R. J., *J. Acoust. Soc. Am.* **89**, 592-597 (1991).
38. Leighton, T. G., and Dumbrell, H. A., *Proceedings of the First Conference in Advanced Metrology for Ultrasound in Medicine (Journal of Physics Conference Series)* (2004, in press).

The Sea Surface Bounce Channel: Bubble-Mediated Energy Loss and Time/Angle Spreading

Peter H. Dahl

Applied Physics Laboratory, University of Washington
1013 NE 40th St., Seattle, WA 98105-6698, USA

Abstract. A model is presented for the energy loss in the sea surface forward bounce channel due to attenuation from wind-speed-dependent bubbles; the model is compared to data from ASIAEX and other archival data sets. At high wind speeds the model predicts an energy loss bound, i.e., no further attenuation with increasing wind speed. Prior to reaching this bound and while there is attenuation, time and angle spreading in the forward bounce path remain largely controlled by the spectral properties of the air-sea interface, i.e., they remain unchanged by the bubbles. Once bounding of energy loss occurs, initiated by the dominance of bubble scattering over air-sea interface scattering, time and angle spreading of the arrival change profoundly.

INTRODUCTION

The process of sound energy arriving via the sea surface forward bounce path, or channel, is loosely classified using the parameter $\chi = 2kH \sin\theta_g$, where k is acoustic wavenumber, H is rms waveheight, and θ_g is the nominal grazing angle corresponding to specular reflection. Reflection is either important or dominant when χ is less than about 1.5, and scattering dominates when χ is larger than 1.5. For natural sea surfaces and typical conditions, a transition from a coherent reflection to an incoherent scattering process occurs for frequencies between about 1 and 10 kHz.

Thus, forward scattering is operative for frequencies of order 10 kHz and above but also at lower frequencies given sufficiently large kH; here, the coherent intensity loss is typically very large and intensity is for the most part incoherent. In this large-χ regime an overall reduction in received incoherent intensity can also happen owing to angular spreading (and time spreading) beyond that which can be measured by the receive aperture (or processing time window), or from use of highly directional sources. Yet in some propagation modeling schemes, losses associated with coherent intensity reduction, or with time and angle spreading of incoherent intensity, have the potential of being mistaken for real energy losses. In contrast, for beam widths and receive time windows sufficiently large to capture this time and angular spreading, a zero-decibel energy loss for sound arriving via the sea surface bounce path can be readily observed in field data (in a transmission ensemble-averaged sense). For example, this has been shown in measurements taken in the O(10)-kHz frequency range [1] and measurements taken between 400 Hz and 1500 Hz [2], but under waveheight conditions such that χ spanned the range O(1-10).

For air-sea conditions that generate sufficiently high concentrations of near-surface bubbles (often requiring wind speeds in excess of 5 m/s) a true energy loss has been observed in surface duct [3] and shallow water [4] propagation measurements at frequencies in the O(1-10)-kHz range. This loss is the result of attenuation from near-surface bubbles. In this paper a model for energy loss in forward scattering from the sea surface due to such attenuation is introduced and compared to recent field measurements from the East China Sea and other archival data. The model and field data reveal that bubbles impact forward scattering from the sea surface in three phases. The first occurs under mild conditions (wind speed less than 5 to 7 m/s); here the pulse forward scattered from the sea surface is extended in time, but only at levels some 30 dB below the peak level, which itself is not attenuated. The second occurs under more vigorous conditions (wind speed 7 to 12 m/s); here a significant energy loss is observed, but time and angle spreading (dominated by rough surface scattering) remain relatively unchanged. The third occurs under still more vigorous conditions (wind speed greater than 12 to 15 m/s). Here, there is near total occlusion of the sea surface, time and angle spreading are manifestly altered, and bubble-mediated energy loss becomes bounded by scattering from bubbles.

ENERGY CONSERVATION IN FORWARD SCATTERING AND MODEL FOR ENERGY LOSS DUE TO BUBBLES

Figure 1 shows the basic geometry for forward scattering from the sea surface for an acoustic source at position P_1 and receiver at position P_2. Scattering is described by the distribution of the bistatic cross-section over the sea surface as a function of position s, with position s_{SP} corresponding to the specular point. These positions are taken to be on the plane associated with the mean sea surface height. The bistatic cross section associated with sea surface roughness $\sigma_r(s)$ is a function of frequency, geometry (source depth, receiver depth, and range), and environment (sea-surface roughness correlation function and χ). We construct $\sigma_r(s)$ using a combination of surface wave measurements and modeling [1, 5].

The property of energy conservation is assumed to apply as:

$$\iint \frac{\sigma_r(s)B(s)dA_s}{(P_1s)^2(P_2s)^2} = \frac{1}{[(P_1s_{SP})+(P_2s_{SP})]^2}, \quad (1)$$

where $B(s)$ is the combined transmit and receive beam pattern weighting, and the integral on the left side over the area of sea surface is computed as a Riemann sum with area interval dA_s. This equality holds for transmit and receive beam patterns

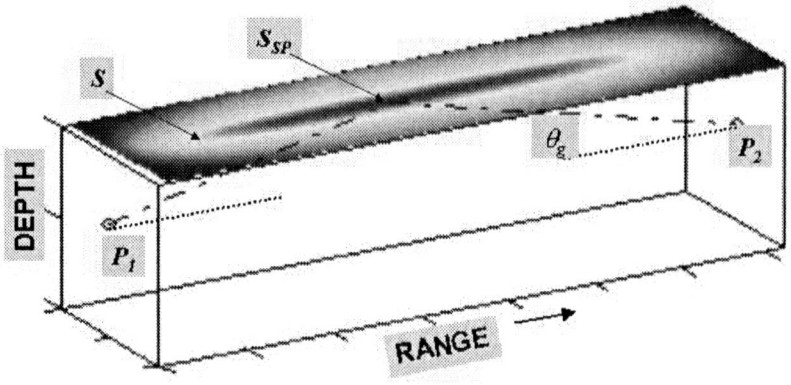

FIGURE 1. Geometry for study of forward scattering from the sea surface; source is at P_1, receiver at P_2, S_{SP} is specular point on the plane corresponding to mean sea surface and S is arbitrary point, with variable shading depicting the hypothetical sea surface bistatic cross section. Path connecting P_1, S_{SP}, and P_2 has grazing angle θ_g.

sufficiently wide to both illuminate the sea surface and receive scattered intensity from areas away from the specular point [6]. As a rough guideline [5] the necessary horizontal angular width for the case of equal source and receiver depths goes as $S_L \sin\theta_g$ and the vertical angular width goes as $S_L \cos\theta_g$, where S_L is the root-mean-square large-scale slope of the sea surface [7]. Taking $S_L \sim 0.15$ as a nominal value, the necessary one-way intensity beam width for transmit and receive is $\sim 20°$. For a more general conditions we take the left side of Eq. (1) as the energy conservation measure to be used subsequently.

Scattering and attenuation from subsurface bubbles contributes to, and modifies, the total bistatic cross section σ as: $\sigma = \sigma_r \alpha_b + \sigma_b$, where α_b (dimensionless) is an attenuation factor and σ_b is the bistatic scattering cross section per unit area sea surface due to bubbles [1]. Both α_b and σ_b depend on the dimensionless parameter β_I, equal to the depth-integrated extinction cross section per unit volume, with β_I entering into α_b as:

$$\alpha_b = \exp(-\beta_I/\sin\theta_i - \beta_I/\sin\theta_s), \qquad (2)$$

where θ_i and θ_s are incident and scattered grazing angles, respectively.

The parameter β_I succinctly describes an acoustically relevant measure of the concentration of near-surface (wind-generated) bubbles; however, an expression for β_I as function of environmental conditions must necessarily be determined empirically. One such expression derived from low-angle backscattering measurements made in the O(10–100) kHz frequency range, that are exceedingly sensitive to the concentration of near-surface bubbles, is:

$$\log_{10}\beta_I = -6.45 + 0.47 U_{10} + 0.85\log_{10}f, \qquad (3)$$

where U_{10} is 10-m height wind speed in m/s and f is frequency in kHz [8].

The above concepts lead to a model for bubble-mediated energy loss in high-frequency ($\chi \gg 1$) forward scattering from the sea surface, which is the following ratio expressed in dB:

$$\frac{\iint \frac{\sigma_r(s)\alpha_b(s)B(s)dA_s}{(P_1 s)^2 (P_2 s)^2} + \iint \frac{\sigma_b(s)B(s)dA_s}{(P_1 s)^2 (P_2 s)^2}}{\iint \frac{\sigma_r(s)B(s)dA_s}{(P_1 s)^2 (P_2 s)^2}}. \quad (4)$$

In the numerator, the left-hand (attenuation) term determines energy loss as a function wind speed (i.e., bubble concentration), and this term dominates at low to moderate wind speeds; in the absence of bubbles $\alpha_b \to 1, \sigma_b \to 0$, and the ratio goes to unity. At high wind speeds, the left-hand term vanishes and the right-hand (scattering) term becomes significant and establishes an *energy loss bound*. This bound is inherently a result of scattering from a two-dimensional surface. When the energy loss reaches the bound determined by bubble scattering, there is in effect total occlusion of the sea surface.

In regards to the attenuation term, we note that $\alpha_b(s) \approx \alpha_b(s_{SP})$, and thus Eq.(4) behaves very nearly as Eq. (2) evaluated at θ_i and θ_s, both set equal to the specular grazing angle θ_g, with implication that energy loss scales with the inverse of θ_g. A typical value for β_I is ~ 0.1 for wind speeds of 8–10 m/s and frequencies near 20 kHz; setting θ_g to 10° puts α_b = 0.32, or an energy loss of about 5 dB per interaction with the sea surface. It is important to keep in mind that because the model applies only to the $\chi \gg 1$ regime, surface decoupling (Lloyds mirror) effects [3, 9] are not operative. For the same example, when the wind speed exceeds about 12 m/s, β_I ~ O(1), and the left side of the numerator in Eq.(4) vanishes. Given sufficiently wide beam patterns (as per above), the energy loss bound is 17 dB, with narrower beams resulting in a higher bound.

FIELD MEAUREMENTS FROM ASIAEX AND EARLIER STUDIES

Measurements of forward scattering from the sea surface were made in the East China Sea as part of the ASIAEX field program [5]. Figure 2 shows the sound speed profile and corresponding ray diagrams for two sets of measurements made simultaneously at frequency 20 kHz. The wind speed during these measurements (0700–0730 UTC 31 May 2001) was 7 m/s ± 0.5 m/s and the rms sea surface waveheight was 0.3 m ± 0.1 m. Based on the sound speed profile (Fig. 2a) the grazing angle associated with the specular path (dashed lines in Fig. 2) for the upper (b) and lower (c) receivers is 6.1° and 10.9°, respectively. Figure 3 shows received multi-path arrival structure for the mean intensity (based on an ensemble average of 20 transmissions) for the 26-m (upper plot) and 52-m (lower plot) receiver depths and model curves corresponding to the mean intensity for the sea surface bounce path. (The overlap between the direct and surface bounce paths for the upper receiver is not addressed by these model curves.) Note that for modeling purposes both the source and receiver are effectively omni-directional. (There were in fact four such receivers separated by 13 cm, 30 cm, and 60 cm at each receive depth to measure vertical spatial coherence [5].)

The model curves are the result of convolving a model for the intensity impulse response [7] with the envelope of the transmit pulse (a 3-ms length boxcar function). The intensity impulse response is set by bistatic cross section σ. The solid curves are based on $\sigma = \sigma_r$ for which an estimate of the 2-D autocorrelation function of sea surface waveheight variation is required (see [5] for additional details on this function), and the dashed curves incorporate bubbles via $\sigma = \sigma_r \alpha_b + \sigma_b$, for which a wind speed of 7.4 m/s is used, putting $\beta_l = 0.01$. Clearly, incorporating bubbles via α_b, σ_b as a uniform distribution over the sea surface is a very simplified representation of the distribution of near-surface bubbles. Yet the two dashed model curves reproduce well bubble scattering phenomena observed fully 20 to 30 dB below peak scattering level and a few dB above the noise. (Model curves are made consistent with the data by adding noise, the level of which is shown in Fig. 3.) Attenuation from bubbles results in a predicted energy loss of 1.14 dB for the shallow receiver and 0.77 dB for the deep receiver; the difference is due to the different nominal grazing angles. The data are consistent with these loss estimates; however, the small difference between the two losses is difficult to verify statistically based on 20 pings.

An estimate of the time spread in forward scattering from the sea surface is made by forming the *time-delay scattering function*, which is a scaled version of the intensity impulse function. Integral measures of the time spread, defined as the characteristic time spread L [7], are noted in Fig. 3 for the cases with and without bubbles; they show that although the pulse extension due to bubbles (seen best with the upper receiver in Fig. 3) appears significant, the overall change in characteristic

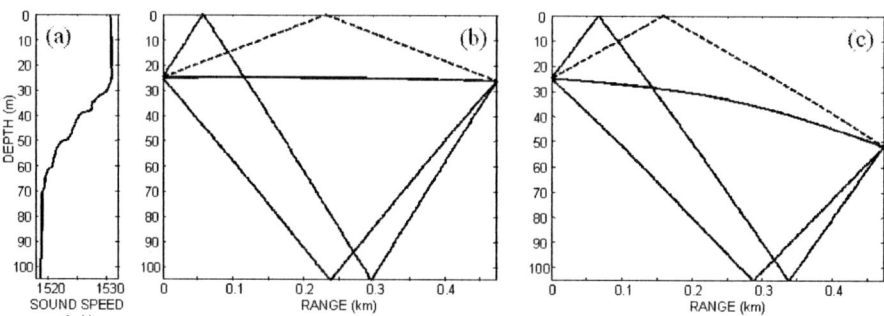

FIGURE 2. (a) Average sound-speed profile corresponding to time of acoustic measurements taken during ASIAEX, East China Sea (b) Ray diagram for 26-m depth receiver and (c) for 52-m depth receiver. Dashed lines in (b) and (c) show rays interacting once with the sea surface.

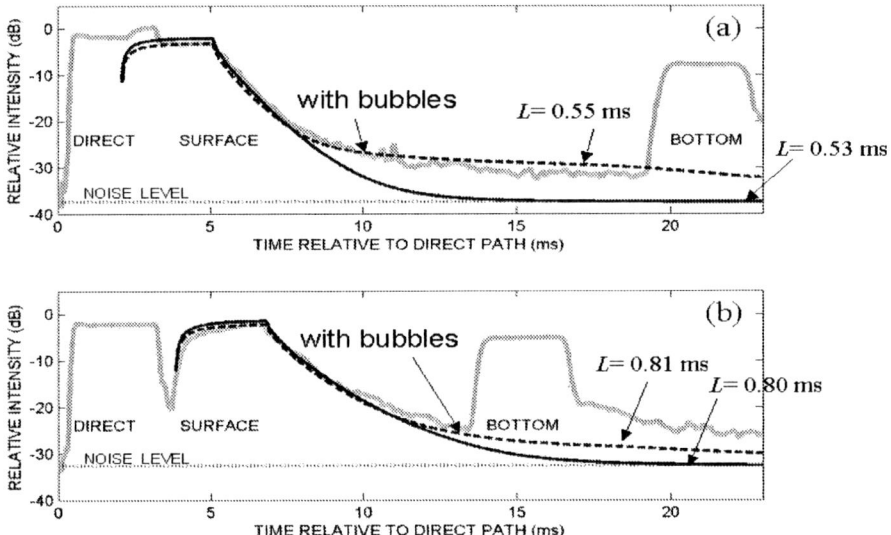

FIGURE 3. (a) Averaged received intensity (gray line) in dB plotted on relative scale (0 dB corresponds to approximately 126 dB re μPa) for the direct, surface-bounce, and bottom-bounce paths corresponding to the geometry in Fig. 2b. Solid, black line is model for mean intensity in the surface bounce path based on a 3-ms length, CW pulse with center frequency 20 kHz. Dashed, black line is same model but includes the effects of bubbles; (b) corresponds to geometry in Fig. 2c. The noise level for each geometry is shown by the dotted, black line.

time spread due to bubbles is small. There is, however, a significant difference in L associated with the different receiver depths, and a model for L [7] predicts results in Fig. 3 reasonably well, giving L = 0.67 ms and 0.83 ms for the upper and lower geometries, respectively.

A somewhat analogous, yet different, situation exists for angular spreading in the sea surface bounce path. In ASIAEX, angular spreading was determined via measurements of vertical spatial coherence [5]. The ($e^{-1/2}$) vertical coherence length d^* at 20 kHz for the 1-m VLA at the 26-m depth is 3.4 wavelengths, whereas this value is 4.0 wavelengths for the VLA located at 52 m. Angular spreading goes as $1/kd^*$, thus vertical angular spreading for the upper receiver is slightly greater than that for the lower receiver. This result is opposite that for time spreading, but consistent with the models for the geometric dependence for both time and angle spreading given in [7]. The analogy is that bubbles also have little influence on angular spreading. Spatial coherence estimates can be modeled well using an approach involving the bistatic cross section and the van Cittert-Zernike theorem [5]. Model results with and without bubbles show no difference, consistent with time and angle spreading in the sea surface bounce path being largely set by properties of rough surface scattering. Measurements made at 30 kHz displaying more substantial loss (3 dB) also suggest that characteristic time and angle spreading in forward scattering from the sea surface are altered little due to scattering from bubbles [7]. We show subsequently that this conclusion changes when bubble concentration is sufficiently high such that total occlusion of the sea surface is in effect.

Figure 4 shows estimates of energy loss due to attenuation from near-surface bubbles (i.e., in excess of spreading and sea-water absorption) for the entire ASIAEX measurement set taken at 20 kHz and similar archived data. The ASIAEX measurements, taken over two continuous 24-h periods (separated by 6 days), represent the largest data set of this kind. There is considerable geographic variety represented in Fig. 4: ASIAEX measurements were taken in the western Pacific littoral; FLIP measurements [7] were taken in the Pacific pelagic zone; Quinault measurements [10] were taken in eastern Pacific littoral; and Whidbey measurements [11] were taken in inland waters of Puget Sound although with an extended fetch to the west. Each measurement represents a careful accounting of losses due to spreading and sea water absorption, for a single interaction with the sea surface. The error bars (not available for the data from [11]) take into account both calibration uncertainty and statistical fluctuations (giving a negative loss in some instances) that depend on the number of transmissions; e.g., uncertainty in the ASIAEX estimates is due largely to the 20 transmissions that enter into the average.

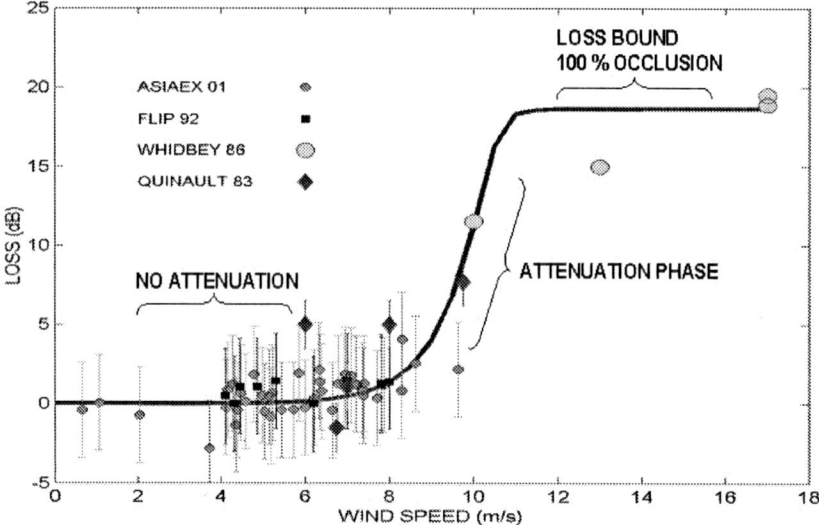

FIGURE 4. Estimates of energy loss in the single surface bounce channel due to attenuation from near-surface bubbles as a function of wind speed. Frequency is 20 kHz and nominal grazing angle is 9°. Results from four experiments (year of experiment identified in legend) are shown. Whidbey measurements were taken between 15 kHz and 25 kHz; see text for further explanation on the grazing-angle-scaling of some of these data points.

The solid curve is the bubble energy loss model based on Eqs. (3) and (4) computed at 20 kHz, with source, receiver, and range geometry such that nominal grazing angle θ_g equals 9°, corresponding to the majority of the ASIAEX data; further $B(s) = 1$ because all measurements in Fig. 4 were made with omni-directional sources and receivers. Note, however, that some measurements from the other experiments were taken at grazing angles different from 9°, e.g., the FLIP measurements represent the range $\theta_g = 4.5\text{--}14.5°$. Therefore, these data have been scaled by the factor $\sin\theta_g / \sin 9°$ where θ_g is the particular grazing angle of the measurements.

The data and model suggest three phases of impact of bubbles on forward scattering from the sea surface: no attenuation, increasing attenuation with increasing wind speed, and an attenuation bound phase (occlusion) at very high wind speeds. Significantly, the data in Fig. 4 also demonstrate that it is very difficult to observe a loss-versus-wind-speed signature in the field measurements of forward scattering until the wind speed exceeds about 7 m/s. This contrasts with low-angle backscattering, which is exceedingly sensitive to wind speed [8], but is also consistent with the model given here, that puts the loss at only 0.45 dB at 7 m/s wind speed. With addition of the larger set of ASIAEX measurements, a transition to the attenuation phase (wind speeds between 6 m/s and 8 m/s) now appears to be displayed by the combined data set. Although the combined data set in Fig. 4 is reasonably consistent with the model, there are, unfortunately, fewer measurements made in the attenuation phase that are also based on a single interaction with the sea surface, such as the measurements in Fig. 4. The measurements of Wille and Geyer [4] show convincingly, however, how an excess total transmission loss in shallow water involving both sea surface and seabed interaction, increases and becomes strongly dependent on wind speed, when in their case the wind speed exceeds about 10 m/s, and thus are somewhat consistent with Fig. 4. (In this case, one component of excess transmission loss is due to scattering into higher grazing angles with energy subsequently lost to the seabed.)

The field measurements reported in McConnell [11] represent intriguing observations apparently made under conditions of total occlusion, for which an energy loss bound is observed. These measurements (plotted on the far right side of Fig. 4) were also interleaved with measurements of vertical and horizontal spatial coherence, the results of which were first given in a 1990 report [12] and re-visited here. Figure 5 shows the estimates of horizontal and vertical coherence compared with model-bands for spatial coherence, computed using the method from [5]. The three versions of model-bands are based on rough-surface scattering equivalent to a wind speed of 17 m/s and fetch of 40 km, plus scattering and attenuation from bubbles for three cases: no bubbles, bubble concentration from Eq. (3) for wind speed of 10 m/s, and for wind speed of 17 m/s representing total occlusion. The model-bands incorporate uncertainties (the bands) in the six receiving beams involved in the measurements, three distributed horizontally and three vertically, and appear nominally consistent with the coherence estimates. (Here the apparent insensitivity of the models for vertical coherence to bubble concentration, is due to the individual beam patterns that compose the vertical receiving array.) Most significant, however, is that horizontal coherence must always exceed vertical coherence for sea surface forward scattering with this acquisition geometry, yet it can be seen in Fig. 5 that horizontal coherence has been knocked down to levels predicted by bistatic scattering from near-surface bubbles and subsequent total occlusion of the sea surface.

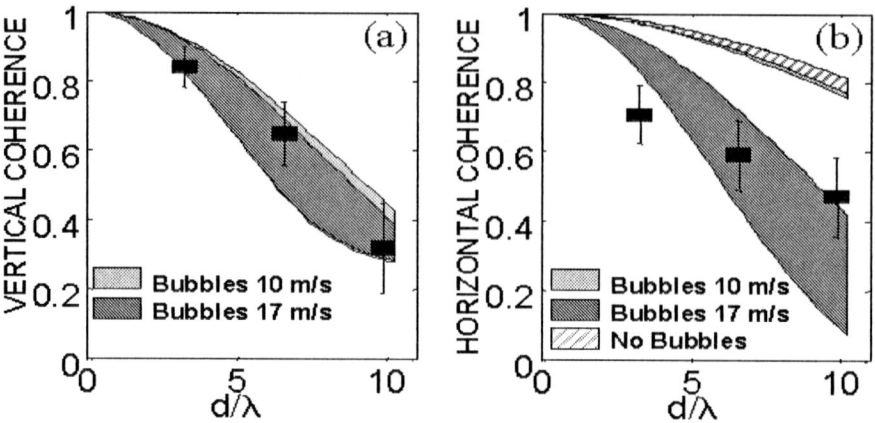

FIGURE 5. Estimates of the magnitude of vertical (a) and horizontal (b) spatial coherence plotted as a function of receiver separation normalized by 15-kHz wavelength. Model-bands are derived using rough-surface bistatic cross section at wind speed 17 m/s and three cases for bubbles: no bubbles, bubble concentration at 10 m/s, and 17 m/s. For vertical coherence the case of no bubbles and bubble concentration at a wind speed of 10 m/s are indistinguishable.

SUMMARY

A model for energy loss in the sea surface bounce path due to attenuation from near-surface bubbles has been presented; it applies to the nominal frequency range $O(10-100)$ kHz and assumes the parameter χ is $\gg 1$. The model compares reasonably well with measurements from the recent ASIAEX experiment and archival data sets. Three phases of impact of bubbles on forward scattering from the sea surface are illustrated: the first is no discernable attenuation, which occurs under mild conditions (wind speed < 5–7 m/s), wherein bubbles extend the pulse forward scattered from the sea surface, but only at levels 30 dB below the peak level, which itself is not attenuated. The second occurs under more vigorous conditions (wind speed 7–12 m/s); here a real energy loss is observed, but time and angle spreading (dominated by rough surface scattering) remain relatively unchanged. The third occurs under still more vigorous conditions (wind speed > 12–15 m/s); here there is near total occlusion of the sea surface, time and angle spreading are manifestly altered, and bubble-mediated energy loss becomes bounded by scattering from bubbles. Although two major effects of total occlusion, the reduction in horizontal spatial coherence and the bounding of attenuation, were demonstrated with field data, additional field measurements of this phenomenon are needed to verify the model presented here.

ACKNOWLEDGMENTS

This work was funded by the Office of Naval Research, Ocean Acoustics Program.

REFERENCES

1. Dahl, P.H., "On Bistatic Sea Surface Scattering: Field Measurements and Modeling," *J. Acoust. Soc. Am.*, **105**, 2155-2169 (1999).
2. Nichols, R.H., and Senko, A., "Amplitude Fluctuations of Low-Frequency Underwater Acoustic Pulses Reflected from the Ocean Surface," *J. Acoust. Soc. Am.*, **55**, 550-554 (1974).
3. Weston, D. E., "On Losses due to Storm Bubbles in Ocean Sound Transmission," *J. Acoust. Soc. Am.*, **86**, 1546-1552 (1989).
4. Wille, P.C. and D. Geyer, "Simultaneous Measurements of Surface Generated Noise and Attenuation at the Fixed Shallow Water Range *NORDSEE*," in: *Proceedings Advanced Research Workshop on Natural Mechanisms of Surface Generated Noise in the Ocean*, June 1987, Lerici, Italy, pp. 295-308.
5. Dahl, P.H. "The van Cittert-Zernike Theorem and Forward Scattering from the Sea Surface," *J. Acoust. Soc. Am.*, **115**, 589-599 (2004).
6. McDonald, J. F. and Spindel, R.C., "Implications of Fresnel Corrections in a non-Gaussian Surface Scatter Channel," *J. Acoust. Soc. Am.*, **50**, 746-757 (1971).
7. Dahl, P.H. "High-frequency Forward Scattering from the Sea Surface: The Characteristic Scales of Time and Angle Spreading," *IEEE J. Ocean. Eng.*, **26**, 141-151 (2001).
8. Dahl, P.H. "The Contribution of Bubbles to High-Frequency Sea Surface Backscatter: A 24-h Time Series of Field Measurements," *J. Acoust. Soc. Am.*, **113**, 769-780 (2003).
9. Tappert, F. D., "Inhomogeneous Absorption and Geometric Acoustics", *J. Acoust. Soc. Am.*, **103**, 1282-1287 (1998).
10. Thorsos, E. I., "High Frequency Surface Forward Scattering Measurements," presented at the 108th meeting of the Acoustical Society of America, October 1984.
11. McConnell, S. O. "Acoustic Measurements of Bubble Densities at 15-50 kHz," in: *Proceedings Advanced Research Workshop on Natural Mechanisms of Surface Generated Noise in the Ocean*, June 1987, Lerici, Italy, pp. 237-253.
12. Dahl, P.H. and McConnell, S. O. "Measurements of Acoustic Spatial Coherence in a Near-Shore Environment," APL-UW TR 9016, August 1990.

On the Relationship between Signal Bandwidth and Frequency Correlation for Surface Forward Scattered Signals

Lee Culver and David Bradley

Applied Research Laboratory and Graduate Program in Acoustics
The Pennsylvania State University, P.O. Box 30, State College, PA 16804

Abstract. The relationship between the signal bandwidth and the correlation of a single surface reflected arrival with the transmitted signal has been investigated experimentally and compared with two theories. The dependence of correlation on signal bandwidth is termed *frequency correlation*. Decorrelation of surface scattered signals is a direct consequence of time spread. Thus the acoustic measurement utilized two pure tone signals, from which time spread has been estimated, and four broadband signals with different bandwidths, from which correlation with the transmitted signal has been calculated. A model developed by Dahl for the ocean surface bistatic scattering cross section was used to predict time spread, which agreed very well with the measured time spread. Next, scattering cross section prediction was employed in two theories that predict frequency correlation. The first, published by Reeves in 1974, compared well with the measurements for bandwidths up to 2 kHz, but under predicted correlation for signal bandwidth between 7 and 22 kHz. In the second, linear systems theory was used to develop a mathematical relationship between time spread and frequency correlation. Predictions made using the linear systems theory agree well with the measured values for signal bandwidths up to 22kHz. Further work is required to evaluate the linear systems theory under higher sea state conditions.

INTRODUCTION

There has been widespread effort in recent years to increase the bandwidth of sonar systems and components (e.g. transducers) in an effort to improve system performance against noise or interference. A relevant question, therefore, is "How much bandwidth will an ocean acoustic channel accommodate without introducing frequency-dependent effects that will degrade coherent processing?" This paper addresses that question for the ocean surface forward scattered acoustic path, and thus applies to signals that have been forward scattered at the ocean surface.

There are many aspects of ocean surface scattering to investigate, understand and model. Fortuin [1] and Ogilvy [2] provide good overviews of ocean surface scattering research. Early efforts were to understand how the mean forward scattered energy varied with sea state, grazing angle, and frequency. More recently, significant progress has been made toward understanding the variation in signal structure caused by changes in transmitter or receiver location, called *spatial coherence*, to understand the limits of acoustic array performance [3]. The focus of the present research is on how the structure of the forward scattered signal is affected by signal bandwidth, which will be referred to as the *frequency correlation*, in order to determine the limits of broadband sonar performance.

The terms *correlation* and *coherence* can take on different meanings, and so for clarity we now state what we mean by these terms. Consistent with Bendat and Piersol [4], the term *correlation* in this paper refers to the operation

$$\Phi(l,\beta) = \left\langle \int p_\beta(t+l)\, q_\beta^*(t)\, dt \right\rangle. \quad (1)$$

where $p_\beta(t)$ is the received signal with bandwidth β, and $q_\beta(t)$ is a replica of the transmitted signal. Here t is time, l is the time lag between $p_\beta(t)$ and $q_\beta(t)$, and the brackets $\langle\ \rangle$ indicate ensemble averaging.

The term *correlation coefficient* will refer to the normalized correlation:

$$\rho(l,\beta) = \left\langle \frac{\int p_\beta(t)\, q_\beta^*(t+l)\, dt}{\left[\int |p_\beta(t)|^2\, dt\right]^{1/2} \left[\int |q_\beta(t)|^2\, dt\right]^{1/2}} \right\rangle. \quad (2)$$

which must be between -1 and 1. The term *frequency correlation coefficient* will refer to the dependence of $\rho(l,\beta)$ signal bandwidth β.

THEORY PREDICTING FREQUENCY CORRELATION

In 1974, Jon Reeves published theory and measurements that related the decorrelation of a single ocean surface forward scattered arrival to the bandwidth of the signal [5]. That theory was based upon the *time spread*, which is the spreading in time of an acoustic signal due to scattering from bubbles and multiple facets at the ocean boundary. Drawing upon earlier work by Martin [6] and Weston [7], Reeves related frequency correlation to the number of sea surface facets, N, expected to contribute to the received signal within an interval corresponding to the temporal resolution of the signal. The total of all contributions, N_T, is proportional to the total temporal elongation (time spread) of the received signal. The number of contributions that are correlated for a particular signal is proportional to the signal time resolution or inverse bandwidth β^{-1}. The correlation estimate is taken as the ratio of the average contributions N/N_T. The theory thus predicts that increasing bandwidth (meaning increased temporal resolution and thus smaller N) results in decreased correlation. Reeves' theory was shown to agree well with correlation measurements using signal bandwidths up to 2 kHz [5].

Dahl [8-10] has developed a detailed model for the bistatic scattering cross section of the ocean surface. The model can be used to compute the intensity impulse response function $I_{imp}(\tau)$, a function which, when convolved with the magnitude squared transmit pulse, produces the ensemble-averaged intensity of a pulse that has been forward scattered from the sea surface. This quantity is known as the time spread. Dahl defines a *characteristic time spread* for the sea-surface bounce path

$$L = \frac{\left[\int_0^\infty I_{imp}(\tau)\,d\tau\right]^2}{\int_0^\infty I_{imp}^2(\tau)\,d\tau} \quad \text{sec.} \tag{3}$$

and postulates that the inverse of the characteristic time spread, L^{-1}, with units cycles/sec (Hz), is the channel *frequency coherence bandwidth*, the bandwidth over which coherent processing can be expected to increase the signal to noise ratio.

Ziomek [11] utilizes linear systems theory to define the *transfer function correlation function*

$$R_H(\Delta f, \Delta t) = E\left[H(f,t)H^*(f+\Delta f, t+\Delta t)\right], \tag{4}$$

which parameterizes how the channel decorrelates the envelopes of different frequency signals, or, more simply, how differently the channel affects signals separated in time Δt and/or frequency Δf. It is the width of $R_H(\Delta f, \Delta t)$ in frequency that determines the channel frequency correlation. If $R_H(\Delta f, \Delta t)$ is broad in Δf, broadband signals will suffer minimal decorrelation; if $R_H(\Delta f, \Delta t)$ is narrow in Δf, correlation will drop off rapidly as signal bandwidth is increased.

Linear systems theory provides a mathematical relationship between the time spread introduced by a channel, e.g. an ocean surface reflection, and the channel transfer function correlation function. The main equations are now reviewed.

Given a source and receiver, the transmitted signal $x(t)$ and the received signal $y(t)$ are related through convolution with the linear time varying channel *impulse response function* $h(\tau, t)$:

$$y(t) = \int x(t-\tau)\,h(\tau,t)\,d\tau. \tag{5}$$

Here the source directionality and receiver spatial response have been absorbed into $h(\tau,t)$. If the scattering process is wide sense stationary[1] and uncorrelated for different delays, a condition referred to as the *wide-sense stationary uncorrelated scattering (WSSUS)* assumption [11], then the mean square received signal can be expressed as

$$E\left[|y(t)|^2\right] = \int |x(t-\tau)|^2\, R_s(\tau,\phi)\,d\phi\,d\tau. \tag{6}$$

Here $R_s(\tau,\phi)$ is the *scattering function*, which parameterizes how the channel spreads acoustic energy in time and frequency. Considering the definition of the intensity impulse response function, $I_{imp}(\tau)$, given above, Equation (6) means that

$$I_{imp}(\tau) = \int R_s(\tau,\phi)\,d\phi. \tag{7}$$

Now, using the transmitted signal as the replica in Equation (1) defines the *replica correlation function* (for zero Doppler shift or time compression):

$$RC(l,\beta) = \left\langle \int H(f,t)X_\beta(f)\,X_\beta^*\,g\,\exp(j2\pi ft)\exp(-j2\pi g(t+l))\,df\,dg\,dt \right\rangle \tag{8}$$

---

[1] The mean and autocorrelation function of a wide sense stationary process do not vary with time [4].

Here $X_\beta(f)$ is the Fourier transform of the signal with bandwidth β and $H(f,t)$ is the channel transfer function. Under the WSSUS assumption, the mean square replica correlator output can be expressed as

$$E\left[|RC(l,\beta)|^2\right] = \left\langle \int R_H(\Delta f, \Delta t) |\Gamma_\beta(\Delta f, \Delta t)|^2 \exp(j2\pi \Delta f\, l)\, \Delta df\, \Delta dt \right\rangle \quad (9)$$

where Δf and Δt are frequency and time separations, respectively, $R_H(\Delta f, \Delta t)$ was defined in Equation (4), and we have introduced a *spectral ambiguity function*

$$\Gamma_\beta(\Delta f, \Delta t) \equiv \int X_\beta(f) X_\beta^*(f + \Delta f) \exp(-j2\pi f \Delta t)\, df. \quad (10)$$

Equation (9) provides a method for predicting the mean square replica correlator output for signals with different bandwidths. The spectral ambiguity function of each signal is calculated using Equation (10). From linear system theory, if the WSSUS assumption is valid, $R_H(\Delta f, \Delta t)$ can be calculated from the scattering function by Fourier transform

$$R_H(\Delta f, \Delta t) = \int R_S(\tau, \phi) \exp(j2\pi(\phi \Delta t - \Delta f \tau))\, d\tau\, d\phi. \quad (11)$$

If we assume that $R_S(\tau, \phi)$ is separable in ϕ and τ, then Equation (11) can be integrated over ϕ and, using Equation (7),

$$R_H(\Delta f, \Delta t) \approx \int I_{imp}(\tau) \exp(j2\pi(\Delta f \tau))\, d\tau. \quad (12)$$

Notice that Equation (9) amounts to multiplying the square of the spectral ambiguity function times the channel transfer function correlation function and integrating over all Δf. Assuming that the functions $|\Gamma_\beta(\Delta f, \Delta t)|^2$ are normalized to the same total energy, the replica correlator output will be approximately constant as long as $|\Gamma_\beta(\Delta f, \Delta t)|^2$ is narrower than $R_H(\Delta f, \Delta t)$. However, once the width of $|\Gamma(\Delta f, \Delta t)|^2$ exceeds that of $R_H(\Delta f, \Delta t)$, the replica correlator output will begin to fall off because the signal bandwidth is wider than the channel bandwidth. This is a key explanation for the linear systems theory based predictions.

TIME SPREAD AND FREQUENCY CORRELATION MEASUREMENTS

Concurrent measurements of time spread and frequency correlation are now described and compared with the theory presented above. An ocean acoustic measurement conducted in August 2002 had, as a primary objective, to directly measure the decorrelation, with increasing signal bandwidth, of direct path, surface reflected, and fully refracted propagation paths through the ocean. Concurrently, ocean surface wave height directional spectra, wind speed and direction, ocean current, and sound speed vs depth were measured in order to investigate the physical mechanisms associated with signal decorrelation. The experiment location was ($32°$ 38.2' N, $117°$ 57.4'W), which is about 2.5 km east of San Clemente Island and about 80 km west of San Diego, California. Water depth is approximately 500 m.

Acoustic Measurement Instrumentation

The measurement geometry is shown in Fig. 1. Signals were transmitted from International Transducer Corporation (ITC) 1001 and 6084 acoustic projectors attached to the riser of a bottom-moored surface buoy. The buoy mooring included an elastic section composed of eight 20 m (unstretched) bungee cords capable of being stretched to 60 m, which served to reduce the buoy watch circle to a few 10's of meters. Signals were received at ITC 6080C hydrophones suspended from the research vessel Acoustic Explorer, that ship being in a three-point moor. Elastic tethers were used to decouple the hydrophones from ship heave.

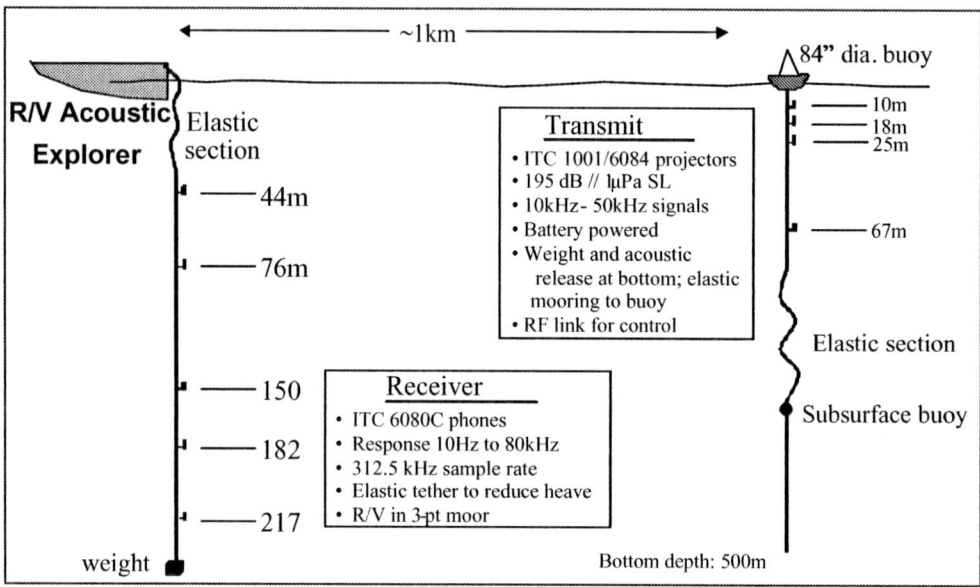

FIGURE 1. Measurement geometry, August 2002 about 2.5 km east of San Clemente Is., California.

Transmit electronics were housed in the surface buoy. A compact PCI based computer was controlled via 900 MHz and 2.4 GHz radio frequency (RF) links with the Acoustic Explorer. Twenty-six 12-volt gel cell marine batteries provided power for several days of continuous operation. Signals were clocked out at 125k samples/sec using a CPCI board designed and built by ARL/PSU. An Instruments Inc. L6 amplifier modified by the manufacturer to accept 48 VDC input power was controlled using the remote interface. Only one projector was active at a time, with projector selection accomplished using high current relays. On board the Acoustic Explorer, received signals were bandpass filtered and sampled at 312.5k samples/sec using boards designed and built at ARL/PSU.

Environmental Measurements

Figure 2 shows a sound speed profile calculated from a CTD drop made during the experiment. It shows a very shallow mixed layer and a strong downward refracting region down to about 100 m, and below that the water is relatively isothermal. A ray-trace made using the Comprehensive Acoustic System Simulation / Gaussian Ray Bundle (CASS/GRAB) acoustic propagation model [12] for the 67 m deep projector and 217 m deep receiver shows slightly refracted direct and a surface reflected paths. For this projector-hydrophone pair, the difference in travel time between the direct and surface reflected paths is about 19 msec.

The measurement site was very much in the lee of San Clemente Island, which significantly affected wind speed and direction and reduced surface wave height. The wind measured at the site of the experiment was from the north-northwest during most of the experiment, and averaged 4 – 6 m/s, corresponding to a sea state 3 on the World Meteorological Organization (WMO) chart [13]. However, a NOAA buoy west of San Clemente Island measured winds from the west, indicating that San Clemente Is. was significantly affecting the wind at the experiment site.

Directional wave height spectra were measured during the experiment using an AXYS Technologies Triaxys wave buoy. A surface waveheight wavenumber spectrum is required as the environmental input to Dahl's bistatic scattering cross section model. For forward scattering geometries, the waveheight spectrum must extend to ocean surface wave numbers of about $k/4$, where k is the acoustic wave number [10]. Using the method developed by Dahl [9], and the "D" wave height spectrum model developed by Plant [14], which uses the wind speed and fetch as inputs, the measured waveheight spectrum was extended to frequencies well above those measured by the wave rider buoy.

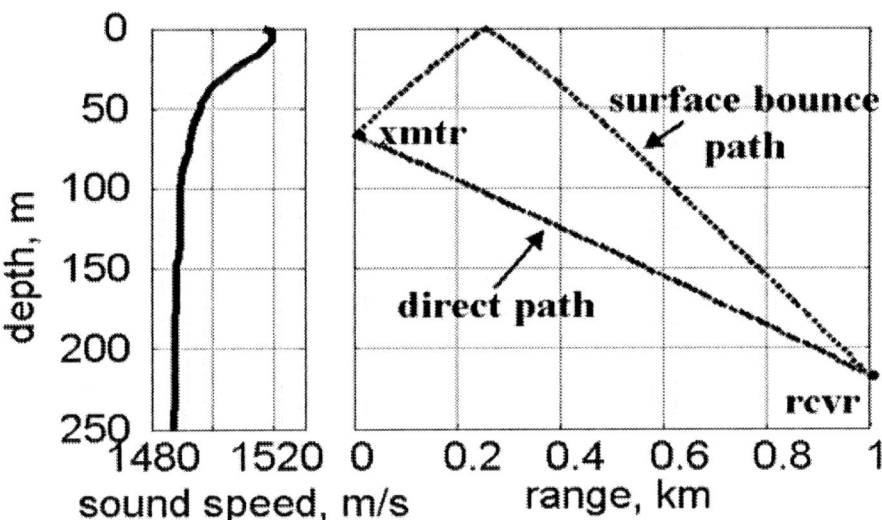

FIGURE 2. Sound speed profile and ray trace between the 67 m deep projector and 217 m deep hydrophone.

Acoustic Measurements

Moving on now to the acoustic data, the transmit signals are summarized in Table 1. Two Continuous Wave (CW) pulses and four Linear Frequency Modulated (LFM) pulses were transmitted at two different center frequencies. We take the separation between sinc function zero crossings (2/T) as the bandwidth of a CW pulse in Table 1.

TABLE 1. Signals transmitted at center frequencies of 20kHz and 40kHz

Signal type	Duration	Bandwidth
CW pulse	0.25 ms	8.0 kHz
CW pulse	1.0 ms	2.0 kHz
LFM	8.0 ms *	1.0 kHz
LFM	8.0 ms *	7.0 kHz
LFM	8.0 ms *	13.0 kHz
LFM	8.0 ms *	22.0 kHz

- Pulse length was 8.0 ms for projectors 1,2 and 4, and 10.0 ms for projector 3.

Each signal was transmitted from a single projector at a time using a 10 Hz repetition rate for 30 s. The short CW signals were designed for estimating the time spread. Figure 3 shows 300 short CW pulses transmitted from the 67 m deep projector and received at the 217 m deep hydrophone (upper: 20 kHz; lower: 40 kHz). The received signals were match filtered to enhance signal to noise ratio. The vertical band in the left half of each panel is the Direct Path (DP) arrival; the second vertical band, about 20 msec later, is the Surface Bounce (SB) arrival.

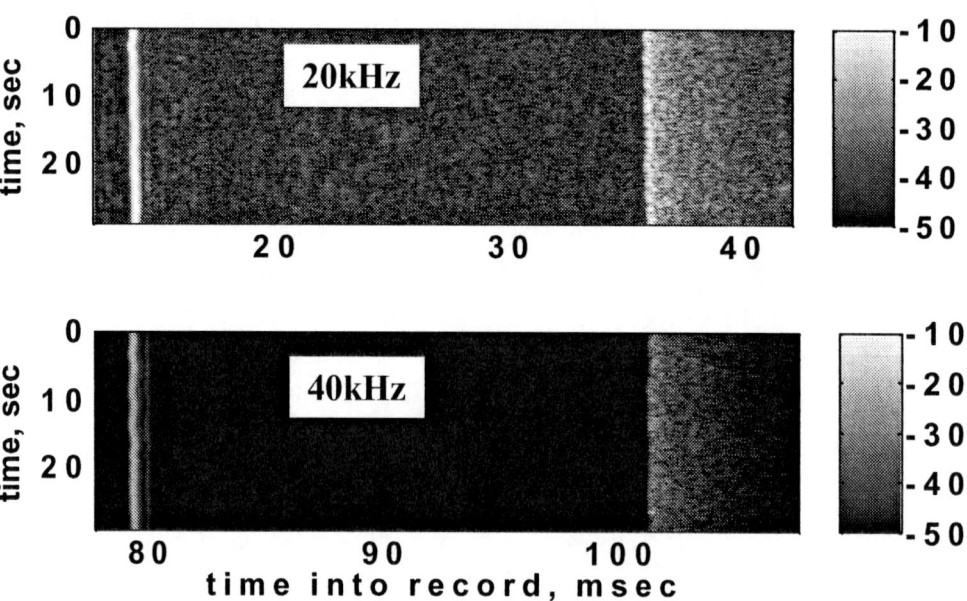

FIGURE 3. Acoustic data recorded at approximately 0200 UTC on 18 Aug 2002. Each panel contains a stack of ~300 short CW pulses transmitted from the 67 m deep projector and received at the 217 m deep hydrophone (upper panel: 20 kHz; lower panel: 40 kHz). Gray scale is level in dB.

Several features are evident. First, there is some jitter in the arrival times of both the DP and SB arrivals. This is due partly to relative movement between the projector and receiver but also to variation in the propagation path. Second, a fully refracted path may be seen just after the DP arrival. Third, the background level is about 10 dB lower in the 40 kHz band than in the 20 kHz band, and fourth, the DP arrival is sharp and distinct, but the SB arrival is following by a smattering of arrivals extending for 8 to 10 msec. These arrivals, which follow the SB arrival, are termed the *time spread*.

MODEL – MEASUREMENT COMPARISON

We now extract the time spread of the surface bounce path from the data shown in Fig. 3. After aligning the SB arrivals by their leading edges, we calculate the ensemble average. The resulting time spread measurements (normalized to 0 dB peak) are solid lines in Fig. 4.

To calculate a time spread prediction, the bistatic scattering cross section model is used to calculate the intensity impulse response function $I_{imp}(\tau)$, which is then convolved with the envelope of the transmitted signal. It was important to use the receive beam pattern in this calculation. In Fig. 4, a noise floor was added to produce the time spread prediction. The agreement is good at both frequencies.

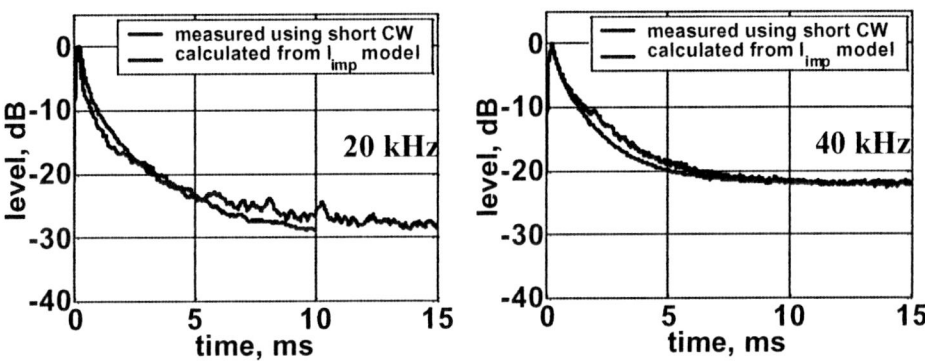

FIGURE 4. Time spread: measured (solid line) and predicted (dotted line) for the short CW pulses transmitted from the 67 m deep projector and received at the 217 m deep hydrophone (left: 20 kHz; right: 40 kHz). Curves are normalized to 0 dB peak.

We now compare measured frequency correlation with the theory. From the data, the frequency correlation coefficient was calculated using Equation (2), separately for the direct path and surface bounce LFM signals. A direct path arrival was used as the replica in both cases in order to account for frequency dependent absorption. The highest direct path and surface bounce path correlation coefficients were extracted from each ping and averaged over all pings to obtain an ensemble average.

The solid lines with square markers in Fig. 5 indicate the correlation coefficient of the DP arrival for the four LFM signals transmitted by the 67 m deep projector and received at the 217 m deep receiver. The correlation of the DP arrival remains close to 1 independent of bandwidth, indicating that propagation has little effect on the correlation over all bandwidths considered. The dashed lines in Fig. 5 indicate the correlation coefficient of the SR arrival. Correlation of the SR arrival decreases with increasing bandwidth in a manner similar to that reported Keranen [15].

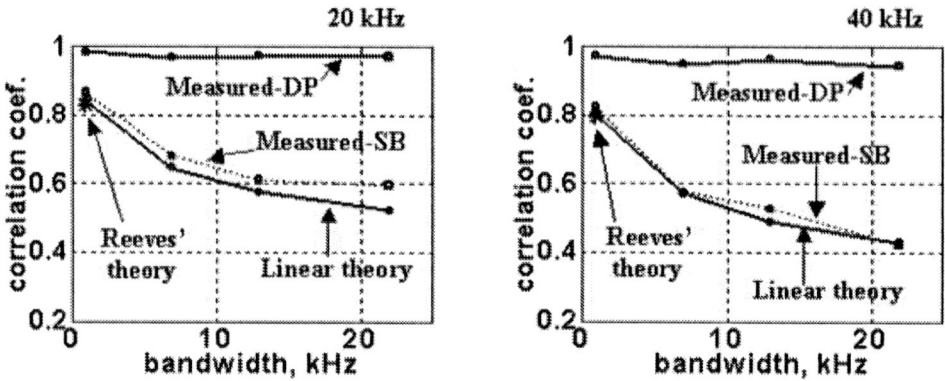

FIGURE 5. Frequency correlation for the 67 m deep projector and received at the 217 m deep hydrophone (left: 20 kHz; right: 40 kHz). Measured for direct path (DP) and surface bounce (SB) path; modeled using linear systems theory (circles); and modeled using Reeves' theory (asterisk).

Next we use the intensity impulse response function and apply the theory due to Reeves, calculating the ratio

$$RC_{REEVES}(\beta) = \frac{\int_0^T I_{imp}(\tau)d\tau}{\int_0^\infty I_{imp}(\tau)d\tau} \quad (13)$$

where the upper integration limit of the numerator is the temporal resolution of the signal. The asterisks in Fig. 5 indicate the prediction based upon Reeves' theory. That theory was developed from measurements made using signals with up to 2 kHz bandwidth, and the theory compares well with the measurements for smaller bandwidth signals.

Applying the linear systems theory, the intensity impulse response function $I_{imp}(\tau)$ is Fourier transformed to obtain $R_H(\Delta f, \Delta t)$ (Equation (11) assuming negligible frequency spread), and Equation (10) used to compute $|\Gamma_\beta(\Delta f, \Delta t)|^2$ for the four LFM signals used in the measurement. Then Equation (9) is used to predict the mean replica correlation coefficient as a function of signal bandwidth. The black lines marked by circles in Fig. 5 denote the prediction based upon linear systems theory. This prediction is in good agreement with the measurements for all bandwidths, although the agreement is better at 40 kHz than at 20 kHz.

SUMMARY AND CONCLUSIONS

We have presented ocean surface forward scatter time spread and frequency correlation measurements made in August 2002, about 2500 m east of San Clemente Island, California, under very modest sea states. Our time spread measurements were found to compare well with predictions calculated using a bistatic scattering cross section model developed by Peter Dahl [8-10]. Our frequency correlation measurements were compared with two different theories. The first is a physics-based theory published by Jon Reeves nearly 30 years ago [5]. Consistent with his own measurements, Reeves' theory is found to match our measured correlation well for signal bandwidths up to 2 kHz. Second, we used linear systems theory [11] to develop the equations connecting frequency correlation and time spread. We find that frequency correlation predicted using the linear systems theory matches measured correlation very well. An important next step in this work is to validate the linear systems theory for higher sea state conditions.

ACKNOWLEDGMENTS

This work was supported by the Office of Naval Research (Dr. J. Tague, Code 321US) under award No. N00014-02-1-0156. The measurement involved significant cooperation from the many people at the Marine Physical Laboratory (MPL) of Scripps Institution of Oceanography, especially Capt. Bill Gaines (USN-Ret) and Dr. Gerald D'Spain. Acoustics graduate students Steven Lutz, Rachel Romond and Tom Weber did superb work at sea and their contributions are appreciated.

REFERENCES

1. Fortuin, L., J. Acoust. Soc. Am **47**, 1209-1228 (1969).
2. Ogilvy, J.A., *Theory of Wave Scattering from Random Rough Surfaces*, Bristol, England: Institute of Acoustics, 1991.
3. Dahl, P.H., J. Acoust. Soc. Am. **115**, 589-599 (2004).
4. Bendat, J.S. and A.G. Piersol, *Random Data: Analysis and Measurement Procedures*, 2nd Ed., New York: John Wiley and Sons, 1986.
5. Reeves, J.C., *Distortion of Acoustic Pulses Reflected from the Sea Surface*, Ph.D. Dissertation, University of California, Los Angeles (1974).
6. Martin, J.J., J. Acoust. Soc. Am **43**, 405-417 (1968).
7. Weston, D.E., J. Acoust. Soc. Am. **37**, 119-124 (1965).
8. Dahl, P.H., J. Acoust. Soc. Am. **100**, 748-757 (1996).
9. Dahl, P.H., J. Acoust. Soc. Am. **105**, 2155-2169 (1999).
10. Dahl, P.H., IEEE J. Oceanic Eng. **26**, 141-151 (2001).
11. Ziomek, L.J., *Underwater Acoustics: A Linear Systems Theory Approach*, Orlando: Academic Press, 1985.
12. Weinberg, H. and R.E. Keenan, NUWC-Newport TR 10,568, Naval Undersea Warfare Center Division Newport, Rhode Island, 1996.
13. Groves, D.G. and L.M. Hurt, *Ocean World Encyclopedia*, New York: McGraw-Hill, 1980.
14. Plant, W.J., J. Geophys. Res. **107(C9)**, 3120, doi:10.1029/2001JC000909, 2002.
15. Keranen, J.G., *Effect of the Ocean Environment on the Coherence of Broadband Signals*, MS Thesis, The Pennsylvania State University, State College, PA (2001).

Modeling Acoustic Signal Fluctuations Induced by Sea Surface Roughness

Robert M. Heitsenrether, Mohsen Badiey

Ocean Acoustics Laboratory, College of Marine Studies, University of Delaware, Newark, DE 19716

Abstract. An empirical fetch-limited ocean wave spectrum has been combined with an acoustic ray-based model to predict the acoustic signal time-angle fluctuations induced by sea surface roughness. Rough sea surface realizations are generated and used as sea surface boundaries with the acoustic model. To validate this model, results are compared against experimental data collected in a fetch limited region. These data includes simultaneous wind speed and acoustic propagation (1-18 kHz) measurements in a fetch limited coastal region. Modeled time-angle fluctuations compare well with field data at lower wind speeds (< 10 m/s).

INTRODUCTION

Surface waves are among several environmental parameters that can have significant influence on the propagation of high frequency underwater acoustic waves. Quantifying the impact of sea surface roughness on the acoustic wave propagation is an important step in both determining performance levels of underwater acoustic instrumentation and developing techniques for using acoustic waves to measure sea surface roughness. This study involves a combined approach based on experimental observation and modeling of both surface waves and acoustic waves in order to assess the detail of acoustic signal interaction with the sea surface.

A high frequency acoustics experiment was conducted during September 22 through September 29, 1997 (HFA97 experiment) in a shallow water region of the Delaware Bay [1]. During the experiment, acoustic signals were transmitted between source-receiver tripods deployed on the sea floor, while highly calibrated environmental data was collected simultaneously from a nearby oceanographic observation platform [2]. Source-receiver tripods were carefully spaced in range so rays with a single surface interaction were easily distinguished in received signals. Extensive analysis of the single surface reflected portion of received signals shows correlation between signal fluctuations and wind speed [1].

In order to further understand the interaction of acoustic waves with the rough air-sea boundary, a combined acoustic-ocean surface model has been employed to simulate the time-angle fluctuations observed in shallow water acoustic transmissions. The model combines the BELLHOP ray-based acoustic model [3] and an empirical wind driven sea surface model [4]. The HFA97 data set is used to guide model development and validate results.

EXPERIMENTAL DATA

The HFA97 experiment was conducted in a central region of the Delaware Bay at 75° 11' West and 39° 01' North. Two bottom mounted tripods, each having an acoustic source and three receiving hydrophones, were placed in 15 m of water and separated by 387 m. On each tripod, the source was located 3.125 m above the sea floor and the three receiving hydrophones were located at 0.33, 1.33, and 2.18 m respectively (Fig. 1). Sources transmitted broad-band chirp signals over the frequency range of 0.6-18.0 kHz.

During the experiment, different pulse transmission rates were used so as to capture the fast and slow temporal variations of the acoustic field driven by different physical ocean processes. In one case, the broad-band chirp signal was transmitted every 0.345 s for a 40-s interval and then repeated every hour for the entire experiment. During these 40-s intervals, each received signal had sufficient time to clear before the next signal arrived so that overlapping did not occur.

Analysis presented here focuses on received signals that result from acoustic waves traveling from the source on one tripod to the three remotely mounted hydrophone receivers, located 387 meters away on the opposite tripod. For these signals, the HFA97 experimental design allowed for examination of the time evolution of ray paths involving only one surface interaction [paths 2-5 in Fig. 1 (a)].

In previous HFA97 analysis, remotely received signals across the three hydrophones were used with a beamforming technique to calculate signal arrival angle as a function of arrival time [1]. By considering the geometry of the HFA97 experimental setup [Fig. 1 (a)], the resulting beamformed plots can be used to easily distinguish the portion of the received signal corresponding to Single Surface Reflected (SSR) ray paths. Also, at lower wind speeds, beamformed plots can be used to distinguish between four individual SSR ray paths [Fig. 1 (b)].

During HFA97, several oceanographic and meteorological measurements were made coincident with acoustic measurements which included, tide height, current profiles, sound speed profiles, air temperature, wind speed, and wind direction.

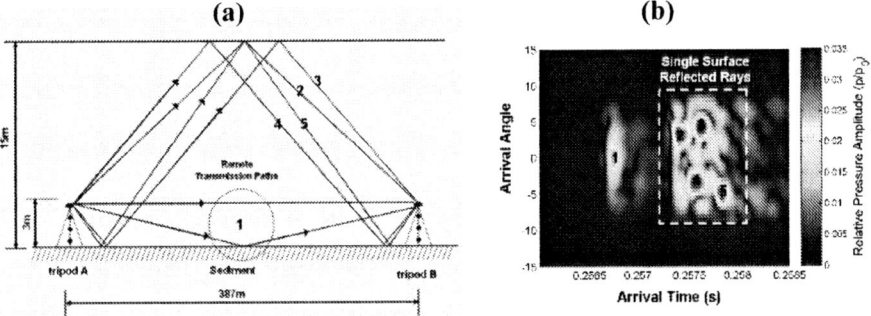

FIGURE 1. (a) HFA97 Experimental setup and ray paths associated with remote transmissions. Single surface reflected ray paths are individually numbered. (b) Remotely received signal arrival angle versus arrival time for a calm period (wind speed of about 2 m/s); single surface reflected ray paths are easily distinguished in the signal [numbers correspond to rays labeled Fig 1(a)].

MODELING METHODS

Ray Theory and Gaussian Beam Tracing

There have been a number of efforts to modify conventional ray theory in order to develop improved methods that provide more accurate results but retain computational efficiency. One such method is Gaussian beam tracing [3]. With this technique, a fan of rays is traced from a point source with trajectories governed by the standard ray equations. The Gaussian beam method associates with each ray a beam with a Gaussian intensity profile normal to the ray. An additional set of equations which govern beam width and curvature are integrated along with the standard ray equations.

The Gaussian beam tracing method has been adapted to the typical ocean acoustics waveguide and has been implemented as a tool called BELLHOP. This model has rigorously been tested and results show excellent agreement with certain full wave models at high frequencies. The method is free of numerical artifacts affecting standard ray models and still retains the computational efficiency of a ray based approach. As the detail of this model is provided in [3], here we refrain from further explanation.

Modeling the Ocean Surface using JONSWAP

In coastal regions, the wind acts on a limited fetch. As a result, the sea will not become fully developed and the large-scale or swell components of the waves will be significantly reduced in amplitude. The JONSAWP spectral model computes a sea surface frequency spectrum, $S(\omega)$, under fetch-limited conditions as function of wind speed [4]. This model is based on an extensive wave measurement program (Joint North Sea Wave Project) carried out in 1968 and 1969 in the North Sea. The JONSWAP spectrum provides a good starting point for modeling surface conditions in the area where the HFA experiments were conducted.

The JONSWAP spectral model takes the form:

$$S(\omega) = \alpha g^2 \omega^{-5} \exp\left[-\frac{5}{4}\left(\frac{\omega}{\omega_p}\right)^{-4}\right] \gamma^\delta \quad (1)$$

where δ is a peak enhancement factor:

$$\delta = \exp\left[-\frac{(\omega - \omega_p)^2}{2\sigma_0^2 \omega_p^2}\right]. \quad (2)$$

The parameters γ and σ_0 are given as $\gamma = 3.3$, $\sigma_0 = 0.07$ for $\omega \leq \omega_p$, and $\sigma_0 = 0.09$ for $\omega > \omega_p$, while α is a function of fetch, X and wind speed, U:

$$\alpha = 0.076\left(\frac{gX}{U}\right)^{-0.22}, \quad (3)$$

and peak frequency ω_p is given as:

$$\omega_p = 7\pi \left(\frac{g}{U}\right)\left(\frac{gX}{U^2}\right)^{-0.33}. \qquad (4)$$

A sea surface height wavenumber spectrum, $W(k)$ can be obtained from the JONSWAP frequency spectrum using the relationship $S(\omega)d\omega = W(k)dk$, and the gravity wave dispersion relation, $\omega = \sqrt{kg}$, where k is the wavenumber of ocean waves.

The spectral method can be used to generate one dimensional, sea surface realizations consistent with the JOWNSWAP spectrum [5,6]. Surface heights are generated at N points with spacing Δx across the horizontal range of length $L = N\Delta x$. Realizations with the desired spectral properties can be generated at points $x_n = n\Delta x$ (n = 1,...,N) with the following expression for surface height function $f(x)$:

$$f(x_n) = \frac{1}{L}\sum_{j=-N/2}^{N/2-1} F(K_j) e^{iK_j x_n} \qquad (5)$$

where for $j > 0$,

$$F(K_j) = [2\pi L W(K_j)]^{1/2} u \qquad (6)$$

and for $j < 0$, $F(K_j) = F(K_j)^*$. In this expression, $K_j = 2\pi j/L$, u indicates an independent sample taken from a zero mean, unit variance Gaussian distribution, and $W(K)$ represents the JONSWAP wavenumber spectrum.

When generating these 1-D surface realizations, surface partition width, Δx, must be selected. For this modeling study, the dominant wavelength predicted by the JONSWAP spectrum at each wind speed will be used to set Δx.

When calculating the JONSWAP frequency spectrum for a chosen fetch and wind speed, the model gives a predicted peak frequency of the spectrum, ω_p (4). At each different wind speed, ω_p can be used to calculate a peak wavelength, λ_p using the deep water dispersion relation and the relationship between wavenumber, k and wavelength, λ (where $\lambda = 2\pi/k$). Here, λ_p represents the dominant wavelength of ocean surface waves for the given conditions. For this modeling case, at each wind speed, Δx will be set to one half of this dominant wavelength. Surface heights between generated points will be linearly interpolated.

When using the JONSWAP wavenumber spectrum to generate 1-D surface realizations, the total wave energy in the spectrum is applied to waves propagating along the x-axis. This may exaggerate surface roughness slightly. In addition, in this process the out of plane scattering of the acoustic field may be neglected. A better approach would be to use 1-D cross sections through 2-D surface realizations in which the wave energy is also distributed in azimuth. However, the focus of this study is to demonstrate the feasibility of the combined acoustic and surface wave modeling approach only.

Integration of BELLHOP and Surface Model

Empirical sea surface models have been combined with acoustic models in past studies of similar nature [6-10]. The modeling approach presented here however is unique in terms of computational efficiency. The concept behind this combined sea surface/acoustic model is the utilization of rough ocean surface realizations and the Gaussian beam tracing model (i.e. BELLHOP [3]). Rough surface realizations are generated using HFA97 wind speed measurements, the JONSWAP wavenumber spectrum, and the spectral method. These surfaces are read into BELLHOP as (horizontal range, surface height) points and become the upper boundary over the water column through which beams are traced. When a beam interacts with the rough surface boundary, the beam trajectory is geometrically reflected from the rough surface, using the beam's angle of incidence and the surface slope at the point of intersection. The resulting model output simulates the fluctuations in arrival angle and arrival time observed in the HFA97 transmissions.

In acoustic wave scattering theory, the scale of ocean surface roughness is usually specified by the surface roughness (Rayleigh) parameter [11] which is defined by, $\chi \equiv 2kh_{rms}\sin(\theta_g)$, where k is the acoustic wavenumber, h_{rms} is the rms sea surface displacement mean level, and θ_g is the grazing angle. For the HFA97 case, using the center frequency of the signal (12 kHz) and the typical h_{rms} for the region considered (0.2-0.4 m), $\chi \approx 2$ which indicates that the SSR portion of received signals consist of incoherent scattering. This combination of high frequency and large scale roughness justifies the approach of geometrically reflecting acoustic ray paths from individual points on the rough ocean surface.

MODEL RESULTS

Acoustic Time-Angle Fluctuations

Time-angle fluctuations of SSR arrivals were measured in the HFA97 data. Time-angle standard deviations were calculated for each hourly, 40-s transmissions consisting of 115 chirp signals.

Beamformed plots [Fig. 1 (a)] can be used to pick out the portion of a received signal that corresponds to a specific ray path. Figure 1 (b) represents a signal that was transmitted during a calm period (wind < 3 m/s) and four individual SSR ray paths can be clearly distinguished. In similar plots for rougher periods, it becomes difficult to distinguish between four individual SSR rays due to the breakup and formation of micro-multi paths resulting incoherent scattering at the rough sea surface. For most rough and calm periods, however, it is feasible to pick out the very first arriving SSR ray path in the second group of arrivals.

Beamformed results were used to track time-angle fluctuations of first SSR arrivals in HFA97 data. Time-angle standard deviations of first SSR arrivals are calculated for the group of signals received during each hourly 40-s transmission interval. Time-angle standard deviations are then plotted against the wind speed recorded at that transmission time.

The BELLHOP/JONSWAP model was used with a Monte Carlo simulation to calculate the standard deviation of arrival time and arrival angle of the first arriving beam with a single surface interaction and no bottom interaction (first SSR beam shown as path 2 in Fig. 1). Separate model runs were made for each one meter/second increment in wind speed (for the range of 1-15 m/s). For each run, 200 surfaces were generated for the given wind speed. A separate BELLHOP beam trace was performed for each of the 200 rough surfaces. Standard deviations of arrival time and arrival angle of the first SSR beams were calculated for each wind speed increment using output from the 200 runs. These standard deviations provide a description of received signal fluctuations which increase with wind speed and surface roughness.

Figure 2 shows comparisons of modeled and measured time-angle standard deviations of the first SSR arrivals. Model results and data agree well for wind speeds of about 9 m/s and less. At lower wind speeds, both time and angle standard deviations show an approximately linear increase with wind speed. Model deviation from HFA97 data at higher wind speeds is a possible indication that increased breaking wave activity occurred at the sea surface at higher wind speeds. The sea surface generator used by this model does not consider the nonlinear hydrodynamics of breaking waves. Therefore, at this point, the combined BELLHOP/JONSWAP model is useful for predicting acoustic signal fluctuations at lower wind speeds.

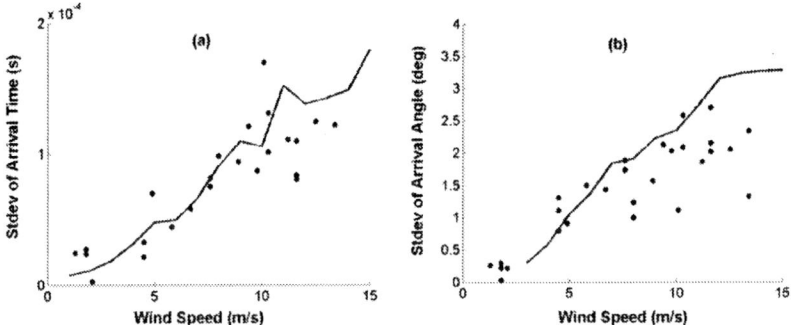

FIGURE 2. Comparisons of BELLHOP/JONSWAP model results obtained from Monte Carlo procedure (solid line) and measured HFA97; fluctuations of single surface bounce beam versus wind speed standard deviation of (a) arrival time in seconds and (b) arrival angle in degrees.

Observed Amplitude Fluctuations

Modeling signal amplitude fluctuations remains to be explored in subsequent work, as open area of research due to complexities stemming from combined sea surface roughness and interactions between acoustic waves and bubbles resulting from breaking waves. Here, observed signal amplitude fluctuations are presented. Remarkably, these amplitude fluctuations show the same trends as the time-angle fluctuations presented above.

As stated earlier, HFA97 experimental setup was designed so that the portion of remotely received signals corresponding to single surface reflected (SSR) rays is easy to distinguish. A method was developed to separate this portion of a received signal in order to calculate mean amplitude across the duration of a SSR portion's arrival time. This average SSR amplitude was calculated for each ping in a 40-s transmission, and then the standard deviation of the group of values was calculated for different wind speeds.

Figure 3 (a) shows a plot of SSR amplitude standard deviation versus wind speed. Similar to the results shown in the time-angle plots above, amplitude fluctuations increase roughly linearly with wind speed and the trend stops after about 9 m/s for this data. Figure 3(b) shows the average SSR amplitude calculated for the whole group of 115 pings at each transmission time. This average SSR amplitude remains close to a single value at lower wind speeds and then suddenly drops off at higher wind speeds. This type of decrease in amplitude of surface reflected acoustic waves typically occurs when there are a significant amount of bubbles in the water column near the sea surface [12]. The trends shown in Fig. 3 (a) and (b) provide another possible indication that an increase in breaking wave activity occurred at the ocean surface during periods of higher wind speeds.

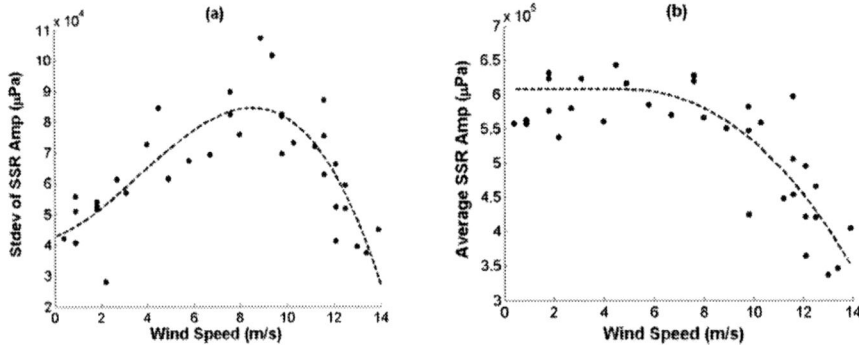

FIGURE 3. HFA97 SSR amplitude fluctuations versus wind speed; (a) measured standard deviation of SSR amplitude for 40-s group of signals (dots) and least squares polynomial fit (line); (b) measured average SSR amplitude 40-s group of signals (dots) and least squares polynomial fit (line).

CONCLUSIONS

Combining an empirical wind driven sea surface model and a ray-based acoustic model presents a unique approach to predicting fluctuations in acoustic signals induced by sea surface roughness. Tracing beams through sea surface height deviations and changing beam direction at surface reflection based on surface slope results in a realistic simulation of time-angle fluctuations in received signal at lower wind speeds. Also, using ray-based acoustic methods makes the model extremely computationally efficient since multiple model runs can be made quickly supporting timely model modification and improvement.

Initial comparisons between this combined model output and HFA97 observations yield good results for lower wind speeds. Data from other high frequency shallow water acoustic experiments will be compared with the model for further validation of this approach. Also, subsequent work will focus on using this modeling approach to predict amplitude fluctuations of acoustic signals induced by fetch limited sea surface roughness.

ACKNOWLEDGMENTS

The authors wish to thank all participants of the HFA97 experiment, particularly Steve Forsythe for his help in signal processing. Special thanks is due to Michael Porter for providing help with the BELLHOP model. This work was supported by the Office of Naval Research, code 321OA and in part by the Sea Grant program.

REFERENCES

1. Badiey, M., Mu, Y., Simmen, J.A., Forsythe, S.E., "Signal Variability in Shallow-Water Sound Channels," IEEE Ocean Eng. 25 (4), 2000, pp. 492-500.
2. Badiey, M., Lenain, L. Wong, K.C., Heitsenrether, R., Sundberg, A., "Long-term Acoustic Monitoring of Environmental Parameters in Estuaries," in Proc.Oceans 2003 Marine Technology and Ocean Science Conference, San Diego, CA.
3. Porter, M.B., Bucker, H.P., "Gaussian Beam Tracing for Computing Ocean Acoustic Fields, "J. Acoust. Soc. Am. 82 (4), 1987, pp. 1348-1359.
4. Hasselmann, D., Dunckel, M., and Ewing, J.A., "Directional Wave Spectra Observed During JONSWAP 1973," Jour. Phys. Ocean. 10, 1980, pp. 1264–1280.
5. Ogilvy, J.A., *Theory of Wave Scattering from Random Rough Surfaces* (Institute of Physics Publishing, Bristol and Philadelphia, 1991) pp. 228-229.
6. Thorsos, E.I., "Acoustic scattering from a 'Pierson-Moskowitz' sea surface", J. Acoust. Soc. Am. 88 (1), 1990, pp. 335-349.
7. McDaniel, S.T., "Composite-Roughness Theory Applied to Scattering From Fetch Limited Seas, "J. Acoust. Soc. Am. 82 (5), 1987, pp. 1712-1719.
8. Dahl, P.H., "On the Spatial Coherence and Angular Spreading of Sound Forward Scattered From the Sea Surface: Measurements and interpretive model," J. Acoust. Soc. Am. 100(2), 1996, pp. 748-758.
9. Dahl, P.H., "On Bistatic Sea Surface Scattering: Field Measurements and Modeling," J. Acoust. Soc. Am. 105 (4), 1999, pp. 2155-2169.
10. Dahl, P.H. "High-Frequency Forward Scattering from the Sea Surface: The Characteristic Scales of Time and Angle Spreading," IEEE Ocean Eng. 26, 2001, 141-151.
11. Clay, C.S., Medwin, H. (1977). *Acoustical Oceanography* (Wiley-Interscience Publications), 49-51.
12. Ostrovsky, L.A., Sutin, M.S., Soustova, I.A., Matveyev, A.L., Potapov, A.I., Kluzek, Z. "Nonlinear scattering of acoustic waves by natural and artificially generated subsurface bubble layers in the sea," J. Acoust. Am. 113, 2003, pp. 741-749.

Mid-Frequency Signal Fluctuations and Target Localization

W.S. Hodgkiss, G.L. D'Spain, and D.E. Ensberg

Marine Physical Laboratory, Scripps Institution of Oceanography, La Jolla, CA, 92093-0701, USA

Abstract. Environmental fluctuations (e.g. water column sound speed perturbations due to internal waves) result in variability of the vertical arrival angle structure observed from an acoustic source at a given range and depth. Experimental data collected by MPL in 2001 with a mid-frequency, vertical aperture receiving array in shallow (~165 m deep) water provided an opportunity to measure both environmental and the resulting acoustic fluctuations. The variability of both shallow and deep 3.5 kHz source transmissions from 4 km and 2.5 km range is summarized statistically using histograms of signal vertical angle of arrival. These appear to indicate that shallow source arrivals fluctuate more than deep source arrivals. In addition, the use of relatively short-range (~1.0-1.7 km) 3.5 kHz source arrivals (direct path and surface bounce) observed on the vertical array to estimate source range and depth via backwards ray tracing is discussed. Shallow and deep source localization is shown feasible with a slowly drifting source being localized to within ~200 m in range and ~10 m in depth.

INTRODUCTION

Sound speed fluctuations result in variability of the vertical arrival angle structure observed at an array. Here we focus on shallow water experimental observations of 3.5 kHz transmissions from both quasi-stationary source locations as well as source tows and specifically consider transmissions from both shallow and deep source depths. First, we will summarize statistically the observed signal fluctuations from 4 km and 2.5 km source ranges using histograms of signal vertical angle of arrival. Second, we will use relatively short-range, direct path and surface bounce arrivals to estimate source range and depth via backwards ray tracing

EXPERIMENT

The experimental data discussed was collected in July 2001 on a shallow ridge known as Fortymile Bank located 67 km (36 nm) west of San Diego and 41 km (22 nm) southeast of the southern tip of San Clemente Island. The R/P FLIP was moored in 165 m deep water and deployed a 64-element vertical receiving array buoyed up from the seafloor. The array consisted of two nested, 43-element apertures each spaced half-wavelength at 3.75 and 7.5 kHz, respectively. Both stationary source and source tow transmissions were made from locations along the ridge axis. As shown in Fig. 1, the bathymetry along the ridge has slight range-dependency. A set of 40 sound speed profiles derived from CTD casts taken during the experiment also are shown.

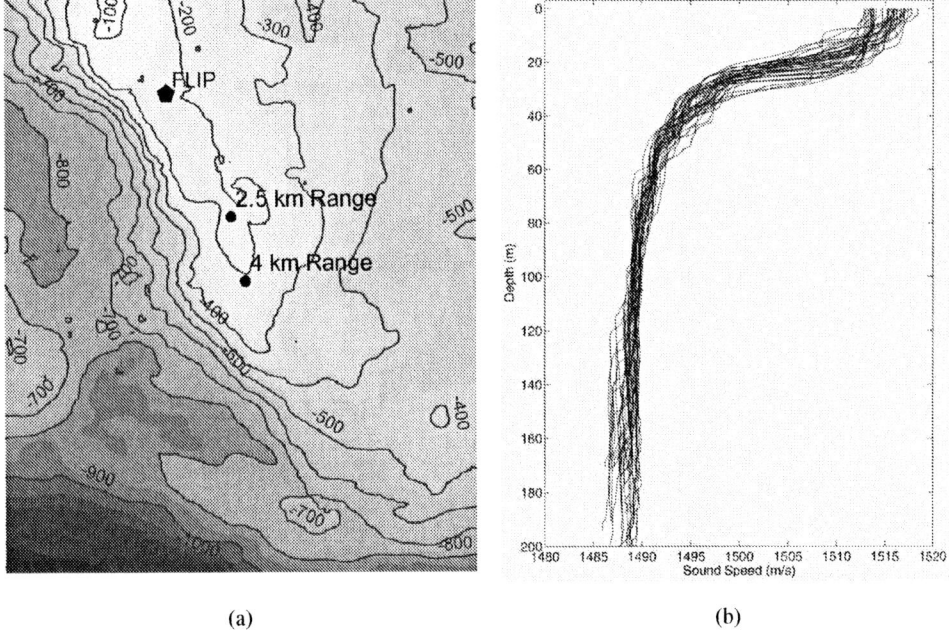

(a) (b)

FIGURE 1. Bathymetry in the vicinity of Fortymile Bank west of San Diego showing the locations of the receiving array deployed from the R/P FLIP and the 2.5 km and 4 km quasi-stationary source stations. Also shown are sound speed profiles derived from 40 CTD casts over the period JD 201-204.

SIGNAL FLUCTUATIONS

Quasi-stationary source transmissions were made from locations 4 km and 2.5 km in range from FLIP. These transmissions were made from several source depths and each was of duration 5 min. Of interest here is a comparison between the arrival structure from shallow (10 m) and deep (70 m) source transmissions.

The observed vertical arrival angle versus time from the 4 km source range is shown in Fig. 2 for these two cases (negative angles correspond to upward looking beams). The presence of near-horizontal arriving energy from the shallow source is due to the range-dependent bathymetry (i.e. slope conversion effects). The largest-level arrivals are in the vicinity of $-6.5°$ and $-2.5°$ for the 10 m and 70 m source depths, respectively. Histograms of these largest-level arrivals are shown in Fig. 3.

Similarly, the observed vertical arrival angle versus time from the 2.5 km source range is shown in Fig. 4 for the shallow and deep source transmissions. The largest-level arrivals are in the vicinity of $-13.5°$ and $-4.5°$ for the 10 m and 70 m source depths, respectively. Histograms of these largest-level arrivals are shown in Fig. 5. Note that a second significant arrival dominates the last half of the deep source observations yielding a bimodal distribution of arrival angles.

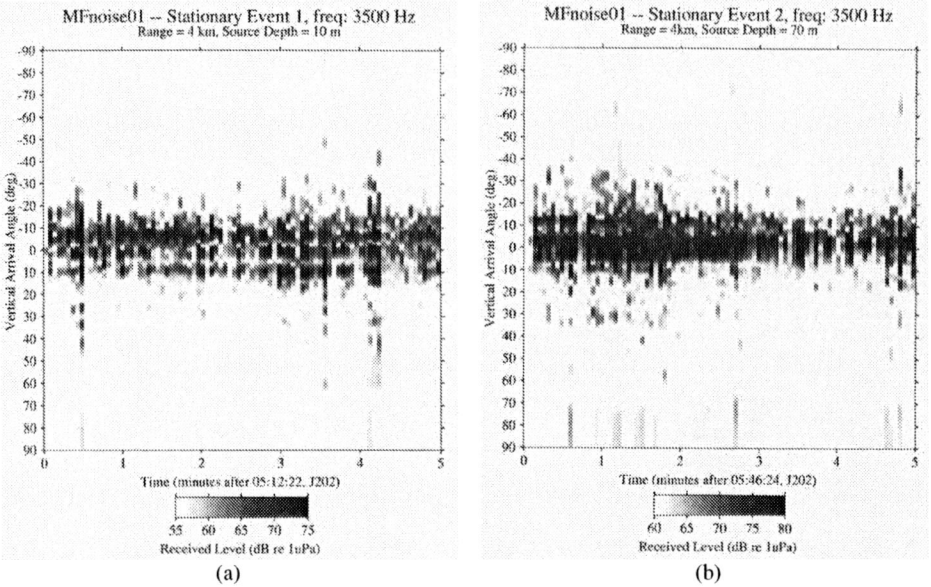

FIGURE 2. Time-evolving vertical arrival structure for the 10 m (a) and 70 m (b) source depths at the 4 km range source station.

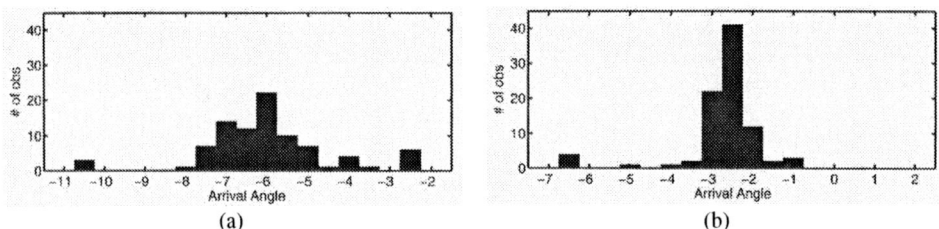

FIGURE 3. Histograms of vertical arrival structure for the 10 m (a) and 70 m (b) source depths at the 4 km range source station in the vicinity of $-6.5°$ and $-2.5°$, respectively.

The histograms appear to indicate that the largest level shallow source arrivals fluctuate more than the corresponding deep source arrivals. Analysis of other source depth transmissions (not shown) indicate similar results (e.g. 15 m and 90 m source depths at 4 km range and 16 m and 89 m source depths at 2.5 km range).

TARGET LOCALIZATION

Using a backwards ray tracing approach, the observed arrival angles from shallow and deep source transmissions at relatively close range were used to localize the source in range and depth. Direct path and surface bounce ray arrivals were identified in the receptions and traced backwards to their crossing points using a measured sound speed profile.

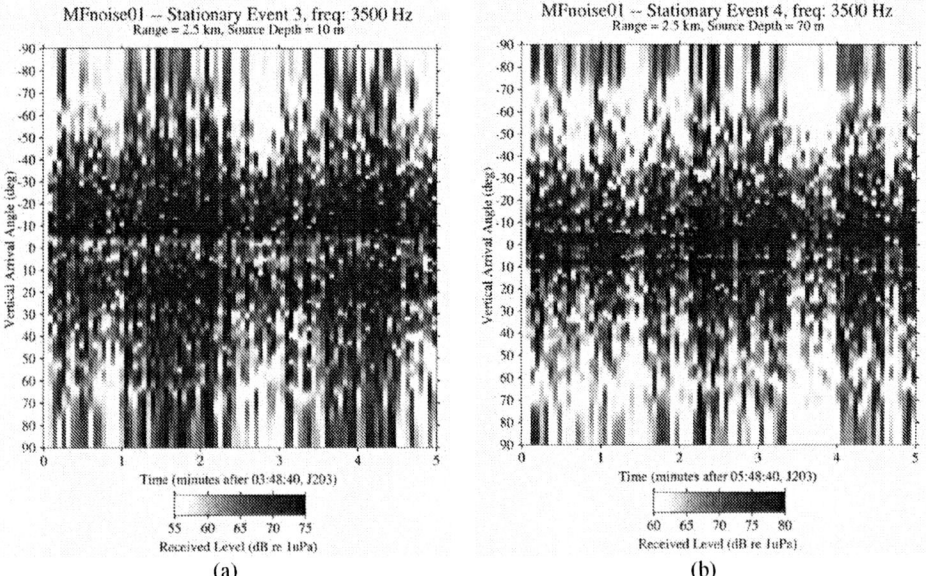

FIGURE 4. Time-evolving vertical arrival structure for the 10 m (a) and 70 m (b) source depths at the 2.5 km range source station.

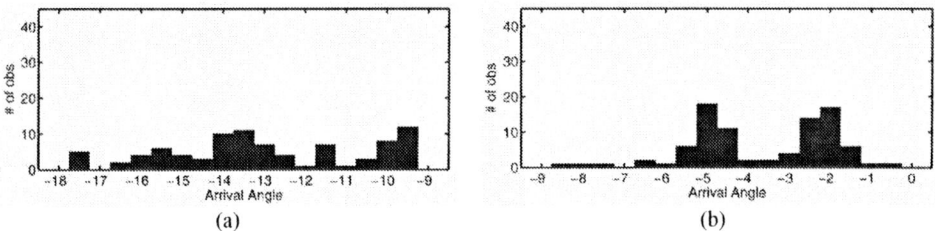

FIGURE 5. Histograms of vertical arrival structure for the 10 m (a) and 70 m (b) source depths at the 2.5 km range source station in the vicinity of $-13.5°$ and $-4.5°$, respectively.

In the first case, the source was drifting slowly at a range of approximately 1.1-1.3 km. The observed vertical arrival angle versus time for shallow (30 m) and deep (70 m) source transmissions is shown in Fig. 6. Examples of carrying out backwards ray tracing from identified arrivals are shown in Fig. 7. Scatter plots of the results are shown in Fig. 8 with the 30 m source transmissions being carried out near a range of 1.04 km and the 70 m source transmissions being carried out near a range of 1.27 km. Although there is some scatter in the results, shallow (30 m) and deep (70 m) source localization is feasible and the sources were localized to within ~200 m in range and ~10 m in depth of their true locations.

In order to investigate the influence of sound speed fluctuations on the results, a simulation was carried out. Representative direct path and surface bounce arrival angles were fixed (-7.5° and $-12.5°$ for the 30 m source and $-4.6°$ and $-12.6°$ for the 70 m source) and traced backwards through the time-evolving sound speed structure based on thermistor string measurements made at FLIP. The resulting scatter plots of

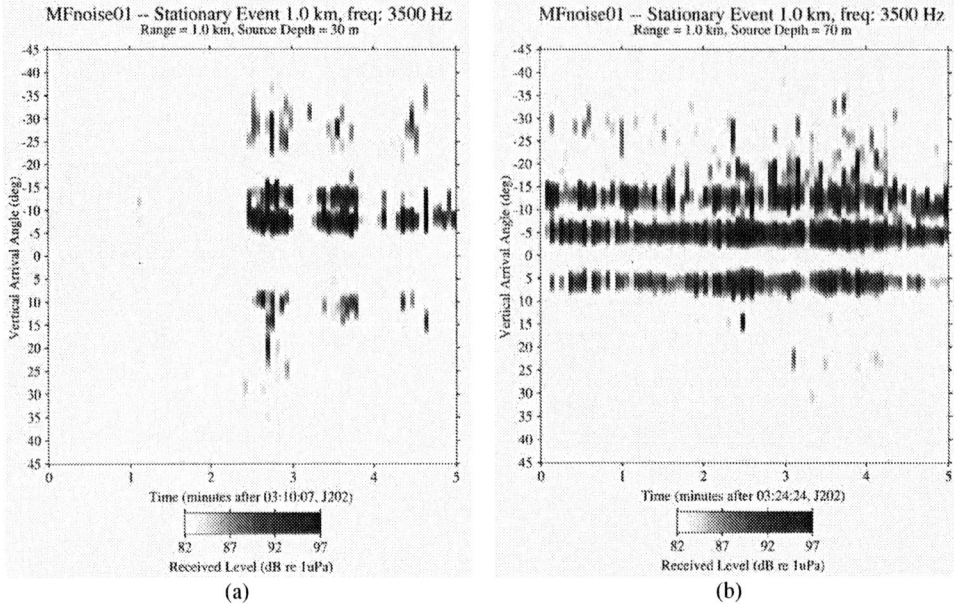

FIGURE 6. Time-evolving vertical arrival structure for the 30 m (a) and 70 m (b) source depths at ranges 1.04 km and 1.27 km, respectively.

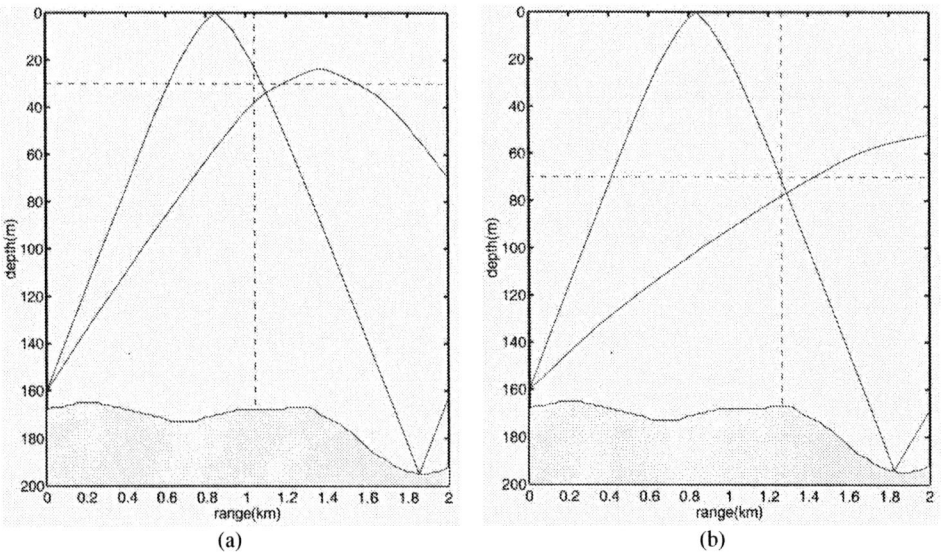

FIGURE 7. Backwards ray tracing of the direct path and surface bounce arrivals observed from the shallow (30 m) and deep (70 m) source transmissions. The dotted lines indicate the known source locations.

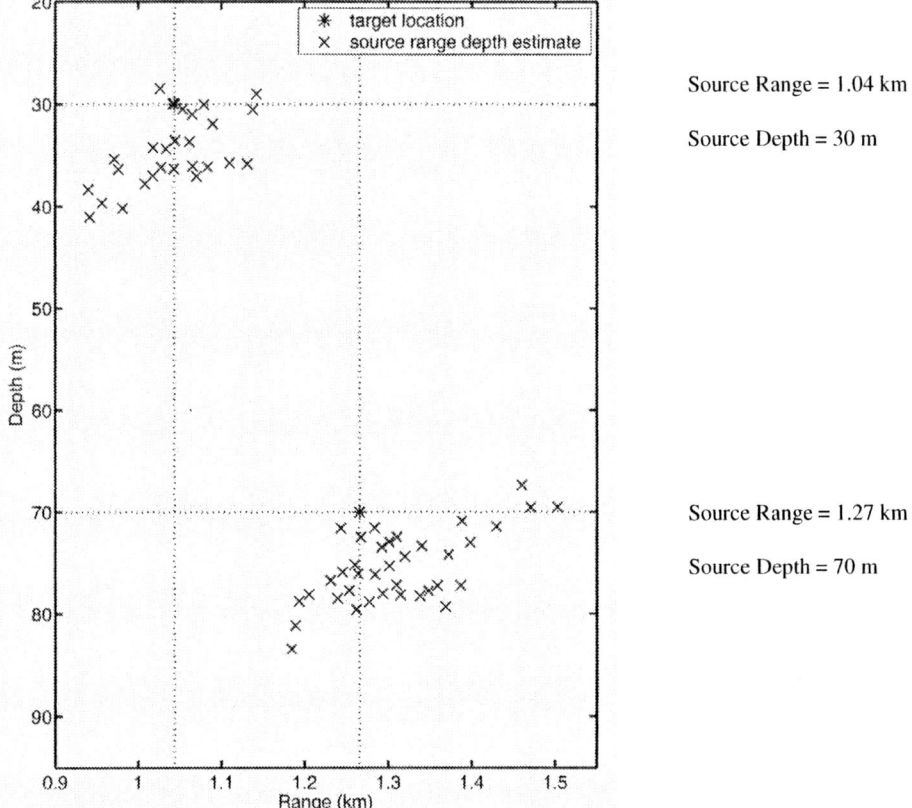

FIGURE 8. Scatter plots showing the clustering of source range/depth estimates based on backwards ray tracing of the observed arrival angles from the shallow (30 m) and deep (70 m) source transmissions.

the estimates of source range and depth are shown in Fig. 9 which includes both sound speed variation over the 5 min observation periods as well as a larger 7 hour period encompassing the source transmissions. The thermistor string observations over the 7 hour period also are included. The larger scatter in Fig. 8 is a result of the effects of several additional sources of variability not simulated: (1) sea surface roughness, (2) source motion, and (3) error in arrival angle estimation.

In the second set of data analyzed, the source was being towed with the range interval of analysis corresponding to approximately 1.1-1.7 km. The observed vertical arrival angle versus time for shallow (17.5 m) and deep (70 m) sources is shown in Fig. 10. Scatter plots of the results from backwards ray tracing are shown in Fig. 11 with the 17.5 m source transmissions being carried out over a range of 1.08-1.66 km and the 70 m source transmissions being carried out over a range of 1.21-1.69 km. As with the quasi-stationary source transmissions, there is scatter in the results but there is a clear distinction between clusters of shallow and deep source depth estimates.

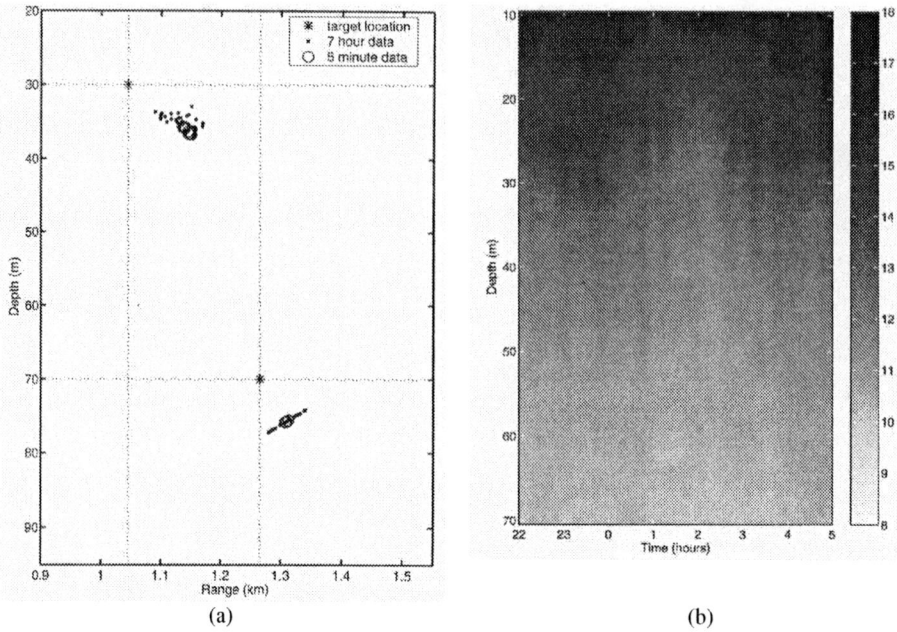

FIGURE 9. Simulations of range/depth scatter based on observed sound speed fluctuations for fixed arrival angles (a). Both sound speed variability over the 5 min observation periods as well as a larger 7 hour period encompassing the source transmissions is investigated. Thermistor string observations over the 7 hour period also are shown (b).

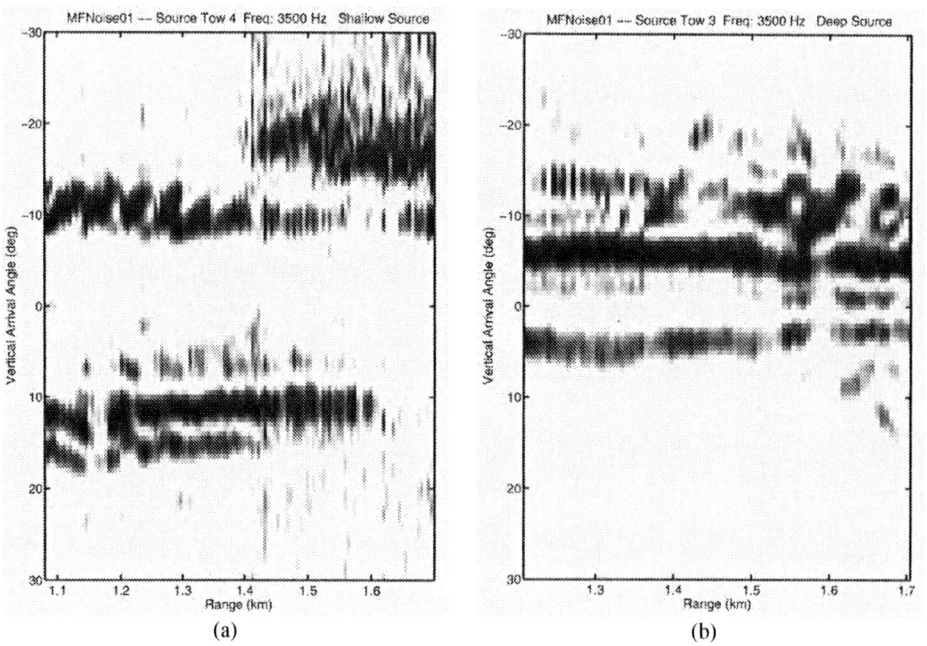

FIGURE 10. Time-evolving vertical arrival structure for the 17.5 m (a) and 70 m (b) source tow transmissions.

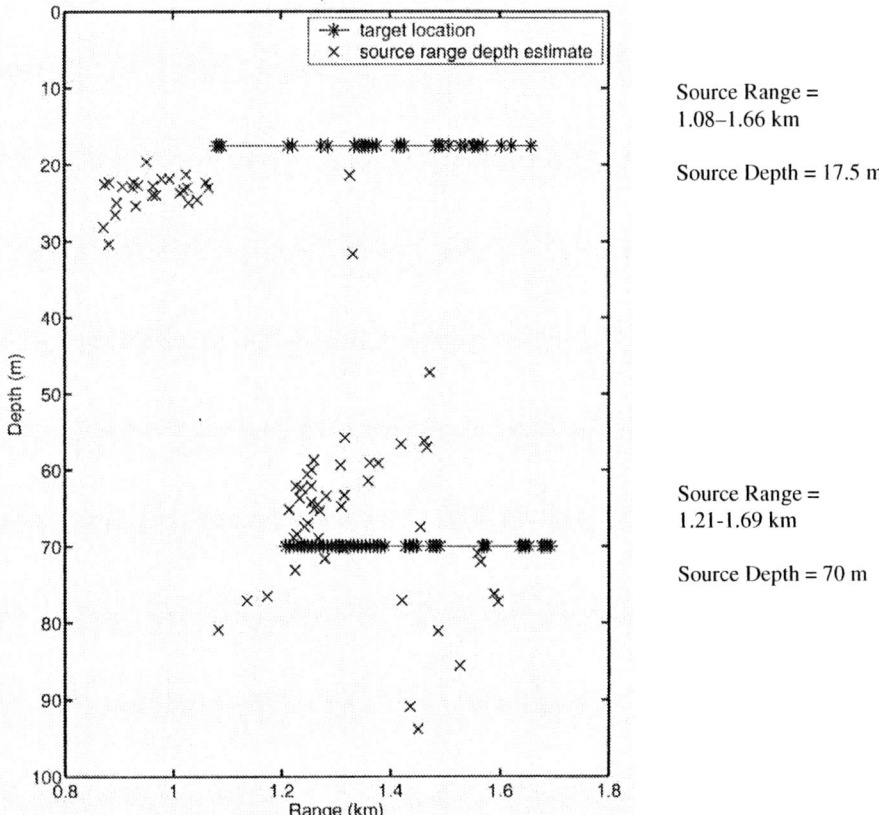

FIGURE 11. Scatter plots showing the clustering of source range/depth estimates based on backwards ray tracing of the observed arrival angles from the shallow (17.5 m) and deep (70 m) source tows.

SUMMARY

Experimental data collected with a mid-frequency vertical aperture receiving array in shallow (~165 m deep) water has been analyzed. The variability of both shallow and deep 3.5 kHz source transmissions from 4 km and 2.5 km source ranges has been summarized statistically using histograms of signal vertical angle of arrival. In addition, the use of relatively short range 3.5 kHz source arrivals (direct path and surface bounce) observed on the vertical array to estimate source range and depth via backwards ray tracking has been demonstrated.

ACKNOWLEDGMENTS

This work was supported by the Office of Naval Research, Contract No. N00014-01-D-0043-D01. We also thank Jeff Skinner, Jim Murray, and Katherine Kim who also participated in the experiment.

HF Doppler Acoustic Imaging of the Ocean Surface and Interior

Robert Pinkel and Jerome A. Smith

Scripps Institution of Oceanography, University of California, San Diego, 9500 Gilman Drive, La Jolla, CA 92093-0213

Abstract. HF phased array Doppler sonar represents a new tool for obtaining Three-dimensional (r,q,t) images of the oceanic surface and interior velocity field. While the capabilities of the approach are unique, the design constraints are also unusual. Examples of both are presented in this work.

INTRODUCTION

With the advent of internally recording instruments in the 1960's, the process of ocean investigation through "time series analysis" began in earnest. A single time (or space) series represents a one-dimensional picture of the four-dimensional world, a solitary light flickering in the darkness. With the advent of satellite remote sensing and Doppler sonar in the 1970's, two-dimensional images became available. While the gains resulting from this advance (and the associated field of image processing) have been enormous, two-dimensional data represent a "tunnel-vision" view of reality. In situations where variability is non-homogeneous/non-stationary or strongly anisotropic, there is motivation to develop three-dimensional sensing systems.

We have developed a series of Phased Array Doppler Sonars (PADS) in an effort to obtain three-dimensional measurements of the oceanic velocity field. These sense the radial component of velocity in a planar sector as a function of range, azimuth and time. To date, the instruments have been used singly, to measure flows in arctic leads, ([1] Figure. 1), upper ocean Langmuir cells [2] and nearshore rip currents. With separated pairs of instruments, the same region can be probed from two perspectives, enabling resolution of two components of velocity (Figure.2). Using this technique x,y,t maps (movies) of the vertical component of vorticity in the nearshore off Duck, N.C. have been formed(Figure. 3). The purpose of this paper is to introduce the technology and to illustrate some of the design constraints unique to this form of HF acoustic remote sensing.

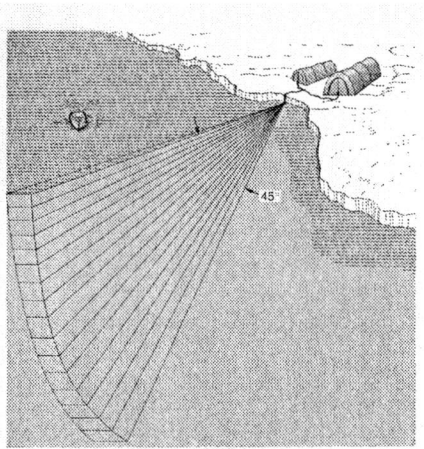

Figure 1. Schematic of the Arctic Leads Experiment (LEADEX) 1992 deployment of the sector scan sonar. The instrument was deployed with the measurement plane oriented vertically to image flows in the mixed layer and upper thermocline.

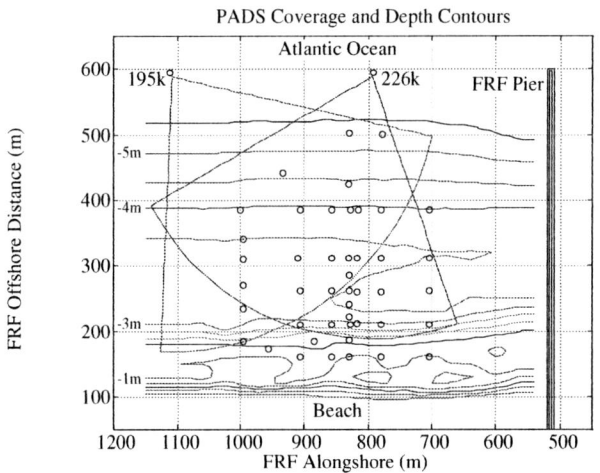

Figure 2. Schematic of the DUCK 97 nearshore deployment of two phased array Doppler sonars. Both components of horizontal velocity can be resolved in the region defined by the overlapping beams. Data were collected over a 60 day period through a variety of conditions.

BACKGROUND

Phased array technology is not new. Acoustic systems have long been used for fisheries research [3], bathymetric mapping [4] and collision avoidance. Jaffe, [5] has explored the feasibility of 4-d volume-time imaging systems. While low-frequency Doppler phased arrays have been developed for military work, application of this technology at ultrasonic frequencies is new.

Our initial system, developed in 1991-92 for arctic research, consisted of a 16-element phased array receiver that operated at 195 kHz with a 10 kHz bandwidth. Data were amplified, demodulated, and digitized within the receiver, and transmitted via optical fiber to the host computer. The transmitter was a single, curved face transducer that produces a 45° by 2° beam. In the arctic deployment (LEADEX), three repeats of a 13-bit Barker code were transmitted, providing a range resolution of 8 m. [6]

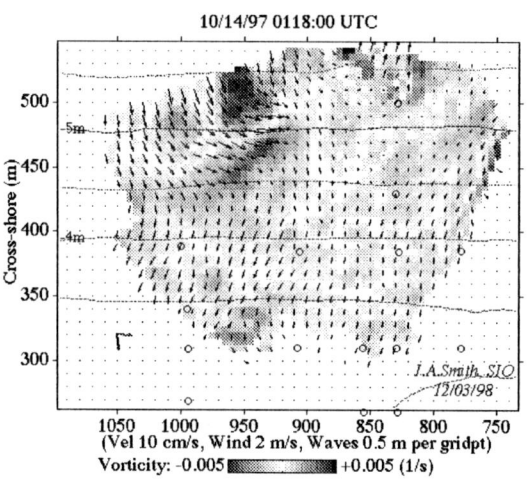

Figure 3. A planar map of the nearshore current field (arrows) at DUCK, as determined by the pair of crossed sonars. The vertical component of vorticity is indicated by the shading. This image represents one "frame" from a 3-d space-time movie of the velocity and vorticity fields.

The data were processed by a National Instruments 2305 Digital Signal Processing Card at a rate of 0.3 Mbyte s^{-1}. Twenty-eight independent beams were formed, spanning a 45° sector. At 2-min intervals, average scattering intensity and radial velocity maps were produced. These were displayed by the host computer and recorded on optical disk.

Subsequent instruments have been developed at 195 and 240 kHz, operating over 90° sectors. Dense 16 element arrays are used in these second-generation devices (Figure. 4).

Figure 4. Second Generation Phased Array Doppler Sonar

DESIGN

There are significant technical challenges associated with the development of these PADS systems. From the assembly perspective, a 10° phase error will result if an individual array element is mis-positioned by .04 cm. Such errors can critically affect beam-forming capability. In terms of data processing, the sonars now produce in excess of 1 Mbyte/s of echo information, steady state. This must be processed in real time with field transportable hardware. Analog challenges include the minimization of acoustic and electrical crosstalk and the matching of phase an amplitude response across the array.

Here we focus on the rather stringent demands placed on the system beam-forming and the resulting beam patterns associated with the volume-scattering application of PADS. Initially, sonar (and radar) systems were developed to detect discrete "targets". The magnitude of the main lobe of a sonar beam relative to its side lobes plays a significant role in the detection of isolated reflectors. If the contrast between main and side-lobe levels is great, false detections will be rare. In detection sonars, beam width is commonly indexed by the "half power point," the angle at which the beam pattern falls 3db below its peak value.

In volume scattering situations, a distributed cloud of targets is encountered. One wishes to detect signals from one region of the cloud while rejecting signals from the rest. Here, the volume of the main lobe relative to the volume in the collective side-lobes is the relevant parameter. This is a function of the shape and shading of a transducer, but NOT its size.

In dealing with multibeam reverberation sonars, one can envision a half-space uniformly populated with scatterers, except in some discrete sector.

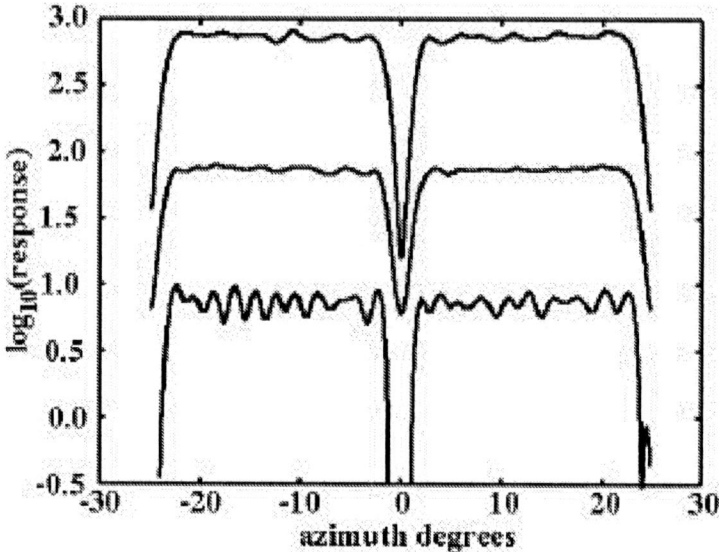

Figure 5. Theoretical "inverse beam patterns" for a 16 element sector scan sonar. A uniform cloud of scatters is assumed except in the region ± 2° in azimuth from broadside. Here an absence of scatters is posited. The array response is given for a Gaussian (top) triangular (middle) and rectangular shading of the receive array. The rectangular window best resolves the edges of the "hole". However the "depth" of the hole is only ~ 10 dB. With increased shading, leakage into the hole is reduced (middle, top) but the apparent width is reduced as well.

The ability of the sonar to image this "hole" in the scattering field, standing off side-lobe leakage from all other beam directions, is a meaningful measure of system performance.

Simulations of this ability are easily conducted (Figure. 5). The results, for a 16-element array, are sobering. A scattering void of width $\pm 2°$ can be detected, given the geometry of our first generation system. However, an un-windowed array (Figure. 5, bottom) sees the void as only 10dB deep. The suggestion is that in each of the energetic look-directions, the receive energy will be 90% signal and 10% leakage noise. With increasing windowing (5 middle, top) the depth of the null response increases. However, the associated width decreases.

For detecting spatial variations in scattering strength, the performance as simulated is acceptable. Isolated hard targets will "leak" into neighboring bands, but the leakage can generally be identified. However, the precision in Doppler frequency estimates degrades rapidly as the signal to noise ratio falls below 10. For these systems, the "self-clutter" noise is proportional to the received signal strength. Leakage prevents the signal-to noise-ratio from significantly exceeding 10, even in the total absence of electronic or acoustic noise. The development of this technology is thus an uphill battle, with significant emphasis placed on array side-lobe suppression.

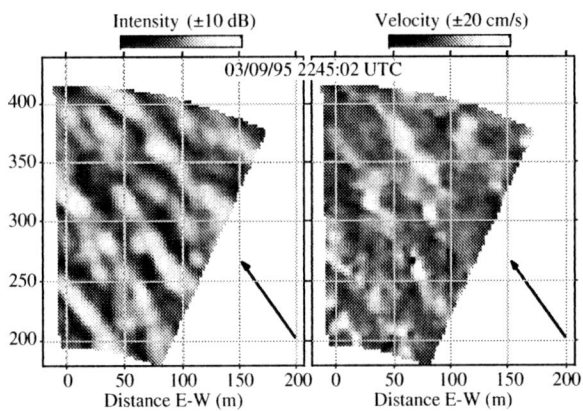

Figure 6. Open ocean observations of Langmuir cells as seen with the phased array. Acoustic intensity (left) is modulated by roughly an order of magnitude as sub surface bubbles (good reflectors) are collected in cell convergences. Corresponding patterns appear in the surface velocity field (right) only after the energetic surface wave motions are averaged out. The cells have a ~ 20 cm s^{-1} signal in this example.

SUMMARY

At present, the major successes of PADS have been in the observation of open-ocean Langmuir cells and near shore current and vorticity structure. It has been found that the spatial patterns of Langmuir cell currents are not aligned with the patterns in scattering strength which result from the collection of sub-surface bubbles in Langmuir convergences (Figure. 6). The Langmuir currents must be detected against a background of surface wave velocities with one or two orders of magnitude more variance. Only PADS technology enables detection of the weak current patterns in the presence of open ocean surface waves. As the processing capability of small computer/DSP cards increases there is significant room for improving the PADS concept. In particular, directionally coding the transmitted pulse will significantly improve angular discrimination, leading to more precise estimates of both velocity and position.

REFERENCES

1. R. Pinkel, M. Merrifield and H. Ramm, *J. Geophys. Res.,* 100(C3). 1995, pp. 4693-4705,.
2. J.A. Smith., *J. Geophys. Res.,* 1998, 92 pp 12,649-12,668..
3. R.B. Mitson, *IEE Proceedings*, 1984, 131-F, No. 3 pp 257-269.
4. C. de Moustier, C., *International Hydrographic Review, LXV* (2), 1988, pp 25-54.
5. J.S. Jaffe and P.M. Cassereau, J. *Acoustical Soc. Am.*, 1988, 83-4, pp 1458-1464.
6. R. Pinkel and J.A. Smith, *J. Atmos. & Oceanic Technol.*, 9, 1992, pp 149-163.

Detection of High Frequency Sources in Random/Uncertain Media.

Leon H. Sibul, Christian M. Coviello, Michael J. Roan

Applied Research Laboratory
The Pennsylvania State University
P.O. Box 30
State College, PA. 16804-0030
Lhs2@psu.edu

Abstract. The results of this paper show how randomness and/or uncertainty of medium, boundary conditions, source characterization, and source and receiver motion affects the probability of detection of a narrow band, high frequency source. Using ray acoustic model, we derive expressions for loss of time coherency and its dual, spectral spreading that is caused motion through a medium with random boundary conditions and inhomogeneities. Spectral spreading decreases probability of detection of a narrowband signals. In this analysis both the usual knowledge and the essential uncertainty are incorporated into problem formulation by separating propagation models and boundary conditions into deterministic and random parts. Maximum entropy method (MEM) is used to incorporate essential uncertainty into model. Maximum entropy method uses what is known in its model, but models what is not known with maximum uncertainty. It does not make any unwarranted assumptions about unknown parameters. MEM is used to calculate confidence intervals and mean values of receiver operating characteristics of high-frequency passive and active sonar detectors when signal to noise ratio is a random variable.

INTRODUCTION

The objective of the research that is presented in this paper to investigate how ever-present randomness and uncertainty in propagation medium and source characterization affects active and passive high-frequency sonar signal processing. Motion of sources and receiver through a propagation medium spatially varying sound speed profiles and random boundary conditions is equivalent to signal propagation through a randomly time-varying or stochastic propagation medium that causes a loss of signal coherence, spectral spreading and an increase of entropy (a measure of uncertainty) of the signal. The difficulty of the performance analysis of sonar is further aggravated by uncertainties in source strength, target models, boundary conditions, background noise characterization, and system specifications. All this suggests that sonar performance analysis should reflect essential uncertainties and randomness of the propagation and system parameters, but the analysis should also be based what is known or specified. This can be accomplished by the *maximum entropy method* (MEM). In this paper we use MEM to derive probability density functions *(PDF)* that are required to calculate confidence intervals and mean values of receiver operating characteristics (ROC) of simple active and passive detectors that operate in random/ uncertain propagation medium. In the case of random medium parameters, the detection statistic is a function of these random medium parameters. Thus, the exceeding of the detection threshold for a given false alarm probability is a random event with a *PDF* $f_D(\theta)$, where θ is a vector of random parameters. In the simple examples presented in this paper, θ is a random signal to noise ratio. Randomness of θ incorporates uncertainty of source strength, propagation loss, and background noise. We feel that plots that show confidence intervals and mean values of the generalized receiver operating characteristics are more effective in displaying the effects of uncertainty and randomness of the medium that so called *"range of the day."* A sonar operator can use our plots to determine, firstly, what are the confidence intervals for a probability of detection at a specified false alarm probability and, secondly, the mean propagation loss to a given range.

We start with a discussion of modeling of high-frequency propagation in time-varying random media, next we will present a brief review of the maximum entropy method for derivation of probability density functions, then we will present derivation of detection statistics for random parameters, and finally we will calculate confidence intervals for the random receiver operating characteristics. The basic propagation problem that we are considering is a high frequency, multipath propagation with random boundary conditions.

MODELING OF HIGH-FREQUENCY PROPAGATION IN TIME-VARYING RANDOM MEDIA

We assume that both the source (or target) and receiver are in motion in a propagation medium with a random boundary and an inhomogeneous sound speed profile. Under these conditions the sound propagates through a time-varying, random or stochastic medium. Propagation in stochastic medium can be characterized by *stochastic Green's functions* [1] or alternatively by *random spreading functions* [2]. Integral transform pairs relate spreading functions and stochastic Green's functions. Spreading functions are more convenient for analysis of signal processing systems and for derivation of ROCs for both active and passive sonar in stochastic media. Spreading functions are random functions that indicate how a stochastic medium spreads a propagating signal in time and frequency. Hence, spreading functions are time-frequency domain characterizations of narrow band signal propagation and scattering. Wideband spreading functions are wavelet transform domain characterizations of propagation and scattering of wideband signals in stochastic media [3]. Scattering functions, which have been widely used for analysis and synthesis of sonar, radar and communication systems, can be calculated from the spreading functions [1]. The receiver input consists of noise $n(t)$ and signal component or echo $y(t)$:

$$y(t) = \iint_{T,\Omega} b(\tau,\omega)\left[U(\tau,\omega)s(t)\right]d\tau\, d\omega, \qquad (1)$$

where $b(\tau,\omega)$ is the *spreading function* and $U(\tau,\omega)$ is a unitary transformation that transforms the signal $s(t)$ as shown below:

$$[U(\tau,\omega)s(t)] = s(t-\tau)\exp(-j\omega t). \tag{2}$$

Here τ is the differential propagation delay and ω is frequency shift, or Doppler shift in the case of active sonar. According to the ray acoustics [4], discrete multipath propagation can be modeled as a sum of individual time-frequency spread signals:

$$y(t) = \sum_i \iint_{T,\Omega_i} b_i(\tau_i,\omega_i)[U(\tau_i,\omega_i)s(t)]\, d\tau_i d\omega_i, \tag{3}$$

In Equation 3 $b_i(\tau_i,\omega_i)$ is the spreading function for the $i-th$ ray. Thus the time – frequency spread signal is a sum of signals that are spread within ray tubes. Under wide-sense stationary, uncorrelated scattering (WSSUS) assumption and under the additional assumption that $s(t)$ is a wide-sense stationary stochastic process, the correlation function of the received signal is

$$R_{yy}(\tau) = R_{ss}(\tau)\sum_i \iint_{T,\Omega_{iii}} S(\tau_i,\omega_i)\exp(j\omega_i\tau)\, d\tau_i\, d\omega_i, \tag{4}$$

where $S(\tau_i,\omega_i)$ is the *scattering function* the is computed by the WSSUS assumption from the spreading functions by Equation 5, below

$$S(\tau_i,\omega_i)\,\delta(\tau_i-\tau_k)\,\delta(\omega_i-\omega_k) = E\{b(\tau_i,\omega_i)b^*(\tau_k,\omega_k)\}. \tag{5}$$

The power spectral density of the received signal is the Fourier transform of the correlation function as expressed by Equation 3:

$$F_y(\omega) = \sum_i \iint_{T,\Omega_i} S(\tau_i,\omega_i)F_s(\omega-\omega_i)\, d\omega_i\, d\tau_i. \tag{6}$$

Equation 4 shows *loss of signal coherency* and Equation 6 shows *spectral spreading* due to multipath effects and motion through inhomogeneous medium with random or rough boundary conditions. Spectral spreading is also an indication of increase of *signal uncertainty or entropy*

$$E(y) = -\int_\Omega F_y(\omega)\,\ell n F_y(\omega)\, d\omega. \tag{7}$$

In this section we have presented a brief review of characterization of stochastic media by spreading and scattering functions. Spectral spreading, loss of signal coherency and increase of signal uncertainty are all factors that cause decrease of probability of detection. To estimate how probability if detection is affected by medium uncertainty we need for expressions probability density functions. Next we will review the MEM for computing probability density functions that can be used for computation of receiver operating characteristics.

MAXIMUM ENTROPY METHOD FOR PERFORMANCE ANALYSIS OF SIGNAL PROCESSING SYSTEMS IN RANOM MEDIA

Maximum entropy formalism for continuous random variables maximizes the entropy [5-7]

$$-\int_a^b f(x)\,\ell n f(x)\, dx, \tag{8}$$

subject to the normalization constraint

$$\int f(x)\,dx = 1,\tag{9}$$

and moment constraints

$$\int_a^b f(x)\,g_r(x)\,dx = a_r \quad r = 1,2,\ldots,m.\tag{10}$$

For constrained maximization, we form the Lagrangian:

$$L = -\int_a^b f(x)\,\ell n(x)\,dx - (\lambda_0 - 1)\left[\int_a^b f(x)\,dx - 1\right] - \sum_{r=1}^m \left[\int_a^b f(x)\,g_r(x)\,dx - a_r\right].\tag{11}$$

This Lagrangian is in the form of a calculus of variation problem

$$L = \int_a^b F[x, f(x), f'(x)]\,dx,\tag{12}$$

where F is a known function. In this case, the integrand is not a function of $f'(x)$ and the Euler-Lagrange equation of the calculus of variations is

$$\frac{\partial F}{\partial f(x)} = 0,\tag{13}$$

which gives

$$f(x) = \exp[-\lambda_0 - \lambda_1 g_1(x) - \lambda_2 g_2(x) - \cdots - \lambda_m g_m(x)].\tag{14}$$

The Lagrange multipliers in the Equation 14 can be determined from the constraint equations

$$\exp(\lambda_0) = \int_a^b \exp\left[-\sum_{j=1}^m \lambda_j g_j(x)\right]dx,$$

$$a_r \exp(\lambda_0) = \int_a^b g_r(x) \exp\left[-\sum_{j=1}^m \lambda_j g_j(x)\right]dx$$

and

$$a_r = \frac{\int_a^b g_r(x) \exp\left[-\sum_{j=1}^m \lambda_j g_j(x)\right]dx}{\int_a^b \exp\left[\sum_{j=1}^m -\lambda_j g_j(x)\right]dx},\tag{15}$$

$$r = 1,2,\ldots,m.$$

These equations can be used to derive maximum entropy distributions [7]. The maximum entropy distributions that we use in this paper are shown in Table 1 [7]. Specifically, we use a gamma distribution to model random signal to noise ratio, and a beta distribution to model conditional (conditioned on random/uncertain signal to noise ratio) probability of detection. Other well-known maximum entropy distributions include multivariate Gaussian, exponential, truncated exponential, beta of second kind, Laplace and Cauchy distributions. All the relevant information that is required for derivation of these distributions is contained in the moments and constraints. Thus, the maximum entropy approach greatly simplifies analysis of signal processing systems in random media. Constraints and moments can be determined from the physics of the media.

TABLE 1. Maximum Entropy Distributions [7]

Range	Constraints	Distribution
$(-\infty, \infty)$	$E\{x\}$ and $E\{x^2\} = \sigma^2 + m^2$	Gaussian: $N(m, \sigma^2)$
$[0,1]$	$E\{\ln(x)\}$ and $E\{\ln(1-x)\}$	Beta distribution of the first kind.
$[0, \infty)$	Arithmetic and geometric means.	Gamma distribution.

MATCHED FILTER DETECTION IN RANDOM MEDIA: A SIMPLE EXAMPLE OF MAXIMUM ENTROPY METHOD

The effects of uncertain target and propagation models, as well as uncertain background noise, manifest in an uncertain signal to noise ratio and time-frequency spreading of the received echo, modeled by Equation 1. Because it is reasonable to assume that, due to multiple scattering random boundaries and propagation through inhomogeneous media, the received signal or echo is a zero mean random process with constrained mean energy. Hence, according to the maximum entropy principle, it is a circularly symmetric (real and imaginary parts have equal variances) complex Gaussian stochastic process. The two hypotheses in this case are

$$H_1: r(t) = \sqrt{E_r} \iint_{T\Omega} b(\tau, \omega)[U(\tau, \omega)s(t)] d\tau\, d\omega + n(t), \quad (16)$$

$$H_0: r(t) = n(t),$$

where E_r is expected received energy and $n(t)$ is Gaussian noise, again using maximum entropy assumption. The detection statistic is

$$\ell(r) = \left| \int_T r(t) s^*(t - \tau_d) \exp(-j\omega_d) \, dt \right|^2. \quad (17)$$

The detection statistic $\ell(r)$ is a χ_2^2 distributed random variable with two degrees of freedom. This follows from the fact that matched filtering operation is a linear operation on a circularly symmetric, complex Gaussian process, $r(t)$; and hence, the magnitude squared of the matched filter output is a sum of squares of the real and imaginary parts [2]. The *pdfs* of the detection statistics under two hypotheses are simple exponential densities:

$$f_{\ell/H_1}(\ell/H_1) = \frac{1}{2\sigma_1^2}\exp\left[-\ell/2\sigma_1^2\right] \quad \ell \geq 0$$
$$= 0, \quad \ell < 0$$
$$f_{\ell/H_0}(\ell/H_0) = \frac{1}{2\sigma_n^2}\exp\left[-\ell/2\sigma_n^2\right] \quad \ell \geq 0$$
$$= 0 \quad \ell < 0,$$
(18)

where $\sigma_1^2 = \sigma_y^2 + \sigma_n^2$. Conditional probability of detection for a given threshold γ for a signal and noise with variance σ_1^2 is

$$f_D(x/\sigma_1^2) = \int_\gamma^\infty \frac{1}{2\sigma_1^2}\exp\left[-\ell/2\sigma_1^2\right]d\ell$$
$$= \exp\left[-\gamma/2\sigma_1^2\right].$$
(19)

Similarly the probability of false alarm at threshold γ and noise variance σ_n^2 is

$$f_{FA}(x/\sigma_n^2) = \exp\left[-\gamma/2\sigma_n^2\right].$$
(20)

If we fix f_{FA} and solve for the threshold, γ, we have an expression for the conditional detection PDF in terms of the signal to noise ratio θ:

$$f_D(x/\theta) = f_{FA}^{\frac{1}{1+\theta}}.$$
(21)

Due to randomness and uncertainty of propagation conditions, target strength and model, we take θ to be a *random variable*. It can be shown that a maximum entropy distribution that satisfies these physical constraints is gamma distribution. The marginal probability of detection is:

$$P_D = \int_\Theta f_\Gamma(\theta) P_{FA}^{\frac{1}{1+\theta}} d\theta.$$
(22)

Equation 22 is used to plot the "mean" of the ROC, however $f_D(x/\theta)$ is a function of random/uncertain signal to noise ratio θ. Since $f_D(x/\theta)$ is constrained to be in range of $[0,1]$, its possible maximum entropy distribution is beta distribution of first kind with moment constraints:

$$E\{\ell n f_D(x/\theta)\} = \int_\Theta f_\Gamma(\theta)\frac{\ell n P_{FA}}{1+\theta}d\theta = k_1,$$
(23)

$$E\{\ell n(1-f_D(x/\theta))\} = \int_\Theta f_\Gamma(\theta)\ell n\left[1-\exp\left\{\frac{\ell n P_{FA}}{1+\theta}\right\}\right]d\theta = k_2.$$
(24)

The parameters for the beta distribution are calculated from constraints specified by Equations 23 and 24:

$$E\{\ln(f_D(x/\theta))\} = \frac{1}{B(m,n)} \int_0^1 x^{m-1}(1-x)^{n-1} \ln(x) dx = f_1(m,n), \quad (25)$$

$$E\{\ln(1-f_D(x/\theta))\} = \frac{1}{B(m,n)} \int_0^1 x^{m-1}(1-x)^{n-1} \ln(1-x) dx = f_2(m,n). \quad (26)$$

The beta density is:

$$f_\beta(x,m,n) = B^{-1}(x,m,n) x^{m-1}(1-x)^{n-1}, \quad (27)$$

where

$$B(m,n) = \int_0^1 x^{m-1}(1-x)^{n-1} dx. \quad (28)$$

The receiver operating characteristic's mean value and confidence intervals are shown on Figure 1. Parameters for the Gamma density can be computed from mean signal to noise ratio and from the mean of the logarithm of the signal to noise ratio, which in turn can be computed from the sonar equation. Details of this computation will be presented in a longer journal paper.

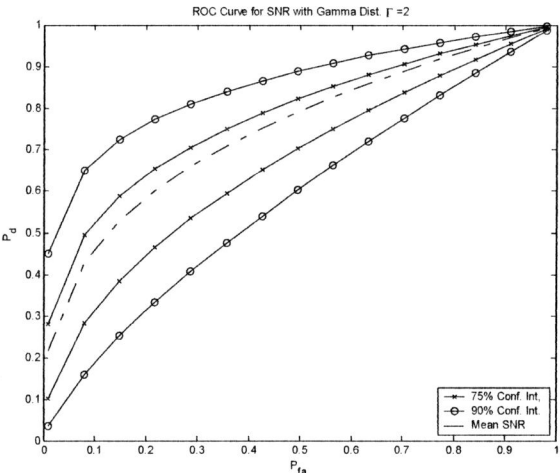

FIGURE 1. Confidence intervals and mean of the receiver operating characteristic of a matched filter detector when signal to noise ratio is a random variable that is Gamma distributed. Gamma=2

CONCLUDING REMARKS

High frequency source and receiver motion through inhomogeneous medium with random boundary conditions causes time and frequency spreading of the received signal. This time-frequency spreading can be modeled by *spreading functions*. In principle, loss of signal coherency, spectral spreading and increase of signal uncertainty can be computed by using the spreading function model. Actual computation requires sophisticated propagation modeling in a media with uncertain and frequently unspecified parameters. In this paper, we presented an example of application of maximum entropy method (MEM) to the performance analysis of a simple matched filter detector. This analysis can also applied to passive sonar detectors. MEM is a constrained optimization problem that maximizes entropy using known moments and range of the random variable as constraints. MEM is maximally uncertain in what is not warranted by data or known models [5-7]. In this performance analysis, uncertainty of propagation and background noise was accounted by treating the received signal to noise ratio as a random variable whose *PDF* is calculated according to the maximum entropy principle. According to the maximum entropy principle, the signal to noise ratio is a Gamma distribution and the conditional probability of detection is a beta distribution of second kind. Parameters of the Gamma distribution can be determined from the mean and logarithmic mean of the signal to noise ratio. The simple example of application of MEM that has been presented in this paper can be extended to more complicated problems.

ACKNOWLEDGEMENTS

This material is based on the work supported by the Office of Naval Research under Contract No.N00024-02-D-6604. Dr. John Tague has suggested this problem.

REFERENCES

1. Adomian, G., "Linear Random Operator Equations in Mathematical Physics I, II III," *Jour. Math.Phys.*, **11**, 1069-1084 (1970), **12**, 1944-1655, (1971).
2. Van Trees, H.L.,, *Detection, Estimation and Modulation Theory*, Wiley, New York, 1971, Chapters 9-13.
3. Sibul, L.H., Weiss, L.G., and Dixon, T.L., "Characterization of Stochastic Propagation and Scattering via Gabor and Wavelet Transforms," *Jour. Comp. Acoustics*, **2**, 345-369 (1994)
4. Jensen, FB., Kuperman, W.A., Porter, M.B., and Smith, H., *Computational Ocean Acoustics*, AIP Press, New York, 1994, pp. 454-459.
5. Jaynes, E.T., "On the Rational of Maximum-Entropy Method," *Proc. IEEE*, **70**, 939-952 (1982).
6. Jaynes, E.T., "Prior Probabilities," *IEEE Trans. on System. Sciences and Cybernetics*, **4**, 227-247, (1968).
7. Kapur, J.N. and Kesavan, H.K., *Entropy Optimization, Principles with Applications*, Academic Press, Boston, 1992.

MARINE MAMMALS

The Dolphin Sonar: Excellent Capabilities In Spite of Some Mediocre Properties

Whitlow W. L. Au

Marine Mammal Research Program, Hawaii Institute of Marine Biology, P.O. Box 1106, Kailua, Hawaii 96734

Abstract. Dolphin sonar research has been conducted for several decades and much has been learned about the capabilities of echolocating dolphins to detect, discriminate and recognize underwater targets. The results of these research projects suggest that dolphins possess the most sophisticated of all sonar for short ranges and shallow water where reverberation and clutter echoes are high. The critical feature of the dolphin sonar is the capability of discriminating and recognizing complex targets in a highly reverberant and noisy environment. The dolphin's detection threshold in reverberation occurs at a echo-to reverberation ratio of approximately 4 dB. Echolocating dolphins also have the capability to make fine discriminate of target properties such as wall thickness difference of water-filled cylinders and material differences in metallic plates. The high-resolution property of the animal's echolocation signals and the high dynamic range of its auditory system are important factors in their outstanding discrimination capabilities. In the wall thickness discrimination of cylinder experiment, time differences between echo highlights at small as 500-600 ns can be resolved by echolocating dolphins. Measurements of the targets used in the metallic plate composition experiment suggest that dolphins attended to echo components that were 20-30 dB below the maximum level for a specific target. It is interesting to realize that some of the properties of the dolphin sonar system are fairly mediocre, yet the total performance of the system is often outstanding. When compared to some technological sonar, the energy content of the dolphin sonar signal is not very high, the transmission and receiving beamwidths are fairly large, and the auditory filters are not very narrow. Yet the dolphin sonar has demonstrated excellent capabilities in spite the mediocre features of its "hardware." Reasons why dolphins can perform complex sonar task will be discussed in light of the "equipment" they possess.

INTRODUCTION

The echolocation system of a dolphin can be divided into three major subsystems: reception, transmission, and signal processing/decision making subsystems. The receiving subsystem consists of the auditory system of the animal, and its capabilities depend on the characteristics of the peripheral and higher auditory centers of the auditory central nervous system. The capability of a dolphin to detect objects in noise and clutter and to discriminate between various objects, and to recognize specific objects depends to a large extent on the information-carrying capabilities of the emitted signals. Also important are the extent to which the dolphin's auditory system can extract pertinent information from the echoes and the animal's cognitive capabilities. In order to make optimal use of acoustical information, the dolphin should have an auditory system that is very sensitive over a wide frequency range. The dolphin should also be sensitive in both quiet and noisy environments and should be able to detect short- and long duration sounds. A good spectral analysis capability is important in discriminating and recognizing predators, prey, and other objects in the environment. Other important characteristics of a good sonar receiver include the ability to spatially resolve and localize sounds, reject externally generated interferences, and recognize temporal and spectral patterns of sounds.

Most of the data that will be discussed here come from the Atlantic bottlenose dolphin, *Tursiops truncatus*. This species is the most common in oceanariums, marine parks, aquaria and other public display facilities. It is also the species that is most common in captivity. Occasionally, data from other species will be used when appropriate.

RECEIVER CHARACTERISTICS

The hearing sensitivity at different frequencies (audiogram) of a bottlenose dolphin was measured in a classic study by Johnson [1]. His results along with those of Au et al. [2] are shown in Fig. 1. The audiogram of the two dolphins indicate that

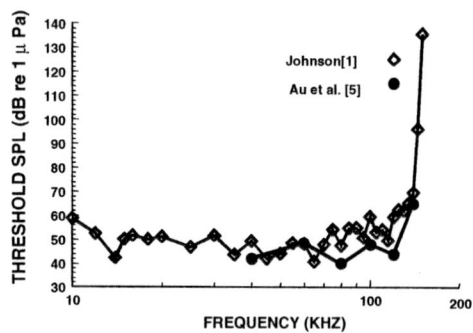

Figure 1. Hearing sensitivity of two Atlantic bottlenose dolphins as a function of frequency (from Johnson [1] and Au et al. [2]).

they have a very broad frequency range of hearing from 100 Hz to 150 kHz, covering approximately 10 octaves. The maximum sensitivity is approximately 40 dB re 1 µPa, which close to sea state 0 when taking into consideration the filter bandwidth in the dolphin's auditory system.

Receiving beam pattern

The receiving beam pattern of a bottlenose dolphin was measured by Au, et al. [3] and their results in both the vertical and horizontal planes for three different frequencies, 30, 60 and 120 kHz, are shown in Fig. 2. The major axis in the vertical plane is pointed between 5 and $10°$ above the horizontal axis. In the horizontal plane, the beam axis is pointed directly in front of the dolphin. The beam patterns are relatively wide in comparison to many technological sonar. For example, the SimRad-Mesotech MS-2000 multibeam sonar has 128 beams in the horizontal plane, each with a beamwidth of $1.5°$ covering a sector of $120°$.

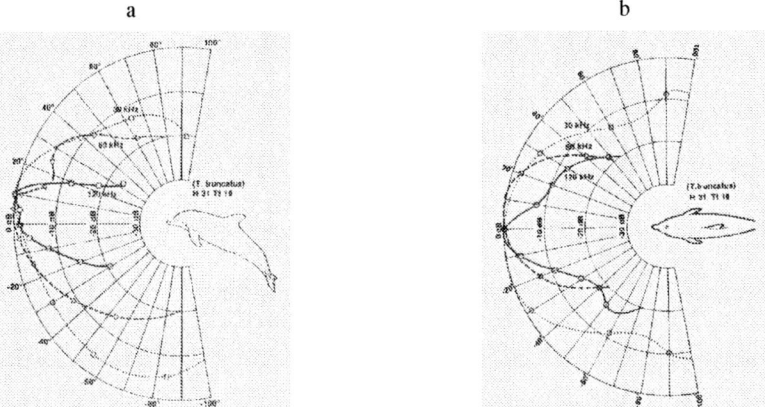

Figure 2. Receiving beam pattern for the bottlenose dolphin for three different frequencies, (a) in the vertical plane, (b) in the horizontal plane (from Au and Moore[3]).

Auditory filter shape

The auditory filter shape of a mammalian subject can be determined by performing a notched noise masking experiment where the tone signal is directly in the middle of the notch. Such a study was performed by Lemonds et al. [4] and their results are shown in Fig. 3 for frequencies of 40, 60, 80 and 100 kHz. Note that the filters are not very narrow. The shapes are similar to that of humans if we normalized

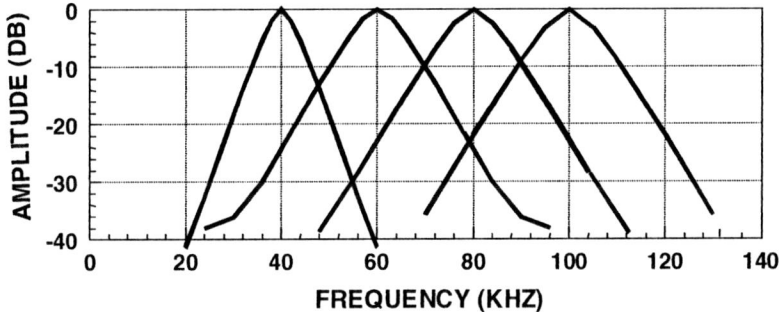

Figure 3. Auditory filter shape for a bottlenose dolphin.

the frequency by dividing by f_o for any given filter. If the 3-dB bandwidth is plotted as a function of the center frequency of each filter, a Q of 8.4 would represent the best constant-Q fit through the bandwidth points.

TRANSMITTER CHARACTERISTICS

Bottlenose dolphins emit short broadband clicks having peak frequencies as high as 120-130 kHz [5]. Signals typically have with 4 to 10 positive excursions and durations that vary from 40 to 70 μs,. Peak-to-peak source levels between 210 and 227 dB re 1 μPa have been measured [1]. Two echolocation signals of the bottlenose dolphin are shown in Fig. 1. Dolphins in tanks naturally emit much lower level signals with lower peak frequency. Examples of echolocation signals are shown in

Fig. 4. The frequency spectrum of transmitted signals is coupled to the output level of the signals having higher frequency as the output level increases. Typical bandwidth is between 40 and 60 kHz.

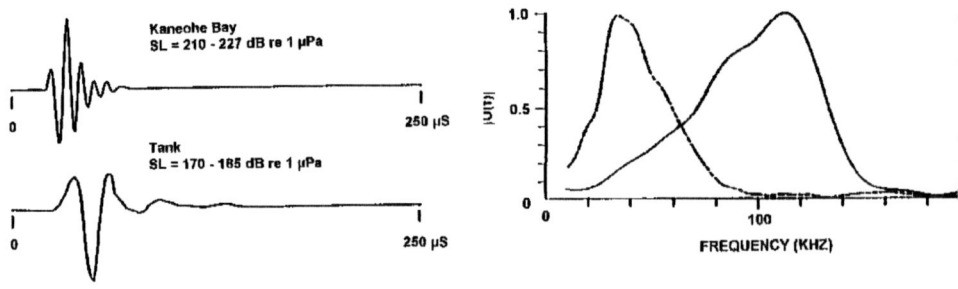

Figure 4. Representative echolocation signals of Tursiops truncates in a tank and in open waters. The waveforms are on the left and the frequency spectra on the right (from Au [5]).

Source levels

The source levels used by a dolphin will vary as a function of the loss involved in a sonar task. Au [5] examined the variation in the source level of five different bottlenose dolphins as a function of the total loss that the animals experienced from two-way spherical spreading loss and target strength and obtained the results shown in Fig. 5. The highest averaged peak-to-peak source level of 224 dB re 1µPa occurred for the dolphins Heptuna and Ehiku searching for a 3-in diameter thin-walled stainless steel water filled sphere at 72.8 m. The target strength shown in the legend is based on energy.

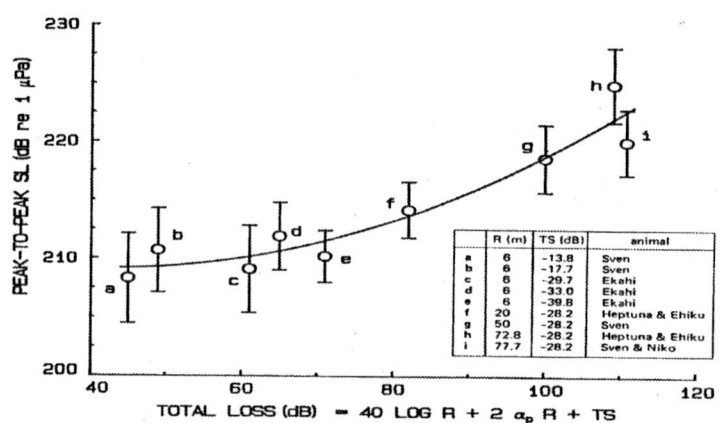

Figure 5. Peak-to-peak source levels used by 5 dolphins as a function of the total loss due to spherical spreading and target strength (from Au [5]).

Although the peak-to-peak source level of the sonar signal can be relatively high the energy flux density is relatively low because of the short duration of the signals. Let us compare the energy flux density of a typical sonar tone burst and that of a dolphin sonar signal by first defining the dolphin sonar signal as p(t) = A s(t), where A is the peak amplitude and s(t) is the normalized waveform. The energy flux density of the dolphin signal can be expressed as

$$E_{dolphin} = SPL_{pp} - 6 + 10\log\left(\int_0^T s^2(t)\,dt\right) \qquad (1)$$

For the echolocation signals shown in the top panel of Fig. 4, the integral term in db is approximately –52 dB [5], so the Eq. 1 can now be expressed as

$$E_{dolphin} = SPL_{pp} - 58 \qquad (2)$$

A similar expression can be written for a tone burst signal of duration T as

$$E_{dolphin} = SPL_{pp} - 9 + 10\log(T) \qquad (3)$$

The difference in the amount of energy is a tone burst over a dolphin signal with the same peak-to-peak source level can now be expressed as

$$\Delta E = E_{TB} - E_{dolphin} = 49 + 10\log(T) \qquad (4)$$

A graph of the amount of energy a tone burst would have over a dolphin sonar signal of the same peak-to-peak amplitude is shown in Fig. 6. A very short tone burst of 100 μs will have about 9 dB more energy than the high-frequency dolphin sonar signal

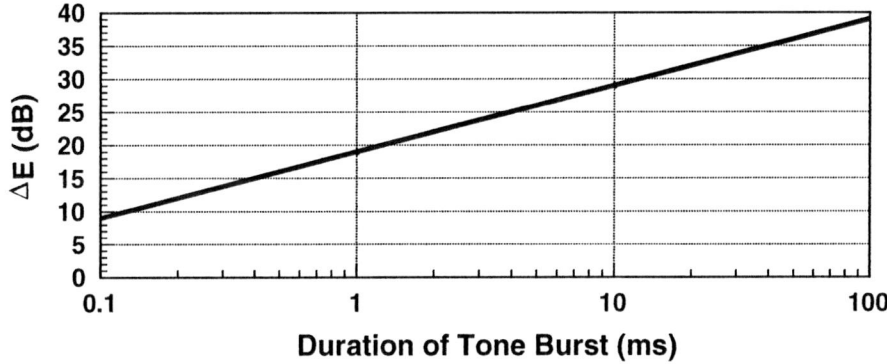

Figure 6. The amount of energy in dB that a tone burst would have over a dolphin echolocation signal of the same peak-to-peak amplitude.

shown in Fig. 4. The excess energy increase logarithmically with the duration of a tone burst.

Transmitting beam pattern

Signals are transmitted in a beam as shown in Fig. 7. The waveform of the signal measured by hydrophones at different angles about the animal's head. The transmit beamwidth of a high frequency sonar signal is 10.2° (horizontal plane) and 9.7° (vertical plane). The receiving beamwidth is slightly wider, 13.7° and 17° in the horizontal and vertical plane respectively. The directional projection and reception characteristics of bottlenose dolphin are poor compared to many technological sonar.

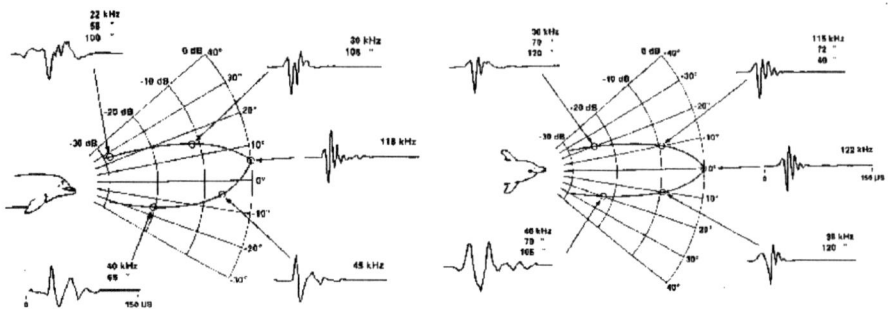

Figure 7 Transmission beam pattern for a bottlenose dolphin in the vertical and horizontal planes (from Au [5]).

One of the properties of the broadband nature of the dolphin sonar signals is the distortion of off-axis signals as can been seen in Fig. 6 for both planes. When a signal is measured at an angle greater than about 5° away from the beam axis, the signals become distorted and the amount of distortion increases as the angle increased.

SYSTEM'S PERFORMANCE

There are many experiments that can be discussed that would highlight the capabilities of echolocation dolphins in performing complex target discrimination tasks. Only three experiments will be discussed here, one on target detection in reverberation and two on target discrimination. Readers who would like to read more on dolphin sonar discrimination experiments should consider Au [5] and Nachtigall [6].

Target detection in reverberation

A sonar system is usually limited by noise or reverberation. Reverberation differs from noise in several aspects. It is caused by the sonar itself and is the total contribution of unwanted echoes scattered back from objects and inhomogeneities in the medium. Murchison [7] studied the effects of bottom reverberation on the target detection capabilities of two bottlenose dolphin in Kaneohe Bay. A 6.35-cm diameter solid steel sphere was used and eventually placed on the bottom. The animals' 50% correct detection threshold ranges for different target depth are shown in Fig.8. The threshold range for the target on the bottom was approximately 70 m. Au [8] used a simulated dolphin sonar signal to measure the scattering strength of the bottom where Murchison performed his experiment. Taking the target strength into consideration and the difference in the transmit and receive beam patterns of the transducer and the dolphin the reverberation form of the sonar equation was used to estimate an echo energy-to-reverberation (E/R) of approximately 4 dB. An example of an E/R ratio of 4 dB are shown in Fig. 9 (Au [8]). The highest highlight of the target echo is clearly detectable; however, the secondary highlights are masked by the reverberation so that the acoustic quality of the echo was altered. The dolphin probably could hear the largest highlight but the echo probably did not "sound" like the sphere they were trained to detect and consequently reported the target as not present.

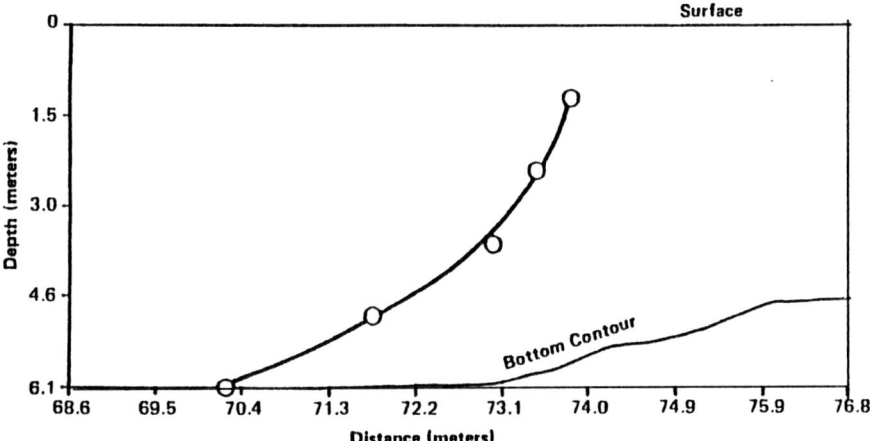

Figure 8. Target detection threshold as a function of target depth. The detection range when the 6.35-cm diameter sphere laid on the bottom was approximately 70 m (from Murchison [7]).

Therefore, it seems that a target detection experiment probably is not purely one of detecting signal in reverberation, but also involves discriminating the features of the echoes from a target. If the lower amplitude highlights are masked by reverberation or noise, the dolphins might hear the larger highlight components of the echo but the echo it probably would not "sound" like the target they were trained to detect. Therefore, target detection in noise and reverberation, also involves target recognition.

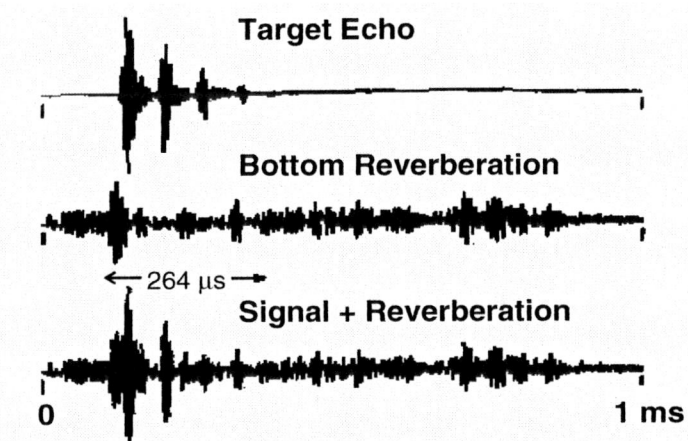

Figure 9. Target echo in reverberation at the dolphin's threshold of detection (from Au [8]).

Discriminating composition and thickness of metallic plates

Evans and Powell [9] demonstrated that a blindfolded, echolocating bottlenose dolphin could discriminate between metallic plates of different thickness and material composition. The dolphin was trained to recognize a 30-cm diameter circular copper disc of 0.22-cm thickness from comparison targets of the same diameter A schematic of the dolphin performing a typical search and the various comparison material and plate thickness are shown in Fig. 10. The dolphins could perform the task well above chance.

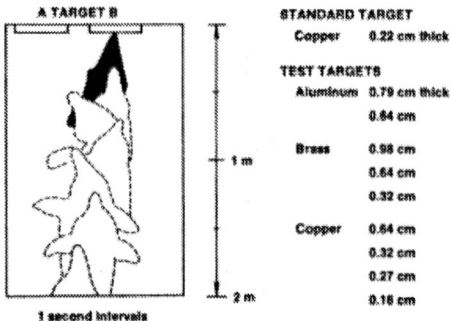

Figure 10. A typical sonar search by the blindfolded bottlenose dolphin and the various comparison targets comparison target used by Evans and Powell [9].

Au and Martin [10] examined the plates used in the experiment of Evans and Powell [9] with an echo ranging system that projected simulated dolphin echolocation signals. Backscatter results at normal incident indicated that virtually no cues for discrimination was present in the echoes. However, when the plates were examined at angles away from the normal, the different plates began to display unique highlight structures. Examples of backscatter at normal incident and at 14° incident are shown in Fig. 11. The echoes from the 14° incident angle are about 20 dB below that of the normal incident, yet the discrimination cues were present for the off-axis backscatter. This implies that dolphin are able to use cues that are at least 20 dB below the maximum amplitude of the echoes at normal incident in order to discriminate targets.

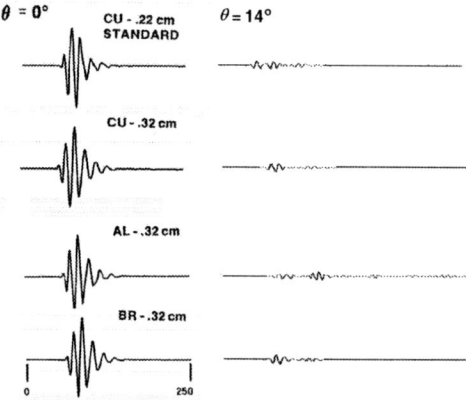

Figure 11. Examples of backscatter from the standard disk and some of the comparison disks used by Evans and Powell [9].

Cylinder wall thickness discrimination

The capability of a bottlenose dolphin to discriminate the wall thickness differences was measured by Au and Pawloski [11]. A dolphin was trained to station in a hoop and echolocate of two targets 8 m away separated by 22° azimuth. The standard target was a 3.81-cm O.D. aluminum cylinder with a wall thickness of 6.35 mm. Comparison targets with wall thickness both thinner and thicker than the standard were used. The comparison targets had incremental differences in wall thickness of ± 0.2, ± 0.3, ± 0.4 and ± 0.8 mm from the standard target. The dolphin was required to echolocate and to respond to the paddle that was on the same side of the center line as the standard target. The dolphin's performance as a function of wall thickness difference is shown in Fig. 11a. The 75% correct response threshold corresponded to a wall thickness difference of –0.23 mm for the thinner targets and +0.27 mm for the thicker targets. Echoes from the standard and the 0.3 mm thinner wall thickness comparison target are shown in Fig. 11b. The echo waveforms are shown in the top two traces, followed by the envelopes of the echo waveforms overlaid on each other and by the frequency spectra in the bottom traces.

The dolphin was able to perform the wall thickness discrimination represented by Fig. 12 b and was below the threshold for the next thinner target. If the animal was using time-domain cues, then the echo data suggest that it could discriminate a 600 ns difference between the arrival of the second highlight for each target. If frequency domain cues were used, than a frequency shift between 3.3 and 3.9 kHz could be detected in the broadband echoes.

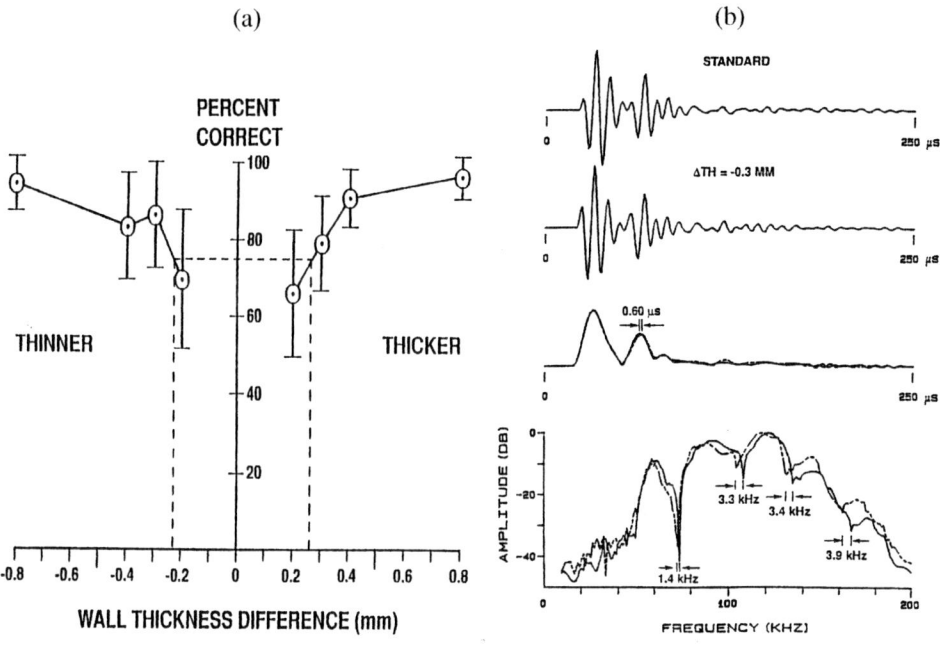

Figure 12. (a) Dolphin wall thickness discrimination performance, (b) Echo waveform, waveform envelope, and frequency spectrum for the standard and comparison target having a wall thickness difference of –0.3 mm. The dashed envelope and spectrum curves are for the comparison target (from Au and Pawloski [11]).

DISCUSSION AND CONCLUSIONS

The various discrimination experiments with echolocating dolphins strongly suggest that these animals possess a sophisticated and well honed sonar system in spite of some fairly mediocre acoustic properties of the sonar system. The transmit and receive beam patterns are not very narrow. The auditory filters are also not very narrow. The amount of acoustic power emitted is not very high when compared to technological sonar. There are many technological sonar with narrower transmit and receive beams and narrower receiver filters that emit more powerful acoustic signals. So we are left with the question: What are important factors that could allow dolphins to have such good sonar discrimination and recognition capabilities?

The first property of the dolphin sonar that contributes to good performance comes from the use of broadband echolocation signals. Shown in Fig. 13 is the envelope of the cross correlation function of the signal shown in Fig. 4 with the echoes from two point source separated by varying time τ. When the point target is separated in time by 15 µs, the two targets begin to be resolvable. When the two point targets are separated by 20 µs, the two peaks in the envelope response are almost completely resolvable. Therefore, the temporal resolution of a dolphin echolocation signal is approximately 15 – 20 µs, which translate to a distance resolution of about 0.015 m or 15 cm. The fine temporal resolution that can be obtained with dolphin sonar signals does not require special processing such as pulse compression making it possible for dolphin to process echoes in the time domain.

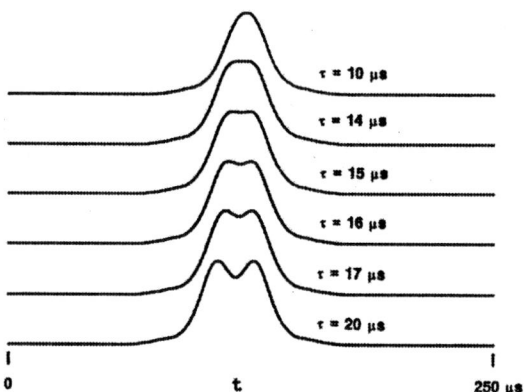

Figure 13. The normalized envelope of the cross correlation function of echoes from two point targets separated in time by τ and the transmitted signal shown in the top left panel of Fig. 4 (from Au [5]).

The use of broadband signals may also allow for the perception of time-separation pitch by dolphins. When a sound consisting of two correlated pulses are projected to humans, a pitch that is equal to the reciprocal of the time delay between the two pulses can be perceived by the human auditory system and may also be perceived by most mammal. If more than two highlights are present in an echo, a time-separation like pitch can still be heard. Therefore, a dolphin may discriminate targets from a pitch-like sound that multi-highlight echoes produce.

However, the short duration of the transmit signal limits the amount of energy within a signal so that the range of the sonar system is not very large. Typically, dolphins seem to be interested in objects that are within 100 m and are hardly concerned about longer ranges. So in a sense, temporal resolution was traded off with maximum range in the evolution process. However, within a 100 m range, there is not a technological sonar that can rival the dolphin in discriminating and recognizing targets. Bottlenose dolphins can even detect and discriminate targets that are buried in ocean sediment.

A second feature of the dolphin sonar system that is used to great advantage by the animals has to do with the dynamic range of its system. The metallic plate discrimination experiment discussed in Section 4.3 suggest that dolphins may gain information on an object by examining echoes that are 20 to 30 dB below the maximum level associated with a particular target. In other words, a dolphin does not seem to only seek out specific orientations to a target that will produce the highest echo levels but orientations that will provide the most information. In the metallic plate discrimination study, the orientation to the target that provided valuable information came from incident angles that were away from the normal to the plates. Therefore, the important parameter for the dolphin may be information level rather than echo level.

A third feature of the dolphin sonar system that is often overlooked is the fact that the sonar is mounted on a very flexible and mobile platform. Dolphins conduct sonar searches in an adaptive manner in that the trajectory of the animal at any given time will be the results of echoes received previously. A dolphin will not be restricted to running preprogrammed track lines or transect but is free to maneuver as the situation dictate. Therefore, a dolphin can approach and search on an object at different orientation and obtain whatever information it needs to recognize a target. The manner in which dolphins conduct sonar searches is another area of research that should be pursued. A system in which the sonar echoes dictate the specific trajectory of a mobile platform at any given time needs to be developed.

The dolphin sonar system has evolved over millions of years as nature's way to optimize an important sensory modality. Humans can take advantage of the natural selection process that has been working in dolphins to improve technological sonar. One obvious direction that should be pursued is the use of broadband signals that imitate the signals used by dolphins. There may be a temptation to adopt a longer broadband signal such as FM signals used by some bats in order to project more energy into the water. I caution against such an approach and suggest that more weight should be placed on using natural section as a guide, and we should strive to first produce a short-range sonar system that can perform as well as the dolphin. Perhaps, after such a system is developed, tested and used should we seek to improve on nature. There are many problems that still need to be solved in terms of processing broadband sonar echoes.

Finally, the ultimate reason for the keen sonar capability possess by dolphins has to do with the entire sonar process being controlled by a mammalian brain that allows for versatility and continuous learning. As a contrast, a neural network is trained to recognize features of specific targets and then the training stops. The "learned" templates are then used to recognize those targets in the field. On the other hand, the mammalian brain continues to learn and in this manner it can adapt to different situations and environment and benefit from previous experiences. Futuristic sonars may need to process signals in a fashion akin to how the brain process signals and control the whole sonar process. This may seem to be a daunting proposition but progress can be made in little steps. For example, it would be useful to develop effective ways to process brief broadband sonar signals, making use of the good temporal resolution inherent in these type of signals. It would also be advantageous to develop techniques in which a sonar on a free-roaming vehicle can control and perform adaptive sonar search patterns. Research should also be done on the process of continuous learning in a sonar function. I believe that we can make considerable progress in developing better sonar by following along the path that has been provided by dolphins.

ACKNOWLEDGMENT

I would like to express my appreciation to all my colleagues that have trained the dolphins used in the various biosonar experiments at SPAWAR Systems Center. In particular, I thank Jeffrey and Debra Pawloski, Ralph Penner, Earl Murchison, Patrick Moore, Norman Chun and Wayne Turl. This is HIMB contribution 1184.

REFERENCES

1. Johnson, C. S. "Sound Detection Thresholds in Marine Mammals". In <u>Marine Bio-Acoustics</u> W. Tavolga (Ed.). Pergamon, New York, pp. 247-260, 1967.

2. Au, W. W. L., Lemonds, D. W., Vlachos, S., Nachtigall, P. E., & Roitblat, H. L. "Atlantic Bottlenose Dolphin Hearing Threshold for Brief Broadband Signals". Journal of Comparative Psychology, Vol. 116: 151-157, 2002.

3. Au, W. W. L., & Moore, P. W. B. "Receiving Beam Patterns and Directivity Indi ces of the Atlantic Bottlenose Dolphin Tursiops truncatus". J. Acoustic. Soc. Am., Vol. 75, 255-262, 1984.

4. Lemonds, D. W., Au, W. W. L., Nachtigall, P. E., Roitblat, H. L., and Vlachos, S. A. "High Frequency Auditory Filter Shapes in an Atlantic Bottlenose Dolphin," J. Acoust. Soc. Am., Vol. 108, 2000, p.2614(A).

5. Au, W. W. L.. The Sonar of Dolphins. Springer-Verlag, New York, 1993.

6. Nachtigall, P. E. ""Odontocete Echolocation Performance on Object Size, Shape and Material". In <u>Animal Sonar Systems</u> R. G. Busnel & J. F. Fish (Eds.). Plenum Press, New York, pp. 71-95, 1980.

7. Murchison, A. E. "Maximum Detection Range and Range Resolution in Echo-locating bottlenose porpoises (*Tursiops truncatus*)". In <u>Animal</u> Sonar "R.G. Busnel & J.F. Fish (eds.) Plenum Press, New York, pp. 65-70, 1980.

8. Au, W. W. L. "Application of the Reverberation-Limited Form of the Sonar Equation to Dolphin Echolocation". J. Acoust. Soc. Am., **92**, 1822-1826, 1992

9. Evans, W. W., & Powell, B. A. "Discrimination of Different Metallic Plates by an Echolocating Delphinid". In <u>Animal Sonar Systems: Biology and Bionics</u> R. G. Busnel (Ed.). Laboratoire de Physiologie Acoustique, Jouy-en-Josas, pp. 363-382, 1967.

10. Au, W. W. L., and Martin, D. "Sonar Discrimination of Metallic Plates by Dolphins and Humans," in <u>Animal Sonar: Processes and Performance</u>, edited by P. E. Nachtigall and P.W.B. Moore, Plenum Press, N. Y. Pp 809-813, 1988.

11. Au, W. W. L., & Pawloski, D. A. "Cylinder Wall Thickness Difference Discrimination by an Echolocating Atlantic Bottlenose Dolphin". J Comp Physiol A, 172, 41-47, 1992.

Biomimetic Signal Processing Using the Biosonar Measurement Tool (BMT)

Ahmad T. Abawi*, Paul Hursky*, Michael B. Porter*, Chris Tiemann* and Stephen Martin⋆

*Center for Ocean Research, Science Applications International Corporation San Diego, CA 92121,
⋆ SPAWAR Systems Center, San Diego, San Diego, CA 92152

Abstract. In this paper data recorded on the Biosonar Measurement Tool (BMT) during a target echolocation experiment are used to 1) find ways to separate target echoes from clutter echoes, 2) analyze target returns and 3) find features in target returns that distinguish them from clutter returns. The BMT is an instrumentation package used in dolphin echolocation experiments developed at SPAWARSYSCEN. It can be held by the dolphin using a bite-plate during echolocation experiments and records the movement and echolocation strategy of a target-hunting dolphin without interfering with its motion through the search field. The BMT was developed to record a variety of data from a free-swimming dolphin engaged in a bottom target detection task. These data include the three dimensional location of the dolphin, including its heading, pitch roll and velocity as well as passive acoustic data recorded on three channels. The outgoing dolphin click is recorded on one channel and the resulting echoes are recorded on the two remaining channels. For each outgoing click the BMT records a large number of echoes that come from the entire ensonified field. Given the large number of transmitted clicks and the returned echoes, it is almost impossible to find a target return from the recorded data on the BMT. As a means of separating target echoes from those of clutter, an echo-mapping tool was developed. This tool produces an echomap on which echoes from targets (and other regular objects such as surface buoys, the side of a boat and so on) stack together as tracks, while echoes from clutter are scattered. Once these tracks are identified, the retuned echoes can easily be extracted for further analysis.

INTRODUCTION

Dolphins have an impressive ability to identify underwater targets using their biological sonar (biosonar) system. They can identify objects based on their shape, exterior wall thickness and exterior and interior material compositions. A variety of reasons contribute to this uncanny ability. From a purely sonar point of view, there are at least two reasons that enable the dolphins to identify underwater targets with such success. One of these reasons is that dolphins can inspect objects by emitting trains or sequence of impulsive sound known as clicks whose frequency content and amplitude as well as inter-click separation can be adaptively controlled. The other reason is that the dolphin sonar operates on a highly mobile platform. They can move around the object of interest, ensonify it at different angles and obtain different "look" directions in much the same way as humans visually inspect objects at different angles. The dolphin clicks are approximately 50-100 µs long with peak frequencies typically ranging between 30-150 kHz and fractional bandwidth between 10%-90% of peak frequency [1,2]. Although the outgoing clicks are brief, echoes reflected from objects can be several milliseconds long rich with information about the object's shape, orientation and composition. The inter-click separation as well as the click amplitude and frequency content can be adjusted depending on the range to and the type of object being interrogated. However, of the three major subsystems that make up the echolocation system, namely reception, transmission and signal processing, despite their impressive performance, the dolphin's reception and transmission subsystems are quite mediocre compared to its signal processing capabilities [3]. It is how dolphins process, integrate and direct sonar functions that gives them the unsurpassed ability to identify underwater objects.

It is the overall performance of the dolphin sonar that has been the subject of extensive research. The objective of many researchers has been to learn how dolphins solve the classification problem and to construct analogous, biomimetic mechanisms. The SPAWAR Systems Center, San Diego (SSCSD) has conducted experiments with dolphins since the 1960's. In these experiments SSCSD researchers have addressed the characteristics of echolocation clicks, mechanisms of click production and echo reception and the adaptive production of clicks relative to the echolocation task performed. To investigate the echolocation strategies of a target-hunting dolphin during a target detection and identification experiment, the team at SSCSD has developed the Biosonar Measurement Tool (BMT) and the Instrumented Mine Simulator (IMS) [4]. The BMT is an instrument that can be held by the dolphin using a bite-plate during an echolocation experiment (see Figure (1)). It records the movement and echolocation strategy of a target-hunting dolphin without interfering with its motion through the search field. The BMT can record a variety of data from a free-swimming dolphin engaged in a bottom target detection task. These data include the three dimensional location of the dolphin, including its heading, pitch roll and velocity as well as passive acoustic data recorded on three channels. The outgoing dolphin click is recorded on one channel and the resulting echoes are recorded on the two remaining channels. The IMS is an instrument that records echolocation clicks at a mine simulator.

In this paper data recorded on the BMT during an echolocation experiment are used to 1) find ways to separate target echoes from clutter echoes, 2) analyze target returns and 3) find features in target returns that distinguish them from clutter returns. This paper is organized as follows: in Section II the experimental setup is described, in Section III results of the data analysis are discussed followed by summary in Section IV.

Figure 1. The picture on the left shows the BMT and the one on the right shows a trainer putting t bite-plate attached to the BMT to the mouth of a subject dolphin.

 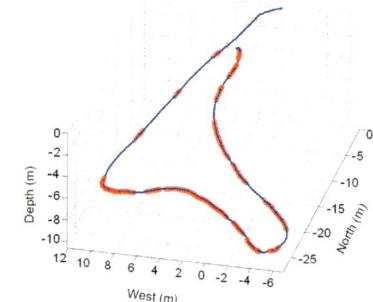

Figure 2. The left picture shows the target used in the trail discussed in this paper. It is composed o sphere attached to a cylindrical post with a rectangular base. The figure on the right shows the dolp search path, as it interrogates the target, issues a positive whistle and returns to the workboat. The dots indicate transmission of the clicks.

EXPERIMENT SETUP

The experimental design used to investigate dolphin echolocation strategies duri target detection and identification is dubbed "hide and seek" [4]. During the experiments one of two dolphin subjects is taken out to a pre-configured location. trail consists of positioning the workboat 20 to 60 meters from one of the surface sw floats. A swim float marks either a positive station (i.e. it has a mine simulator wit 3 to 30 meters located nearby) or a negative station (i.e. a mine simulator is located nearby). The dolphin is trained to station on the port side of the workboat a take the BMT into its mouth during the start of the trial. The dolphin is trained swim towards the surface float while conducting an acoustic search for the bott targets. It reports positive "target present" by whistling at the end of the search. assistant listens to the response with a hydrophone and headset and a bridge signa provided to the dolphin if the response is correct. The dolphin returns to the workb immediately after issuing the positive whistle response. The positive whistle respor is also recorded by the acoustic sensors on the BMT. If the dolphin does not fin target, it is required to swim to and around the surface float before returning to the workboat. Typical trials range in time anywhere between several seconds to 90 seconds.

In the trial of interest to this paper a target (shown in left panel of Figure 2) was placed near the surface swim float on the bottom and a dolphin subject named Flip was used to echolocate the target. The right panel in Figure 2 shows the dolphin search path as it interrogates the target. The red dots along the swim path of the dolphin show the transmissions of clicks. Note that the frequency of click transmission increases as the dolphin gets near the target, located at approximately 10 meters below the surface. After the dolphin issues a whistle, reporting that a target is present, he turns around and swims back up towards the workboat. He continues to click during his return path, presumably to echolocate the boat. During this trial, which lasted a little over 40 seconds, over 1100 clicks were transmitted. The transmitted clicks and the associated echoes were recorded on the BMT. These data will be analyzed in the next section.

DATA ANALYSIS AND RESULTS

For every transmitted click the BMT records tens and tens of echoes from features in the ensonified field. Figure 3 shows a sample of these echoes. As it can be seen in Figure 3, it is impossible to be able to tell whether these echoes belong to the target or

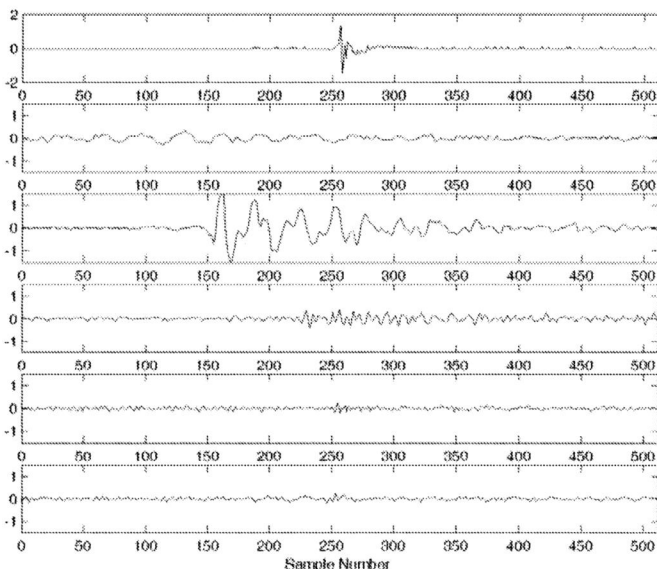

Figure 3. The top panel shows a typical dolphin click. The bottom five panels show echoes from various features in the ensonified field.

clutter.

To be able to group echoes, we plotted the time for each transmitted click along the vertical axis and the corresponding time for the returned echoes along the horizontal axis. The time for the returned echoes is measured from the time that the corresponding click was transmitted. This plot, which we refer to as the echomap, is shown in Figure 4, where each point represents the location of the peak of an echo time series.

The remarkable feature of the above plot is the appearance of tracks, which correspond to echoes that consistently line up regardless of the location of the dolphin. The echoes from the target are shown in red. Note that as the dolphin approaches the target, the time separation between the clicks and the echoes decreases. At the end of the track the dolphin whistles, indicating that he has found the target. The vertical track, which crosses the target track, corresponds to echoes from the ocean surface. Observe that after the dolphin makes the positive identification and is on its way toward the surface, the time separation between the clicks and the surface echoes decreases, indicating that the dolphin is approaching the surface. The other tracks on the echomap cannot be identified easily, as they may correspond to echoes from the surface floats, the workboat or discarded objects on the ocean bottom. Nevertheless, the echomap provides the means to be able to divide the received echoes into at least two groups: those that appear on tracks and those that do not.

Once a point or a series of points are selected for further analysis, their corresponding echo time series can be extracted. Figure 5 shows ten randomly selected returned echo time series along each of the three colored tracks shown on the top left panel. The top right panel shows echoes along the red track, the bottom left panel shows echoes along the magenta track and the bottom right panel shows echoes along the blue track. The echoes along the blue and the red tracks are those of the target. They exhibit the two dominant returns from the front and the back of the target, as is typical of returned echoes from curved shells. The echoes along the magenta track belong to an unknown target. The echo time series for points that do not lie along a track exhibit completely different characteristics. A few echo time series representing these points are shown in Figure 3.

Before analyzing the received echoes, it is useful to look at the transmitted clicks since differences in the received echoes cannot be fully explained without accounting for their corresponding transmitted clicks. Figure 6 shows a correlation matrix of all transmitted clicks, where each click was correlated with every other click. As can be seen in Figure 6, the early clicks (<370) are highly correlated. This is the period of time when the dolphin is in the process of searching and identifying the target. The gap at around click number 370 occurs when he stops clicking and issues a whistle, announcing that he has identified the target. Although the dolphin keeps clicking until he returns to the boat, his later clicks, particularly those above 800, do not correlate well with his earlier clicks when he was searching for the target.

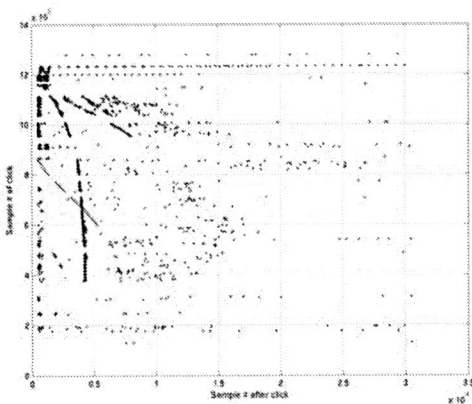

Figure 4. This figure is a plot of the time of received echoes, shown in terms of sample number along the horizontal axis, versus the corresponding outgoing clicks along the vertical axis. The time along the horizontal axis is measured from the time that the corresponding click was transmitted, i.e. all clicks lie along the vertical axis.

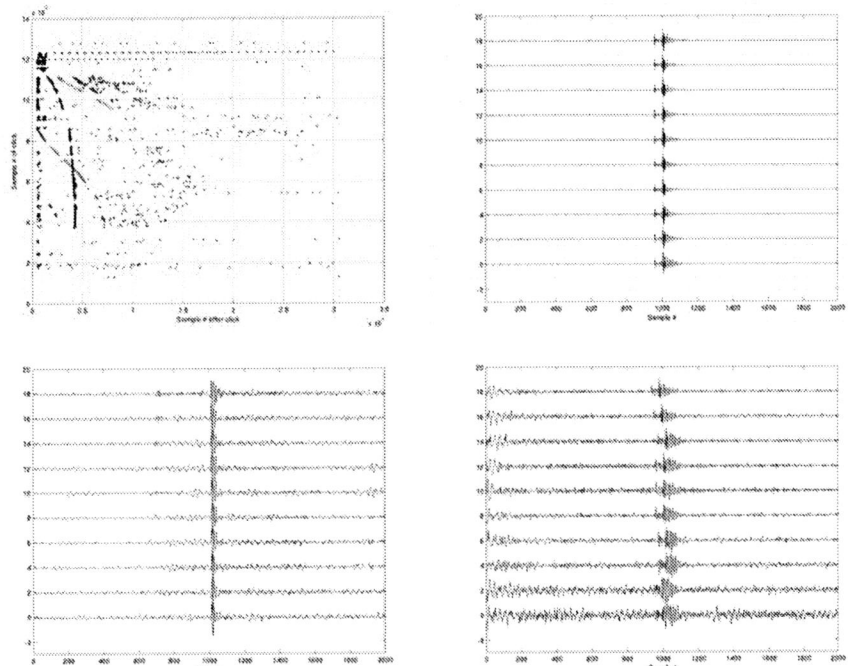

Figure 5. The echo time series corresponding to ten randomly selected points along the colored tracks shown in the top left panel. The top right panel shows the time series for points along the red track, the bottom left and bottom right panels show the same for points along the magenta and blue tracks, respectively.

This suggests that dolphins may use different types of clicks depending on the type of task they have to perform.

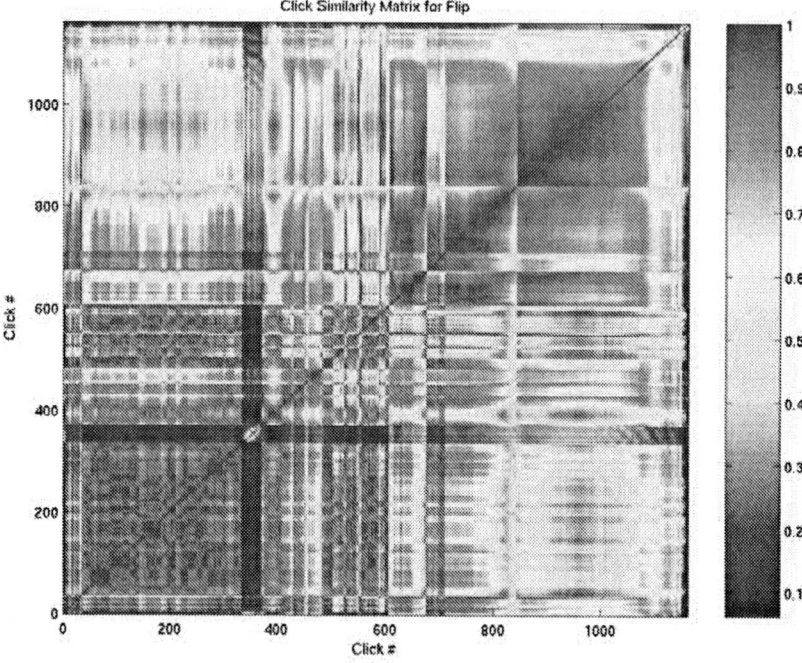

Figure 6. The correlation matrix for the transmitted clicks shows that earlier clicks (<370) are highly correlated. The later clicks, particularly those >800 do not correlate well with the earlier clicks.

As a way to quantitatively study the differences between received echoes we correlated the echoes from the target with all other echoes. An experimental time series echo model was constructed by averaging 31 consecutive returns from the target. This corresponds to the middle of the red track in Figure 4. We also correlated the envelope of the averaged time series and a boxcar model with all received echoes. The boxcar model represents the grossest features of the time series envelope, namely the two dominant peaks. The top left panel in Figure 7 shows the three experimental models used. The remaining three panels in Figure 7 show color-coded echomaps of the correlation between each experimental model and all the received echoes. The color in each echomap represents the amplitude of the correlation, with red representing large values and blue representing small values. The purpose of plotting these color-coded echomaps is to see the distribution of correlation on the echomap.

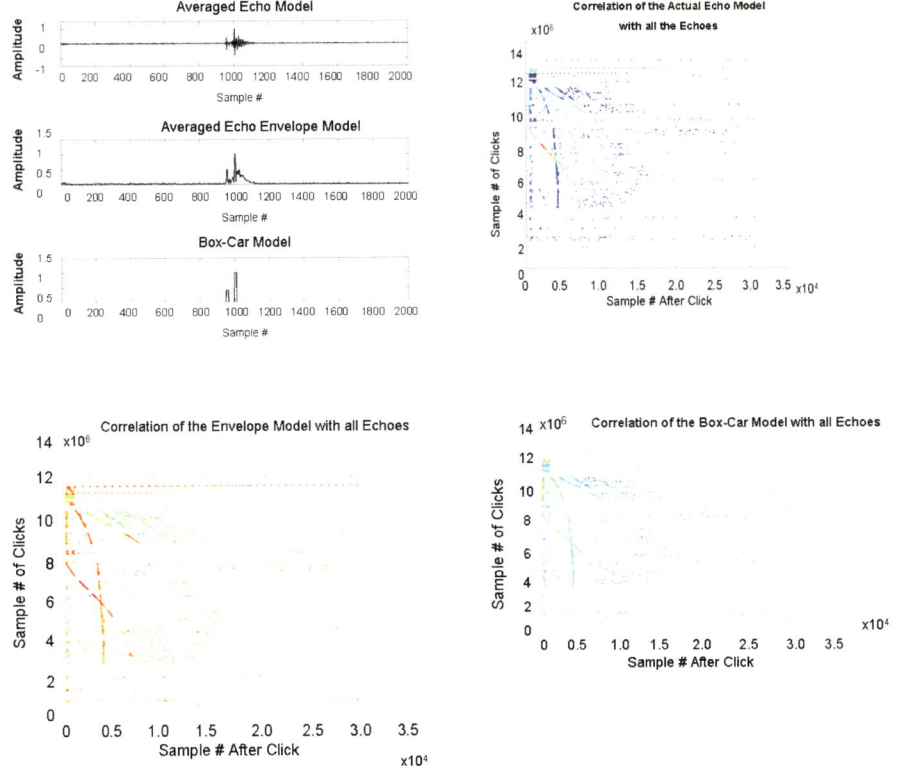

Figure 7. The top left panel in this figure shows the three experimental target echo models: the top panel shows the averaged time series model, the middle one shows the envelope of the averaged time series and the bottom one shows the boxcar model, which has the grossest features of the envelope model, namely the two dominant peaks. The other three panels show color-coded echomaps, which display the distribution of high (red) and low (blue) correlation of the three echo models with all the received echoes. The top right panel is a color-coded echomap showing the distribution of correlation of the averaged time series model with all the other received echoes. The bottom left and bottom right panels show the same for the envelope and box-car models, respectively.

As is evident from the top right panel in Figure 7, the average time series model correlates well with points on the echomap, which belong to echoes from the target (the red track in Figure 4). This is to be expected, since this model was constructed by averaging 31 time series selected from points along the same track. However, observe that this model does not correlate well with echoes that are not located on the target track. Therefore, a threshold correlation value can be found to separate echoes belonging to the target track from the rest of the echoes.

The envelope model can separate echoes that are located on tracks from the rest of the echoes (scattered echoes). The boxcar model shows a slightly better performance. However, to be able to compare the performance of these models quantitatively, the echoes on the echomap were divided into two major groups: those that lie on tracks and those that do not. The ones that lie on tracks were divided into five subgroups color-coded in red, blue, green, magenta and yellow. The red and the blue subgroups represent the target echoes. The experimental models were correlated with the echoes from each group and the color-coded correlations were plotted in Figure 8.

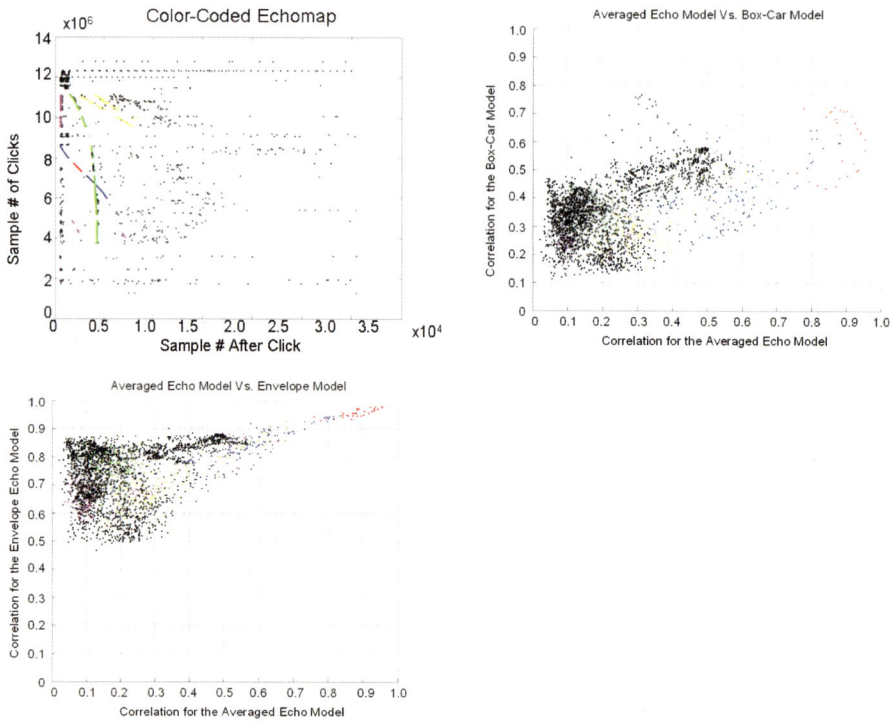

Figure 8. A quantitative comparison of the performance of the experimental echo models using feature space plots. The top left panel shows the division of the echoes in the echomap into six groups, each designated by a different color. The top right panel shows a feature space plot of the box-car model correlation versus the averaged time series model correlation. The bottom left panel shows the same type of plot for the envelope model. The feature space plots have the same color convention as the color-coded echomap.

The top right and bottom left panels in Figure 8 are feature space plots comparing the performances of the three experimental echo models. The top right panels compare the performance of the averaged time series model with that of the boxcar model. Observe that the averaged time series model is able to separate echoes located on the target track (red and blue) from the rest of the echoes very well. The correlation values for echoes located on the target track and the rest of the echoes are well apart and a threshold correlation value of about 0.6 can separate them. The boxcar model cannot separate the target echoes as well and a relatively high threshold correlation value of approximately 0.8 is required to do this. Hence, its performance is not as good as the averaged time series model. The bottom left panel in Figure 8 compares the performance of the envelope model with that of the averaged time series model. Note that the envelope model correlates well with both the target and non-target echoes, as all the correlations have values larger than 0.5. It does separate the target echoes from the rest of the echoes, but a large threshold correlation value of almost 0.9 is required to do this. Therefore, the envelope model has the worst performance of the above three models.

Correlation techniques provide one way of comparing target echoes with non-targets echoes. Pattern recognition techniques can also be employed to discriminate targets and non-targets by looking at the time-frequency response of each class. Figure 9 shows a series of spectrograms selected from points along the target track, where time is along the horizontal axis and frequency is along the vertical axis. The two strong arrivals, due to scattering from the front and the back of the target, are clearly visible. Note that the second arrival is stronger and spans over a wider band of frequency. The two arrivals and the ensuing ringing give the time-frequency response of the target a unique L shape, which is not present in the time-frequency response of the non-target echoes shown in Figure 10. Based on the differences between the two sets of time-frequency responses, in principle it is possible to design a pattern recognition-based classifier to discriminate the target echoes from those of non-targets.

SUMMARY

In this paper data recorded on the BMT were used to analyze the outgoing dolphin clicks and returned echoes during a dolphin echolocation experiment. To accomplish the main objective of the paper, which is to find ways to distinguish target echoes from those of clutter, the peaks of the returned echoes as a function of time and click number were mapped on to what is referred to as the echomaps. After analyzing the returned echoes further, it was verified that on an echomap echoes from regular objects (planar, curved, etc.) appear as tracks and those from irregular objects (rough surfaces, rocks, etc.) appear as scattered points. This property of the echomap was used to divide echoes from the ensonified field into separate categories of target, target-like, clutter-like and clutter. The time series for three target echo models were correlated with the echoes from each one of the above categories. The three target echo models consisted of an experimental echo time series, obtained from averaging 30 echoes from the target, a boxcar model, which consisted of two rectangular pulses collocated at the two prominent returns of the experimental echo time series and the envelope of the experimental echo time series. Various techniques, including feature space plots were used to determine how well each one of the above models could separate target echoes from clutter echoes. It was shown, perhaps not surprisingly, that the experimental echo model had the best performance followed by the boxcar model. Finally, spectrogram matching using pattern recognition techniques is proposed as a potentially more robust method for discriminating target echoes from non-target echoes.

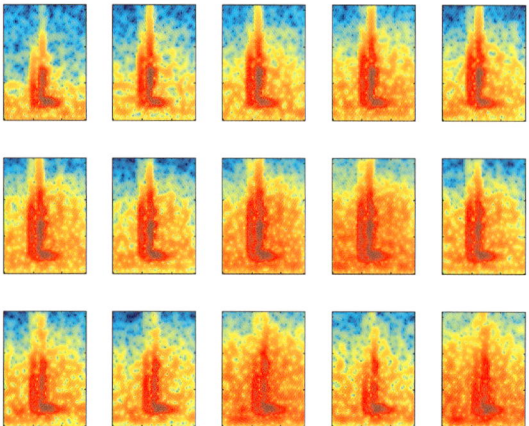

Figure 9. The spectrograms for echoes selected from the target track. The horizontal axis represents time and the vertical axis represents frequency.

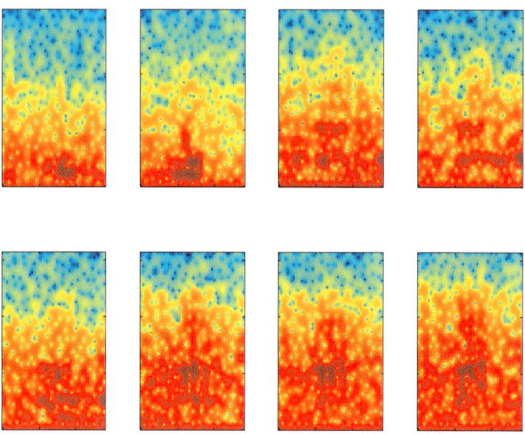

Figure 10. The spectrograms for echoes selected randomly from scattered points on the echomap.

REFERENCES

1. Au, W. W. L., "Echolocation Signals of the Atlantic Bottlenose Dolphins (*Tursiops truncatus*) in Open Waters," in Animal Sonar Systems, edited by G. Busnel and J. F. Fish (Plenum, New York), pp. 251-282, 1980.
2. Houser, D. S., Helweg, D. A. and Moore, P. W. "Classification of Dolphin Echolocation Clicks by Energy and Frequency Distributions," J. Acoust. Soc. Am. 106, 1579-1585, 1999.
3. Au, L. W. W., "The Dolphin Sonar: Excellent Capabilities In Spite of Some Mediocre Properties", this Volume.
4. Klappenback, S. and P. W. Moore, "Biosonar Measurement Tool – Training a Free Swimming Dolphin to use a Computer for Tracking and Research", Proceedings of the 30th Annual IMATA Conference, Orlando, Fl, p.24, 2002.

Active sonar and the marine environment

Erik M. Sevaldsen & Petter H. Kvadsheim

FFI (Norwegian Defence Research Establishment), Horten, Norway

Abstract. A study of the effects of active sonar transmissions on fish and marine mammals in Norwegian waters has been launched following the ordering of new frigates by the Royal Norwegian Navy (RNoN). The frigates will be equipped with active sonars operating at lower frequencies than those of the sonars currently in operation in the RNoN. Lower frequency sonar transmissions are believed to be potentially more harmful to marine life than higher frequency transmissions. The objective of the study is to acquire knowledge about the effects of active sonar on marine life, and produce a set of recommended rules for naval sonar operations in Norwegian waters based on scientific grounds.

INTRODUCTION

We started work on project LFAS (Low Frequency Active Sonar) and the Marine Environment in March 2003. The project, which is sponsored by the RNoN, aims at investigating the effects of lower frequency active sonars on marine mammals and fish. The RNoN has ordered five new frigates, the first one to be delivered in early 2006. The frigates will be equipped with towed array sonars, hull-mounted sonars and helicopter-operated dipping sonars. Frequency range of the sonar system will cover 1 to 8 kHz.

The main purpose of the project is to enable the RNoN to operate its lower frequency active sonar equipment in an environmentally safe way with as few operational restrictions as possible.
To achieve this result we have to
- establish what is known about the effects of sonar transmissions on fish and marine mammals, particularly at lower frequencies,
- create a forum for cooperation among military and civilian experts to discuss and agree on what we know and do not know about the effects of active sonar transmissions on marine mammals and fish found in Norwegian waters,
- conduct field studies recommended by the cooperative forum
- set up a national environmental database containing available information on the presence of fish and marine mammals with their known sensitivities to sonar transmissions, and
- produce a set of recommended rules for naval sonar operations in Norwegian waters based on scientific arguments.

THE EXPERT GROUP

Project work started with a search for available information on fish and marine mammals in Norwegian waters. We must know about the most important species found in Norwegian waters, economically and biologically, how they are distributed and what is known about their sensitivity to acoustic influences, noise in general and active sonar transmissions in particular.

To achieve these ends we have to cooperate nationally and internationally. The first step in this process was to establish a national forum for collaboration and discussions among military and civilian experts, "the Expert Group on Sonar Effects on Marine Life". The Expert Group held its first meeting in Horten 7. October 2003. Experts from the following institutions were invited:

- Department for Arctic Biology, University of Tromsø,
- Department for General Biology, Institute of Biology, University of Oslo,
- OLF/OGP (Association of Norwegian Oil Producers/Oil and Gas Producers International, Stavanger)
- Institute of Marine Research (IMR), Bergen
- The Norwegian College of Fishery Science (NCFS), University of Tromsø.
- Norwegian Polar Institute, Tromsø
- The Norwegian School of Veterinary Science, Tromsø/Oslo
- Norwegian Naval Training Establishment (KNMT), Haakonsvern, Bergen
- FFI (Norwegian Defence Research Establishment), Horten

The Expert Group agreed that we do have considerable information on abundance and distribution of various species in Norwegian waters, particularly the economically important types of fish like cod, herring etc., but also whales and seals. This information is stored at IMR in Bergen. The abundance estimates of fish are based on acoustic methods (echo integration) and trawl sample data. Marine mammal distribution estimates are based on random observations from ships, whereas population estimates of whales are based on the recognized line transect method [1].

The Expert Group also agreed that there is very little information available regarding fish or marine mammal sensitivity to sonar signal exposure. Some knowledge is available on hearing in fish, seals and dolphins.

We do not know much about the great whales' hearing. Their vocalization is known, and it is generally assumed that the mammals have their most sensitive range of hearing in the frequency range of their vocalization. However, since predators like the killer whale vocalize or echo-locate at considerably higher frequencies, it is possible that the great whales are able to hear sounds also in this range of frequencies.

Based on the available information on distribution of important species in Norwegian waters and the insufficient information on sensitivity to acoustic signals of these species, the Expert Group agreed to recommend that the following field studies be undertaken by the project:

1. Effects of sonar signals on survival and development of fish fry (spawn). This study can be continued using the same infrastructure to study behavioral and physiological effects of sonar exposure on adult herring and cod
2. Behavioral effects of sonar exposure on killer whales and minke whales
3. Effects of sonar exposure on seals

These proposals are based on a scientific evaluation of what we know and do not know and how we can harm animal life in the sea most by not knowing the effects of what we do.

RESOURCE DATA BASE

The information on marine life collected from areas within the Norwegian economic zone (see Fig 1) must be stored in a data base accessible for the RNoN. This data base shall contain resource data and data on sonar signal exposure sensitivity for species found in Norwegian waters.

In more detail the data base should contain all available information on

1) which species (of fish and marine mammals) are statistically found in given areas within the Norwegian economic zone at a specific time period of the year. The time period could be season (winter, spring, summer, fall), month or any other given period,
2) what are the animals actually doing in the area
    migration
    feeding
    spawning/breeding
    calving etc.
3) the best available information on sensitivity to sonar signal exposure for the various animals and fish related to activity and time of the year

This information shall be made available for planning of sonar exercises and for the naval vessels actually performing the exercises. Tactical decisions concerning the environment must be made based on the available statistical information found in the resource data base, updated information from external sources like the Institute of Marine Research (IMR) and actual observations from the chosen exercise area. Such observations (of marine mammals) could be visual or from passive sonar or active (whale) sonar. Active sonar exercises may be started by going through a ramp-up sequence of transmitted sonar signal levels if this is found to be a suitable procedure.

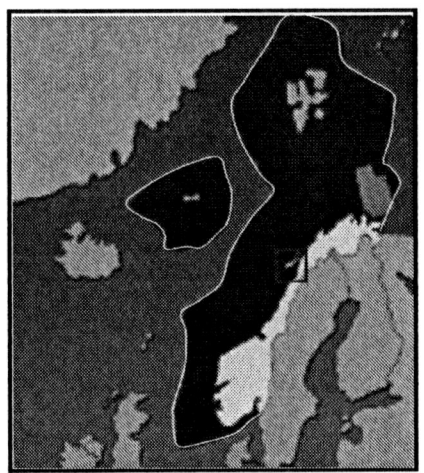 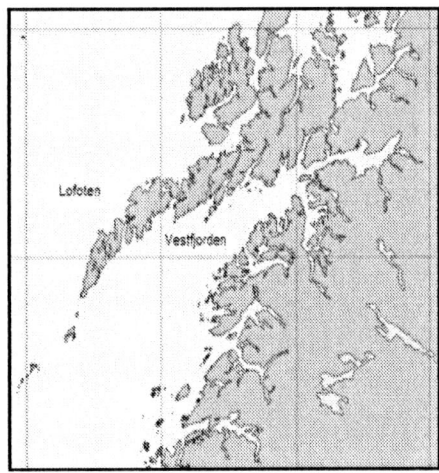

FIGURE 1. Map of the Norwegian Sea with Norway and the Norwegian economic zone. The section extract shows the Lofoten-Vestfjorden area where we intend to do behavioral tests on whales

The structure and format for the resource data base have not been decided on yet. Our solution must comply with the systems to be installed onboard the new frigates. We tend to favor a web-based system connected to a server via satellite telephone. A Web Map Service (WMS) system from Open GIS Consortium Inc. presents a geographical display system with additional layers. It can handle dynamic data with large variability. Our environmental data are semi-static.

EFFECT STUDIES

The first meeting of the Expert Group on Sonar Effects on Marine Life recommended that the three effect studies listed above should be undertaken by the project:

Study of Fish Fry and Fish

Fish fry (spawn): The background for this study is that many types of fish fry (and larvae) are pelagic and contrary to adult fish do not have the ability to escape from unpleasant or harmful sound sources. It is likely that fish fry's swim bladders have resonance frequencies that may be excited by one or more of the frigate sonars. At resonance, the swim bladders may absorb much of the acoustic energy in the impinging sound wave. The resulting oscillations may harm the swim bladder itself if the vibrations become too strong. The swim bladder is also part of a system that amplifies the vibrations which reach the fish's hearing organs, and at resonance the vibrations may become so intense that the hearing organs may be injured. Additionally, species like herring have a thin duct from the swim bladder expanding into an air-filled bullae close to each ear (Fig. 2). The hearing organ may therefore be particularly vulnerable in herring.

When aquatic organisms are being exposed to strong sound pressure waves, shear forces may develop between different types of tissue having unequal density and acoustic impedance. Tissue under development and growth, as in fish fry, will be more vulnerable than tissue in adult fish. Injury to nerve tissue can hamper development of the fish and make the adult fish less likely to survive.

If sonar exposure turns out to be harmful for fish fry, one should avoid sonar exercises in areas of high fish fry density. On the other hand, if it turns out that fish fry are not harmed by sonar exposure, it is very likely that adult fish are not harmed either. The reason for this is that fish fry is believed to be much more sensitive than adult fish to this kind of influence.

Adult Fish: Can active sonars scare away fish from fishing grounds?
To be scared by a sound signal, it is a prerequisite that the sound signal be audible.
The frigate sonars operate in the frequency band 1 – 8 kHz. Fig. 3 shows hearing curves for some species of fish. Most fish do not hear sounds above 1 kHz, but herrings have this ability. The herring has two air-filled sacks (bullae) coupled to the hearing organs, and the sacks are connected to the swim bladder via two narrow air canals (Fig. 2). This fact causes herring to have particularly good hearing, up to 2-3 kHz (Fig. 3). But this again causes herring to be vulnerable to hearing damages over a wider frequency band than other fish, which implies that herring might be scared away and even harmed by the sonar signals.

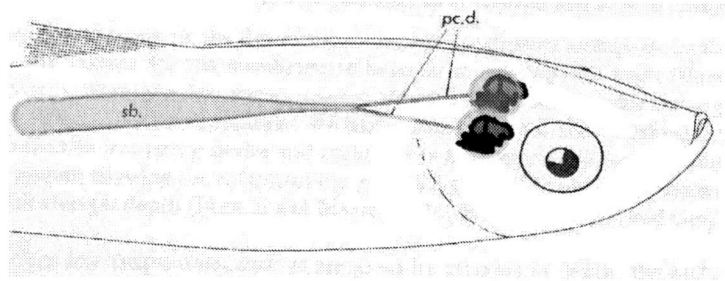

FIGURE 2. Swim bladder with connection to the hearing organs in herring (modified from Blaxter *et al.* (3))

It is not likely that herring will be scared away permanently from a geographical area by a mobile sonar. However, because of its importance, biologically and economically we will include herring in our sonar effect study.

Fish and Fish Fry Studies: These studies will be carried out in collaboration with the Norwegian College of Fisheries Sciences, Institute of Marine Research and University of Oslo. Several groups of fish will be kept in fish net cages, fish fry will be kept in bags in the net cage. Some groups of fish will be exposed to sonar signals. Control groups will be treated the same way, but without sonar exposure.

Sonar exposure will be conducted at several frequency bands and at several exposure levels. Behavior will be recorded acoustically and by video. Fish showing signs of harm will be dissected to examine for histological changes.

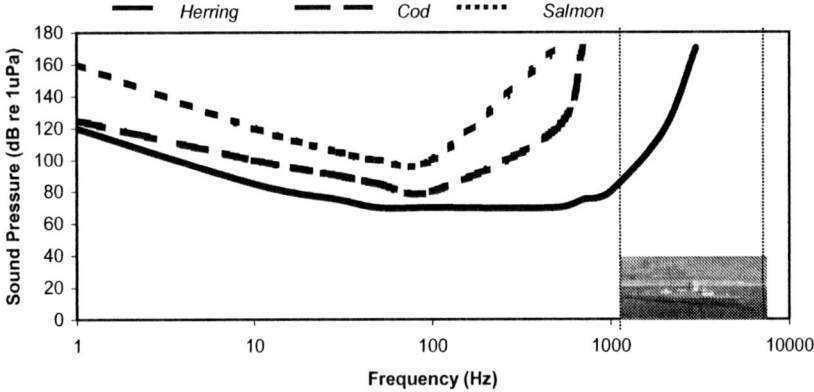

FIGURE 3. Audiograms for some important species of fish (data from Mitson (2)). Frequencies of frigate-sonars are also indicated.

Whale Behavioral Studies

Mass strandings of whales (beaked whales) in Greece 1996, Bahamas 2000 and the Canaries 2002, have been linked with naval sonar exercises. This has resulted in a massive international focus on how active sonars can harm whales.

Some knowledge exists from previous studies on blue whale, humpback whale, fin whale and grey whale. These studies indicate that (LF) sonars do not have any vital negative influence neither on migration nor foraging or vocalization behavior. For these reasons, we find it important to focus on other species. Minke whales and killer whales are very numerous in Norwegian waters and often populate the same area, like the Vestfjorden area (Fig. 1). The two species are biologically very different; one solitary, the other social, living in groups, one is prey, the other predator, one is very vocal, the other less vocal, one is a baleen whale, the other a toothed whale. Because both these very different whale species appear in the same area, we find it interesting to study them both. Fig. 4 shows a killer whale in Vestfjorden.

Behavioral studies of whales are difficult. It is not easy to study behavioral change when we really do not know their normal behavior. Changes in behavior are not harmful to whales, provided that the changes are not biologically vital, i.e. related to migration, foraging or reproduction. Whale behavioral studies are planned to be performed in collaboration with Woods Hole Oceanographic Institution, USA, and IMR, Bergen, Norway, by placing a sensor package containing a data logger which will record movements, incoming sonar signals, depth, time etc., and a VHF transmitter which makes it possible to track the whale. After a period of recording of normal behavior sonar transmissions starts. The animal's reactions are now recorded together with the sonar signals. Finally, the data logger can be remotely released and recovered by localizing the VHF transmitter.

FIGURE 4. Killer whale in Vestfjorden, Norway

Seal Studies

Seal studies will be carried out in collaboration with Department of Arctic Biology, University of Tromsø. Seals will be studied for their own sake and as a testable marine mammal species.

Hearing in Seals: Like fish, we will study seals in fish net cages with and without sonar exposure. The purpose is to observe behavioral changes with increasing sonar exposure level for several frequency bands. If strong behavioral changes are observed, we will proceed to examine if physiological injuries to the animal's hearing organs or brain have occurred.

Seals and Diver's Disease (the bends): From the Canary Island whale stranding in 2002, which occurred during a naval sonar exercise, it has been reported pathological symptoms in the dead whales which indicate that the whales may have suffered from diver's disease [4]. Two theories have been put forward to explain this phenomenon:
1) The animals getting scared by the sonar signals while diving and then surfacing too quickly.
2) The sonar sound waves cause existing micro bubbles in the blood to grow to harmful sizes.

The phenomenon described in Jepson *et al.* [4] is not well documented. However, from Fig. 5, excerpted with permission from [Fig. 2a from KJ Falke et al., SCIENCE 229:556-558]. Copyright [1985] AAAS, it does not seem obvious that rapid surfacing of seals (and marine mammals in general) should increase the risk for diver's disease. On the contrary, because of lung collapse at a certain depth (30 m in Fig. 5), nitrogen content in the blood will be limited, and in fact, be less at the surface after a rapid ascent following a deep dive than after a normal more slow one. This discussion indicates that the first theory is not very likely. The second theory is not well explained and understood.

However, it cannot be ruled out that sonar signals can cause diver's disease to develop in marine mammals. Therefore we will conduct diver's disease experiments on seals. The seal will be put in water in a pressure chamber and pressurized and depressurized to simulate alleged risky diving behavior such as accelerated ascent rates. In addition, a possible physical effect of sonars will be tested by exposing the animals to sonar sound during diving. As in the other tests the exposure will be at several frequency bands and several sound levels. Formation of gas bubbles in the body, *i.e.* development of decompression sickness, will be monitored during the trials using ultrasound Doppler [6].

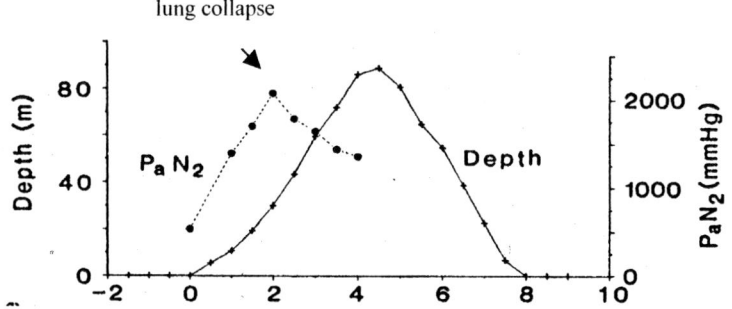

FIGURE 5. Weddell seal diving profile and content of Nitrogen in the seal's blood after Falke et al., reprinted with permission from [5], © 1985 AAAS.

CONCLUSION

After one year of project work, we have established that the resource situation in Norwegian waters is quite well known, and that very little is known about sonar signal effects on marine life. To increase our knowledge on sensitivity to sonar transmissions field tests are planned on biologically and economically important species. Our intention is to start with tests on fish fry (spawn) this spring. Initial tests of sonar effects on whale behavior are planned for late summer/fall 2004, and tests with seals are scheduled for spring 2005. The remaining tests will take place during 2005.

REFERENCES

1. Skaug, H.J., Øien, N., Schweder, T., Bøthun, G. "Current Abundance of Minke Whales in the Northeastern Atlantic; Variability in Time and Space" in *Can. J. Fish. Aquat. Sci.* (in press, 2004).
2. Mitson, R.B. "Underwater Noise of Research Vessels" in *ICES Coop. Res. Rep.:* **209**: 61p (1995)
3. Blaxter, J.H.S., Denton, E.J., Gray, J.A.B. "Acousticolateralis System in Clupeid Fishes" in *Hearing and sound communication in fishes*, edited by Tavolga W.N., Popper, A.N., Fay, R.R., Springer-Verlag, New York (1981).
4. Jepson, P.D., Arbelo, M., Deaville,, R., Patterson, I.A.P., Castro, P., Baker, J.R., Degollada, E., Ross, H.M., Herraez, P., Pocknell, A.M., Rodriguez, F., Howiell, F.E., Espinosa, A., Reid, R.J., Jabert, J.R., Martin, V., Cunningham, A.A., Fernandez, A.. "Gas-bubble Lesions in Stranded Cetaceans", *Nature* **425**: 575-576 (2003).
5. Falke, K.J, Hill, R.D., Qvist, J., Schneider, R.C., Guppy, M., Liggins, G.C., Hochachka, P.W., Elliott, R.E., Zapol, W.M. "Seal Lung Collapse During Free Diving: Evidence from Arterial Nitrogen Tensions" in *Science* **229**: 556-558 (1985).
6. Nishi, R.Y., Brubakk, A.O., Eftedal, O.S. "Bubble Detection" in *Physiology and Medicine of Diving,* edited by Brubakk, A.O., Neuman, T.S. Elsevier Science, London (2003).

Predicting the Environmental Impact of Active Sonar

Alec J. Duncan, Robert D. McCauley, and Amos L. Maggi

Centre for Marine Science and Technology, Curtin University of Technology, GPO Box U1987, Perth, WA 6845, Australia

Abstract. The effect of active sonar on marine animals, particularly mammals, has become a hot topic in recent times. The Australian Environmental Protection and Biodiversity Conservation Act 1999 obligates Defence to avoid significant environmental impacts from Navy activities including those which produce underwater sound such as active sonar. It is in the interests of all parties that these effects be modeled accurately to facilitate both the quantitative evaluation of the consequences of any proposed sonar trials, and the identification of suitable mitigation procedures. This paper discusses the received signal parameters that are of importance when predicting the effect of sonar systems on marine animals and techniques for modeling both the expected values of these parameters and their statistical fluctuations.

INTRODUCTION

Relationships between acoustic signal parameters and the effects of the signals on marine animals are still unclear, as practical and legal difficulties have led to a paucity of experimental data. In the case of marine mammals the similarities between their inner ear mechanisms and those of land mammals [1] leads to an expectation that similar signal parameters will be important in both cases. Consideration of the data that does exist for marine mammals [2] together with data for land animals [3] suggests that the total received energy is likely to be the most important parameter at long range because it determines the loudness of the sound perceived by the animal, and hence the likelihood of a behavioral response. By contrast, at short range the peak signal pressure is likely to be the most important parameter because, for the short bursts of sound typical of sonar systems, it determines the likelihood of physiological damage.

This paper considers methods for predicting these parameters for hull mounted and sonobuoy based active sonars, which generally operate in the frequency range 2.5 kHz to 10 kHz. These systems are capable of transmitting a variety of signal waveforms, but the two most common are the single frequency tone burst, used for initial detection and Doppler determination, and the swept frequency burst used for accurate ranging.

Sound propagation in the ocean is subject to random fluctuations due to the presence of inhomogeneities in the water column and interactions of the sound with the rough sea surface and seabed. When dealing with environmental impacts it is, therefore, desirable to determine the probability distributions of the signal parameters of interest so that the probabilities of exceedence of appropriate thresholds can be computed.

The approach taken in the work described here was to simulate an ensemble of received signals for a typical scenario using a standard high frequency propagation model and the assumption of uncorrelated random fluctuations in the relative arrival times of signals that have traveled by the different ray paths. This was carried out for both a tone burst and a swept frequency signal. The statistics of the received signal energy and signal peak pressure were then computed and fitted to appropriate theoretical probability density functions.

PREDICTION OF RECEIVED SIGNAL PARAMETERS

Theory

The theory presented here is based on the assumption that samples of the envelope of the received signal can be treated as random variables with variance σ_x^2 and a Rayleigh probability density function (pdf) [4]:

$$f_R(x) = \frac{x}{\sigma_x^2} e^{-\frac{x^2}{2\sigma_x^2}}. \tag{1}$$

The equation for this pdf, and the ones that follow, only apply to positive values of their arguments. They are implicitly assumed to be zero for negative arguments.

The samples are assumed correlated with an autocorrelation function that is zero for lags greater than τ_c. There are thus $N = T_r/\tau_c$ independent samples of the signal envelope available in a received signal of duration T_r.

Received energy

The received energy is proportional to:

$$E = \int_0^{T_r} \frac{x(t)^2}{2} dt \approx \frac{T_r}{N} \sum_{i=1}^{N} \frac{x_i^2}{2} \tag{2}$$

where the x_i represent samples of the continuous time envelope function $x(t)$.

By using the usual rules for determining the probability distributions of functions of random variables [4,5] and assuming the samples are independent it is straightforward to show that $y_i = x_i^2/2$ has an Exponential pdf:

$$f_e(y) = \frac{1}{\sigma_x^2} e^{\frac{-y}{\sigma_x^2}} \tag{3}$$

and that E has a Gamma pdf:

$$f_\gamma(E) = \frac{E^{N-1}}{\beta^N \Gamma(N)} e^{-E/\beta} \tag{4}$$

with $\beta = \dfrac{\sigma_x^2 T_r}{N} = \sigma_x^2 \tau_c$ and $\Gamma()$ representing the Gamma function.

The population mean and variance of E are:

$$\mu_E = N\beta = \sigma_x^2 T_r, \text{ and } \sigma_E^2 = N\beta^2 = \dfrac{\sigma_x^4 T_r^2}{N}. \quad (5)$$

These equations can be combined to yield an expression for the number of independent samples:

$$N = \left(\dfrac{\mu_E}{\sigma_E}\right)^2. \quad (6)$$

For sufficiently large N (say > 30) the Central Limit Theorem will apply and the pdf of E will be well approximated by the Normal (Gaussian) pdf with the same mean and variance.

Peak pressure

A pdf for the peak pressure can be derived by recognizing that for independent samples of the signal envelope the probability that the maximum of these samples, $z = x_{max}$, is less than p is given by:

$$\int_0^p f_z(z)\,dz = \Pr(z < p) = \prod_{i=1}^{N} \Pr(x_i < p) = \left[\int_0^p f_R\, x\, dx\right]^N. \quad (7)$$

Evaluating the expression on the right hand side of Equation (7) and differentiating with respect to p results in the following expression for the pdf of the maximum (peak) of the envelope:

$$f_z(z) = \dfrac{Nz}{\sigma_x^2} e^{-\dfrac{z^2}{2\sigma_x^2}} \left(1 - e^{-\dfrac{z^2}{2\sigma_x^2}}\right)^{N-1} \quad (8)$$

Simulation of received signals

The scenario considered here was a 150 m deep isovelocity water column, with a sound speed of 1,500 m.s^{-1} and a density of 1,024 kg.m^{-3}. The seabed was a fluid half-space with a sound speed of 1,750 m.s^{-1}, a density of 1940 kg.m^{-3}, and an attenuation of 0.8 dB per wavelength. The source depth was 6 m, the receiver depth 10 m, and the source to receiver range varied from 2 km to 3 km. Two transmit signals were considered: a tone burst with a frequency of 7.5 kHz, a duration of 1s, and a 10% cosine amplitude taper on each end; and a burst of the same duration and amplitude taper, but with a linear frequency sweep from 6.8 kHz to 8.2 kHz.

The Gaussian beam-tracing model Bellhop ([6], [7]), was used to determine the amplitudes and delays of the various arrivals, with only arrivals with amplitudes greater than one percent of the maximum being used in the received signal reconstruction. This resulted in 18 and 26 arrivals being summed at 2 km and 3 km range respectively. Received signals were generated by summing replicas of the transmit signal delayed and scaled by the computed amounts. This process was carried out in the frequency domain so that phase shifts due to boundary reflections could be incorporated and the delays could be included without interpolation error. Different received signal realizations were obtained by perturbing the computed delays by random amounts prior to calculating the received signal. The perturbations were taken from a Gaussian random number generator with a standard deviation of 200 µs. A sensitivity test showed that the results were indistinguishable for perturbation standard deviations greater than a quarter of a period of the lowest frequency present (37 µs).

Envelopes for five realizations of the received signal for each transmit signal are shown in Fig 1. The received signals due to the different transmit signals have a very different character, with the swept frequency signal envelope having larger and more rapid fluctuations than the tone burst envelope, reflecting its much wider bandwidth.

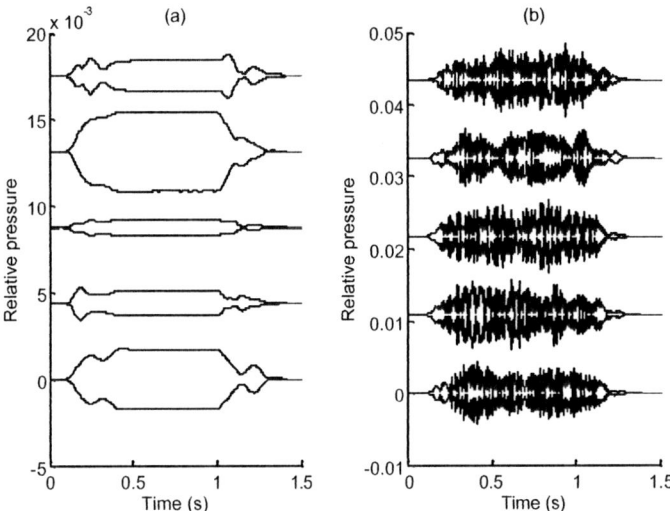

FIGURE 1. Envelopes of five realizations of the received signal for (a) tone burst and (b) frequency sweep.

Results

Results are presented here for a receiver range of 2 km. Limited space precludes including the results for other ranges, which were qualitatively very similar to those presented here.

Received energy

Figure 2 shows a comparison between the pdf of the received energy for the tone burst estimated from the simulation and the Gamma pdf with the same mean and standard deviation.

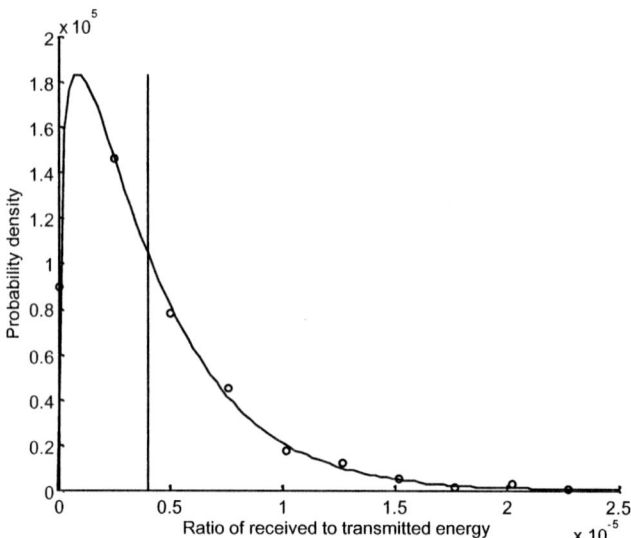

FIGURE 2. Circles are probability density of received energy estimated from 1000 signal realizations for tone burst signal. Solid line is gamma probability density function Equation (4) with $\mu_E = 4.1\times10^{-6}, \sigma_E = 3.6\times10^{-6}$. Vertical broken line is expected signal energy computed from the sum of the squares of the arrival amplitudes, vertical dotted line (indistinguishable from broken line) is mean energy of signal realizations.

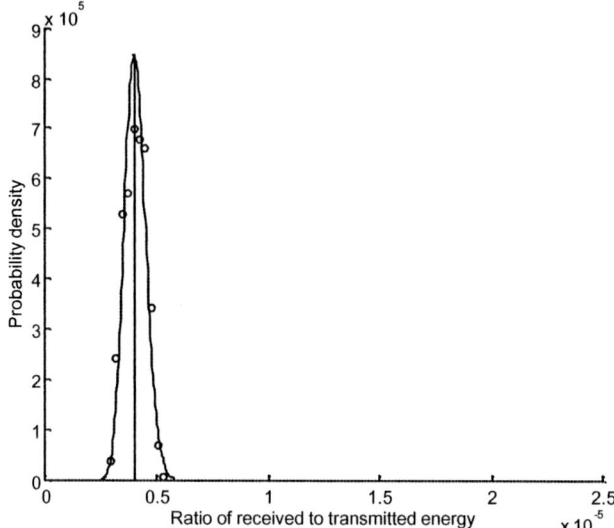

FIGURE 3. As for Figure 2, but swept frequency transmit signal. $\mu_E = 4.1\times10^{-6}, \sigma_E = 4.7\times10^{-7}$.

The equivalent plot for the swept frequency signal is shown in Fig 3. In each case, a histogram of the energies of 1,000 received signal realizations was calculated. This was then normalized for comparison with the theoretical pdf by dividing by the product of the number of samples and the bin spacing.

In both cases the mean energy agreed with the value computed from the sum of the squares of the absolute values of the arrival amplitudes computed by Bellhop, but the tone burst produced a much greater spread in received energy than the sweep. This was a direct consequence of the wider bandwidth of the sweep, which resulted in a shorter envelope correlation time and, therefore, a greater number of independent samples being available in each receive signal. The values of σ_x and N estimated from the data using Equations (5) and (6) are given in Table 1. To make subsequent comparisons easier, the tabulated values of σ_x have been renormalized relative to the peak pressure of the transmit signal, z_{Tx}, rather than the transmit energy, E_{Tx}, by multiplying the values calculated using Equation (5) by a factor of $\sqrt{E_{Tx}}/z_{Tx}$.

TABLE 1. Fitted signal parameters

Estimated Using	Normalized Envelope Standard Deviation, σ_x	Number of independent samples, N
Tone-burst, Energy Pdf	1.4×10^{-3}	1.3
Tone-burst, Peak Pdf	1.3×10^{-3}	2.0
Sweep, Energy Pdf	1.4×10^{-3}	73
Sweep, Peak Pdf	1.4×10^{-3}	258

Peak pressure

Similar results for the pdfs of the peak received signal pressures are shown in Figs 4 and 5. Here the incoherent sum of the arrival amplitudes gives a reasonable estimate of the mean value of the signal peak for the tone-burst, but underestimates it by a factor of 2.5 (8dB) for the sweep. The theoretical pdf given by Equation (4) provides a good fit to the data for both signal types when the parameters are adjusted to minimize the least square error. The fitted envelope standard deviations agree with those calculated using the energy pdfs but the numbers of independent samples differ (Table 1).

Note that the pdfs are clearly skewed, even for the sweep with 258 samples, demonstrating that the Central Limit Theorem does not apply in this case.

DISCUSSION AND CONCLUSIONS

The probability density functions based on the simplifying assumptions of a Rayleigh pdf for the signal amplitude and statistically independent samples provided good fits to the simulation results.

The received energy pdfs for the tone burst and sweep had the same mean, which also agreed with the values computed via an incoherent transmission loss calculation. However, the energy pdf for the tone burst had a much higher standard deviation than that for the sweep, with a consequent higher probability of extreme values.

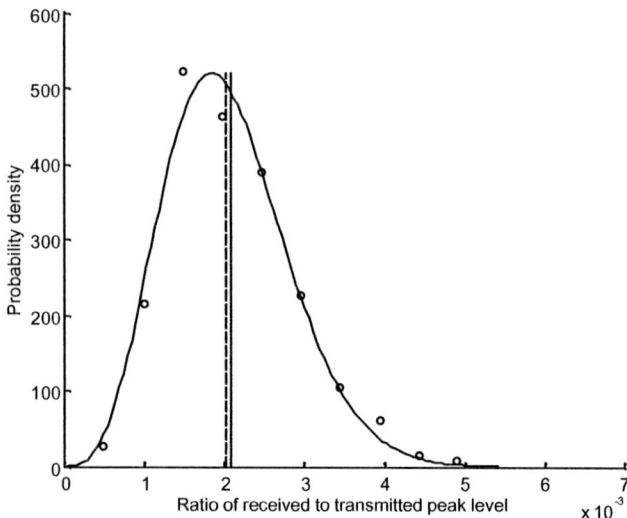

FIGURE 4. Circles are probability density of ratio of peak received pressure to peak transmit pressure estimated from 1,000 signal realizations for tone burst signal. Solid line is fitted theoretical probability density given by Equation (8) with $\sigma_x = 1.26 \times 10^{-3}, N = 2.0$. Vertical broken line is expected ratio computed from incoherent sum of arrival amplitudes, vertical dotted line is mean ratio of signal realizations.

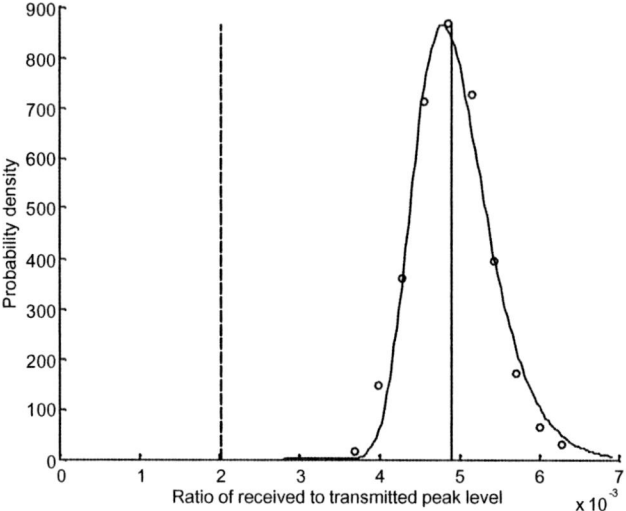

FIGURE 5. As for Figure 4, but swept frequency transmit signal. $\sigma_x = 1.43 \times 10^{-3}, N = 258$.

The peak signal pdf for the tone burst had a mean value close to the value expected from the incoherent transmission loss calculation, but the peak signal pdf for the sweep was shifted to right, with a mean corresponding to a transmission loss 8 dB lower than the incoherent calculation. This is particularly significant in the context of estimating environmental impacts.

These effects are explained by the much wider bandwidth and hence shorter correlation time of the sweep compared to the tone burst, resulting in the sweep signal envelope having a larger number of independent samples than the tone burst envelope.

There was, however, quantitative disagreement between the numbers of independent samples estimated from the energy pdfs and those estimated from the peak pdfs (Table 1). This discrepancy was particularly apparent for the sweep signal and is likely to be due to a breakdown of the assumption that the received signal envelope can be treated as a random variable when there are a limited number of arrivals. The authors are investigating this further.

ACKNOWLEDGMENTS

This work has been made possible by financial support provided by the Defence Science and Technology Organisation, Australia (DSTO).

REFERENCES

1. Darlene R Ketten *Bioacoustics* **8**, 103 (1997).
2. James J Finneran, Carolyn E Schlundt, Randall Dear, et al. *Journal of the Acoustical Society of America* **111** (6), 2929 (2002).
3. Robert Lataye and Pierre Campo *Journal of the Acoustical Society of America* **99** (3), 1621 (1996).
4. Alexander D. Poularikas, *The Handbook of Formulas and Tables for Signal Processing*. (CRC Press, 1999).
5. Ronald E. Walpole and Raymond H. Myers, *Probability and Statistics for Engineers and Scientists*, 3rd ed. (Macmillan, 1985).
6. Michael B Porter and Homer P. Bucker *Journal of the Acoustical Society of America* **82** (4), 1349 (1987).
7. Michael B Porter, Acoustic Toolbox (Atpii_F95) (1999).

Acoustic Propagation Studies For Sperm Whale Phonation Analysis During LADC Experiments

Natalia A. Sidorovskaia*, George E. Ioup[§], Juliette W. Ioup[§], and Jerald W. Caruthers[¶]

*Physics Department, The University of Louisiana at Lafayette,
Lafayette, LA 70504-4210, USA
[§]Physics Department, The University of New Orleans,
New Orleans, LA 70148, USA
[¶]Department of Marine Science,
The University of Southern Mississippi, Stennis Space Center, MS 39529, USA

Abstract. The Littoral Acoustic Demonstration Center (LADC) conducted a series of passive acoustic experiments in the Northern Gulf of Mexico and the Ligurian Sea in 2001 and 2002. Environmental and acoustic moorings were deployed in areas of large concentrations of marine mammals (mainly, sperm whales). Recordings and analysis of whale phonations are among the objectives of the project. Each mooring had a single autonomously recording hydrophone (Environmental Acoustic Recording System (EARS)) obtained from the U.S. Naval Oceanographic Office after modification to record signals up to 5,859 Hz in the Gulf of Mexico and up to 12,500 Hz in the Ligurian Sea. Self-recording environmental sensors, attached to the moorings, and concurrent environmental ship surveys provided the environmental data for the experiments. The results of acoustic simulations of long-range propagation of the broad-band (500-6,000 Hz) phonation pulses from a hypothetical whale location to the recording hydrophone in the experimental environments are presented. The utilization of the simulation results for an interpretation of the spectral features observed in whale clicks and for the development of tracking algorithms from single hydrophone recordings based on the identification of direct and surface and bottom reflected arrivals are discussed. [Research supported by ONR.]

INTRODUCTION

Studies of the acoustic vocalizations and phonations of marine mammals have become one of the hot topics of underwater acoustic research in the last few years. There has been increasing anthropogenic noise in the ocean and an overlap between the spectral content of naval sea operations and the vocalizing and phonating frequencies of deep-diving marine mammals. There is concern that this is a potentially disturbing factor in the mammals' habitat. Unlike visual surveys, biological sampling, and radio-tagging, passive acoustic recordings offer a variety of advantages for the investigation of free moving large marine animals (such as different types of whales): they are unobtrusive and do not change the social behavioral pattern of an animal, they can contain simultaneous information about many individuals at different distances from a receiving system, and they can provide the continuous monitoring of species spending most the their time under water. When using passive acoustics, studying the vocalization/phonation patterns of different animals and species and trying to discern acoustic sound of a specific individual is the way to gain the knowledge (which does not exist at present time) about social acoustic communication of large marine mammals. A considerable amount of the mammals' broadband acoustic data has been collected by the Littoral Acoustic Demonstration Center (LADC).

The Littoral Acoustic Demonstration Center (LADC) is a consortium of the University of New Orleans (UNO), the University of Southern Mississippi (USM), the University of Louisiana at Lafayette, and the Naval Research Laboratory at Stennis Space Center (NRL-SSC), with guidance and support from the Naval Oceanographic Office (NAVOCEANO). It was formed to perform and analyze underwater acoustic measurements of ambient noise and marine mammal phonations, namely endangered sperm whales. The first experiment was conducted in the Gulf of Mexico (GoM) in the summer of 2001 from 17 July to 21 August. Figure 1 shows the LADC study area, which is the same area used in summer 2002. The black dots indicate oil platforms and the whale symbols indicate sperm whale sightings. The second set of measurements was made in conjunction with the Saclant Centre (SACLANTCEN) exercise Sirena02 in the Ligurian Sea from 01 July to 23 July in the summer of 2002. The third set of measurements was made during the late summer and early fall of 2002 in GoM. All the acoustic measurements were accomplished with vertically moored Environmental Acoustic Recording System (EARS) buoys from NAVOCEANO. Each of the EARS buoys had a single omni-directional hydrophone and an instrument package which autonomously recorded the acoustic signals up to 5,859 Hz in the GoM and up to 12,500 Hz in Europe. The hydrophones for these buoys were suspended 50 m from the bottom. The remainder of the mooring, spanning almost all the rest of the water column, was instrumented with self-recording environmental oceanographic sensors which provided time series data of temperature, conductivity, and pressure. In both the summers of 2001 and 2002, the moorings were deployed along an approximately straight line at the 600, 800, and 1,000 m contours, 5.3 km (between 600 m and 800 m moorings) and 13 km (between 800 and 1,000 m moorings) apart [1]. Additional oceanographic data were gathered on various cruise legs to augment the mooring measurements and give a more complete description along the study tracks. In the summer of 2001, a chirp sonar survey along the study tracks was also performed. The data have been inverted to give sound speeds and densities in the bottom [2].

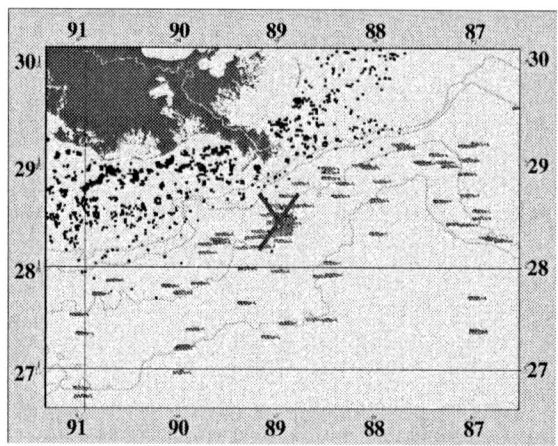

FIGURE 1. Ship track locations for oceanographic sensing. Oil platforms are black dots and whale sightings are whale symbols. Louisiana coast is at upper left. (Reproduced from [1])

SPECTOGRAMS OF THE LADC DATA

The analysis of the LADC recordings reveals a considerable amount of anthropogenic noise, as well as well-identifiable (by aural analysis) phonations of marine mammals (predominantly sperm whales), in approximately 600 Gbytes of recorded data. Distant shipping noise is generally dominant in a frequency range from 10 Hz to 300 Hz with a peak in the spectrum near 50 Hz. More local shipping effects often include many tonal lines superimposed upon the distant shipping spectra [3]. Seismic exploration sources are widely present under 300 Hz in addition to the ship noise. Figure 2 shows a representative four-second fragment of a 60 sec segment of LADC data from the 800 m buoy. It contains very clear recordings of sperm whales. The recording begins on Julian Day 213, Zulu 0 hr, 9 min, 37 sec. In this figure, the top graph shows the structure of the time signal, originally recorded by the mooring. The bottom two graphs in the figure are spectrograms. The first one shows all frequencies up to 5,859Hz while the bottom spectrogram contains only frequencies to 1,000 Hz. Broadband transform lines in the middle figure are sperm whale clicks and closely spaced clicks sound like creaks, although they are distinct from the spectrogram patterns which correspond to what are generally known as creaks in the literature [4-7]. The seismic exploration source (107 km away) is clearly visible as the red peak in the bottom spectrogram of Fig. 2. The low frequency noise without the seismic source present, which still dominates the spectrum, is visible before the seismic pulse arises.

Figure 2 and extensive spectrogram analysis reveals well-defined null patterns in the phonation spectrograms, as, for example, the sequence of clicks at the 49-second mark. More detailed information can be gathered by overplotting the magnitude of the spectra in a single group of clicks. Fifteen distinct groups were identified in the previously mentioned 60-second segment based on the temporal grouping of the clicks. The results of the comparison for five groups are presented on Fig. 3. Some groups show the unique and stable null patterns despite the considerable variations in amplitude in-between. A legitimate question to ask is if the interference pattern in the mammal's spectrograms is caused by propagation effects, or if it can be associated with an individual animal. The second part of this paper presents the results of modeling the propagation effects of the LADC experiment.

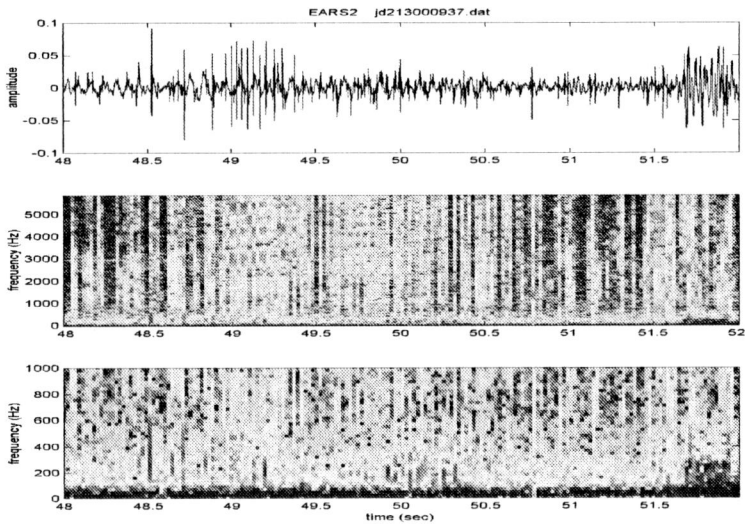

FIGURE 2. 800 m mooring LADC time data extracted from a 60 sec segment beginning at Julian Day 213, Zulu 0 hr, 9 min, 37 sec., and their spectrograms.

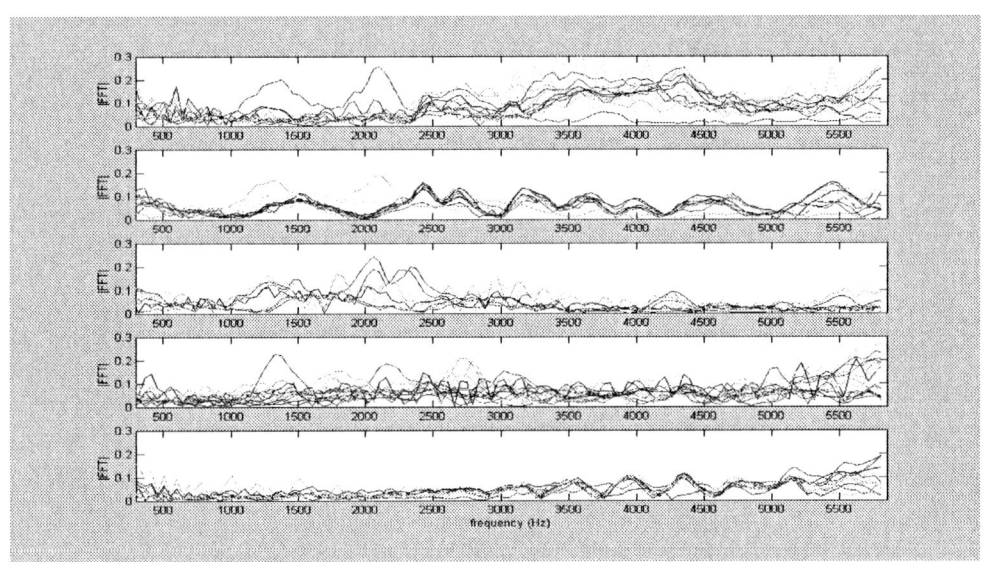

FIGURE 3. Spectrograms for 5 different groups of clicks.

PROPAGATION MODELING FOR LADC DATA

The environmental data collected as a part of the LADC experiment in Summer 2001 (Fig. 4) were input into the Range-dependent Acoustic Model (RAM) by Michael Collins to simulate the broad-band acoustic response on a receiving hydrophone. The hypothetical animal depth is 700 m which corresponds to common foraging depths of sperm whales. The receiver position is 740 m. The source spectral function is assumed to be flat over the frequency range between 500 and 5,859 Hz to study only the effects of the waveguide propagation.

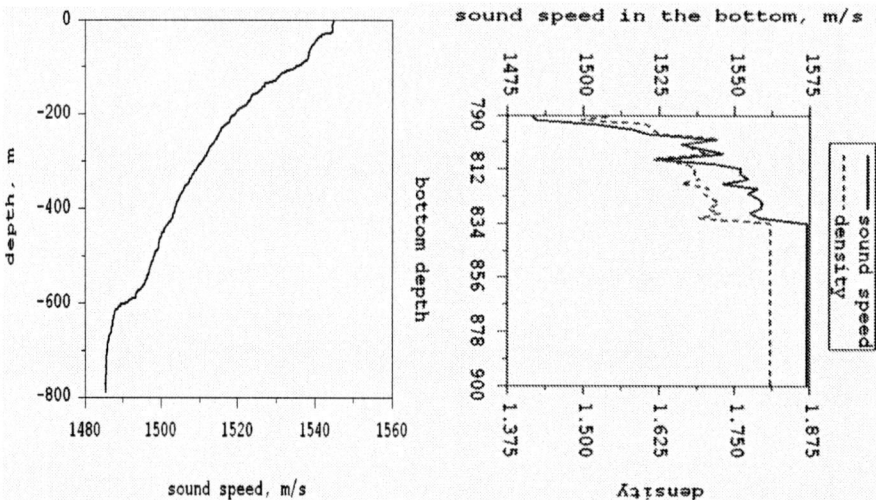

FIGURE 4. Environmental input into the RAM model: sound speed profile in the water column and bottom sound speed and density functions.

Figure 5a represents the frequency dependence of the amplitude of the modeled waveguide transfer functions. The color coding indicates the different horizontal distances between the hypothetical source location and a receiver. We can see the "sine-like" behavior of the transfer function that indicates only direct and bottom reflected pulses are responsible for the most energy transfer between the source and receiver. The 10 Hz frequency step size in the simulations does not account for the fine-scale structure due to surface and multiply reflected arrivals. The frequency of oscillations is decreasing as a hypothetical animal swims away from the buoy. At the horizontal separation of 3 km, the structure of the transfer function becomes more irregular indicating the partial time overlapping between direct and bottom reflected arrivals and the contribution of the surface reflected one. Using the Fourier synthesis procedure, the time domain response can be obtained. The time-domain structure of the transfer function, which is color-coded in accordance with the horizontal separation between a source and receiver, is shown on Figure 5b.

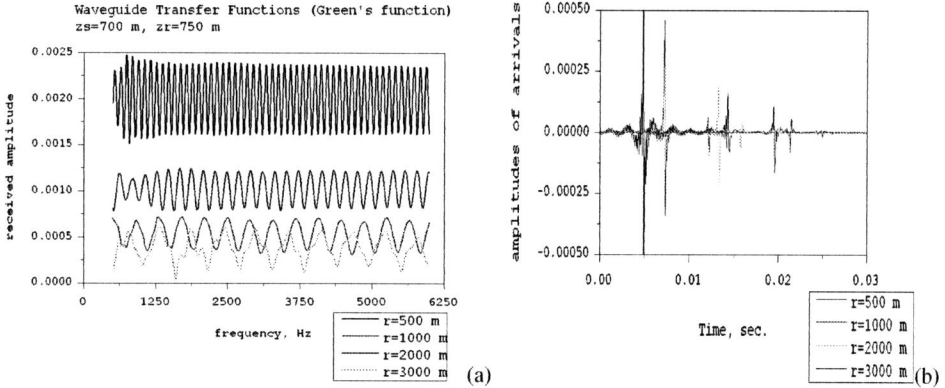

FIGURE 5. (a) - Frequency-domain structures of the waveguide transfer functions for different horizontal distances between the source and receiver for a receiver at a depth of 740 m and a source depth of 700 m; (b) - Temporal structures of the waveguide transfer functions for different horizontal distances between the source and receiver for a receiver at a depth of 740 m and a source depth of 700 m.

From the analysis of time domain response, we can conclude that the temporal delay between the direct and bottom reflected arrival is about 15 msec for a horizontal separation of 500 m and gradually decreases with increasing range. The temporal window for the spectrograms in Figs. 2 and 3 was 4 msec, which corresponds to the average duration of on-axis sperm whale clicks [7]. Applying this windowing function to separate the direct pulse arrival for r=500 m followed by the Fourier transform, we obtain a nearly flat spectrum of the windowed transfer function (Fig. 6). Based on propagation modeling, we can hypothesize that for the short-time spectra and relatively close position of an animal to an EARS buoy the animal phonation apparatus may be responsible for prominent null patterns in the clicks spectrograms presented in Figs. 2 and 3.

The series of sound pressure level maps in Fig. 7 show the distribution of acoustic energy with depth and time for the fixed horizontal separation between source and receiver. The simulation results clearly indicate that the detectability of the acoustically active foraging animals by a surface array increases with an increase in the horizontal separation. The surface reflected arrivals are delayed by more than 80 msec at bottom hydrophones and should not overlap with direct or bottom-reflected arrivals. The direct and bottom-reflected arrivals can be successfully resolved for narrow directional clicks when an animal is up to 2 km away from the receiver. The temporal delay between these two arrivals can be utilized for developing a tracking algorithm based on single hydrophone recordings [8].

CONCLUSIONS

The results of acoustic propagation modeling for the LADC environment suggest the interpretation of null patterns in click spectrograms as being due to distinct features of a mammal phonation apparatus. Many more groups of clicks should be similarly analyzed, and the spectra of individual isolated clicks should be compared before definite conclusions can be drawn. The development of an algorithm, which can provide the identification of individual clicks in continuous recordings, at least with some degree of statistical confidence, should also be addressed in future research.

ACKNOWLEDGMENTS

This research is supported by ONR, Program Officer Melbourne Briscoe. The authors wish to acknowledge with gratitude help of Joal Newcomb, Robert Fisher, Robert Field, and other scientists of NRL, Grayson Rayborn, Stan Kuczaj, Christopher Walker of USM, Mike Wild, Mark Snyder and other scientists and engineers of the Naval Oceanographic Office. They have also benefited from their interactions concerning sperm whales and sperm whale acoustics with specialists including David Mellinger, Aaron Thode, Patrick Miller, Mark Johnson, Peter Tyack, Anthony Martinez, Keith Mullen, Jonathan Gordon, Nathalie Jaquet, Bill Lang, Carol Roden, Sarah Tsoflias, and Bob Gisiner.

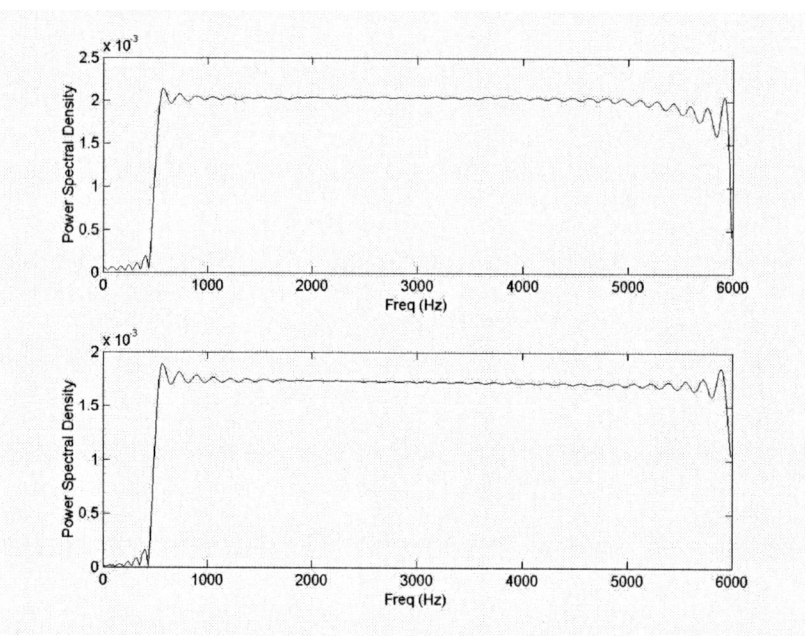

FIGURE 6. Spectrum of the windowed direct and bottom-reflected arrivals.

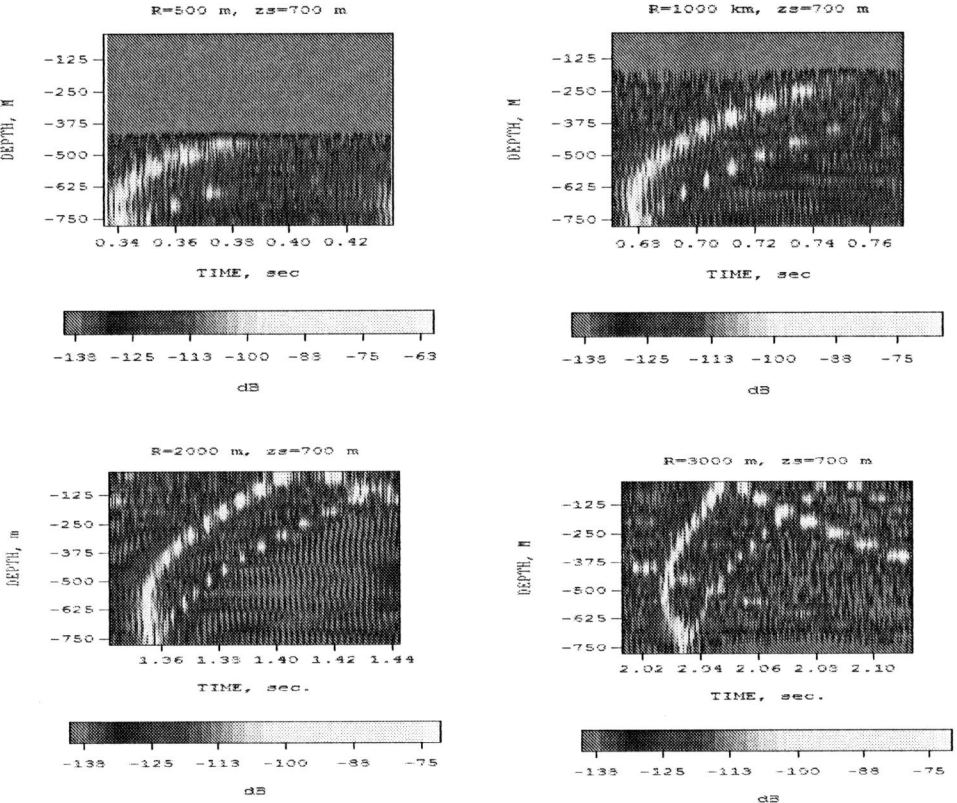

FIGURE 7. Depth-time sound pressure levels for the omnidirectional source for four fixed ranges between the source and the receiver.

REFERENCES

1. Newcomb, J., Fisher, R., et al., "Using Acoustic Buoys to Assess Ambient Noise and Sperm Whale Vocalizations," in Proceedings of MMS ITM 2003, January 2003, New Orleans.
2. Turgut, A., McCord, M., Newcomb, J., and Fisher, R.,"Chirp sonar sediment characterization at the northern Gulf of Mexico Littoral Acoustic Demonstration Center experimental site," MTS/IEEE OCEANS2002 Proceedings, 29-31 Oct 2002, Biloxi, MS, pp. 2248-2252
3. Newcomb, J., Fisher, R., et al., "Modeling and measuring the acoustic environment of the Gulf of Mexico," Proceedings of MMS ITM 2002, January 2002, New Orleans
4. Goold, J.C., and Jones, S.E., "Time and frequency domain characteristics of sperm whale clicks," J. Acoust. Soc. Am, 98(3), 1995, pp. 1279-1291.
5. Gordon, J.C.D., "Evaluation of a method for determining the length of sperm whales, Physeter catodon, from their vocalizations," J. Zool. London, 224, 1991, pp. 301-314.
6. Weilgart, L.S., and Whitehead, H., "Coda communication by sperm whales (Physeter macrocephalus) off the Galapagos Islands," Can. J. Zool., 71(4), 1993, pp. 744-752.
7. Møhl, B., et al., "The monopulsed nature of sperm whale clicks," J. Acoust. Soc. Am, 114(2), 2003, pp. 1143-1154.
8. Thode, A., Mellinger, D.K., et al., "Depth-dependent acoustic features of diving sperm whales (Physeter macrocephalus) in the Gulf of Mexico," J. Acoust. Soc. Am., 112(1), 2002, pp. 308-321.

Underwater Ambient Noise and Sperm Whale Click Detection during Extreme Wind Speed Conditions

Joal J. Newcomb*, Andrew J. Wright**, Stan Kuczaj***, Rachel Thames***, Wesley R. Hillstrom****, and Ralph Goodman*****

*Naval Research Laboratory, Code 7180, Stennis Space Center, MS 39529
**Previously at the University of Wales, Bangor, UK
***University of Southern Mississippi, Psychology Dept, Box 5025, Hattiesburg, MS 39406
****Naval Oceanographic Office, Stennis Space Center, MS 39522
*****University of Southern Mississippi, Dept of Marine Sciences, Stennis Space Center, MS 39529

Abstract. The Littoral Acoustic Demonstration Center (LADC) deployed three Environmental Acoustic Recording System (EARS) buoys in the northern Gulf of Mexico during the summers of 2001 (LADC 01) and 2002 (LADC 02). The hydrophone of each buoy was approximately 50m from the bottom in water depths of 645m to 1034m. During LADC 01 Tropical Storm Barry passed within 93nmi east of the EARS buoys. During LADC 02 Tropical Storm Isidore and Hurricane Lili passed within approximately 73nmi and 116nmi, respectively, west of the EARS buoys. The proximity of these storm systems to the EARS buoys, in conjunction with wind speed data from three nearby NDBC weather buoys, allows for the direct comparison of underwater ambient noise levels with high wind speeds. These results are compared to the G. M. Wenz spectra at frequencies from 1kHz to 5.5kHz. In addition, the impact of storm conditions on sperm whale clicks was assessed. In particular, although the time period during the closest approach of TS Barry tended to produce lower click rates, this time period did not have the greatest incidence of non-detection at all the EARS buoys. It follows that storm-related masking noise could not have been responsible for all the observed trends. The data suggest that sperm whales may have left the vicinity of the deepest EARS buoy (nearest TS Barry's storm track) during the storm and possibly moved into the shallower waters around the other EARS buoys. It also appears that sperm whales may not have returned to the deepest EARS area, or did not resume normal behavior immediately after the storm, as the click rate did not recover to pre-storm levels during the period after TS Barry had dissipated. Results of these analyses and the ambient noise analysis will be presented. (Research supported by ONR).

INTRODUCTION

The Littoral Acoustic Demonstration Center (LADC) is an Office of Naval Research funded consortium consisting of the University of New Orleans, the University of Southern Mississippi, the Naval Research Laboratory, and the University of Louisiana at Lafayette. LADC deployed three Environmental Acoustic Recording System (EARS) buoys in the northern Gulf of Mexico (GoM) during the summers of 2001 (LADC 01) and 2002 (LADC 02) to study ambient noise and marine mammals. The LADC EARS buoy (developed by the Naval Oceanographic Office) is an autonomous, self-recording buoy capable of more than 66 days continuous recording of a single channel at a 11.7kHz sampling rate. The hydrophone of each buoy was approximately 50m from the bottom in water depths of 1034m to 645m along an upslope track. The buoys were labeled EARS 1 (deepest), EARS 2, and EARS 3 (shallowest). Oceanographic data (CTD and XBT data) were obtained during each deployment along a longer upslope track and along a cross slope track centered at EARS 1. Bottom information for these tracks was obtained during LADC 01 from side-scan sonar data.

During LADC 01, Tropical Storm Barry passed within 93nmi east of the EARS buoys. During LADC 02 Tropical Storm Isidore and Hurricane Lili passed within approximately 73nmi and 116nmi, respectively, west of the EARS buoys. The proximity of these storm systems to the EARS buoys, in conjunction with wind speed data from three nearby National Data Buoy Center (NDBC) weather buoys (Fig. 1), allows for the direct comparison of underwater ambient noise levels with high wind speeds.

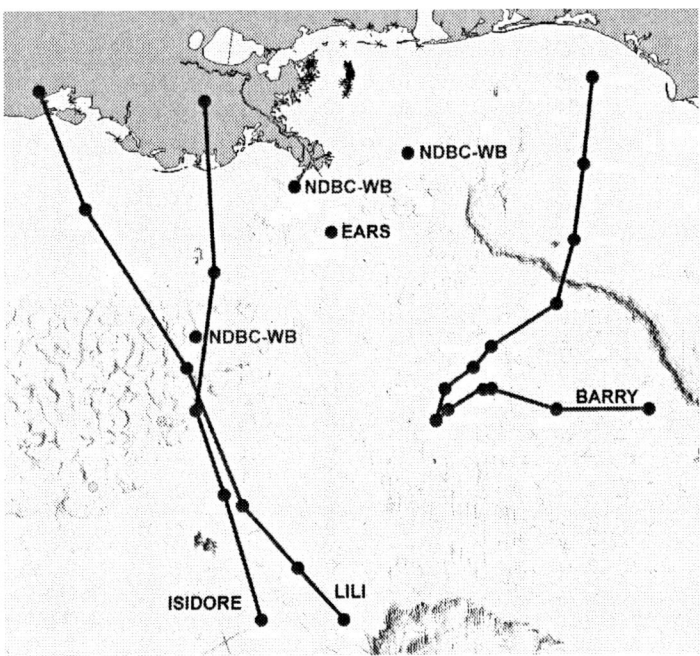

FIGURE 1. Chart of storm system tracks during the summer of 2001 and 2002 relative to the EARS deployment site and the NDBC weather buoys. The storm systems' tracks were obtained from positional data published by the National Hurricane Center (NHC) indicated by the solid circles connected with straight lines.

AMBIENT NOISE RESULTS

The storm systems' tracks were obtained from positional data published by the National Hurricane Center (NHC). Wind speed estimates at the EARS location were obtained from an examination of wind speed and wave height measurement data obtained from the NDBC weather buoys. Ambient noise results were obtained from a spectral analysis of the EARS raw acoustic data.

Hurricane Lili (October 2002)

Hurricane Lili at closest approach passed about 116nmi west of the EARS buoys. All three NDBC buoys were used to estimate wind speed for each time period except for the time period where Hurricane Lili passed very close to NDBC buoy 42041.

Figure 2 illustrates the ambient noise results for Hurricane Lili for EARS 1 from 1kHz to 5.5kHz. The closely grouped spectra are the noise levels for the time periods corresponding to published positions of Hurricane Lili and represent estimated wind speeds of 20 to 47kts. The much lower level spectrum represents the results for a time period during which there were no storm systems in the GoM and is presented only for comparison. As expected, when Hurricane Lili approached EARS 1, the wind speeds increased and the corresponding noise levels increased. Superimposed upon these results are the Wenz curves [1] for Beaufort Wind Forces (BWF) of 5 and 8 (solid black lines) and the projected upper limit of prevailing noise. As can be seen from Fig. 1, the measured spectra agree well with the slope and levels of the BWF 8 curve for wind speeds of 34kt to 40kt. Results from the EARS 3 buoy have overall higher spectral levels, but otherwise are very similar.

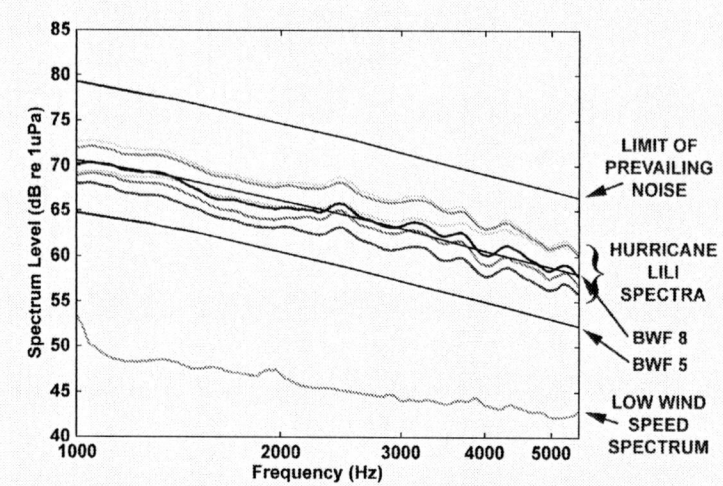

FIGURE 2 Illustration of the ambient noise results for EARS3 during Tropical Storm Isidore superimposed upon the results from Wenz [1] for Beaufort Wind Force 5 and 8. Also shown are the upper limit of the prevailing noise from Wenz [1] and a low wind speed spectrum for comparison.

Tropical Storm Isidore (September 2002)

Tropical Storm Isidore was less organized and weaker than Hurricane Lili. She was also a very large storm geographically. As can be seen from Fig. 1, TS Isidore passed approximately 43nmi closer to the EARS buoys than Hurricane Lili at closest approach (73nmi west of EARS buoys). Consequently, the spectrum levels for TS

Isidore are about the same as the spectrum levels for Hurricane Lili. All three NDBC buoys were used to estimate wind speed for each time period except for the time period where TS Isidore passed very close to NDBC buoy 42041.

Figure 3 illustrates the ambient noise results for TS Isidore from 1kHz to 5.5kHz for EARS 3. The closely grouped spectra correspond to the published positions of TS Isidore and represent estimated wind speeds of 21 to 55kt. The much lower level spectrum represents the results for a time period during which there were no storm systems in the GoM and is for comparison only. As expected, and similarly to the results for Hurricane Lili, when TS Isidore approached EARS 3, the wind speeds increased and the corresponding noise levels increased.

Superimposed upon the results for TS Isidore are the Wenz curves for Beaufort Wind Forces (BWF) of 5 and 8 (solid black lines) and the projected upper limit of prevailing noise. As can be seen from Fig. 3, the measured spectra agree well with the slope and levels of the BWF 8 curve for wind speeds of 34kt to 40kt. Results from the EARS 1 buoy have overall lower spectral levels (except for the EARS 1 anomalous time period), but otherwise are very similar.

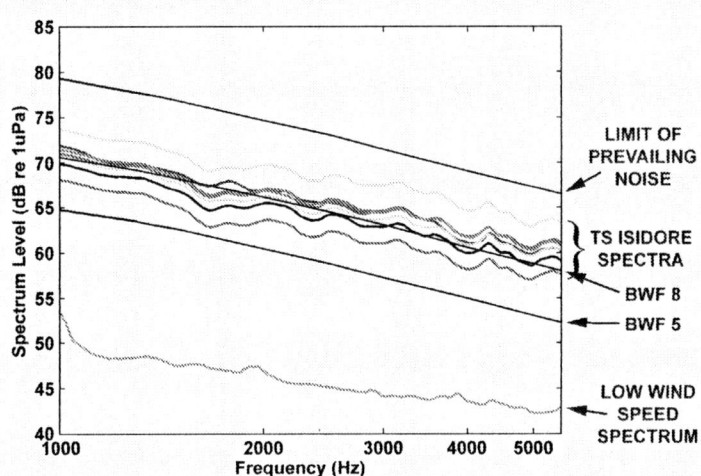

FIGURE 3 Illustration of the ambient noise results for EARS3 during Tropical Storm Isidore superimposed upon the results from Wenz [1] for Beaufort Wind Force 5 and 8. Also shown are the upper limit of the prevailing noise from Wenz [1] and a low wind speed spectrum for comparison.

There were some anomalous time periods during TS Isidore where the noise levels were much higher than those expected from the wind speed estimates. In those cases, acoustic data from 30 minutes before or after the original time period yielded results in line with wind speed estimates. It is postulated that these anomalies might be due to the banding nature of tropical storm systems. In these bands localized severe weather conditions are possible (e.g., rain squalls, high wind speeds, tornadoes, etc.). If these anomalies were due to banding effects, it would be expected that they occur periodically throughout the passage of the storm. This is currently being explored.

Tropical Storm Barry (August 2001)

Tropical Storm Barry passed about 93nmi east of the EARS buoys at closest approach. An acoustic noise analysis of the data and wind speed estimates, similar to the analysis for the 2002 storm systems, is currently underway. A generic spectral analysis of the acoustic data has been performed. Figure 4 illustrates the results of that analysis for a 24-hour period before TS Barry and a 24-hour period during the passage of TS Barry. The broadband lines seen on both spectrograms have been identified as shipping traffic. As can be seen from Fig. 4, the period before TS Barry has more shipping than the period during TS Barry. Also evident is the increase of noise levels in the frequency band from 1kHz to 4kHz during TS Barry. A close examination of the results during TS Barry indicates that the increase of levels is periodic, perhaps due to banding effects.

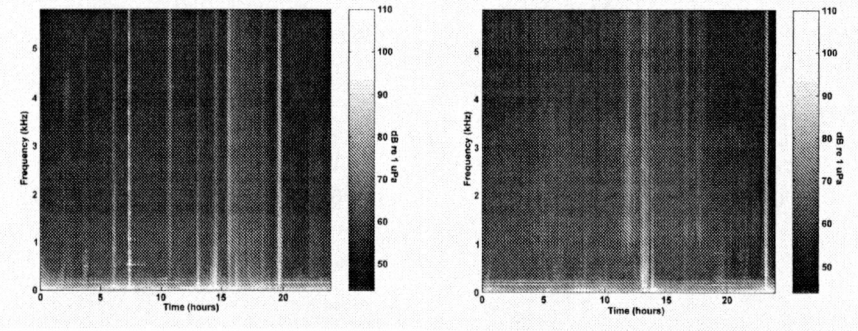

FIGURE 4. Spectrograms of the ambient noise before (left) and during (right) Tropical Storm Barry.

CLICK PRODUCTION

Ishmael [2] version 1.0 was employed to detect clicks automatically in the acoustic data. Ishmael generates noise-equalized spectrograms and calculates an average spectral energy. It then records detections after applying energy threshold criteria for frequencies over 2kHz. The Ishmael output, combined with the use of Raven [3] version 1.0, allowed anomalous sounds to be identified in the data, including pulse noises (e.g., those generated by seismic airguns) and more consistent noise (e.g., shipping noise). Furthermore, the unique multi-pulsed structure [4] of the clicks was periodically confirmed in order to verify that a sperm whale had produced them. Data containing detection events generated predominantly by non-sperm whale sounds were removed from subsequent analyses, as were those containing considerable consistent noise.

The whole study period was broken down into three major sections (Table 1) based on the 2001 NHC advisories for TS Barry. The storm period was defined by the time of the first and last advisories (1 and 17). Statistical significance was determined using Kruskal-Wallis tests to test the null hypothesis that storm activity had no impact on sperm whale vocal behavior.

TABLE 1. Definition of the time periods used in the analyses. Times determined by the NHC advisories for Tropical Storm Barry.

Time Period	NHC Advisory Number(s)	Start of Period			End of Period		
		Date	Julian Day	Time (GMT)	Date	Julian Day	Time (GMT)
Pre-storm	Pre-1	26-Jul	207	00:00	2-Aug	214	18:00
During	1,17	2-Aug	214	19:00	6-Aug	218	08:00
Post-storm	Post-17	6-Aug	218	09:00	13-Aug	225	23:00

Results

The mean click rate was highest during the pre-storm period at all EARS buoys either with or without the zero-files analyzed; zero-files are acoustic data files in which Ishmael detected no clicks. Additionally, the click rate was lowest during the storm period for all analyses with the exception of the EARS 2 data with the zero-files still included (i.e., the unadjusted click rate).

Although patterns also can be seen in the proportion of zero-files for each period at all three EARS buoys (Fig. 5), the variation was not quite significant ($p=0.026$, $p=0.004$ and $p=0.006$ at EARS 1, 2 and 3 respectively). EARS 1 displayed the highest incidence of zero-files during TS Barry, while EARS 2 and EARS 3 had their lowest proportion of zero-files during the storm confirming that the results do not simply reflect the impact of increased levels of masking noise. Further evidence for this is provided by the fact that the mean adjusted click rate for hours 4-8 at EARS 2 was higher during the storm than at any other time (Fig. 6). It is worth noting that this may also suggest that the whales are increasing the strength of their clicks in response to higher levels of ambient noise, since sperm whales are known to be capable of regulating the sound pressure of their clicks independently of their environment [5].

Produced pneumatically [6], it is possible that louder clicks requires the use of more air which may, in turn, reduce the number of clicks that can be produced before some air recycling must take place [7]. Alternatively, greater pressures might be required placing more physical strain on the tissues involved in production [6]. However, the full repercussions of a potential increase in signal strength, physiological or energetic, can not be determined from this data, although any such compensatory efforts would have implications for regions of heavy vessel activity.

The diurnal distributions at EARS 1 before and after the storm were very similar, although the absolute values after the storm were significantly lower (Fig. 6). Sperm whales are known to live in very stable groups [8], which have been found to extend to foraging behavior at depth and not just surface activities [9]. Consequently, this result, together with the different pattern observed during the storm, suggests that the whales from this area may have moved away during TS Barry and were slowly returning to the site after the storm had passed.

EARS 2 and 3 displayed visibly different diurnal patterns in adjusted click rate during each period (Fig. 6), although the post-storm and during-storm periods were possibly most similar, especially at EARS 2. It is likely that the very different daily peaks were created by the presence of whales that were not recorded in the area before the storm. Storm-related mixing could be expected to simply reduce any diurnal trends in diving behavior rather than producing an alternative trend since prey would be more

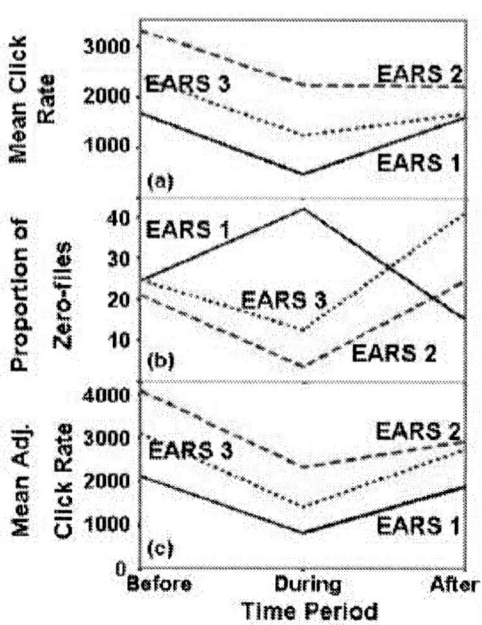

FIGURE 5. (a) Mean click rate, (b) mean proportion of zero-files and (c) adjusted click rate for the three time periods and each EARS buoy.

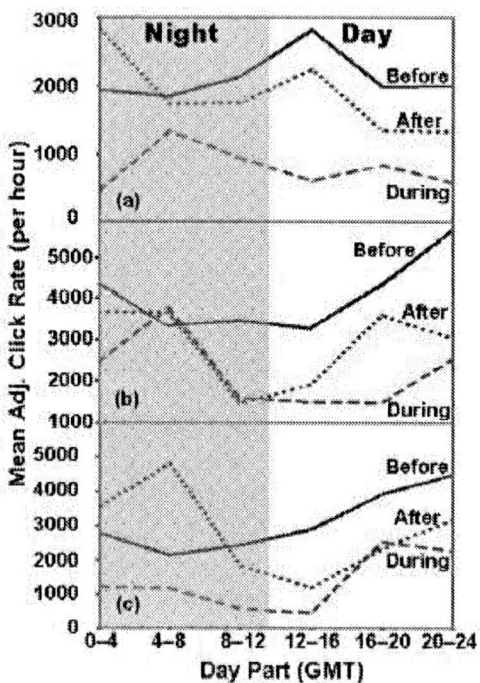

FIGURE 6. Mean adjusted click rate over the parts of the day at (a) EARS 1, (b) EARS 2 and (c) EARS 3 for the three time periods.

evenly distributed throughout the water column. Furthermore, the original trends were not revived after the storm suggesting that the migrations observed were more permanent than the remaining time included in this study (approximately one week).

It is not known whether these movements were due to an avoidance of deep water during the storm or more simply of the storm itself. Neither is it possible to determine if all the observed responses were due to increased ambient noise levels or changes in the Sea State. However, this preliminary study clearly shows that the distribution and behavior of the whales are affected by the presence of near-by tropical storms.

SUMMARY

Ambient noise results in the frequency range of 1kHz to 5kHz from the LADC 2002 experiment during Hurricane Lili and TS Isidore are in good agreement with expected wind trends. They also agree well with the Wenz curves. A similar analysis is planned for the TS Barry acoustic data from LADC 2001. A separate analysis of the data from the LADC 2001 experiment, showed that TS Barry had a definite impact upon the detected sperm whale click rates. A similar analysis of the LADC 2002 data for Hurricane Lili and TS Isidore is planned.

ACKNOWLEDGMENTS

This research was supported by ONR. The authors would like to thank the School of Biological Sciences, University of Wales, Bangor and the Marine Mammal Cognition and Behavior Lab, University of Southern Mississippi for co-operation during the analysis of the sperm whale data. Thanks are also due to John Goold, Robin Paulos, David Mellinger, Leslie Walsh, and all of the crews involved on the deployment and recovery cruises.

REFERENCES

1. Wenz, G. M., "Acoustic Ambient Noise in the Ocean: Spectra and Sources," J. Acoust. Soc. Amer. **34**, 1936-1956 (1962).
2. Mellinger, D.K., *Ishmael 1.0 User's Guide*, Technical Report OAR-PMEL-120, 2001.
3. Cornell Lab of Ornithology, *Raven 1.0 User's Manual*, Bioacoustics Research Program, Cornell Lab of Ornithology, 2002.
4. Goold, J. C., and Jones, S. E., "Time and Frequency Domain Characteristics of Sperm Whale Clicks," J. Acoust. Soc. Amer. **98**(3), 1279-1291 (1995).
5. Madsen, P. T., Payne, R., Kristiansen, N. U., Wahlberg, M., Kerr, I., and Møhl, B., "Sperm Whale Sound Production Studied with Ultrasound Time/Depth-Recording Tags," Journal of Experimental Biology **205**, 1899-1906 (2002).
6. Cranford, T. W., Amundin, M, and Norris, K. S., "Functional Morphology and Homology in the Odontocete Nasal Complex: Implications for Sound Generation," Journal of Morphology **228**, 223-285 (1996).
7. Wahlberg, M., "The Acoustic Behavior of Diving Sperm Whales Observed with a Hydrophone Array," Journal of Experimental Marine Biology and Ecology **281**, 53-62 (2002).
8. Christal, J., Whitehead, H. and Lettevall, E., "Sperm Whale Social Units: Variation and Change," Canadian Journal of Zoology **76**, 1431-1440 (1998).
9. Whitehead, H., "Formations of Foraging Sperm Whales, Physeter Macrocephalus, Off the Galápagos Islands," Canadian Journal of Zoology **67**, 2131-2139 (1989).

EXPERIMENTAL AND MEASUREMENT TECHNIQUES

The Kauai Experiment

Michael B. Porter, Paul Hursky, Martin Siderius[1], Mohsen Badiey[2], Jerald Caruthers[3], William S. Hodgkiss, Kaustubha Raghukumar[4], Daniel Rouseff, Warren Fox[5], Christian de Moustier, Brian Calder, Barbara J. Kraft[6], Keyko McDonald[7], Peter Stein, James K. Lewis, and Subramaniam Rajan[8]

[1]*Center for Ocean Research, SAIC, San Diego, CA*
[2]*University of Delaware, Newark, DE*
[3]*University of Southern Mississippi, Stennis Space Center, MS*
[4]*Marine Physical Laboratory, Scripps Institution of Oceanography, La Jolla, CA*
[5]*Applied Physics Laboratory, University of Washington, Seattle, WA*
[6]*Center for Coastal and Ocean Mapping, University of New Hampshire, Durham, NH*
[7]*Space and Naval Warfare Center, San Diego, CA*
[8]*Scientific Solutions, Inc., Nashua, NH*

Abstract. The Kauai Experiment was conducted from June 24 to July 9, 2003 to provide a comprehensive study of acoustic propagation in the 8-50 kHz band for diverse applications. Particular sub-projects were incorporated in the overall experiment 1) to study the basic propagation physics of forward-scattered high-frequency (HF) signals including time/angle variability, 2) to relate environmental conditions to underwater acoustic modem performance including a variety of modulation schemes such as MFSK, DSSS, QAM, passive-phase conjugation, 3) to demonstrate HF acoustic tomography using Pacific Missile Range Facility assets and show the value of assimilating tomographic data in an ocean circulation model, and 4) to examine the possibility of improving multibeam accuracy using tomographic data. To achieve these goals, extensive environmental and acoustic measurements were made yielding over 2 terabytes of data showing both the short scale (seconds) and long scale (diurnal) variations. Interestingly, the area turned out to be extremely active with a large mixed layer overlying a very dynamic lower channel. This talk will present an overview of the experiment and preliminary results.

OVERVIEW

The site of the experiment at the Pacific Missile Range Facility was selected primarily to take advantage of a network of over 200 hydrophones distributed over an offshore area larger than the state of New Mexico. As discussed later, this hydrophone array provides a unique capability for HF acoustic tomography. With that decided, the various participants agreed on a 6 km track following the 100-m isobath as shown in Fig. 1. Experience in other tests has shown that this length of track in this water depth typically provides an interesting combination of propagation conditions, including simple 2-path propagation in the near field and transitioning to order 10 echoes downrange (depending on bottom reflectivity). Similarly, communications work in the 8-16 kHz band has typically shown that standard sources (about 185 dB) start to fade out at the extreme of this range.

The University of New Hampshire, Center for Coastal and Ocean Mapping conducted the first stage consisting of mapping an area of approximately 100 km^2 in water depths of 30-900 m northwest of Kauai (Fig. 1). This was done with a combination of multibeam echo-sounders both capable of 150 deg swath widths: a RESON SEABAT 8111 sonar system operating at 100 kHz and used for water depths from 30 m to 400 m, and a Kongsberg EM120 sonar system operating at 12 kHz for depths greater than 400 m.

The deepwater system was permanently mounted on the hull of the ship. The shallow water sonar was installed temporarily for the purpose of this experiment, and it was deployed about 1 m deeper than the ships draft (about 5 m), which is also the baseline of the deepwater sonar, and the depth at which sound speed variations can have the most impact on beam pointing accuracy. Both sonar systems shared the same shipboard attitude and navigation sensing equipment.

The recorded soundings (pairs of (travel time, angle) for each beam for each ping) were processed and ray-traced into depths (z) and horizontal offsets (x, y) using sound speed vs. depth profiles derived from CTD and XBT casts taken throughout the survey area. The bathymetric grid shown in Fig. 1 was produced at a cell size of 20x20 m^2, with the WGS84 horizontal datum and the mean lower low water (observed tides) vertical datum.

There is currently only limited understanding of the role that different environmental factors, such as ocean structure, ocean variability, sea state, bottom roughness, currents, etc. play on noncoherent and coherent communications. Furthermore, the high frequencies imply wavelengths on the order of centimeters, which raises the concern that immeasurable quantities may play an important role.

To address these concerns, numerous environmental sensors such as an ADCP, waverider, CT sensors and thermistor strings were deployed in or near the propagation path as shown in Fig. 2. Besides the above described multibeam echosound survey, a sidescan survey was conducted after the experiment. To obtain a detailed picture of the ocean thermal structure, the thermistor strings were deployed in a curtain along the propagation path. In practice, it would have been difficult to produce a denser sampling; however, the spacing between strings took into consideration the typical correlation scale of internal waves. We will discuss the acoustic sensors in more detail below; however, receiving arrays were also deployed regularly along the propagation path to obtain a detailed sampling of the acoustic field.

ENVIRONMENTAL MEASUREMENTS

Sea state is one of the key drivers of HF propagation and communications performance. There are at least 3 factors at work here. First, a (hypothetical) static sea surface causes scattering with eventual losses of the steeper paths into the seabed. Second, the true sea surface is obviously dynamic, inducing coherence losses. Third, a rough sea leads to bubble injection and the resulting bubble clouds can cause strong attenuation in the 40 – 50 kHz band (depending on bubble size).

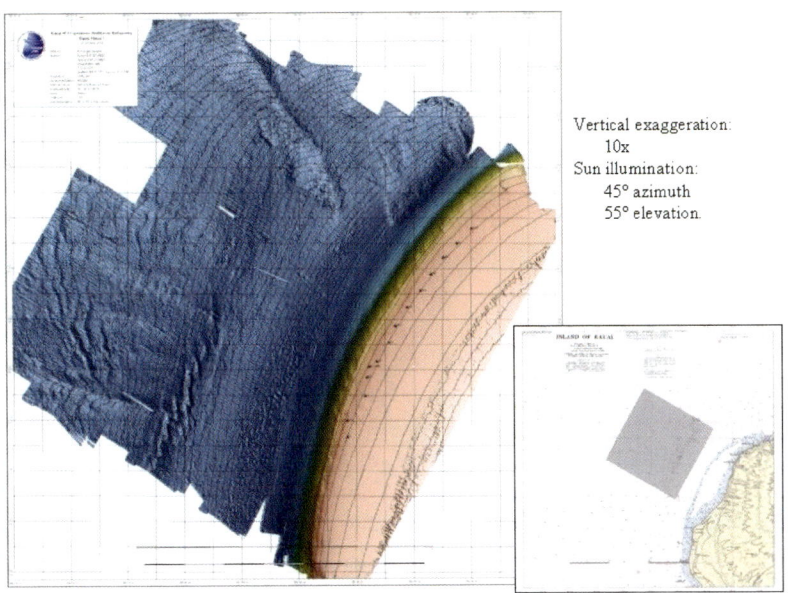

FIGURE 1. Multi-beam echo sound survey of the site and location relative to the island of Kauai (inset).

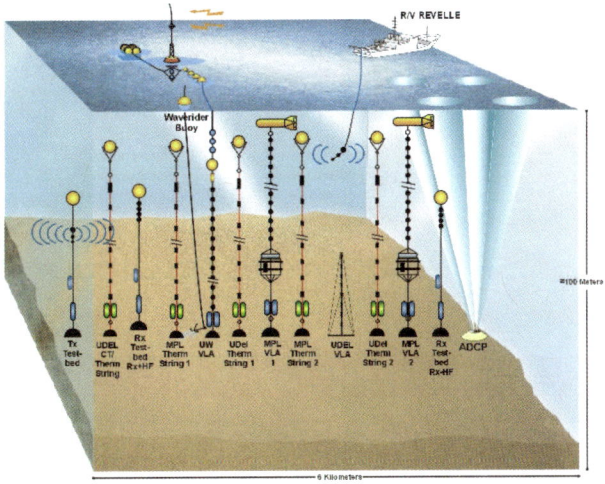

FIGURE 2. Equipment lay down for KauaiEx. The principal sound source located at the southern end (left) broadcast to VLA's distributed along the track. Acoustic sensors were interleaved with environmental sensors including thermistor strings along the entire propagation path.

The relative role of these effects is poorly understood. Indeed treatments of surface reflection are generally naïve on the role of the time scale that controls whether the surface should be treated as static or dynamic (or in between). Consider that a typical MFSK or DSSS communications system may process a 25-msec tone or symbol. This tone acoustically flashes the ocean surface capturing in a freeze frame much of the important dynamics. The treatment of these effects is an open research question; however, it was clear for the Kauai Experiment that an effort should be made to monitor the surface wave spectrum. The results in Fig. 3 show that data over the 2 weeks of the experiment.

To understand this plot it is useful to understand the surface wave behavior near Kauai. There are essentially 3 types of waves that occur: 1) long-period swell from the Southern hemisphere during the southern winter, 2) long-period swell from the northern hemisphere during the northern winter, 3) local short-period chop due to local waves. With regard to the local chop, wind on this side of Kauai follows a fairly regular pattern in which the eastern trade winds begin to pick up around noon producing a well-developed sea that then lies down in the late evening. The trade winds also have a tendency to clock around slightly to the north. The site of the experiment was purposefully selected so that the island of Kauai would cast a wind shadow on the site, which was gradually uncovered as the trade winds clocked to the north. Figure 3 shows both long and short period wave action associated with these different mechanisms.

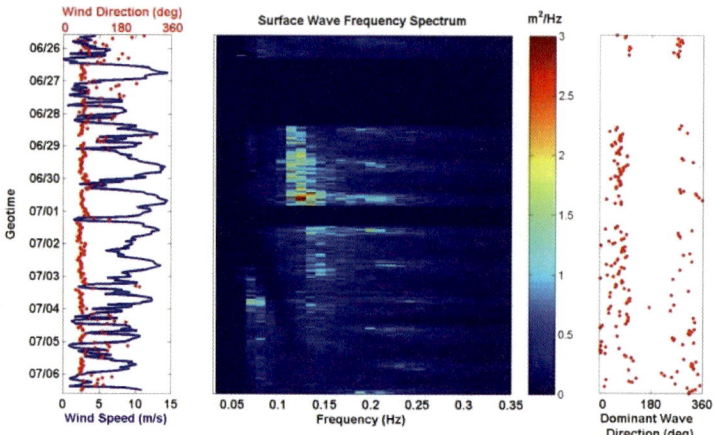

FIGURE 3. Waverider data taken over the duration of the experiment showing the daily wind cycles and associated wave spectrum.

TELESONAR TESTBEDS

The Telesonar Testbeds, developed at SPAWARSYSCEN San Diego and funded by ONR, played a central role in the experiment by providing two sound sources and two receivers. These unique, high fidelity, modular, reconfigurable, autonomous, wideband instruments were designed for high-frequency acoustic propagation and

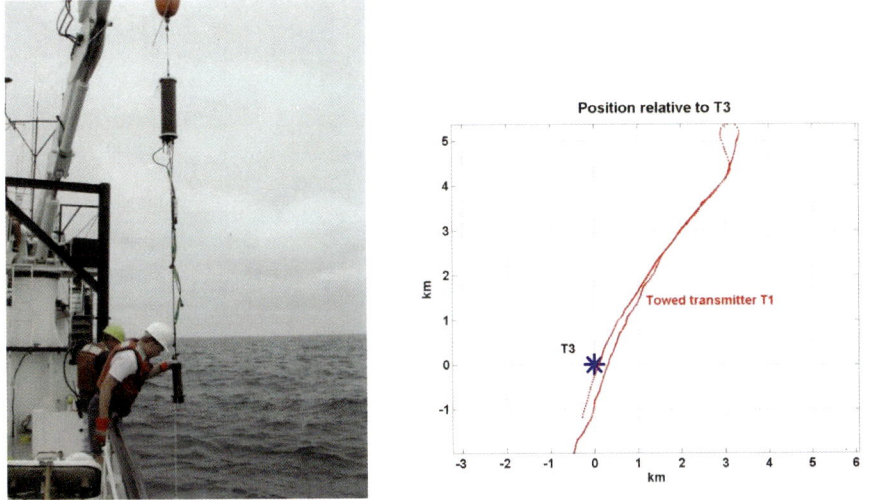

FIGURE 4. Deployment of the telesonar testbed (left). Track of the towed transmitter (T1) relative to the receiver testbed (T3) (right).

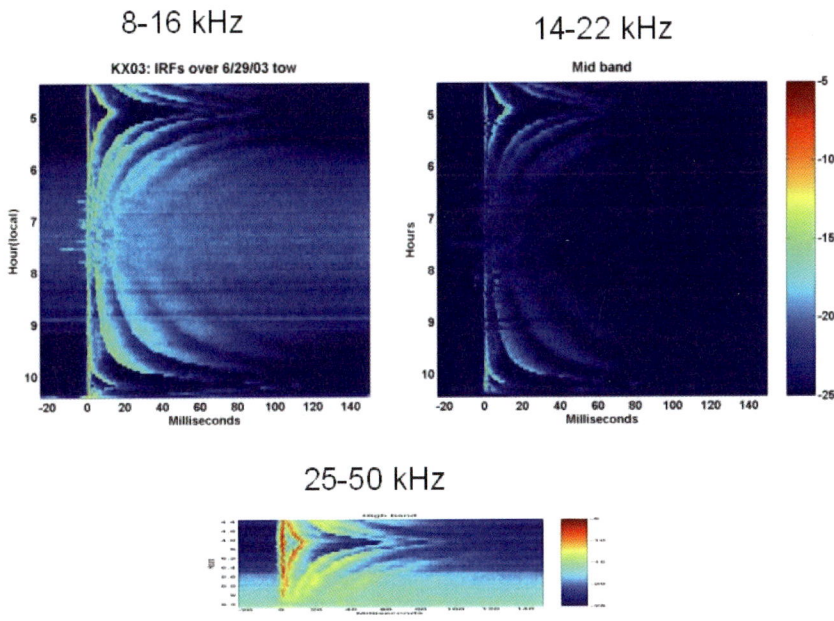

FIGURE 5. Matched-filter response or replica correlogram showing the impulse response of the channel in 3 separate frequency bands.

communication research. Each Testbed is configurable as a transmitter, receiver or both. The transmit-configured mooring is 6 meters long and has a mass of 46 kg in air 19 kg in water. This extremely small size and lightweight allow frequent field tests from an array of surface vessels down to 7 meters in length.

The Testbed transmitters are capable of sourcing arbitrary waveforms at 183 dB in three frequency bands, i.e. 8-16, 14-22, 25-50 kHz. The receiver-configured Testbeds are instrumented with a 4-channel 1.5 – 22 kHz, and a 2 channel 1.5-50 kHz receive arrays. Inside the electronics canister, a microcontroller coordinates the mission. A single-board computer running a robust real-time operating system under DOS orchestrates the sourcing and recording of data from and to a hard-disk drive. A MFSK modem is incorporated into the instrument to provide a link for remote control and status. The microcontroller, robust real-time operating system, and the modem all combine to ensure reliable mission execution. The paper by McDonald, et al. [1] provides additional information on the testbeds and their use.

Figure 4 shows the deployment of one of these systems with a PC104 computer housed in the aluminum pressure case linked to a 3-band transmitter system. The principle source unit or *diva* was deployed at the southern axis of the propagation track with the remaining VLA's providing the audience. An additional unit was towed in various patterns around the main propagation path to look at off-axis propagation and multi-user communications. Towed sources are also critical for understanding modem sensitivity to Doppler effects. As a corollary modem researchers view these measurements as either worthless or critical depending on whether they are interested in fixed or mobile networks.

An example of one such source tow is shown in Fig. 4 in which a testbed was towed to a range of about 6 km and back. Range here is measured with respect to the receiver testbed (T3) shown as the third mooring from the left in Figure 2. This sort of measurement is very useful to understand the range-dependence of the multipath structure. In addition, because the telesonar testbeds have an unusually large bandwidth, we are able to examine changes in both propagation physics and communications performance across that same large band.

Figure 5 shows the expansion and contraction of the multipath pattern as the source moves out in range then back derived by matched-filter processing. For those not familiar with the standard 'matched-filter' or 'replica correlation' procedure, we briefly review that. A chirp (also known as a linear frequency modulated sweep or LFM) is transmitted in the band of interest. The received waveform is then correlated with the transmitted waveform. Since the transmitted waveform of the chirp has a flat power spectrum, its autocorrelation produces a much shorter impulsive function (technically a sinc-pulse). The channel is considered as a linear filter so that the process of correlation can be done on the received waveform with the same effect as if it had been done to the chirp before it was transmitted. Thus, correlating the received waveform with the transmitted waveform is equivalent to transmitting an impulsive sinc function and yields an estimate of the channel impulse response.

The individual panels show the different frequency bands (low, mid, and high). We find many researchers not familiar with this frequency band are very surprised to observe a clear multipath structure involving many boundary interactions. These results clearly show this feature, even in the highest band. This multipath pattern

FIGURE 6. The right panel shows the Data Acquisition Unit (DAU) with three batteries mounted on the base of tripod supporting the vertical line array in center. The left panel shows the array during a deployment.

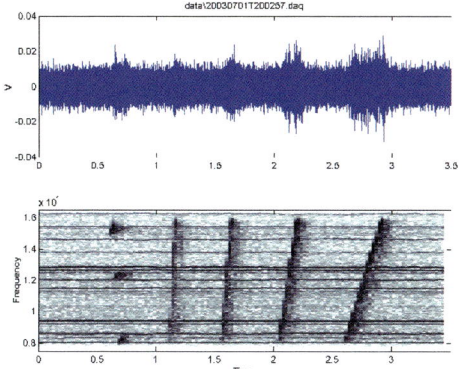

FIGURE 7. Received signal for low frequency transmissions, 8-16 kHz 320 msec. (a) Amplitude versus arrival time, (b) Frequency-time diagram for 8-16 kHz for a set of transmissions preceded by three marker tones.

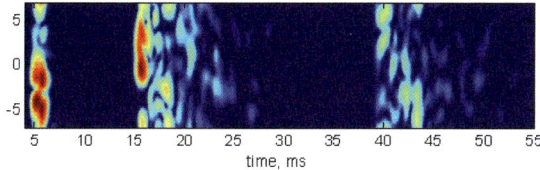

FIGURE 8. Incoming energy as a function of time and arrival angle. Note three distinct arrivals around 5-7 msec., 15-25 msec., and 40-50 msec. during 7/2/04 at 12:32 PM.

provides a unique fingerprint of the source location and can be used to track a source to many kilometers using a single phone.

MODULAR RECEIVE ARRAY

The University of Delaware designed and built a new VLA for KauaiEx using commercial off-the-shelf components. This array provided a unique capability in having enough closely spaced elements to facilitate beamforming and thereby study fluctuations in the arrival time and angle. In addition, since the array is rigid and mounted on the sea bottom, it provides a stable receiver for higher frequency signals where the effects of small array motion can influence the signal fluctuations. In this array, three batteries feed an eight channel Data Acquisition Unit (DAU) with the sampling frequency of 98 kHz. The array is lightweight and easy to deploy, requiring a winch. Figure 6 shows the VLA, the DAU and supporting tripod during one of the deployments.

To capture the effect of variability of the ocean environment on the propagation of broadband signals, a set of four LFM sweeps with variable duration were transmitted over various frequency bands every half hour. The varying durations are designed to help characterize surface variability over a very short time scale by providing different times over which to average the surface reflected signals. A single transmission consists of 9 sets of the FM sweeps 40-320 msec. in duration as shown in Figure 7. A set of three narrowband FM sweeps was used as a marker at the beginning of each transmission set; a set consists of FM sweeps over various frequency ranges. The transmissions are spaced about 3 seconds apart for a total duration of about 30 seconds.

The received signals were matched filtered and beamformed. Figure 8 shows a sample of beamformed signal depicting three distinct arrivals. The detail of the variability on the arrival energy for short and long times are shown in Ref. [2] in this volume.

AUTONOMOUS ARRAY SYSTEM

The Marine Physical Laboratory brought two water-spanning 16-channel VLA's, which provided an invaluable sampling of the acoustic field (Fig. 9). These autonomous systems are composed of a PC running Labview for data acquisition, recorded to large onboard disk drives.

Channel probes were transmitted regularly throughout the experiment to measure directly the channel impulse response. Both chirps (with pilot tones) and m-sequences were used as these each have different advantages for estimating the impulse response. An example of these receptions on the MPL array is shown in Fig. 10. Many researchers look at a single chirp and that may be said to represent a purer version of the channel impulse response. However, simply stacking the results of 20 chirps provides a vastly clearer picture of the echoes. On the other hand, if one stacks too many chirps then the temporal variations in the waveguide cause the arrivals to be smeared out. Thus, the Goldilocks solution is the figure in the middle panel.

FIGURE 9. The MPL Autonomous Array Systems laid out on deck before deployment.

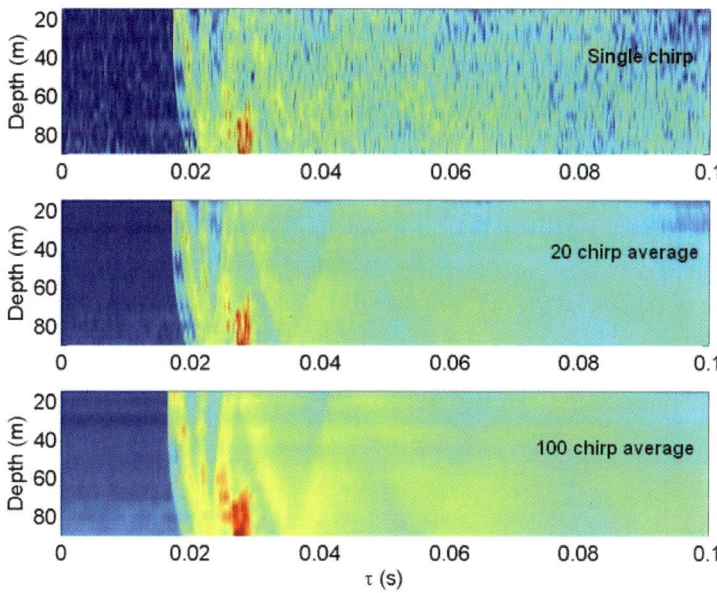

FIGURE 10. Impulse response obtained by matched-filtering the data on the MPL VLA. Increasing (top to bottom) the number of chirps averaged to get the impulse response progressively draws out weaker arrivals but smears out their arrival time. Note the bottom-ducted energy showing up as a bright spot near the bottom.

In general, we may describe this as an accordion pattern where each fold in the accordion represents an additional surface or bottom reflection. The quiescent period at the beginning ends as the direct path arrives forming the first broad arc. Since the source is near the bottom, a bottom reflected path comes almost immediately behind it. There is also a significant blob of energy creeping along the bottom and arriving at about 25 msec on this time scale. These are the rays trapped in the lower sound channel that some have called the *bouncing ball* paths.

Of course, a key goal of this experiment is to understand more clearly how the propagation physics effects acoustic communications performance. Again, the MPL array provides a particularly valuable resource since it allows us to examine bit-error rates (BER) at the same time, throughout the channel.

Figure 11 shows the results from one of many schemes transmitted during the experiment. This is a multiple-frequency shift keyed system running at 2400 bps. This is arguably the simplest scheme for transmitting data and may be compared nearly precisely to the process of striking a chord on a piano. The received data is interpreted by recognizing the chords (a spectrogram) and the transmission rate is limited principally by the ocean reverberation. This in turn requires that the sound of one chord dies down before the next is played. Practical MFSK systems typically yield a bandwidth efficient of ¼ bps/Hz. Here we are using a 4800 Hz band and pushing the scheme to 2400 bps which generates somewhat high BER. However, it should be noted that in practical operation we include a half-rate convolutional coder that reduces the data rate in half but reduces the error rate to virtually zero. Since channel coding reduces the errors by a huge amount, gaining useful statistics would require many more transmissions. Therefore, we typically prefer to interpret sensitivity of the modem scheme without the channel coding.

The role of the channel on the modem performance here is obvious to the most casual observer. Large numbers of channel errors are generated in the surface mixed layer and progressively decreasing for deeper receivers. The ocean thermal structure seen in the lower panel reveals the warm mixed layer in the upper half of the water column and the cool duct at the bottom.

To discuss these effects one should be aware of numerous factors. First, the SNR is higher in the lower part of the water column, partly because of the lower duct. Ambient noise plays into that also; however, it is seen to not vary a large amount with depth. In addition, all paths in the upper mixed-layer are surface interacting causing both static and dynamic surface losses. In contrast, the bouncing ball energy in the bottom duct is strong and steady. Finally, one must consider the overall multipath spread. Increased multipath aids SNR, which is good, but causes the reverberant effect (intersymbol interference), which is bad. The paper by Siderius, et al., [3] in this volume discusses these effects in more detail.

A final point of interest in an array such as this is how multiple channels can be exploited to improve modem performance. There are many ways to think of this problem. For instance, one can imagine beamforming to a particular eigenray so that multipath spread is reduced. One can also view the receivers as being robust against fading since they fade independently (spatial diversity). A trivial way to take advantage of this spatial diversity is to process each of the channels and sum them in a sort of voting process to estimate the particular tones played. Figure 12 shows how summing increasing numbers of channels provides a major improvement to the system performance.

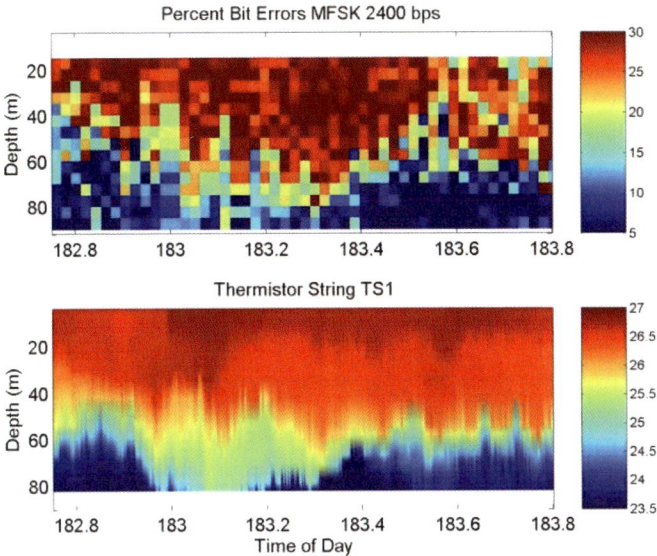

FIGURE 11. Bit-Error-Rate over depth and time (upper panel) versus the ocean temperature structure (degrees Celsius) (lower panel).

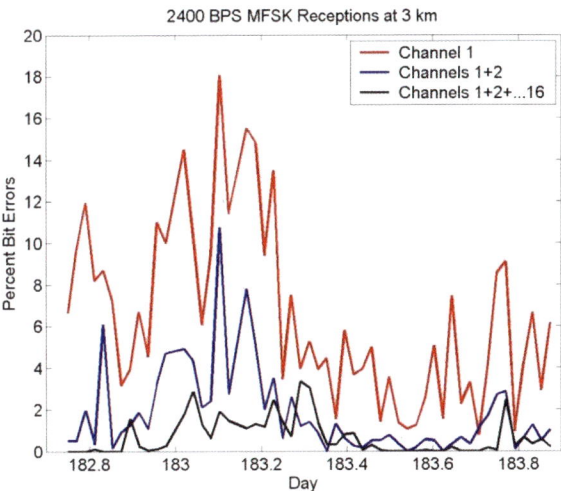

FIGURE 12. Bit-error-rates plotted vs. time and combining varying numbers of channels. Increasing the number of receivers provides a significant improvement in performance.

UW ARRAY

The UW-APL array provided another dimension to the experiment. Its 8 channels sampled a smaller slice of the water column but with greater density. In addition, the wireless access via radio buoy allowed real-time monitoring of the transmissions. The position of the array within the water column is shown in Fig. 13. One can see from the ray trace that it is poised to receive the direct and surface paths, and depending on the mixed layer depth may also get a hint of leakage energy from the bouncing ball paths.

Figure 14 shows the arrival pattern as observed down the array. Because the source is sitting on the bottom, the arrivals tend to arrive in pairs. The spacing between the array elements is 2 m and the arrival pattern shows significant structure even with this relatively dense sampling. The figure represents an average over 10 seconds so the late arriving paths that have multiple interactions with the sea surface are more diffuse than the distinct early arrivals.

HF ACOUSTIC TOMOGRAPHY

As mentioned above, a motivator for doing the experiment in this site was access to hydrophones in the Pacific Missile Range Facility. Figure 15 shows the network of sources and receivers that were used for this phase of the testing. Matched-filter output from these phones was equally successful in extracting the various boundary echoes and preliminary model runs (Figure 16) show that the arrivals for phones in deeper parts of the range are also well modeled. The paper by Lewis, et al. [4] in this volume discusses these data in greater detail.

SIDESCAN SONAR SURVEY

Finally, a sidescan survey was conducted by USM at the end of the experiment to characterize more fully the nature of the bottom. The results shown in Figure 17 are images of a small region of the bottom surveyed simultaneously with two frequencies (150 kHz on the left and 300 kHz on the right). The slant range on each side of each frequency image is 100 m and the depth, represented by the darker band down the middle of each, is 20 m. (The other dark vertical bands are nulls in the sidelobes. The wider dark band on the right side of the 300-kHz image is just incidental to the gain setting.) The speckled structures are the echoes from two fish schools (partly directly under the boat, but also off to the right side). Further to the right (offshore) and left (in shore) are sand ripples (not too clear in these figures), eventually giving way to the larger-scale ridge structures on the shoreward side also seen in the multibeam survey.

The paper by Caruthers, et al. [5] presents a fuller discussion of the sidescan survey.

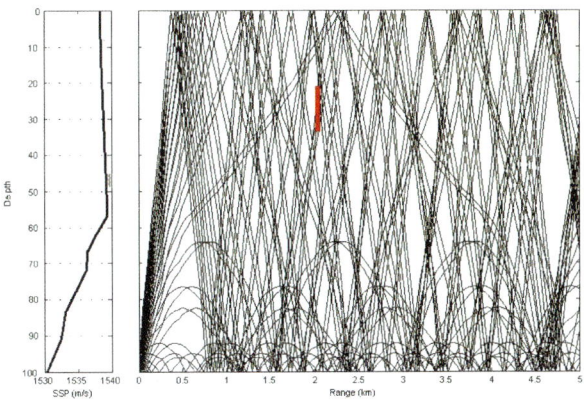

FIGURE 13. Sound-speed profile and ray trace for KauaiEx.

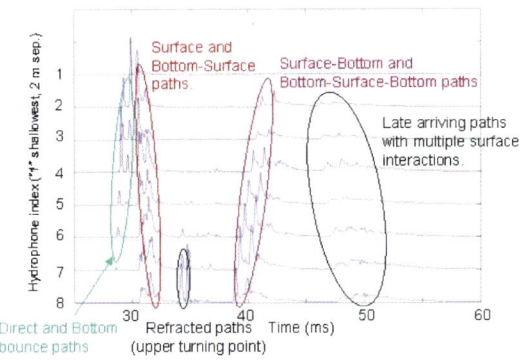

FIGURE 14. Matched-filter response for the UW array. Various arrivals crossing the array are clearly identified.

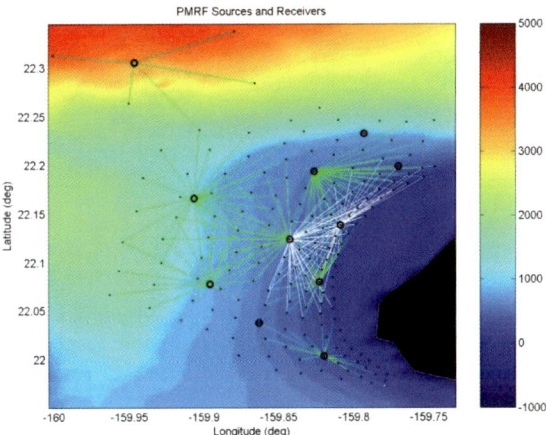

FIGURE 15. Network of sources and receivers provided by the Pacific Missile Range Facility and used for tomographic imaging of the ocean structure.

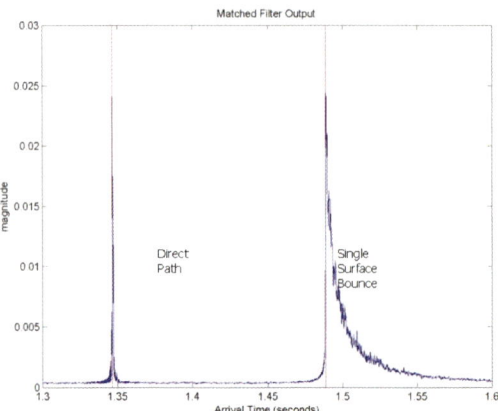

FIGURE 16. Matched-filter output from PMRF showing the two dominant arrivals in a deeper part of the range. The red curves are modeled arrival times and closely overly the blue curves derived from the data.

FIGURE 17. Sidescan survey images of the propagation path as seen at 150 kHz (left panel) and 300 kHz (right panel). The band down the middle represents the water depth. A school of fish is visible near the top of the plot and sand ripples along the inshore side (left).

ACKNOWLEDGMENTS

The High-Frequency Experiment (HFX) was supported by ONR OA321. The acoustic communications experiment (SignalEx) was supporter by ONR SS. The HF tomography component was supported by CEROS. Partial support for APL-UW was from ONR under the ARL program.

REFERENCES

1. McDonald, V., Hursky, P., and the KauaiEx Group, "Telesonar testbed instrument provides a flexible platform for acoustic propagation and communication research in the 8-50 kHz band," this volume.
2. Badiey, M., Forsythe, S., Porter, M., and the KauaiEx Group, "Effects of environmental variability on the bottom mounted vertical line array in KauaiEx," this volume.
3. Siderius, M., Porter, M., and the KauaiEx Group, "Impact of thermocline variability on underwater acoustic communications: results from KauaiEx," this volume, 2004
4. Lewis, J., Stein, P., Rajan, S., Rudzinsky, J., Vandiver, A., and the KauaiEx Group, "High-frequency tomography using bottom-mounted transducers," this volume.
5. Caruthers, J., Quiroz, E., Fisher, C., Meredith, R., Sidorovskaia, N., and the KauaiEx Group, "Side-scan sonar survey operations in support of KauaiEx," this volume.

Ocean Variability Effects on High-Frequency Acoustic Propagation in KauaiEx

Mohsen Badiey[1], Stephen E. Forsythe[2], Michael B. Porter[3], and the KauaiEx Group

[1]*College of Marine Studies, University of Delaware, Newark, DE 19716*
[2]*Naval Undersea Warfare Center, Newport, RI, 02841*
[3]*Science Applications International, San Diego, CA 92121*

Abstract. During the Kauai experiment in summer of 2003 a bottom-mounted vertical line array containing 8 hydrophones spaced 0.6 meter apart was deployed in a 100-m shallow water region near the Pacific Missile Range Facility. The acoustic source was placed about 2 km away on a flat sea bottom at 95 meter water depth. The element spacing was sufficiently small to allow measurements of the temporal variability of time-angle intensity fluctuations of the acoustic energy. Measurements were made simultaneously of the broadband acoustic pulse transmissions (8-50 kHz) and environmental parameters. The latter measurements included current, temperature and salinity profiles, directional surface wave spectra, as well as wind speed and direction above the sea surface. Arrival time-angle fluctuations were found to be correlated with the environmental variability due to ocean dynamics in this region. It is shown that variations of the sea surface dynamics exhibit different temporal effects than those occurring within the water column.

INTRODUCTION

Variability of ocean physical parameters can cause significant fluctuations in the arrival of broadband acoustic signals in shallow water. Arrival time and angle information of a pulse is a useful quantity in different applications of underwater acoustics because it can be a good indicator of the dynamics of the ocean volume or boundaries. The arrival time of energy following a particular ray path depends on sound speed and current of the ocean through which the ray passes and on roughness of the ocean boundaries with which reflects or scatters.

In shallow water, where the energy traveling along several ray paths may contribute to an arriving signal, significant signal bandwidth is necessary to distinguish (temporally) the arrivals corresponding to individual ray paths. Even with sufficient signal bandwidth, on a windy day having a rough sea surface it may be difficult to identify an arrival time for an individual ray that intersects the sea surface, due to the time and angle spreading and collective interference of the scattered energy. To address these issues a highly calibrated acoustic experiment was conducted during the summer 2003 at a shallow water location near the Kauai Island, Hawaii.

The Kauai Experiment (KauaiEx) was conducted from June 22 to July 9, 2003 with the objective to study high-frequency (8-50 kHz) acoustic propagation in a shallow water waveguide. In contrast to much of the previous literature, emphasis was placed on multipath arising from multiple boundary interactions. The main theme of this experiment was the role of the environmental physical parameters on high-frequency acoustic signals applicable to underwater communications. A great deal of effort was made to characterize the environment including the surface wave spectrum, 2-D temperature structure along the propagation path, salinity, currents, and bottom properties. Using autonomous instruments, most of these parameters were measured continuously over the two weeks of the experiment providing information on the diurnal cycles. At the same time, extensive acoustic measurements were made using a variety of vertical line arrays some of which spanned the entire water column. Detailed description of the experiment and the overview is provided in [1]. During the course of the experiment there were actually three different deployments of acoustic source and receiver arrays. In this paper, the results are presented from the second deployment pertaining to a fixed vertical line array placed on the sea floor 2 km from the acoustic source. A brief description of the oceanographic data is presented followed by the acoustic measurements and data analysis.

MEASUREMENTS

Oceanographic Measurements

Detailed oceanographic measurements made during the experiment included directional surface waves, water column temperature and salinity profiles at different points along a 6-km propagation path. The current profile, wind speed and direction were also measured at single points. Figure 1 shows the schematic diagram of the acoustic and the environmental measurement arrays. The sea surface waves and temperature profile near the bottom mounted receiver array for a period of 47 hours (from 11:00 AM July 1 through 10:00 AM July 3, 2003) are presented here. During this time the temporal variability of acoustic signals is shown to be correlated with the environmental variability.

Five thermistor arrays were placed along the propagation track to measure temporal and spatial distribution of the sound speed during the experiment. Measurements of the salinity profile at different points showed a negligible variation. Therefore, the salinity is considered constant for calculation of the sound speed profile. For data discussed here, the temperature profile (UDEL Themistor String in Fig. 1) near the receiver array at 2 km from the sound source is shown in Fig. 2. This time corresponds to the same time for which both surface waves and the acoustic propagation measurements were made.

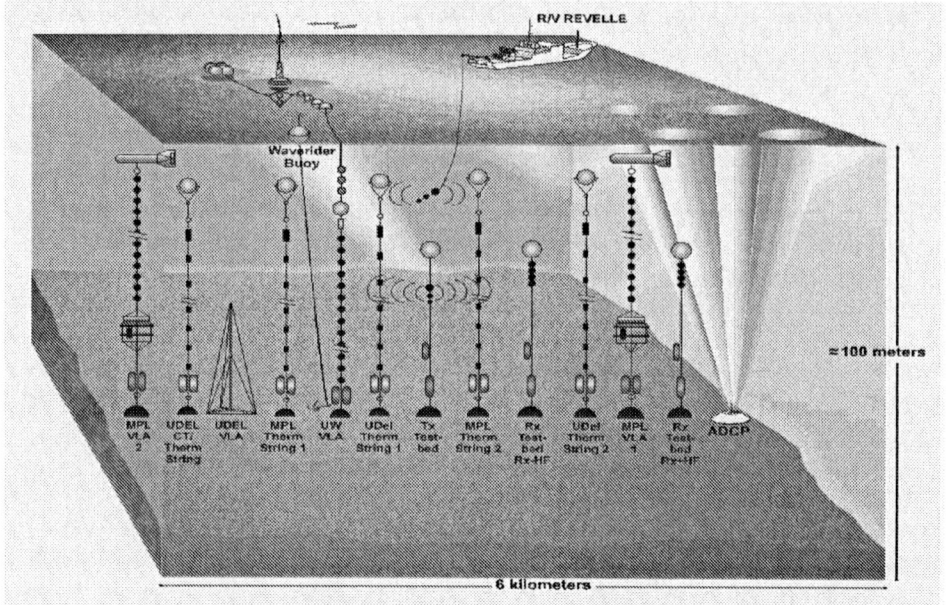

FIGURE 1: Schematic diagram of the Kauai Experiment. Data from UDEL Vertical Line Array (VLA) and UDEL-CT Thermistor String are discussed here.

It is noticed that for most of this time the water column is a very well mixed layer down to about 50 m depth. A cold layer (about 4-5 degrees C lower than the mixed layer) emerges at nearly tidal cycles. Variations of this layer pertaining to oceanographic features are repeated during the entire experiment.

Hourly measurements of wave frequency spectra were made during the entire experiment using a directional buoy. The wave spectrum and the wind speed for the period from 11:00 AM July 1 through 10:00 AM July 3, 2003 are shown in Fig. 3. The right plot shows hourly measurements of the non-directional wave frequency spectra. These plots show that changes in spectral level correspond to changes in wind speed. Changes in high-frequency spectral levels are abrupt while changes in low frequency spectral levels occur gradually. It is noticed that surface wave energy changes in three different distinct frequency bands. Open ocean swells show up on (0.05 – 0.1 Hz) which are low frequency waves traveling in from the open ocean from far distances. Then the wind generated surface waves arrive at two different bands. First, there are larger scale waves formed after the wind has blown in the same direction for some duration of time (0.1 -0.2 Hz), and then small scale surface chop (0.2-0.35 Hz) that appears almost immediately after wind speed increases and disappears shortly after wind speed decreases or changes direction (this is referred to as the land breeze effect).

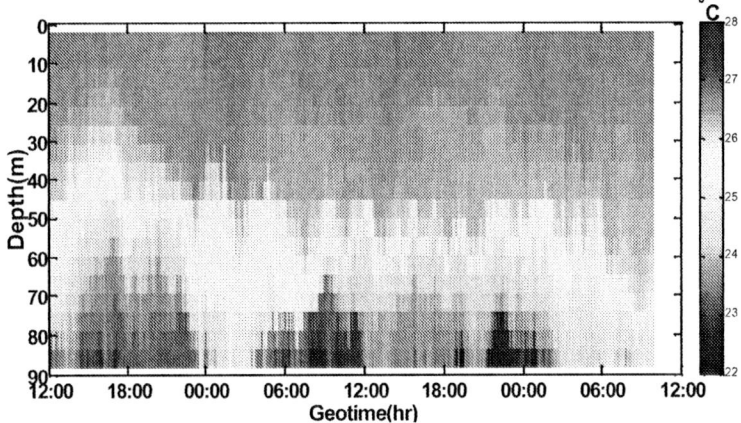

FIGURE 2: Temperature profile for the UDEL-CT/Thermistor String mooring from 11:00 AM on July 1 through 10:00 AM on July 3, 2003.

Changes in spectral levels for the 0.2 – 0.3 Hz small scale, surface-chop frequency show an immediate increase and decrease corresponding to changes in wind speed. Spectral levels for the 0.1 - 0.2 Hz larger scale wind-generated waves show the same correspondence to changes in wind speed, but spectral level changes occur on longer time scales. Energy remains in this frequency band for sometime after wind speeds decrease.

For both days covered in this period, wind speeds undergo a late morning increase and a late night decrease. Changes in spectral levels for the 0.2 – 0.3 Hz small scale, surface-chop frequency show an immediate increase and decrease corresponding to changes in wind speed. Spectral levels for the 0.1 - 0.2 Hz larger-scale wind generated waves show the same correspondence to changes in wind speed, but spectral level changes occur on longer time scales. Energy remains in this frequency band for some time after wind speeds decrease and generally the dominant wave direction corresponds to wind direction. This indicates the majority surface wave energy is coming from wind-generated waves.

In an acoustic measurement the sea-surface fluctuations may induce fast fluctuations in the acoustic signal propagation while temporal variability of the sound speed profile may induce large-scale fluctuations. To resolve both these scales, we consider different sampling of the ocean on both short and long geophysical time scales.

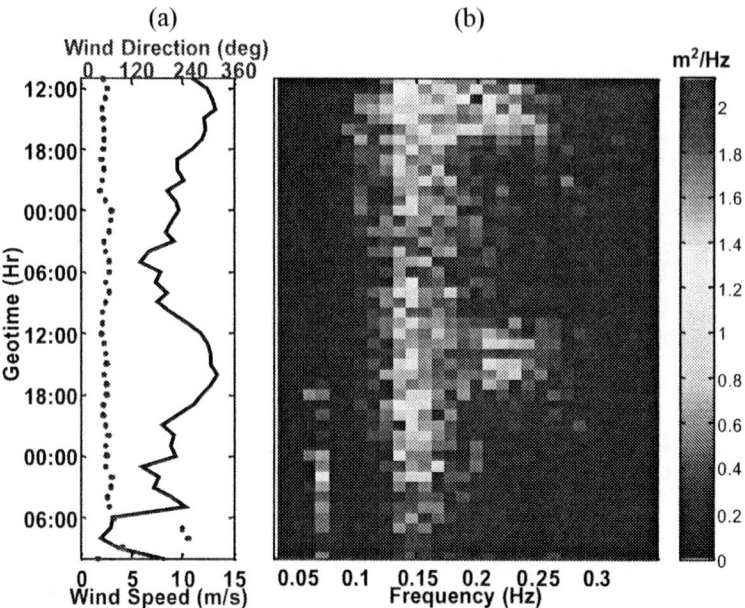

FIGURE 3: (a) Hourly wind speed and direction; (b) Measured sea surface wave frequency data from 11:00 AM July 1 through 10:00 AM July 3. This period corresponds to 48 hours of acoustic data recorded on the bottom mounted vertical acoustic line array.

ACOUSTIC MEASUREMENTS

A set of four LFM sweeps with variable duration was transmitted in three frequency bands (low 8-16 kHz, mid 14-22 kHz, and high 25-50 kHz). The duration of these chirp signals was 40, 80, 160 and 320 milliseconds (msec.) respectively. These four sweeps were transmitted nine times with a 3-second time interval in between transmissions, for a total duration of 27 seconds. This group of nine transmissions is used to estimate variability over a short time scale. Groups of nine were transmitted with the time interval of 30 minutes throughout the experiment period. The variable durations are designed to help characterize surface variability over a very short time scale by providing different times over which to average the surface reflected signals. In this paper we describe the signals transmitted in the 8-16 kHz frequency band. The impulse response for the water column shown in Fig. 4 is obtained by correlating the arrivals at the receiver array with the original transmitted signal.

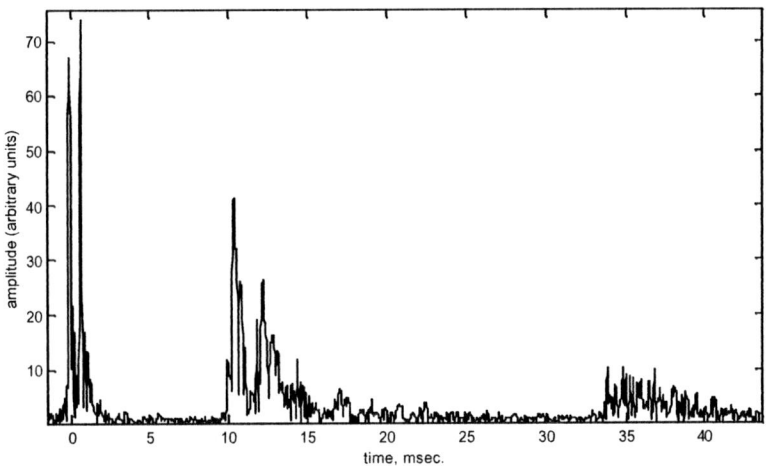

FIGURE 4 - Matched-filtered signal for a typical pulse arrival for geotime 7/2/04 12:31 PM.

The distinct early arrivals (around 0 msec.), which are bottom interacting, are subject to one or more bottom bounces. Surface-interacting ray groups arrive starting at around 10 and 34 msec. Since the original transmitted signal is broadband (8-16 kHz), sub-bands of the signal can be used to provide a time-averaged value of the arrivals. Here a 2 kHz bandwidth has been used to give a resolution of about 0.5 msec. for 3 dB down from max peak.

The 8-hydrophone receiver array mounted in a vertical stave has roughly a 4 m aperture. This array is used in a beamforming algorithm designed to provide better resolution of the ray arrivals by showing the acoustic field arrival in both time and angle. Beams are formed by taking the Fourier transform of the 8 channels' digitized time series and multiplying each channel by a complex phasor to "correct" the expected phase shift (as a function of frequency and arrival angle) between the vertically-deployed hydrophones in the array. The 8 corrected Fourier transforms are then summed and the inverse Fourier transform applied to give a single time series. This signal is effectively spatially filtered to look only at the incoming energy from the selected arrival angle. The process is repeated for all arrival angles between -20 and 20 degrees. The envelope is then formed using a Hilbert-transform, and displayed as a false-color image in Fig. 5. The correction formula is $\phi_{ij} = k_i z_j \sin\theta = \frac{2\pi f_i}{c} z_j \sin\theta$, where ϕ_{ij} is the phase correction to the ith spectral amplitude of the jth channel of the 8 hydrophones' time series, z_j is the vertical displacement of the jth hydrophone from the center of the array, and θ is the selected angular direction for the spatial filtering operation. While the element spacing was fairly small, it was not sufficiently small to eliminate grating-lobe effects that appear as repeating patterns in the angular direction of the time-angle plot. As an example, energy arriving from 0 degrees (horizontal) needs no phase correction, since $\sin\theta = 0$ for that angle. However, if the product,

$\frac{f}{c} \Delta z \sin \theta$, (where Δz is the spacing between adjacent elements) is equal to 1, the phase shift is also 0, which is indistinguishable from the horizontally incident case. So, the angle at which aliased energy arrives horizontally is $\theta = \sin^{-1}\left(\frac{c}{f \Delta z}\right)$. A sample time-angle plot for a pulse transmission during July 1, 2003 at 12:31 PM is shown in Fig. 5.

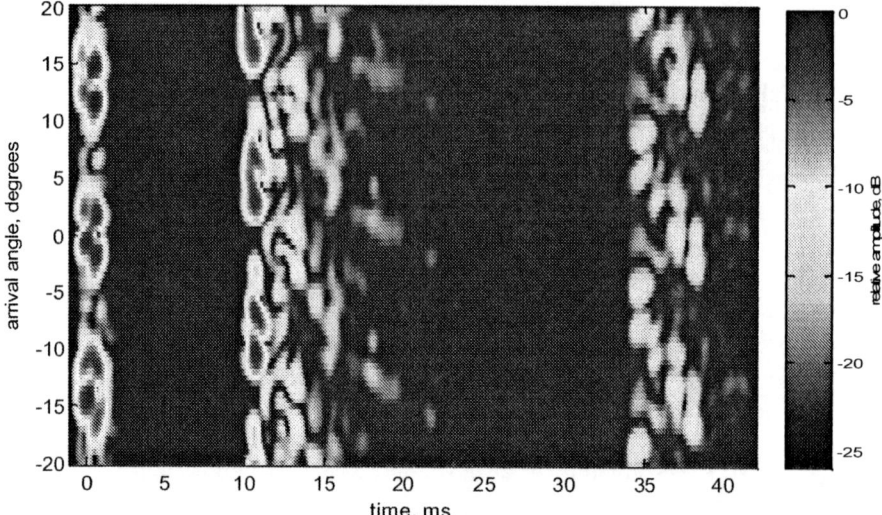

FIGURE 5 - Incoming energy as a function of time and arrival angle for geotime 7/2/04 12:31 (center frequency of energy is 11 kHz and bandwidth is 2 kHz).

The vertical repetition is noticed in the series of arrivals at 0 and 10 msec. From the above formula, the aliasing angle at 11 kHz and 0.6 meter element spacing at which incoming energy will be aliased (for small arrival angles around 0°) is calculated to be 13.2°. We take half of this angle on each side (i.e. 6.6°) for the angular region of analysis.

Based on the above analysis, it is difficult to distinguish the aliased energy peaks from the actual ones. It is possible however to appeal to the known ocean conditions (source depth, receiver depth, and sound speed profile) to eliminate some regions of the arrived energy as being inconsistent with known transmission characteristics. This is a topic of future analyses. In this paper we are concerned with variation from set to set (i.e. 3 second time separations) and from transmission to transmission (i.e. 30 minute time separations). The following plots will concentrate on the angular region from -6.6° to 6.6°, recognizing that energy arriving from larger angles (e.g., via surface interactions) will be aliased.

SIGNAL VARIABILITY

Variability over Long Geotime

To show the effects of ocean variability on the acoustic pulse propagation, two time scales are considered. We refer to these time scales as the short and long geophysical times (abbreviated here as "geotimes") corresponding to the transmission intervals of 3 seconds and 30 minutes respectively. The beamformed time-angle plots shown in Fig. 6 depict the pulse arrivals for four separate geotimes.

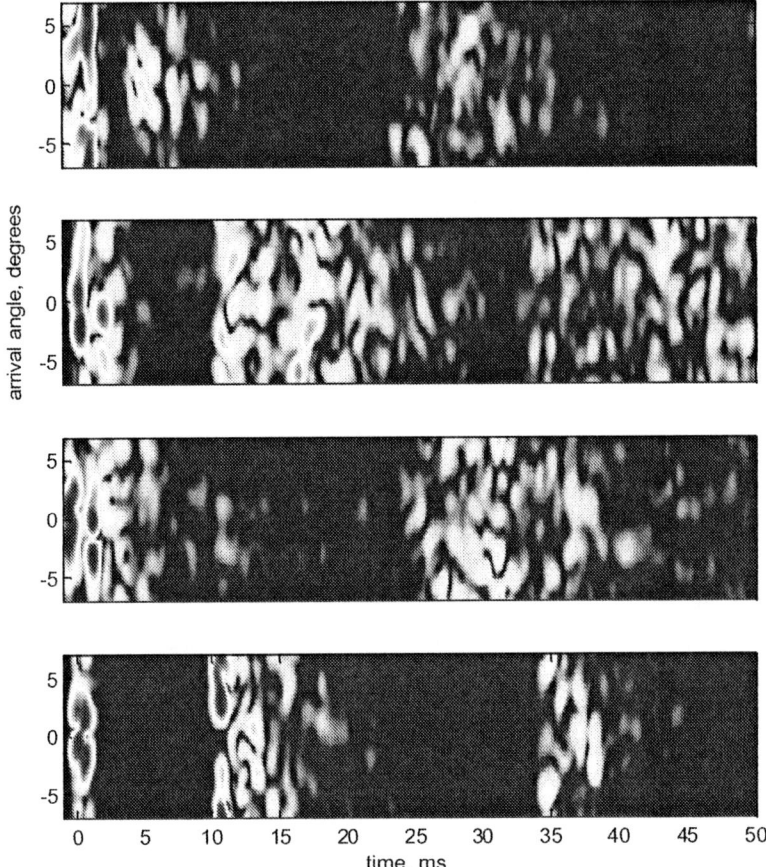

FIGURE 6 - Time-angle plots for showing a large variation in the macro-structure of the ray arrivals over few hours during 7/2/2003. (A) 00:01 AM, (B) 04:01 AM, (C) 06:00 AM, (D) 12:32 PM.

Based on the temperature variations shown in Fig. 2, the sound speed profile over this period changes radically, with the sound speed near the bottom varying by about 10 meters/sec. However, as shown in Fig. 3, the surface is relatively calm over this period.

Variability over Short Geotime

Next the variation of the time-angle fluctuations over signal transmissions separated by 3-second time intervals is considered. An example of short geotime variations of the beamformed results is shown in Fig. 6.

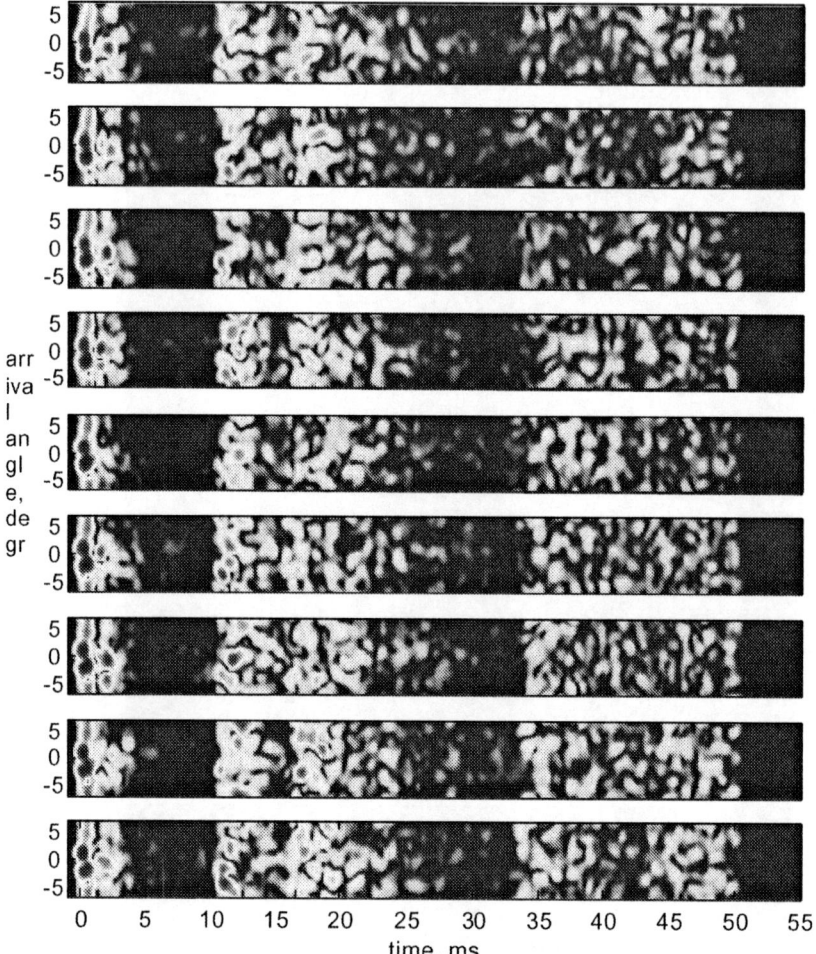

FIGURE 7 - Time-angle plots for 9 transmissions over 27 seconds starting at 7/2 04:01 AM.

At first glance, the frame-to-frame stability of the received signal is noted for the energy arriving from 2-4 msec. The coherence of the first surface arrival (11 msec.) is also noted in Fig. 7. This arrival time corresponds to an almost flat, calm sea surface (Fig. 3 for surface waves and wind conditions at 7/2/2003 around 04:00 AM). The surface-reflected energy is somewhat stable, corresponding to the observed surface wave energy spectrum for this geotime (long period waves, not much chop). It is also noted that the later surface-interacting arrivals (34-50 msec.) are not as coherent as the earliest arriving energy. To interpret these results, a ray tracing calculation for the experimental source-receiver geometry is shown in Fig. 8.

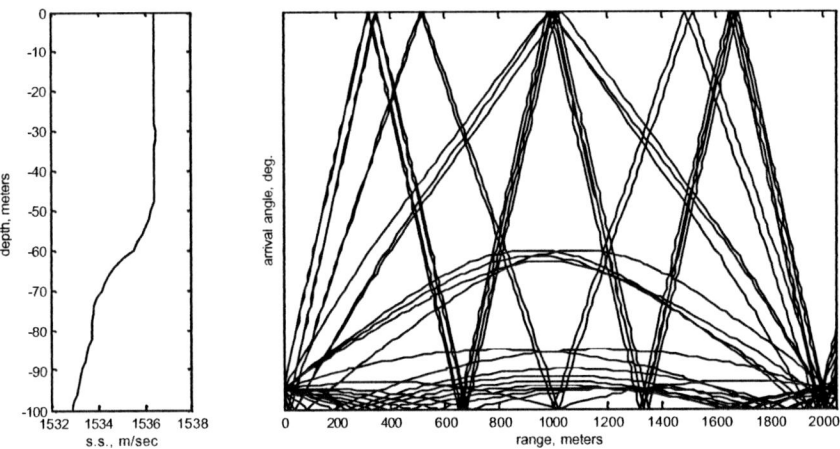

FIGURE 8 - Eigenrays for 07/02/04 04:00 AM showing rays arriving from distributed sources (94-100 meters depth) at 2000 meters range. The receiver is placed at range=0, depth 95 meters.

The number of eigenrays included in this display has been arbitrarily limited by restricting the ray fan emitted from the source. In practice, this limit would be controlled by the bottom critical angle; however, we are only interested in broad qualitative features here. Since the source depth for the experimental data was at 95 m, we consider a range of source depths between 94 and 100 meters and calculate the eigenrays for different depths. The following diagrams in Figs. 9 and 10 show an angular distribution of possible significant eigenrays as a function of arrival time and angle for different combinations of surface and bottom interactions.

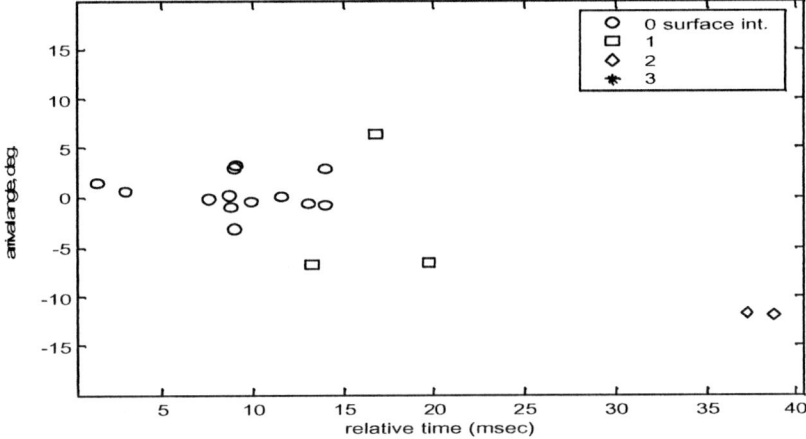

FIGURE 9 - Calculated ray arrival times versus arrival angle showing the number of surface interactions for the sound speed profile on 7/2/2003 at 04:00 AM.

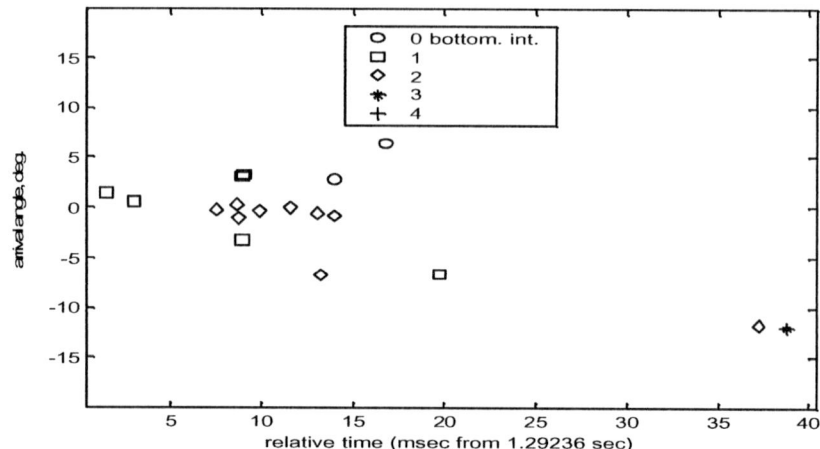

FIGURE 10 - Calculated ray arrival times versus arrival angle showing the number of bottom interactions for the sound speed profile on 7/2/2003 at 04:00 AM.

The spread in time-angle arrivals in these two figures indicates that bottom-interacting energy of the signal could be over a period of about 13 msec. In the set of arrival diagrams shown in Fig. 11 (covering a period of 27 sec) the early arrival shown (i.e. 0 msec.) corresponds to bottom-bounce energy only. The later arrivals are surface interacting. This is known *a priori* from the geometry and sound speed profile at that geotime.

Finally, the effect of rough sea surface is shown in Fig. 12 during 9 transmissions (a period of 27 sec) starting at 7/1 12:31 PM. Note the lower amplitude and the lack of coherence of the surface reflected signals. The energy shown in the plots from 4-10 msec. is bottom interacting, based on the ray tracing calculations. Surface interacting rays are absent. This is due to high-frequency surface waves (the choppy surface shown in Fig. 2) during this period. The low amplitude and lack of coherence in the bottom-interacting signals that arrive around 5 msec. and the signals arriving between 25 to 30 msec. are not fully resolved at this time.

SUMMARY

High-frequency acoustic signals are affected by the small and large spatial and temporal scale variations of the ocean environment processes. These processes can be due to the ocean volume or the dynamic boundary condition roughness. Concurrent oceanographic and acoustic observations were made near a shallow water region of Kauai Island, Hawaii in summer 2003. A subset of the data collected during the experiment is presented here to examine the correlation between the oceanographic variability and the high frequency acoustic wave propagation. Results show a direct relationship between sound speed changes, the surface wave spectrum, and the acoustic wave propagation in this shallow water region.

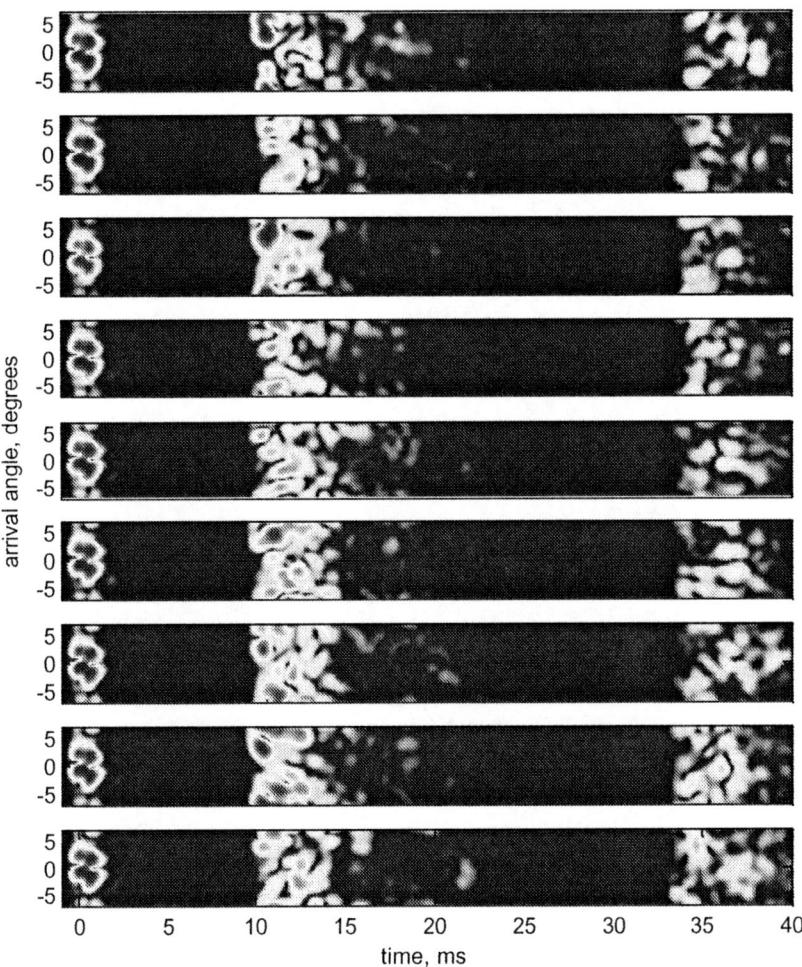

FIGURE 11 - Time/angle arrival picture for geotime 7/2/04 12:32, showing significant arrivals (time scale start is arbitrary, and is set so that t=0 is approximately the time of first arrival).

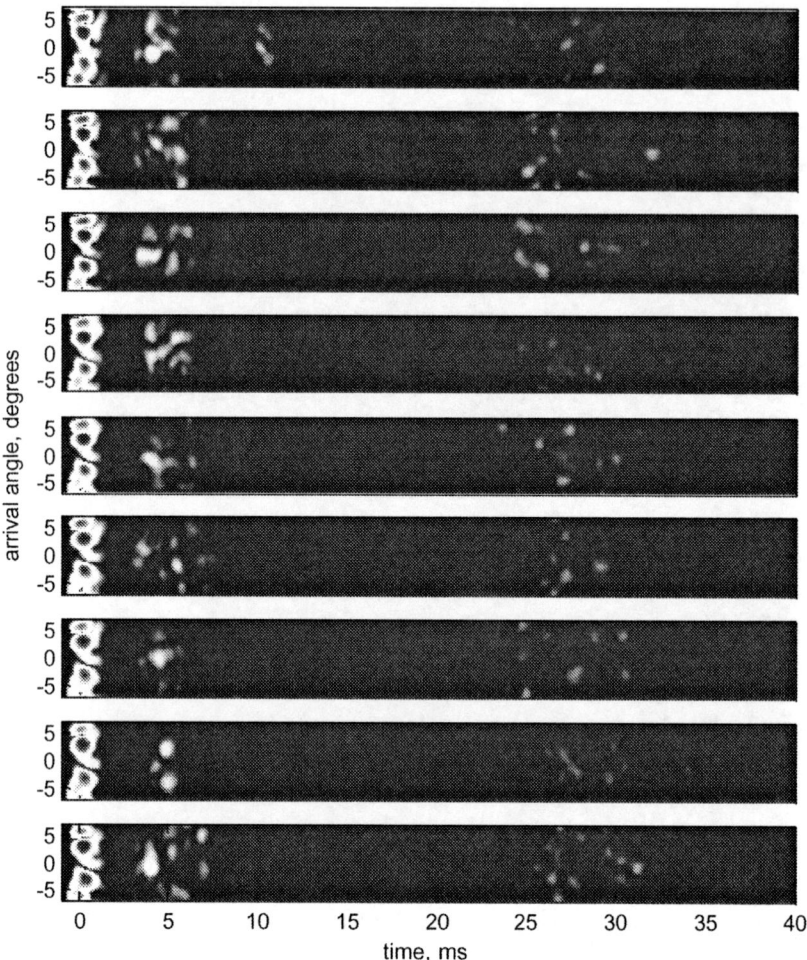

FIGURE 12 - Time-angle plots for 9 transmissions starting at 7/1/04 12:33. Note that during this period the surface was very rough (see Fig. 3).

ACKNOWLEDGMENTS

This work was supported by the Office of Naval Research under grant N00014-00-D-0115. Partial support provided by the Sea Grant Program is acknowledged. The authors wish to extend their special thanks to Art Sundberg for his contributions to the Kauai Experiment and to Robert Drake for his help in the Data Acquisition System. The KauaiEx group members who participated in the field experiment were, Art

Sundberg, Luc Lenain, and Robert Heitsenrether (UDel), Paul Hursky, Martin Siderius (SAIC), Jerald Caruthers (USM), William S. Hodgkiss, Kaustubha Raghukumar (SIO), Daniel Rouseff and Warren Fox (APL-UW), Christian de Moustier, Brian Calder, Barbara J. Kraft (UNH), Keyko McDonald (SPAWARSYSCEN), Peter Stein, James K. Lewis, and Subramaniam Rajan (SSI).

REFERENCES

1. Porter, M. B., et al., "The Kauai Experiment," this volume.
2. Badiey, M., Mu, Y., Forsythe, S., Simmen, J., Lenain, L., *"Arrival Time Fluctuations for High-Frequency Pulse Transmissions in Shallow Water,"* 5th European Conference on Underwater Acoustics: ECUA2000, France, 2000.
3. Carbone N., and Hodgkiss, W., "Effects of Tidally Driven Temperature Fluctuations on Shallow-Water Acoustic Communications at 18 kHz," *IEEE Jour. Oceanic Engr.*, Vol. 25(1), 2000.
4. Badiey, M., Lenain, L., Wong, K., *"High Frequency Acoustic Propagation in the Presence of Oceanographic Variability,"* published in book entitled Impact of Littoral Environmental Variability on Acoustic Predictions and Sonar Performance, edited by N. Pace and F. Jensen, Kluwer Academic Publishers, pp.35-42, 2002.

Telesonar Testbed Instrument Provides a Flexible Platform for Acoustic Propagation and Communication Research in the 8 – 50 kHz Band

Vincent K. McDonald[*], Paul Hursky[†], and the KauaiEx Group

[*] Space and Naval Warfare Systems Center, San Diego, CA. 92152
[†] Center for Ocean Research, SAIC, 10260 Campus Point Drive, San Diego CA, 92121

Abstract. The underwater channel remains a difficult medium for transmitting communication signals. Frequent field tests are required for validating models, testing new waveforms and coding schemes, developing link protocols, designing an adaptive multi-mode modem, testing third-party prototype modems, testing a new directional transducer, and developing new DSP-efficient algorithms. A modular, flexible, autonomous instrument was designed to easily and inexpensively conduct such field tests. This instrument, called the Telesonar Testbed, was originally designed five years ago to specifically support the Telesonar Program at the Space and Naval Warfare Systems Center, San Diego. It has since taken on a wider role supporting the High-Frequency Initiative, the SignalEx project, and a new project testing Multiple Input/Multiple Output (MIMO) systems. Over the last five years several major design changes have been made, making it smaller, lighter weight, more reliable, and acoustically commandable. This paper will describe the design and features of this instrument which has been the centerpiece of 10 experiments to date. This paper will also apply six carrier estimation and six symbol timing estimators to data received by the Telesonar Testbeds in the KauaiEx and ElbaEx Experiments.

INTRODUCTION

The Telesonar Testbed is a unique, high-fidelity, modular, reconfigurable, autonomous, wideband instrument for high-frequency acoustic propagation and communication research. Central to its design is its ability to provide ample experimentation opportunities at low cost. Autonomous operation and a lightweight and small package allows for a meaningful experiment using a craft as small as 7 meters in length. It has been the workhorse of many experiments including the coastal waters off of California, Massachusetts, Hawaiian Island of Kauai, and most recently in the northwest Italian coastal waters near Elba Island. This paper will discuss the design and features of this instrument and present the results of applying carrier and symbol synchronization techniques to data collected by a Telesonar Testbed deployed at the KauaiEx and ElbaEx Experiment sites.

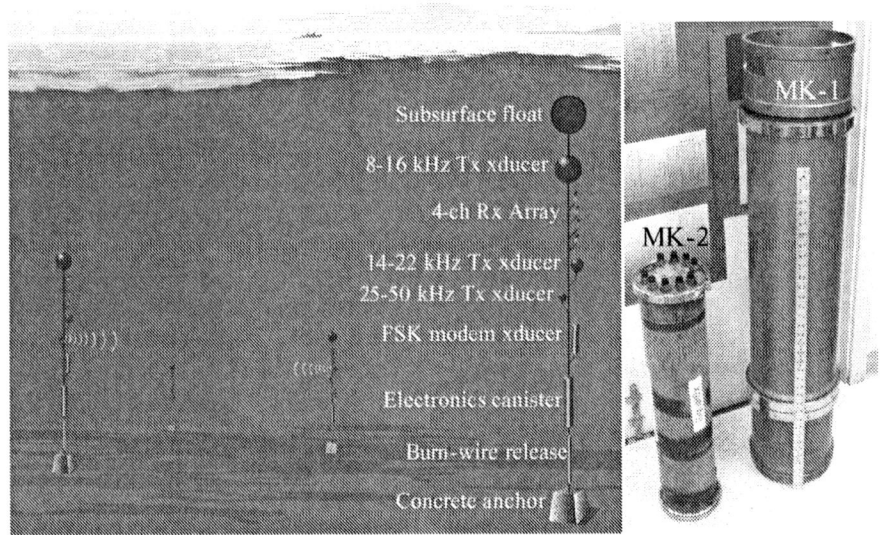

FIGURE 1. Telesonar Testbed bottom-deployed vertical configuration (left). Electronics canisters for MK-1 and MK-2 (right).

The Telesonar Testbed design and development was originally funded in 1997 by the Office of Naval Research to support the Telesonar Program in addressing underwater acoustic communication issues such as shallow-water model validation, link protocol development, multimode adaptive modem research, assessment of transmission security, spatial diversity testing, and ocean impulse response observations to name a few. Since 1997 the instrument has undergone two significant design overhauls, each time reducing the size and power consumption while increasing its capabilities. Figure 1 (right) depicts the original Testbed (MK-1), and the current instrument (MK-2).

TESTBED INSTRUMENT DESCRIPTION

Instrument Mooring

The instrument mooring is typically a serial, in-line configuration shown in Fig 1 (left). Starting from the bottom and working upwards, one to three commercially available concrete footings each with a mass of 16 kg in water are used to anchor the instrument to the seafloor. Above that is a pair of ultra-lightweight (0.2 kg in water) releases that upon an acoustic command will "burn" a small wire through an electrically accelerated anodic dissolution process. Next is the instrument itself with internal electronics and an alkaline battery pack. If the instrument is configured as a transmitter, an additional dual-chemistry (alkaline and NiCad) transmit battery pack is strapped to the electronics canister to provide the necessary additional battery capacity and current sourcing capability. A set of transmit and receive transducers are located above the electronics canister. Finally, a subsurface float is used to keep transducers stable and the instrument mooring vertical in the water column in addition to providing the necessary buoyancy for instrument recovery once released from the seafloor.

The Space and Naval Warfare Systems Center, San Diego finished building four Testbed instruments in May of 2003. The instruments are configurable as a transmitter, receiver or both; however, typically one transmits while the other three, deployed at various ranges, receive and record the acoustic signals. While testing multi-access signaling methods during the KauaiEx and the ElbaEx Experiments, two instruments were configured as transmitters and two as receivers. In addition to being bottom moored, the instrument can be towed at speeds up to five knots while mounted in a custom-built tow body.

Mission Description

Before continuing with the design and features of this instrument, a mission description within the context of an experiment is in order. While the instruments are within the lab or sealed and on deck, mission parameters are loaded through a serial connection or via an RF link. The instrument is then placed in the mooring string and necessary mechanical and electrical connections made before it is lowered to the bottom or is deployed in a freefall manner. Once on the bottom or lowered to its operating depth within the tow body, a commercially-available FSK (frequency shift keying) modem enables checkout of the instrument.

At a prescribed time, the transmitter begins transmitting a probe signal followed by a communication waveform. The receiver, being cognizant of the mission plan, selects the appropriate receive array, sample rate, and opens and closes files commensurate with each short (≤ 1min) transmission for ease of postmortem analysis.

During the KauaiEx Experiment, an interleaved, multi-band, transmission schedule was attempted for the first time. In an attempt at maximizing our data collection efficiency/density, the transmissions were time-division multiplexed from the tow-body and the bottom mounted instrument. That way, we obtained transmissions from a fixed transmitter to a fixed receiver and a moving transmitter to a fixed receiver. Further complicating matters, transmissions would cycle through the three transmit bands, *i.e.* 8-16, 14-22, and 25-50 kHz. The clock accuracy required for this interleaving, required the late addition of very accurate real-time clocks in each instrument coupled with edge-triggered hardware circuitry.

Mechanical Description

The Testbed electronics and external battery canisters measure roughly 15-cm diameter by 78-cm length. The mass of the entire receiver mooring minus the concrete anchors (*i.e.* transducers, acoustic releases, canister, etc.) is 20 kg and 6 kg in air and water respectively. The mass of the entire transmit mooring minus the concrete anchors is 46 kg and 19 kg in air and water respectively. The mooring lengths for both receive and transmit configurations are roughly six meters.

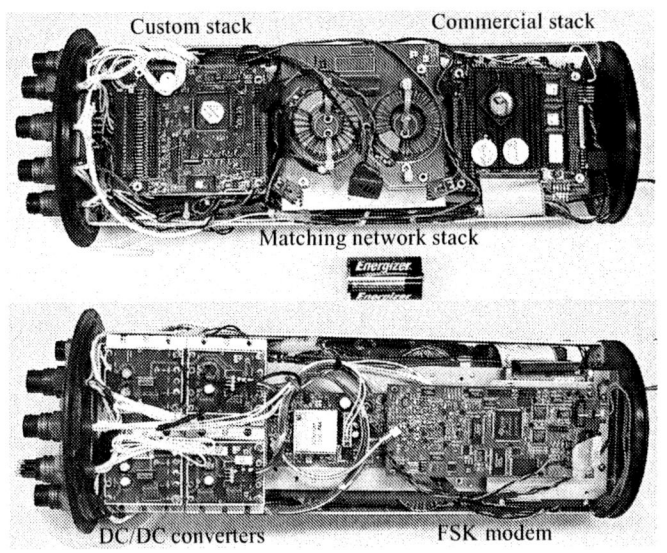

FIGURE 2. Top and bottom view of the electronics within the Telesonar Testbed. A D-cell battery is shown for scale.

Electrical Description

The electronics within the canister can be divided up into six main subsystems: 1. custom circuit card stack, 2. matching-network stack, 3. commercial circuit card stack, 4. DC/DC converters, 5. amplifier, and 6. FSK modem (Figures 2, and 3).

The custom stack is comprised of four cards sandwiched via two 50-pin connectors on each board. The master controller (MC) board coordinates mission execution; the power-distribution board (PDB) routes power to the subsystems under MC control; the RF board allows control of the instrument via RF link when the instrument is sealed but not deployed; and finally, the amplifier board sets the transmitter gain and routes the signal for transmission to the appropriate matching network/transducer combination.

A matching network is required for each of the three transmit bands. The matching networks are not simply transformers for boosting signals fed to the transducers and canceling the capacitive reactance of the transducer, but rather they are Norton transformations made up of two or more toroidal inductors and several high-voltage/current capacitors providing favorable load characteristics to the amplifier and equalizing the overall response for a wider useable transmit bandwidth, Fig 4.

Three PC-104 compliant boards comprise the commercial board stack: single-board computer (SBC), 12-bit A/D, and a 12-bit D/A board. The SBC coordinates the sourcing of hard-disk-drive stored data to the D/A for transmission. It also coordinates the digitizing of received waveforms for storage to the hard-disk drive.

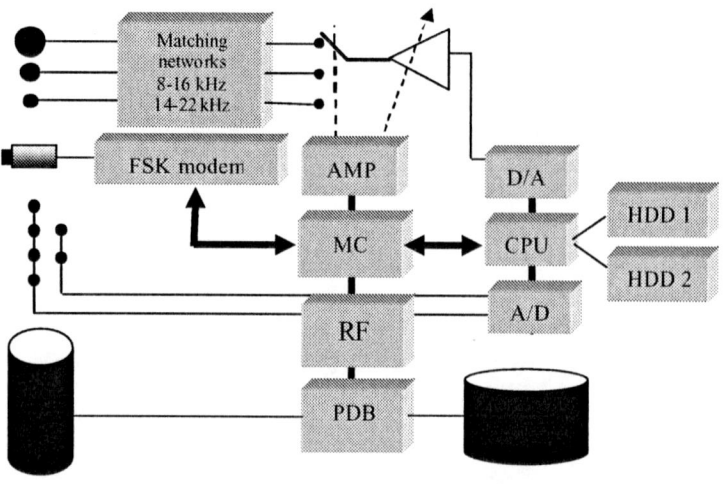

FIGURE 3. Block diagram of the Telesonar Testbed.

Finally, the SBC logs all activity and error messages to file for post-mortem mission-execution analysis.

There are four Vicor Inc. DC/DC converter modules that provide ±48 volts to the amplifier, 5 volts to the custom and commercial circuit card stacks, and 28 volts to the FSK modem.

The 200-watt, class A/B amplifier by APEX is mounted to the endcap for heat dissipation.

The commercial FSK modem by Benthos Inc. is used for remote control and status once the instrument is deployed.

Instrument Reliability

Reliability is crucial for autonomous, ocean-deployed systems in which experimental delays are expensive. There are several features of the Testbed instrument that ensure successful mission execution. The mission coordinator is an 8-bit microcontroller (PIC17C756A) residing on the MC board. The simplicity of the hardware architecture and the C-code that runs on it, virtually eliminates unexpected code execution. A GNU general public licensed Real-Time Operating System (RTOS) TICS is required to allow the instrument to simultaneously execute multiple tasks. This simple, yet powerful RTOS which is run under DOS (Disk Operating System) on the SBC, provides a robust, reliable software platform for carrying out mission commands from the MC board issued over a serial RS-232 connection between them. This serial link is also used by the MC to verify correct command execution. The MC reboots the SBC if it suspects it has "hung" or is improperly executing commands. As a last resort, the FSK modem can be used to remotely toggle a hardware line that resets the microcontroller on the MC board if topside scientists believe the microcontroller is not functioning properly.

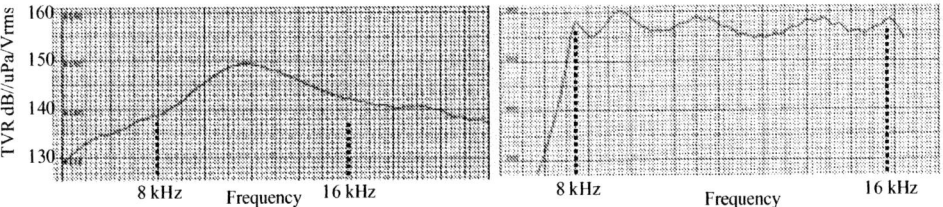

FIGURE 4. Transmit voltage response of the ITC-1007 (International Transducer Corp.) uncompensated (left), with Norton transformation matching network, (right).

Salient Features

The testbed instrument is capable of sourcing arbitrary waveforms at a maximum of 183 dB re 1μ Pascal @ 1 meter, continuously, in the three frequency bands mentioned above. A single external transmit battery pack provides 24 hours of transmit time at the maximum source level. Strapping on another similar-sized battery canister doubles the transmit time to 48 hours. The FSK modem link allows for command submission and is also used to monitor system status including disk-space usage, proper mission execution, and battery voltages. Long-term experiments for studying seasonal variations in acoustic signal propagation and communication performance is made possible by a low-power microcontroller and the sleep mode of the FSK modem. These are the only two devices that are continuously powered. Lastly, ground work has been laid for adding real-time modem development on a general-purpose Texas Instruments' 32-bit floating-point DSP processor.

DATA ANALYSIS

As mentioned above, the Telesonar Testbeds have been used for a variety of experiments to address different issues in underwater acoustic communications. Here we discuss one particular application in which we studied various techniques for carrier estimation and symbol synchronization. In particular, the next few sections apply six carrier estimation and six symbol synchronization techniques to QPSK (quadrature phase shift keying) signals collected by receiver-configured Testbeds during the KauaiEx and ElbaEx Experiments.

The highly-distortive nature of the underwater medium [1] has governed the development of synchronization methods used there [2, 3]. Typically, power and time is expended on a preamble for estimating Doppler, symbol synchronization, and equalizer initialization. Contemporary free-space RF systems cannot afford this luxury and thus synchronization techniques that rely on secondary properties of the communication signals have been developed. We were curious as to how well these techniques would perform against signals transmitted through the underwater channel. A complete description of the techniques described next can be found in a yet to be published thesis [4].

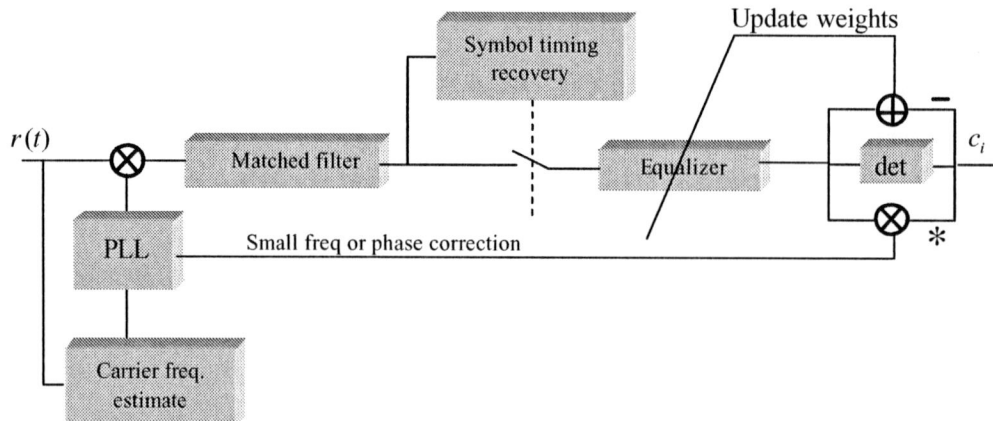

FIGURE 5. Simple receiver block diagram.

Synchronization Basics

In order to baseband a received passband signal for subsequent processing, receivers in phase-coherent systems are required to estimate the frequency and phase of the carrier. In older systems, these tasks were assisted by aids that were sent along with the modulated signal. These and other synchronization aids have generally been phased out due to disadvantages they each posses. These aides either require more bandwidth or power, or they decrease throughput, all precious commodities. Therefore, most receivers are expected to extract both carrier and symbol timing from the received modulated signal by exploiting secondary signal properties.

In most modems, large frequency offsets are taken care of before other synchronization processes are started. Therefore, methods operating on the passband signal directly without access to detected symbols must be used. Once the frequency of the carrier has been estimated and is within 0.1 of T (symbol period), a symbol-timing loop can be started. Symbol timing estimation can be carried out in the presence of small carrier frequency error; implying that, obviously, phase-lock has not been achieved. Nevertheless, symbol-timing detector performance generally improves with decreased carrier frequency offset.

Once symbol sychronization has been achieved, symbols can then be delivered to a slicer, via an adaptive equalizer, that forms a maximum likelihood decision on the delivered symbol. Then, the angle between the delivered symbol and the intended location provide the carrier frequency estimator a phase-error signal for phase locking to the carrier, and provides the error signal for updating the equalizer weights. Figure 5 shows a simple high-level block diagram of a receiver.

The balance of this paper will now briefly cover the six carrier frequency estimators and six symbol-timing synchronizers. All six carrier frequency estimation methods discussed operate independent of symbol timing and vice versa. Furthermore, the symbol timing methods all are able to operate with some residual rotation of the in-

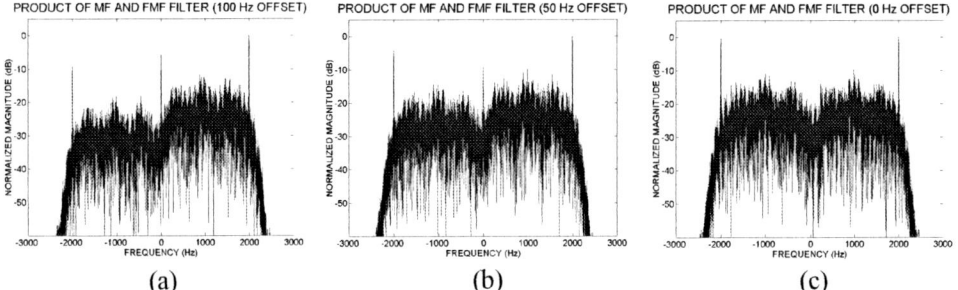

FIGURE 6. Product of the matched filter and frequency matched filter. Plot (a) and (b) plots show a DC line proportional to a 100 and 50 Hz offset between the up- and down-convert frequencies. Notice that the DC line is absent when the correct down-convert frequency is used (plot (c)).

phase and quadrature symbols. Lastly, because phase-locked loops (PLLs) are at the heart of most synchronization systems, at least one, second-order proportional-plus-integral PLL [4] was integrated into each synchronization technique's closed-loop implementation. Lastly, there is little latitude in the open-loop implementation of the methods; however, the closed-loop solutions leave more room for implementation flexibility and thus may bias the performance comparisons. Nevertheless, every effort was made to put the techniques on common footing.

Carrier Frequency Estimation Introduction

In QPSK communication systems, the information is carried by the phase modulation of a carrier. It is therefore essential to replicate exactly the frequency and phase of the carrier so that the in-phase and quadrature projections that were impressed upon the carrier can be determined. Franks in [5] offers a nice tutorial on symbol and carrier synchronization. With few exceptions, synchronizers for frequency-selective fading channels are adopted from methods using the AWGN (additive white Gaussian noise) noise models; not because of their superior performance but rather because of a shortage of more optimal solutions. Most carrier-frequency estimators for the AWGN channel rely upon the spectral symmetry of the received signal to provide a center-of-gravity estimate of the carrier frequency. The underwater channel's frequency-selective fading, frequency dependent absorption, and the transmit and receive instrument's non-flat transmit and receive response respectively, all couple together to present a non-symetric spectrum that biases center-of-gravity techniques.

Technique Descriptions for Carrier Frequency Estimation

The six techniques for carrier frequency estimation can be further segregated into the following broad categories: 1. maximum likelihood, 2. center-of-gravity, 3. spectral line.

The maximum likelihood frequency estimator forms the product of two filters, the matched filter and the frequency matched filter (MF-FMF) [4, 6]. The DC term of this product is proportional to the frequency offset between the up- and down-convert

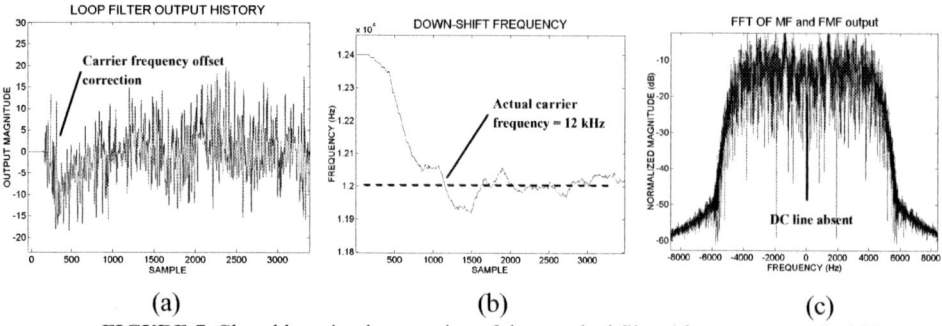

FIGURE 7. Closed loop implementation of the matched filter / frequency matched filter

frequencies. Figure 6 shows the output of these two filters as a function of the residual frequency offset. Figure 7 depicts the dynamic operation of the closed-loop solution.

The conjugate product estimator directly estimates the error in the down-convert or basebanding frequency. The underlying idea is rather simple. Suppose that the initial down-convert frequency is in error. If the basebanded signal is oversampled, the correlation from sample-to-sample will be a function of the sinusoidal modulation of the in-phase and quadrature data caused by this frequency mismatch. To find the rotation rate and thus the frequency error, the angle between successive samples is determined. The modulation contributes to the noise in this estimate and therefore, the conjugate product is fed to an averager before taking the arctan.

Another non-data-aided frequency estimator uses a bank of frequency-translated matched filters. The idea here is again simple. The received passband signal is fed to a bank of frequency-shifted matched filters that span the expected range of possible frequency shifts due to Doppler or differences in transmitter and receiver clocks. The frequency spacing of each match filter is dictated by the amount of residual frequency offset a down-stream carrier PLL can tolerate. In other words, this maximum-searching estimator provides an estimate of the carrier frequency but not the phase. The magnitude squared of each matched filter output is then summed over L symbols and is fed to a peak detector. If the symbols are highly oversampled, then this technique provides both the frequency offset and symbol timing as depicted by the ambiguity surface in Fig 8.

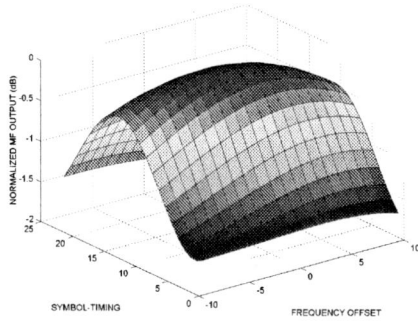

FIGURE 8. Ambiguity surface created by frequency-translated matched filter. Symbol timing and down-convert frequency can be gleaned from plot.

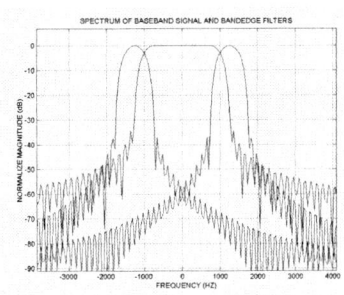

FIGURE 9. Spectrum of band-edge filters plotted on top of signal spectrum

The FFT-window technique is a center-of-gravity method that compares spectral energy in the signaling band for a collection of down-convert frequencies. The carrier frequency associated with the maximum power in the signaling band is then chosen as a first-order approximation of the down-convert frequency. Subsequent processing will eliminate any residual frequency offset and will phase-lock the local generated carrier to incoming passband signal.

Another center-of-gravity technique that relies upon the signaling-band symmetry, compares the energy in the right and left band edges. The generation of the band-edge filters is discussed in [4]. The filters are shown in Fig 9. These so-called band-edge filters are created by the frequency-domain product of the matched filter and the derivative matched filter. Two advantages of this technique are its rapid estimation of the frequency offset, and its operation without a cross product of any kind. Unfortunately, it will provide a biased estimate if the channel distorts the signal leaving unequal energy in the two band edges.

The spectral-line method is the only technique that operates off of the carrier directly. The modulation of a QPSK signal can be stripped by sending it through a 4th power non-linearity. This method can provide both the frequency and phase of the carrier for sufficient receive signal-to-noise (SNR). At reduced SNRs this method provides the best estimate of the carrier frequency if the output of the PLL is averaged over many symbols. Figure 10 shows the PLL cycle slipping as it attempts to lock

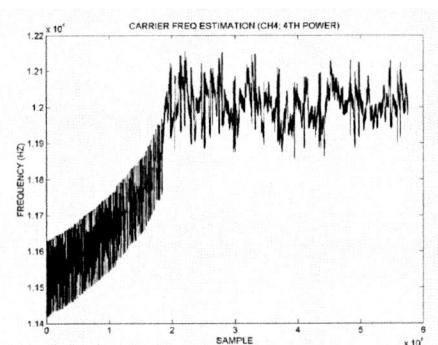

FIGURE 10. Fourth–power carrier frequency estimation. Notice non-linear cycle slipping before loop tracks close to the actual carrier frequency of 12 kHz.

TABLE 1. Carrier estimation techniques

Technique	North Elba	South Elba	HFX
4th power	+3	-10	+7
Frequency offset MF	+40	-120	-240
FFT window	+40	-120	-240
Band-edge filters	+40	-120	-240
Conjugate product	-20	-245	-221
FMF-MF	-40	-240	-300

onto the carrier of 12 kHz. Once the PLL leaves a non-linear operating region (i.e. after cycle slipping), it tracks frequencies at or nearby 12 kHz.

Table 1 shows the results of the carrier frequency estimation techniques applied to data collected by the Telesonar Testbed during the KauaiEx and ElbaEx Experiments. As described above, the non-flat received signal spectrum significantly biases the result of all techniques except the 4th-power method. Since the 4th power technique is the only one that operates off of the carrier directly, it is insensitive to a non-flat receive spectrum and performs well. The actual carrier frequency was 12 kHz, the symbol rate was 2K symbols/second, and the sample rate was 48 ksps.

Symbol Timing Estimation

The function of any symbol timing synchronizer is to maximize the SNR of the symbol delivered to down stream signal processing blocks in a receiver. Typically the received matched filter is a copy of the pulse shape that was sent. Therefore, the matched filtering operation on the received data is a correlation process that produces peaks when the received pulse is time aligned with the matched filter. These peaks represent the symbol estimate with the maximum SNR. Therefore, the timing recovery algorithm must provide sampling instants with the correct frequency and phase for sampling these peaks. If Nyquist pulses are used (e.g. raised cosine) then in an ideal channel exhibiting only a time shift, sampling at the peak will provide the best, inter-symbol interference free, estimate of the symbol. Tutorials on symbol timing are rare but can be found in [5, 7-9].

There are two main classifications of techniques for estimating symbol timing: Decision Directed (DD) or Data Aided (DA), and non-decision directed (NDD). This paper will deal only with NDD methods or those that estimate sampling instants without the aid of detected symbols. The following few paragraphs will describe the 6 techniques applied to data distorted by underwater channels.

The function of a timing recovery algorithm boils down to estimating τ. This parameter accounts for the time lag between transmitter and receiver, and is unknown but non-random, and therefore Maximum Likelihood (ML) techniques direct us toward an optimal timing recovery algorithm. An approximation to the ML algorithm is the so called matched filter/derivative matched filter (MF-DMF) technique [10]. This method can be justified heuristically by observing that it will attempt to move the sampling phase until the derivative of the baseband signal is zero, which occurs at the peaks of the signal.

TABLE 2. Performance of symbol timing techniques.

Technique	North Elba	South Elba	HFX
Gardner	0.036	0.11	0.05
MF-DMF	0.28	1.3	1.9
MF-FMF	0.46	0.36	1.6
Early-late Gate	0.75	2.5	1.9
Square with BE filter	0.53	2.6	1.9
Square without BE filter	5.6	6.2	5

The early-late gate symbol synchronizer [11] is a ML approximation that exploits the symmetry properties of the matched filter output. This technique requires three samples and the timing is adjusted so that the first and the third samples are equal to one another. That way, the middle sample should be at or close to the peak.

Another very popular symbol timing estimator was presented by F. M. Gardner in [12] and is a minimum-likelihood method since it operates on the zero-crossings instead of the peaks of the basebanded signal. This algorithm adjusts the timing so that the sample between the symbols is zero. As Table 2 shows, this algorithm works surprisingly well, outperforming the ML techniques.

The MF-FMF described above not only produces a DC line proportional to the error in the down-convert frequency, but also produces spectral lines at the symbol rate. A PLL can then be locked to this line and the sampling instants generated by picking the peaks of the PLL output.

Passing a QPSK signal through a squaring non-linearity produces a spectral line at the symbol rate. These lines are created by the convolution in frequency of the spectrum of the received signal with a copy of this spectrum. Remember that the product in time is equivalent to convolution in frequency. Therefore, when the spectrum perfectly overlaps a DC term forms. Also, when the band edges overlap a spectral component at the symbol rate will form.

Since the symbol timing information is contained in the band edges, the squaring algorithm can be improved if we reject the signaling band through the use of a band-edge filter. Table 2 demonstrates the improved performance of this modification to the squaring algorithm.

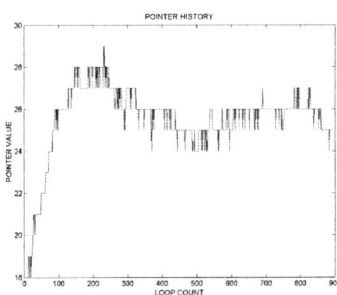

FIGURE 11. Dynamic operation of the Gardner zero-crossing method

Each of the six techniques just described were implemented in a closed-loop form using a second-order loop filter. The performance metric was the variance of the pointer value that the loop produced. Since there are 24 samples available per symbol, these tracking loops converged to one of 24 possible pointer values. Figure 11 shows the dynamic operation of the Gardner loop. Table 2 ranks the performance of the six techniques tested.

CONCLUSIONS

It is clear from Table 1 that all of the carrier frequency estimation techniques except for the 4^h-power method, perform poorly when working on signals that have been distorted by an underwater channel. On the other hand, Table 2 indicates that the NDD symbol timing techniques appear to be adequate for estimating symbol locations.

The Telesonar Testbed will continue to be utilized for underwater communication and high-frequency acoustic propagation research

ACKNOWLEDGMENTS

The KauaiEx Group is: Michael B. Porter, Paul Hursky, Martin Siderius (SAIC), Mohsen Badiey (Univ. Delaware), Jerald Caruthers (Univ. Southern Miss.), Daniel Rouseff, Warren Fox (Univ. Washington), Chritian de Moustier, Brian Calder, Barbara J. Kraft (Univ. New Hampshire), Keyko McDonald (SPAWARSYSCEN), Peter Stein, James K. Lewis, and Subramaniam Rajan (Scientific Solutions).

We would like to extend our gratitude to the SACLANT Undersea Research Centre and all involved SACLANT scientists and engineers who made the Elba Experiment a success. A special thanks to the Chief Scientists Finn Jensen and Mark Stevenson.

REFERENCES

1. Rice, J. A., "Acoustic Signal Dispersion and Distortion by Shallow Undersea Transmission Channels," Proc. NATO SACLANT Undersea Research Centre Conf. on High-Freq. Acoustic in Shallow Water, Lerici, Italy, pp. 435-442, July 1997.
2. Stojanovic, M., Catipovic, J. A., and Proakis, J. G., "Adaptive multi-channel combining and equalization for under water acoustic channels," Journal of the Acoustical Society of America, vol. 94, pp. 1621-1631, 1993.
3. Stojanovic, M., Catipovic, J. A., and Proakis, J. G., "Phase-coherent Digital Communications for Underwater Acoustic Channels," IEEE Journal of Oceanic Engineering, vol. 19, 100-111, January 1994.
4. McDonald, V. K., Carrier and Symbol Synchronization Techniques for Phase-coherent Communication Applied to Shallow Water Ocean Channels, 2004 (yet to be published).
5. Franks, L. E., "Carrier and Bit Synchronization in Data Communication – A Tutorial" Review, IEEE Transactions on Communications, vol. Com-28, no. 8, August, 1990.

6. Harris, F. J., "Band edge Filtering and Processing for Carrier and Timing Recovery," COMCON-7 Athens Greece, 28 June – 2 July 1999.
7. Bennett, W. R., "Statistics of Regenerative Data Transmission," BSTJ 37 pp. 1501-1542, Nov. 1958.
8. Gitlin R. D., and Hayes, J. F., "Timing Recovery and Scramblers in Data Transmission," BSTJ 54(3), march 1975.
9. Aaron, M. R., "PCM Transmission in the Exchange Plant," BSTJ 41 pp. 99-141, January 1962.
10. Saltzberg, B. R., "Timing Recovery for Synchronous Binary Transmission," BSTJ, pp. 593-622, March 1967.
11. Gitlin R. D., and Salz, J., "Timing Recovery in Pam Systems," BSTJ 50(5) p. 1645, May and June 1971.
12. Gardner, F. M., "A BPSK/QPSK Timing-Error Detector for Sampled Receivers," IEEE Transactions on Communications, Vol. COM-34, pp. 423-429, May 1986.

Channel Effects on Direct-Sequence Spread Spectrum Rake Receiver During the KauaiEx Experiment

Paul Hursky*, Vincent K. McDonald[†], and the KauaiEx Group

* *Center for Ocean Research, SAIC, 10260 Campus Point Drive, San Diego, CA 92121*
[†] *Space and Naval Warfare Systems Center, San Diego, San Diego, CA 92152*

Abstract. Acoustic communications using a direct-sequence spread spectrum modulation with a RAKE receiver was tested during the KauaiEx experiment (in Hawaii). The oceanography was measured along the communications path during this test, so that the propagation of the communications transmissions could be subsequently modeled. Such simultaneous measurements provide a continuous picture of how the ocean waveguide evolved during the transmissions, enabling us to isolate what was influencing communications performance. This paper presents an analysis of DSSS Rake receiver performance at this shallow water coastal site.

INTRODUCTION

The SignalEx experiments (see [1]) have been conducted to gain an understanding of how various oceanographic phenomena influence the "channel" seen by underwater acoustic communications systems, and to further measure how these phenomena impact the performance of these systems. The ocean "channel" is a challenging one, because it is a fluctuating waveguide, with a rough surface in motion, an irregular bottom, and a volume in which wave speed can vary with depth and range. The severity of these effects and the relatively limited bandwidth in the underwater channel renders many of the techniques developed for terrestrial wireless or wire channels less effective.

A large repertoire of modulation schemes and receiver designs can be deployed to combat these channel impairments, all with parameters that potentially need to be tuned. This flexibility is also a curse, for it is difficult to predict in advance what particular setting suits a given environment. A goal of the SignalEx work is to develop the ability to predict communications performance, given a description of the oceanography at the targeted site and using acoustic propagation models, so that acoustic communications systems can be deployed at their optimal settings, turn-key, and with reasonably predictable performance.

Acoustic propagation modeling at low frequencies and for a "frozen" ocean (i.e. ignoring time-varying phenomena, such as surface wave motion), even with arbitrary range and depth dependencies, is well developed and successful in predicting complex propagation environments (e.g. witness the many successful demonstrations of matched field processing). However, at higher frequencies, where ocean dynamics due to surface motion and water column sound speed fluctuations, boundary roughness, and air bubbles all have a more dramatic effect, modeling is not as mature. In fact, it is not clear what needs to be modeled to predict acoustic communications performance.

After briefly reviewing the DSSS-Rake receiver, we present results of testing this receiver in three different shallow water ocean sites. We conclude by demonstrating a channel simulator that demonstrates the sort of time-varying phenomena that must be included in a communications performance prediction, especially for receivers having tracking loops (the DSSS-Rake receiver has a delay-locked loop for tracking the symbol timing).

DIRECT SEQUENCE SPREAD SPECTRUM (DSSS) AND RAKE RECEIVER

Spread spectrum modulation techniques, as their name implies, use more bandwidth than strictly necessary to transmit a particular information bit rate, using their redundant bandwidth to either provide gain or to allow multiple users to share the same band. Conceptually, using extra bandwidth provides more "slots" for information than is needed for a given information bit rate. Frequency shift keying over a sparse "frequency hopping" sequence (with different users assigned different hopping patterns) is one form of spread spectrum modulation. Phase shift keying according to a pseudo-random noise (PRN) sequence (with each user assigned a different sequence) at a higher rate than the information rate is another form, called direct sequence spread spectrum (DSSS), or code division multiple access (CDMA). In this paper, we will focus on the DSSS/CDMA receiver in a single user, point-to-point configuration, testing a particular receiver design (described in [2]). Improvements to this basic receiver have also been described in [3].

In DSSS/CDMA, a single carrier is continuously transmitted, phase modulated by a bi-polar "spreading" sequence (of 1's and –1's, called chips). If L is the "spreading factor", L chips are transmitted for every information bit. An information bit is conveyed by setting the sign of the block of L chips used to transmit it. In the next section, we will be presenting results of transmitting 4000 chips/second in an 8-16 kHz band with information bit rates of 50 bps (80 chips/bit), 100 bps (40 chips/bit), and 200 bps (20 chips/bit).

Much effort has gone into discovering PRN sequences that have good auto and cross correlation properties (see [4]), so that delayed versions of the same sequence (e.g. in the presence of multipath) or multiple different spreading sequences (e.g. transmitted by multiple users with distinct spreading sequences) will not interfere with each other. We will use Gold sequences, although the receiver we are using does not reap the full benefit of these sequences because each information bit is spread over only a subset of the entire sequence (ultimately, a consequence of the limited bandwidth we are using, relative to the information bit rates we have chosen to test).

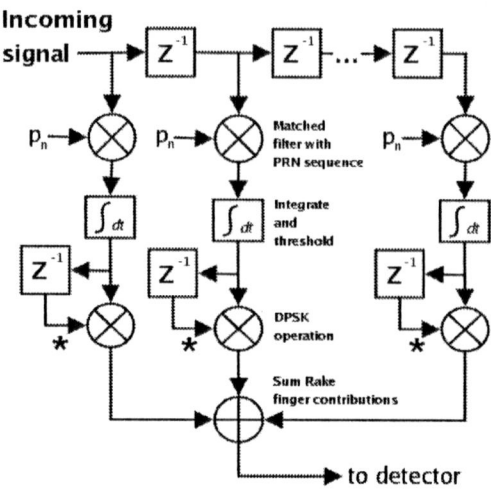

Figure 1. Diagram of DSSS-Rake receiver.

Rake receivers (see Figure 1) were designed to exploit the above-mentioned good correlation properties of spread spectrum sequences by combining the copies of the transmitted signal, which arrive at different multipath arrival times. Because the signal has been "spread" (by a sequence with "good" auto-correlation properties), the overlapping arrivals only interfere with each other to a limited extent.

The Rake receiver consists of a bank of correlators (called "fingers" of the Rake receiver), each matched to the known PRN sequence. Each correlator is applied at a different delay in order to recombine or despread the multipath arrivals before "detecting" which information bit value was sent. The RAKE fingers are spaced to cover the time spread of the channel, or at least the subset of it that we seek to recombine.

Because each symbol period (i.e. information bit) contains only a subset of the sequence, and we cycle through the entire sequence as we move through multiple bits, we must initially synchronize our copy of the sequence with the received sequence. This is done by detecting three repetitions of a subset of the sequence. After this initial acquisition, a delay-locked loop (DLL) is used to track any subsequent drift, by adjusting the timing to minimize the amplitude difference between early and late versions of a dominant Rake finger output.

The receiver we are testing uses differentially coded information bits. This means we do not have to track carrier phase, but only detect how the phase changes from one symbol to the next (i.e. whether the current RAKE output is plus or minus the previous RAKE output, where outputs are produced at the bit rate). The taps of the RAKE receiver will constructively add to the detection output if their corresponding multipath arrivals have the same phase across two bit periods.

Note the recombination of the RAKE taps is only possible because overlapping delayed copies of the spreading sequence do not interfere with each other. Overlapping copies of a single carrier system that has not been spread would interfere with each other.

To summarize our expectations of the receiver being tested: we expect that the DLL will track a slowly varying time offset between the transmitter and receiver, and we expect that the multiple taps of the RAKE receiver will provide some gain in a multipath environment, if the multipath arrival structure is stable over a symbol period. We have put these expectations to the test using a receiver array in shallow water spanning most of the water column, so that we get a comprehensive picture of how performance varies over depth, time, and the accompanying oceanographic changes.

PERFORMANCE AND ANALYSIS OF EXPERIMENT DATA

Figure 2 shows the experiment configuration off the coast of Kauai and the time-varying impulse response at three depths, measured by applying a matched filter to LFM chirps (sweeping from 8-16 kHz in 50 ms, repeated every 250 ms). The strongest arrivals were observed near the bottom, with the second group of arrivals showing progressively more compact structure as we move from the surface to the bottom.

Figure 3 shows the array geometry, the sound speed profile, and bit error rate versus depth and time for three bit rates (50, 100 and 200 bps). The performance was measured over two diurnal cycles and nearly the entire water column.

The performance was uniformly better near the bottom. As Figure 2 shows, most of the energy is in the second group of arrivals, and the concentration only gets stronger as we move from the ocean surface to the ocean bottom. As seen from looking at the sound speed profile in Figure 3 (and as described in [5]), this concentration of energy is due to a thermocline that extends almost to the bottom, where a strong duct is formed, focusing these later arrivals.

The concentration of errors peaking at hour 30 seems to be correlated with the changes in sound speed profile, in particular the degree of ducting near the bottom. When the duct is weaker, due to mixing in the water column, the second group of arrivals is no longer so concentrated, and the receiver seems to have less of a dominant signal to lock onto.

All the bit error rates shown are without error correction (which would probably recover fully from bit error rates up to 12 percent or so). Arguably, the 100 and 200 bps tests yielded marginal performance (too many instances of bit error rates beyond any error correction repair). This is a consequence of the relatively small spreading factors (40 for 100 bps and 20 for 200 bps), which determine the matched filter gain. The amount of multipath in the channel in which the receiver was tested requires more gain to overcome the multipath interference. Increasing the bandwidth would allow longer PRN sequences (with more gain) to be used, without sacrificing the information bit rates. Adding an equalizer and a chip-rate timing loop, as suggested in [3], are other potential remedies, although they are more computationally demanding.

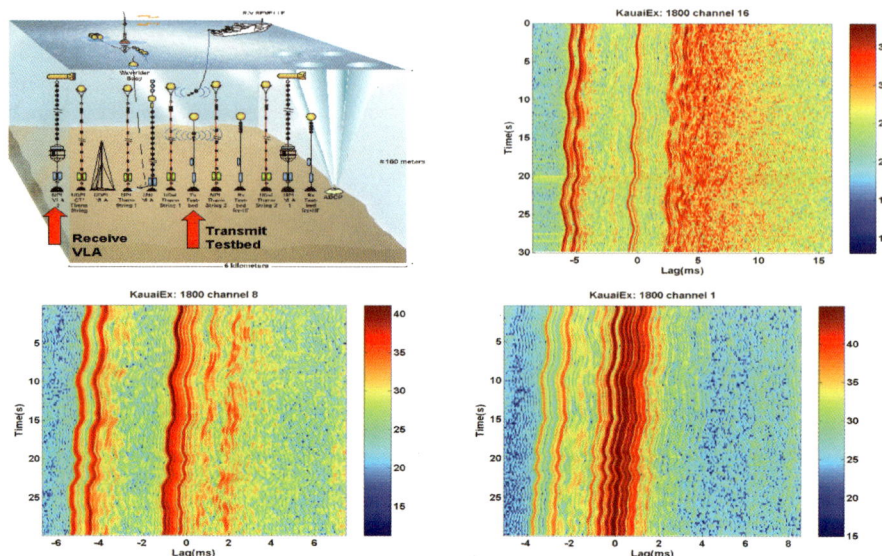

Figure 2. Kauai experiment configuration (upper left), and measured impulse response functions at three depths: near surface (upper right), middle of water column (lower left), and near the bottom (lower right). Only a small portion of the impulse response, around the earliest arrivals, is shown so that the fluctuations in arrival time can be observed.

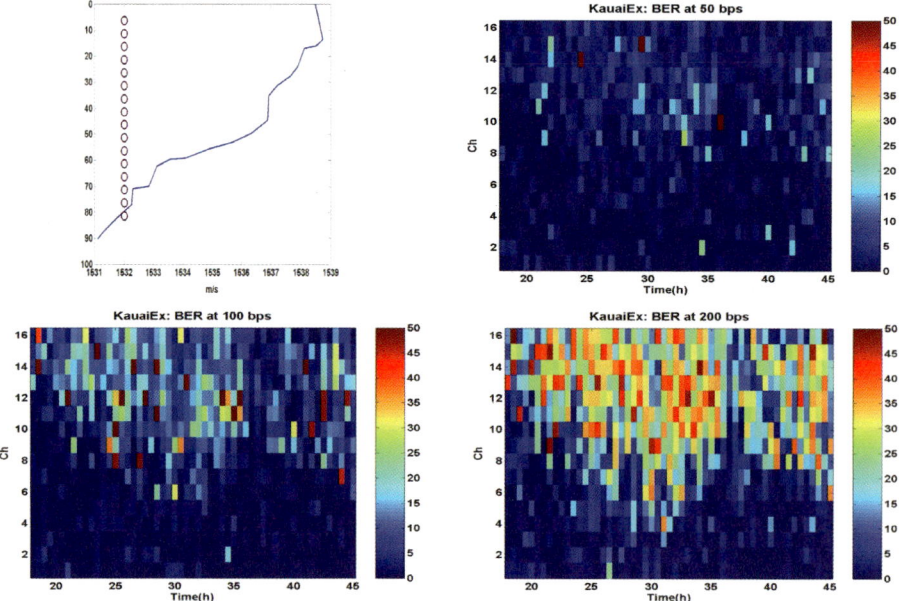

Figure 3. Upper left figure shows array element placement relative to sound speed in the water column. Bit error rates versus array element and time (45 h total) are also shown: at 50 bps (upper right), 100 bps (lower left), and 200 bps (lower right) in Kauai.

CHANNEL SYNTHESIZER FOR PERFORMANCE PREDICTION

Our approach for a channel synthesizer is to start with a frozen ocean model, using the best available sound speed profile for the targeted site, and to calculate arrival times and complex amplitudes using the Bellhop gaussian beam propagation model. We have had good success modeling the high frequency channel with such an approach, demonstrating matched field processing in the 8-16 kHz band (see [6]). Besides time of travel, amplitude and phase information, ray (and gaussian beam) models typically provide information about how many times individual arrivals have interacted with the boundaries. This additional information enables us to add fluctuations the otherwise static paths, for example striving to reproduce fluctuations caused by interactions with a moving surface. As the measured impulse response functions in Figure 2 show, sometimes this is as simple as providing a sinusoidal variation whose amplitude and frequency match the wave motion of the ocean surface (as in the impulse response near the bottom). Other times, such surface interactions are clusters of arrivals with no easily identifiable track (as can be seen in the impulse response near the surface).

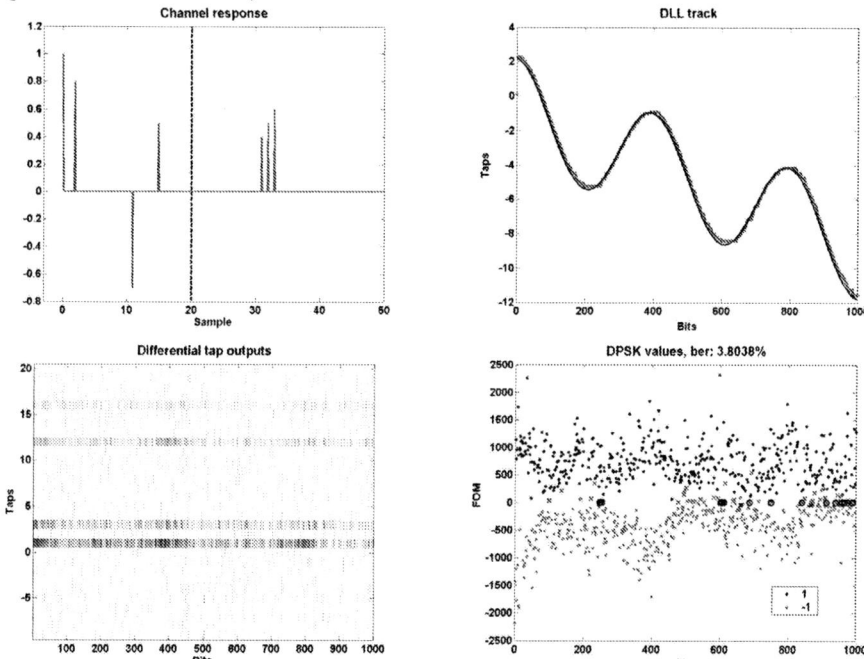

Figure 4. Channel simulation: "frozen ocean" impulse response function (upper left), simulated and estimated time-of-arrival showing Doppler shift caused by source and surface motion (upper right), differential output of Rake fingers (lower left), DSSS constellation color coded to indicate known values of information bits (lower right). At higher Doppler rates, when surface motion reinforces source motion, we see fading on the Rake finger outputs and the resulting closing of the eye pattern in the constellation as the high phase rate causes time spreading of the matched filter outputs.

Figure 4 shows an example of testing the DSSS Rake receiver against a synthetic channel, in which the transmitter drifted toward the receiver (this constant Doppler produces a mild ramp in the arrival times), and the path being tracked by the timing acquisition loop (i.e. the DLL) corresponded to a surface path having a sinusoidally varying arrival time. The superposition of these two motions resulted in arrival times having a downward trajectory with a sinusoidal oscillation. Note how the constellation (shown in the lower right subplot) expands and contracts, following the Doppler acceleration shown in the tracking loop. This causes bit errors when the SNR drops, due to the Rake matched filters not being able to "follow" the Doppler accelerations.

This is a simple example of why realizations (and not just averaged quantities, such as spectra) must be synthesized for the multipath fluctuations, especially for simulations of receivers that have tracking loops. Furthermore, each path may have a distinct Doppler process, depending on which part of the water column it has traveled through.

Our channel synthesizer relies upon resampling the transmitted waveform, realizing a true time-varying time dilation (and not just a phase shift: our 8-16 kHz band does not qualify for a narrrowband approximation) corresponding to a Doppler trajectory. Furthermore, each path arrival is given a potentially distinct Doppler trajectory, so that arrivals that have interacted with the ocean surface at different locations have suitably independent trajectories, and arrivals that have interacted with the surface more than once have greater "spread" than paths that have interacted only once.

What remains to be done is constructing more realistic models for the surface, volume and bottom interactions (i.e. changing bathymetry as the communicating platforms move around the site) that can be driven by measurements of the prevailing oceanography and weather (e.g. wind speed, sea surface temperature, etc).

CONCLUSIONS

We have tested the DSSS-Rake receiver at depths spanning the water column for several tidal cycles while simultaneously measuring the prevailing oceanography along the propagation path from source to receiver. Not surprisingly, we can correlate bit error rates with many of the oceanographic phenomena at both sites. We have discussed how using PRN spreading sequences having the touted "good cross-correlation" properties in the DSSS-Rake receiver does not realize the gains implied by these propereties, because only subsets of the full sequences are used for any signle information bit. As a result, given the typical channel spreads in underwater channels, only modest rates can be reliably achieved with the DSSS-Rake receiver, unless the bandwidth is dramatically increased.

We have demonstrated a channel synthesizer that accurately reproduces the prevailing "frozen ocean" multipath and also captures a variety of time-varying channel effects, including continuously and rapidly varying Doppler at surface-interacting paths and discontinuous times of arrival due to volume and rough boundary scattering. As an example of why it is important to use actual realizations of time-varying channels (and not just spectral averages that do not capture the detailed phase trajectories of the various multipath components), we have shown how the Doppler due to surface motion can cause sporadic outages in the DSSS-Rake receiver. Note that a "frozen ocean" channel would not stress the various tracking loops that are typically used in receivers for various phase-modulated signaling schemes, especially higher-rate phase-coherent schemes.

ACKNOWLEDGMENTS

This work was supported by ONR through the ONR High Frequency Initiative and the SignalEx project.

The original code for the DSSS-Rake receiver was obtained from Ethan Sozer and John Proakis with whom we had many fruitful discussions on acoustic communications. We also benefited from later work on this receiver by Michael Porter.

The data used to test this receiver was collected during the KauaiEx experiment, with Michael Porter as chief scientist, and the KauaiEx group, all contributing ideas and resources. The KauaiEx Group is: Michael B. Porter, Paul Hursky, Martin Siderius (SAIC), Mohsen Badiey (Univ. Delaware), Jerald Caruthers (Univ. Southern Miss.), Daniel Rouseff, Warren Fox (Univ. Washington), Christian de Moustier, Brian Calder, Barbara J. Kraft (Univ. New Hampshire), Keyko McDonald (SPAWARSYSCEN), Peter Stein, James K. Lewis, and Subramaniam Rajan (Scientific Solutions).

REFERENCES

1. M. B. Porter, V. K. McDonald, P. A. Baxley, and J. A. Rice, "SignalEx: Linking environmental acoustics with the signaling schemes," *Proceedings of MTS/IEEE OCEANS'00 Conference*, 2000, pp. 595–600.
2. E. M. Sozer, J. G. Proakis, M. Stojanovic, J. A. Rice, A. Benson, and M. Hatch, "Direct sequence spread spectrum based modem for under water acoustic communication and channel measurements," *Proceedings of MTS/IEEE OCEANS'99 Conference*, pp. 228 - 233, September 1999.
3. M. Stojanovic and L. Freitag, "Hypothesis-feedback equalization for direct-sequence spread-spectrum underwater communications," *Proceedings of MTS/IEEE OCEANS'00 Conference*, pp. 123 - 129, September 2000.
4. D. V. Sarwate and M. B. Pursley, "Cross-correlation properties of pseudorandom and related sequences," *Proc. IEEE*, vol. 68, no. 5, pp. 593-619, 1980.
5. Martin Siderius, Michael Porter and the KauaiEx Group, "Impact of thermocline variability on underwater acoustic communications: results from KauaiEx," *Proceedings of HF Acoustics Conference*, La Jolla, California, March 1-5, 2004.
6. Paul Hursky, Martin Siderius, Michael B. Porter, and Vincent K. McDonald, "High-frequency (8-16 kHz) model-based source localization", *J. Acoust. Soc. Am.*, vol. 115, no. 6, pp. 3021-3032, June 2004.

Impact of Thermocline Variability on Underwater Acoustic Communications: Results from KauaiEx

Martin Siderius, Michael Porter* and the KauaiEx Group [†]

Center for Ocean Research, SAIC, 10260 Campus Point Drive, San Diego, CA, 92121
[†]*SAIC, UD, UW-APL, SIO, USM, UNH, SSI*

Abstract. In July 2003, the KauaiEx, high-frequency acoustic experiments were conducted off the coast of Kauai, Hawaii. Both acoustic communications signals and probe signals (to measure the channel impulse response) were transmitted in the 8-50 kHz band. These signals were transmitted over several days from fixed and moving platforms and were received at multiple ranges and depths using vertical arrays and single hydrophones. Extensive environmental measurements were made simultaneous to the acoustic transmissions (e.g. measurements of the water column temperature structure, wind speed and surface wave heights). The experimental site has a relatively reflective seabed made up of sand that was combined with highly variable oceanographic conditions which led to communications performance closely tied to source/receiver geometry. In this paper, the correlation between environmental factors and communications performance will be discussed. The focus is on communications signals in the 8-13 and 14-19 kHz frequency bands at source receiver range of 3 km. Results show the performance in the higher band was approximately the same as for the lower band. Results also show a strong dependence on receiver depth with the deeper hydrophones having fewer bit errors. The ocean sound speed structure at this site appears to have a large impact on the communications performance and the time variability.

INTRODUCTION

During June and July of 2003, the KauaiEx series of experiments were conducted off the island of Kauai, Hawaii. These experiments were designed to measure the environment and simultaneously transmit acoustic communications waveforms over a period of several days. These tests involved both towed and fixed sound sources. In this paper, two topics are addressed: 1) the performance of underwater communications in the 8–13 kHz band as compared with the 14–19 kHz band and 2) performance over time to determine if environmental factors have a significant influence. The first topic, comparing frequency bands, is important since the ability to use higher frequencies implies having a larger available bandwidth (for obtaining higher data rates) and a decrease in the physical dimensions of sources and receiver arrays. The higher data rates are obviously important for many applications, but the smaller dimensions can be also be important as these systems begin to be installed on smaller platforms (e.g. autonomous underwater vehicles). The second topic, determining the impact of the environment on performance, is important for predicting when and where underwater communications systems will fail. Many of the environmental factors that influence performance can be either measured *in situ* or obtained through archival data. Knowing that particular environmental factors will cause a failure of communications systems may influence how, when or where a system is deployed.

In this paper, Multi-Frequency-Shift-Keying (MFSK) signaling is considered. MFSK is both simple and robust and for that reason it is currently used in commercially available modems. MFSK is robust because it is a non-coherent method (uses intensity and not phase of the signal) and this makes it valuable in uncertain operating environments. The main attraction of the coherent methods is their potential to more efficiently use the available bandwidth (i.e. obtain higher data rates). However, this comes at the price of more complex processing to overcome channel variability. In addition to being valuable in its own right, the simple and robust nature of MFSK signaling makes its performance a useful yardstick against which to measure other methods.

The first section of this paper describes one deployment from the KauaiEx experiments and the data that is used for the analysis. Both the environmental and acoustic communications signals are described. The next section shows the relationship between the water column temperature structure, the wind speed and communications performance.

KAUAIEX

Details of all the experiments during KauaiEx are described in Ref. [1]. In this paper, only the second deployment, which took place from June 30 to July 3, 2003 will be considered. The geometry for the experiment is shown in Fig. 1. Data analyzed here is from the testbed transmissions (Tx Testbed near the middle of the track) with the source located about 5 m from the seabed. The receiver array (MPL-VLA2) is about 3-km away had 16 hydrophones spaced 5 m apart with the first channel about 8.5 m from the seabed.

Environmental measurements

As can be seen from Fig. 1, there were extensive environmental measurements including: five strings of thermistor sensors to measure water column properties along the acoustic track, a waverider buoy to measure wave-heights, and an Acoustic Doppler Current Profiler (ADCP). Other geophysical measurements such as grab samples, seismic profiling and multibeam mapping were also made to help characterize the seabed. The entire data set is too large to consider in this paper so only a small amount of data will be discussed here with the main focus on the water column variability in the vicinity of MPL-VLA2.

The water column sound speed generally showed a region near the surface with a high degree of mixing due to the often windy conditions. The depth where the mixed layer ended and the thermocline began varied with location and time. In the left panel of Fig. 2 are 5 measured sound speed profiles taken during the 2nd deployment (on July 1, 2003). The mixed layer depth is 40–50 m for 4 of the profiles and decreases to about 20 m for one. In many locations around the world's oceans, the sound speed near the surface is highly variable responding to surface heating; however, in Kauai the mixing

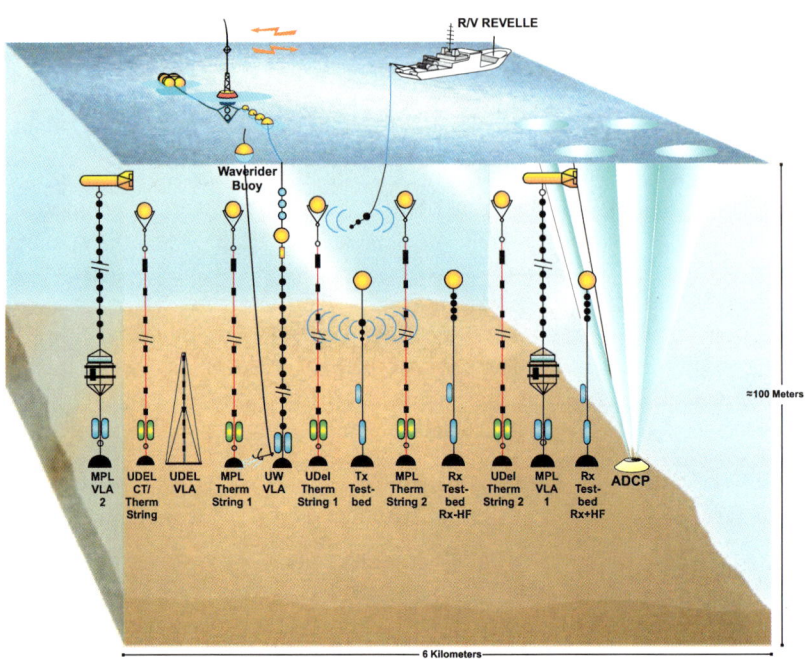

FIGURE 1. Experimental setup for the 2nd KauaiEx deployment (June 30–July 3, 2003). Data from the MPL-VLA2 (about 3 km from the moored source) is analyzed here along with measurements from the UDEL-CT/Thermistor string located about 500 m away. The VLA has 16 equally spaced hydrophones and spanned depths of 17–92 m.

causes the water near the surface to be more uniform with a high degree of variability occurring at greater depths. These sound speed profiles give a sense of the structure and variability, but the thermistor strings give a time history. In the right panel in Fig. 2, the data from the thermistor string nearest MPL-VLA2 is shown (labeled UDel CT/Therm. String in Fig. 1). There were 13 thermistors located at depths between 4 and 82 m. There is a clear, regular pattern evident in the thermistor data showing the thermocline depth moving up and down in the water column over time. The impact of these variations on the acoustic communications signals will be discussed in the next section.

MFSK transmissions

The MFSK signals considered here use two bands each with 128 frequency components spaced 40 Hz apart from 8 to 13.2 kHz and 14 to 19.2 kHz. We will refer to these as the low and mid bands. (An additional high band covering 25-50 kHz was also included in the experiment but is not discussed here.) The upper and lower 4 tones in each

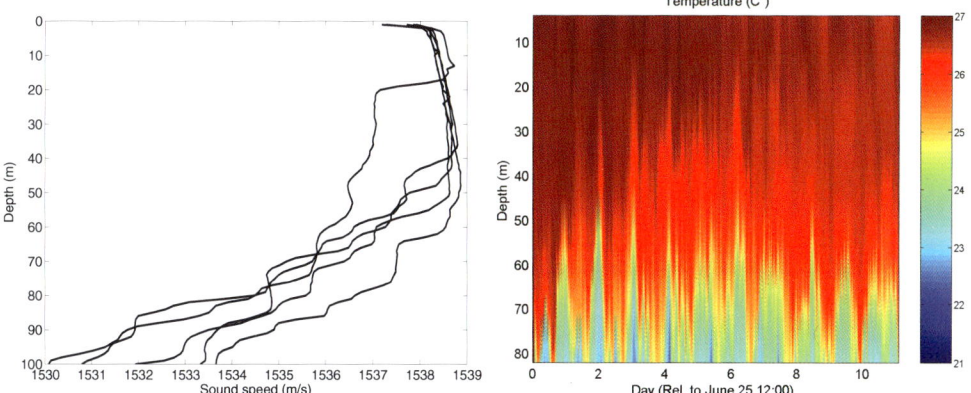

FIGURE 2. Left panel shows measured sound speed profiles taken on July 1, 2003 near the experimental site. Note the change in depth of the mixed layer. The right panel shows a time history of the ocean temperature during the experiment. This was from the UDel CT/Therm. string located near MPL-VLA2.

band are reserved for pilot tones to compensate for Doppler. The information is passed using a subset of the 128 frequencies that can be modified every 25 ms. One detail of the modulation scheme is the use of 1 of 4 coding. This means 4 tones are used to encode 2 bits of data. The advantage of this is that in decoding, only a decision about which of the 4 tones is loudest is needed to determine if the transmission is a 0-0, 0-1, 1-0 or 1-1. This method is less sensitive to intensity variations than having the decoder decide if a tone is a 1 (on) or 0 (off). Based on the frequency band used here, the maximum data rate in each band is 60 bits in 0.25 ms, or 2400 bits per second (bps) over each band (4800 bps total). To transmit at lower data rates, the time duration of the tones is increased (e.g. 1200 bps is achieved by holding the tones on for 50 ms).

Preceding the MFSK transmissions is an m-sequence that is used to determine the signal start. To decode the data, the receptions are matched filtered (with the replica m-sequence) to acquire the start of the signal and frame the MFSK transmission. A spectrogram is then taken of the MFSK portion of the time series using a non-overlapping boxcar window of 25-ms duration. The highest tone in each of blocks of 4 tones is then determined. Although errors can be reduced by coding the transmissions prior to transmissions (at the expense of data rate) this has not been done here and all errors reported are the raw bit errors. Channel coding improves the communications performance but then requires many more transmissions to collect the statistics required to interpret the environmental effects.

The overall performance of the different data rates in the two frequency bands is shown in Fig. 3. The left six panels shows the bit errors versus signal to noise ratio (SNR) for the low band and the right six panels for mid band. Within each band there are different data rates in each of the six panels. In general, both bands and all data rates follow the trend that better SNR leads to lower bit errors. On the other hand, there is also considerable scatter in the plots. As expected, the lower data rates show fewer errors for the same SNR.

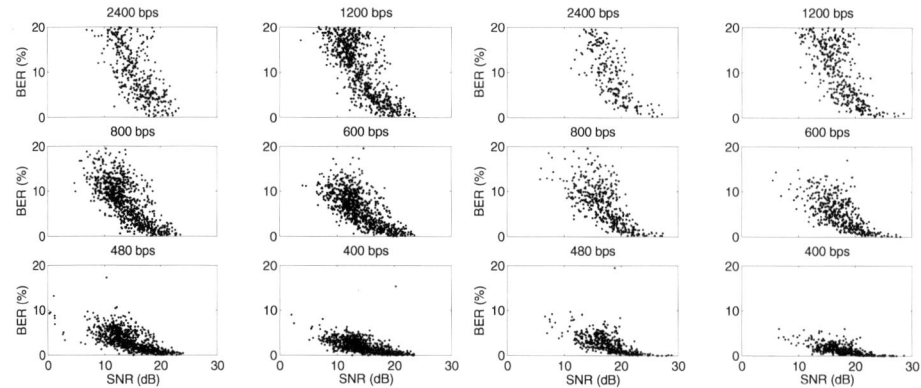

FIGURE 3. Bit errors as a function of SNR for different data rates in two frequency bands. The left six panels show the low (8–13.2 kHz) band at data rates of 2400, 1200, 800, 600, 480 and 400 bits/sec. The right six panels are the same for the mid (14–19.2 kHz) band.

One might expect that the higher frequency band would have poorer performance since volume attenuation tends to increase the transmission loss at higher frequencies. However, an analysis of the statistics shows that we actually obtained better performance in the higher band. There are many other factors at work that explain this. First, the source level was 1–2 dB stronger in the higher band. Second, the ambient noise is lower. Third, scattering losses are higher. That latter effect decreases the signal level but simultaneously decreases the intersymbol interference associated with multipath. Ultimately, a reliable channel simulator is key to predicting the performance. However, it is noteworthy that the higher band has both better performance and provides a more compact system (smaller projector).

Another interesting lesson from these tests was the performance improvement with hydrophone depth. This can be seen in the left panel of Fig. 4 where the bit errors as a function of depth are averaged over about 1 day (at the data rate of 2400 bits/s for both the low and middle bands). The improvement in performance with depth is mainly due to higher SNR at deeper depths although there was no indication the ambient noise level was strongly depth dependent. The deepest hydrophone at about 91.5 m shows about 5% bit errors while the most shallow hydrophone at about 16.5 m shows about 30%. This trend is seen both on the short and the long time scale. Shown in the right panel of Fig. 4 is the ambient noise at the center frequency of the two bands averaged over the 1 day of transmissions. There is only a weak depth dependence of ambient noise although the much reduced ambient noise level for the higher band is evident.

IMPACT OF OCEAN THERMOCLINE ON COMMUNICATIONS PERFORMANCE

We can gain insight into the temporal variability of performance by looking at the time history of the bit errors taken over 1 day (again for 2400 bps). The 24-hour period

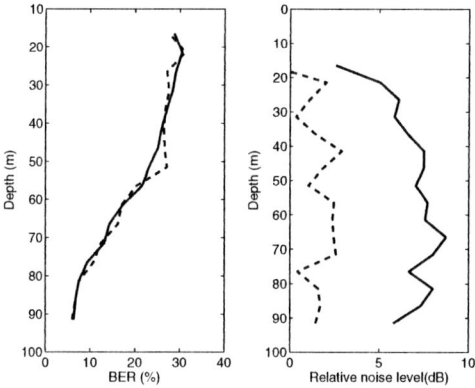

FIGURE 4. The left panel shows the depth dependent percent bit errors averaged over about 1 day of transmissions for the low band (solid) and the middle band (dashed). The right panel shows the corresponding ambient noise as a function of depth averaged over the same time period.

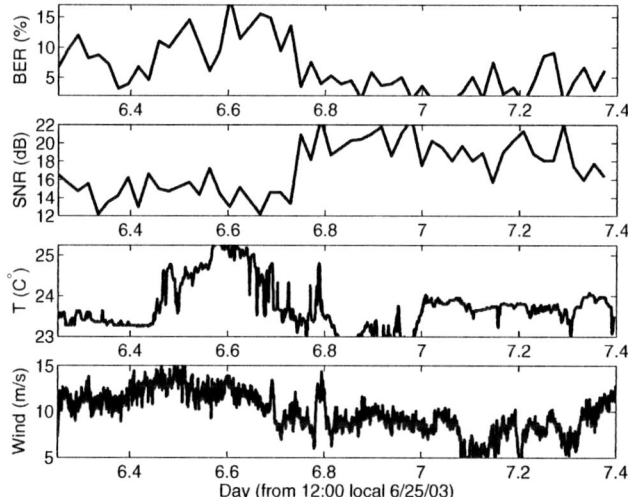

FIGURE 5. Top panel shows the percent bit errors for 2400 bps transmissions in the low band over about 1 day on the deepest hydrophone channel (91.5 m). Below is the corresponding SNR. The temperature at a thermistor located at 82 m depth about 500 m away is shown in the third panel. The lowest panel shows the wind speed during the same period. Days are relative to 12:00 on June 25, 2003 (local time).

is important since there are diurnal events both with the oceanography and with the winds. The top panel of Fig. 5 shows a time history of the bit errors on the deepest hydrophone channel (91.5 m). The panel below shows the corresponding SNR. The period between about day 6.4 and 6.8 shows a marked increase in bit errors. There is a rough correspondence with SNR especially near day 6.8 when the SNR increases and the performance improves. The lower two panels in Fig. 5 show the temperature and wind

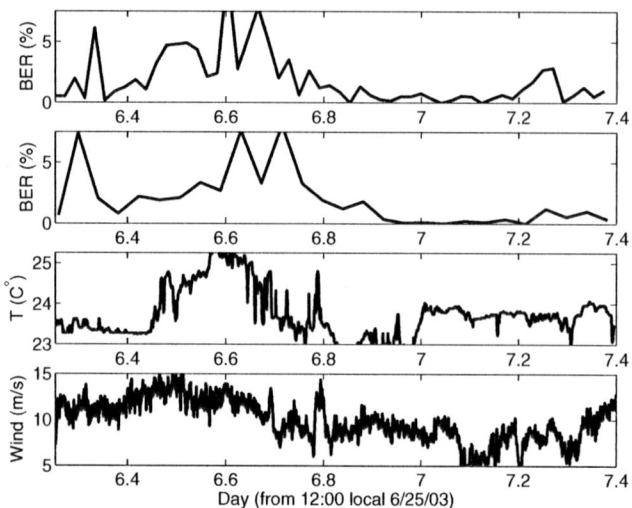

FIGURE 6. Top panel shows the percent bit errors for 2400 bps transmissions in the low band over about 1 day using a coherent average of the two deepest hydrophone channel (86.5 and 91.5 m). Below is the same for the mid band. The temperature at a thermistor located at 82 m depth about 500 m away is shown in the third panel. The lowest panel shows the wind speed during the same period. Days are relative to 12:00 on June 25, 2003 (local time).

speed during the same time period. During the period with increased bit errors there is a sharp increase in the water temperature and there is also some indication that the wind is changing during the same period. Wind increases generally cause the noise level to increase and this could be responsible for the decrease in SNR. However, if wind was responsible it is reasonable to assume the errors would show the same characteristics regardless of depth and this is not the case. For hydrophone channels in the mid to upper part of the water column where the water is well mixed there is no indication of bit errors increasing in this time period. Another example of the increases in bit errors during the period of day 6.4 to 6.8 is shown in Fig. 6. In this figure, channels 1 and 2 are combined coherently before decoding and the bit errors are therefore reduced. The top panel shows the results for the low band and below the mid band. The lower panels are again the temperature and wind speed. Although not conclusive, there appears to again be a correspondence between the increase of bit errors and the change in temperature. Also, note the end of the plot near day 7.4 where the wind speed begins to increase but there is no indication of increasing bit errors.

CONCLUSIONS

We have shown the performance of MFSK transmissions over 3 km in the 8–13.2 and 14–19.2 kHz bands. The two bands had about the same bit error rates for the same SNR, however, the higher band often had overall lower numbers of bit errors. This was due to a slightly higher source level of about 1–2 dB, but also due to a much lower ambient noise level.

There was also a marked difference in the performance in both bands as a function of receiver depth. The receivers near the sea-surface had the worst performance and, again, this was closely tied to differences in SNR as a function of depth. The ambient noise level did not seem to increase with depth and the loss of SNR is attributed to lower signal received at the shallower depths. In addition to correlation with SNR, there are indications that during certain periods when the water column becomes mixed throughout the water column the performance decreases on the deepest hydrophones. Future work will include acoustic modeling to determine if the observed time changes in the ocean sound speed structure can account for the observed changes in signal level over depth and time.

ACKNOWLEDGMENTS

This work was supported by the Office of Naval Research under contract N00014-00-D-0115. The KauaiEx Group is: Michael B. Porter, Paul Hursky, Martin Siderius (SAIC), Mohsen Badiey (Univ. Delaware), Jerald Caruthers (Univ. Southern Miss.), William S. Hodgkiss, Kaustubha Raghukumar (Scripps Inst. of Oceanography), Daniel Rouseff, Warren Fox (Univ. Washington), Christian de Moustier, Brian Calder, Barbara J. Kraft (Univ. New Hampshire), Keyko McDonald (SPAWARSSC), Peter Stein, James K. Lewis, and Subramaniam Rajan (Scientific Solutions)

REFERENCES

1. Porter, M. B., and the KauaiEx Group, "Results from the KauaiEx experiments," in *Proceedings of the High-Frequency Ocean Acoustics Conference*, AIP, 2004.

Side-Scan Sonar Survey Operations in Support of KauaiEx

Jerald W. Caruthers[*], Erik Quiroz[*], Craig Fisher[*], Roger Meredith[⊥], Natalia A. Sidorovskaia[¶], and the KauaiEx Group[1]

[*]*Department of Marine Science, University of Southern Mississippi, Stennis Space Center, MS 39529*
[⊥]*Naval Research Laboratory, Stennis Space Center, MS 39529*
[¶]*Department of Physics, University of Louisiana at Lafayette, Lafayette, LA 70503*

Abstract. In support of the high-frequency channel characterization experiment (KauaiEx), three days of Side-Scan Sonar (SSS) surveys were conducted off the northwest coast of Kauai, Hawaii. The SSS used in this survey was a specially modified Marine Sonic Technology, Ltd, system operating alternately at 150 and 300 kHz and producing high-resolution digital data as well as standard tiff images of the seafloor. This paper is, in part, a summary of work reported as the initial report "Side-Scan Sonar Survey: Narrative of Operations and Initial Data Report" which was based on analyses of the standard image data, and can be found at ftp://moray.dms.usm.edu/Caruthers/sidescan/Kauai/. Along the primary paths of transmission of the underwater-communication experiment there appears to be no obstructions or outcroppings, such as coral, at scales smaller than the KauaiEx multibeam bathymetry. However, several small-scale variations in the texture of the bottom are present, e.g., sand ripples with crests running approximately parallel to the depth contours and wavelengths of about 1 m and globular-like inhomogeneities with a scale near 3 m. In the southeast corner, some larger, more rugged, ridge-like structures are suggested. (This work is supported by the Ocean Acoustics Program of the Office of Naval Research.)

INTRODUCTION

From the 8th to the 9th of July 2003, a Side-Scan Sonar (SSS) survey was conducted off the Pacific Missile Range Facility (PMRF) at Kauai, Hawaii, in support of the High-Frequency Channel Characterization Experiment (HFX, and also known as KauaiEx).[1] The survey was conducted with a specially modified Marine Sonic Technology, Ltd, (MSTL) SSS. The sonar operated at alternating dual frequencies of 150 and 300 kHz and simultaneously in a standard mode (producing tiff images) and a digital mode (producing high-resolution digital waveform data). The technical characteristics of this sonar and its use in bottom-scattering research are discussed in a previous report.[2] The original three-day survey in the KauaiEx range could not be accomplished according to plan owing to bad weather and high seas on the first day. There were also problems with the deep towing of the sonar that were compounded by the high seas. With continued weather and sea state problems in the deeper waters of the range, survey operations conducted on the second day were in 35 m of water in a

---

[1]Michael B. Porter, Paul Hursky, Martin Siderius (SAIC), Mohsen Badiey (UD), Jerald Caruthers (USM), William S. Hodgkiss, Kaustubha Raghukumar (SIO), Dan Rouseff, Warren Fox (APL-UW), Christian de Moustier, Brian Calder, Barbara J. Kraft (UNH), Keyko McDonald (SPAWARSSC), Peter Stein, James K. Lewis, Subramaniam Rajan. (SSI).

more protected area. These data were of excellent quality verifying the capabilities of the sonar under such more favorable conditions. Herein are displayed some of the images acquired on that day and discussions of some of the results are presented. At the end of that series of shallow-water runs, tests were run at the 100-m depth contour (still in somewhat protected waters) and deep-water techniques were re-evaluated and modified before returning to the HFX range the next day. With modifications to the operational plan and to the equipment, and with improved weather in the morning of the third day, a survey in the range was accomplished. This survey was limited, however, due to increased wind and sea later in the afternoon.

RESULTS

The data discussed here are in the form of tiff images in an MSTL format designated with the extension '.mst'. Software that can display these images and associated files can be downloaded from the MSTL web site http://www.marinesonic.com/ and is labeled there as "Sea Scan PC Review Software". All the 'mst' files collected can be found at the FTP site ftp://moray.dms.usm.edu/ Caruthers/sidescan/Kauai/.[3] A few of those images are included here as figures discussed in this text.

Shallow Water Work of July 9

As mentioned previously, this work has little relevance to the HFX experiment. It is presented here, in part, to demonstrate the quality of the data when taken under better circumstances and, in part, to provide a benchmark to check data in the HFX range if found to have similar character. (But the latter turned out certainly not to be the case.) Finally, and probably most importantly, this work adds to the knowledge of bottom scattering in varied bottom types. The general location of this site is 22° 1' N 159° 47' W.

There had been prior comments about coral in the region around and near the HFX site. The structures found in shallow water appear to leave the question open, however. Good images of such features at each sonar frequency are seen in Figs. 1 to 3. (Some of the striations in the images are caused by yaw of the fish or turns in the survey.) These features appear to stand a meter or two above a sandy seafloor. (Some patchiness of sand ripples may suggest the presence of muddy regions as well.) The structures are believed to be primarily lava with coral in isolated locations.

In these figures, patches of sand (and possibly mud or sand with different characteristics) are clearly visible (the smooth darker regions of the image between the more solid formations). The sand in the region is likely to be calcareous sand resulting from the breakdown of the coral and bits of shell, or volcanic sand resulting from the breakdown of lava. Such sands, both calcareous and volcanic, are observed on various beaches along the Kauai coastline.[4]

Several localized patches of sand ripples are visible in Figs. 2, and 3 and other images not displayed here. Particularly notable regions of sand ripples are seen in these figures. In Fig. 1 (the lower left side of both frequencies) speckle patterns can be seen. Often, this is indicative of a school of larger fish, but here they appear to be

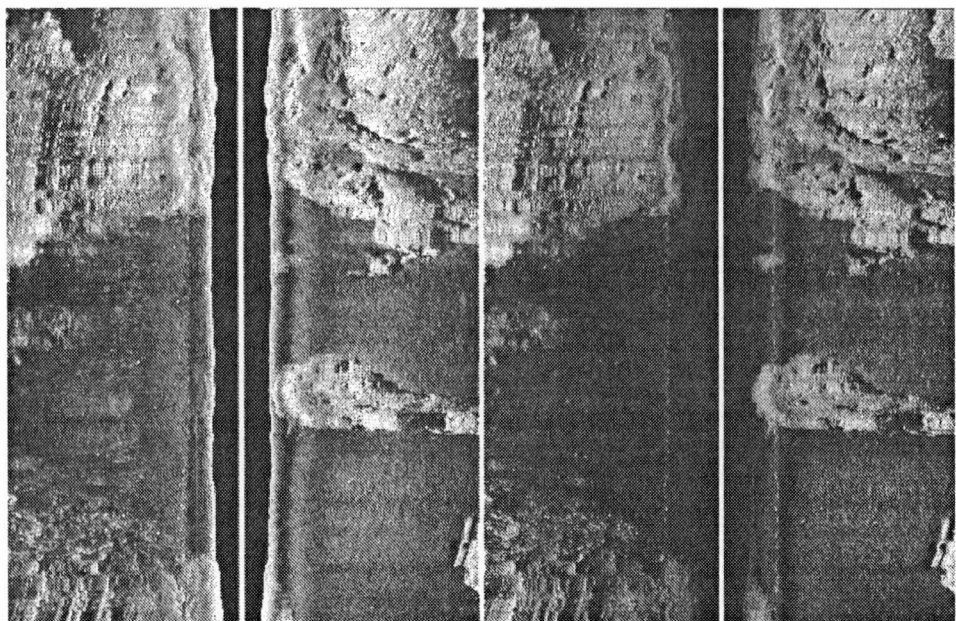

FIGURE 1. Sonar images L0709019.mst and H0709019.mst. (For this four-part image and the sonar images that follow, the left pair is for 150 kHz and the right pair is for 300 kHz. Within each pair, the left image is for the left side and the right image is for the right side.)

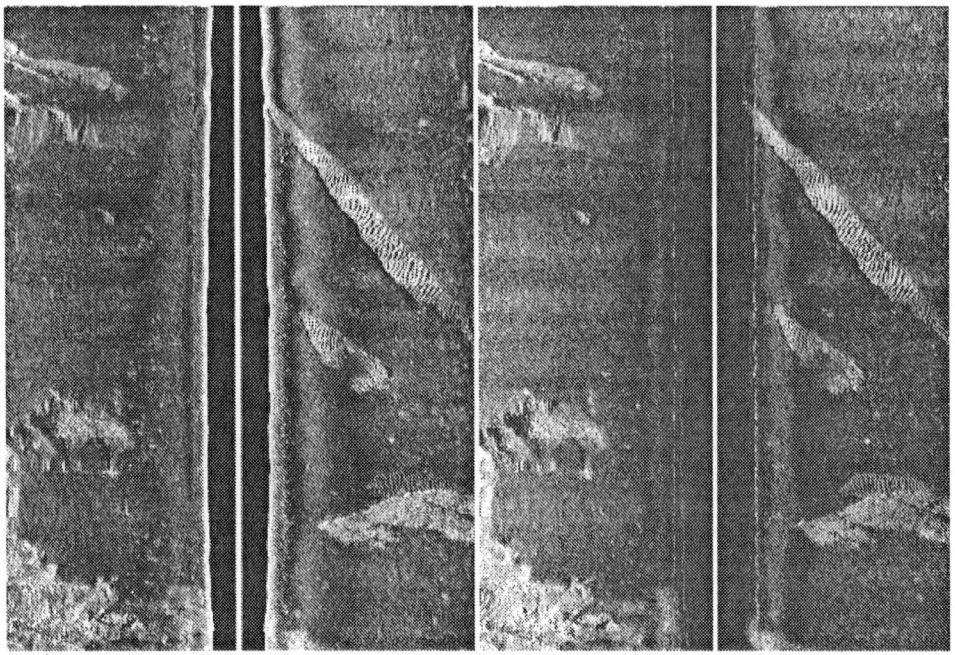

FIGURE 2. Sonar images L0709038.mst and H0709038.mst.

FIGURE 3. Sonar images L0709043.mst and H0709043.mst.

stronger reflections (vs. scattering) from facets in the coral or lava structures. Such reflections are usually somewhat coherent and, if present, may affect the structural details of communication signals. On the other hand, what may be facets at wavelength scales may not appear at scales of the HFX lower frequency signals.

Work in HFX range of July 10

Anticipating stronger winds in the afternoon, the survey on the 10th of July was begun very early in the morning. The seas were rough even then, but the work was manageable. Moving up the range against a light wind (initially) the boat speed-over-the-ground was held at about 1.8 kt and the fish was held to within 20 m off the bottom and the range scale was set to 100 m. This line was made essentially over the line previously occupied by the HFX moorings. Figures 4 and 5 show some of the results during this run. They also show the large excursions in depth that could not be controlled. There are cross-range striations at small scale that are likely sand ripples with wavelengths of about a meter running parallel to the depth contours. There appear to be additional inhomogeneities in the images that are difficult to see among the sand ripples. These inhomogeneities have a globular appearance at a scale of about 3 m. Other than these small-scale features the bottom is generally smooth without any outcropping such as ridges, coral heads, or lava flows.

For the run back to the south (which was about a 100 m east of the run to the north), the minimum speed over the ground that could be maintained was about 2.5 kt. This resulted in the fish being about 40 m off the bottom, but the depth held the

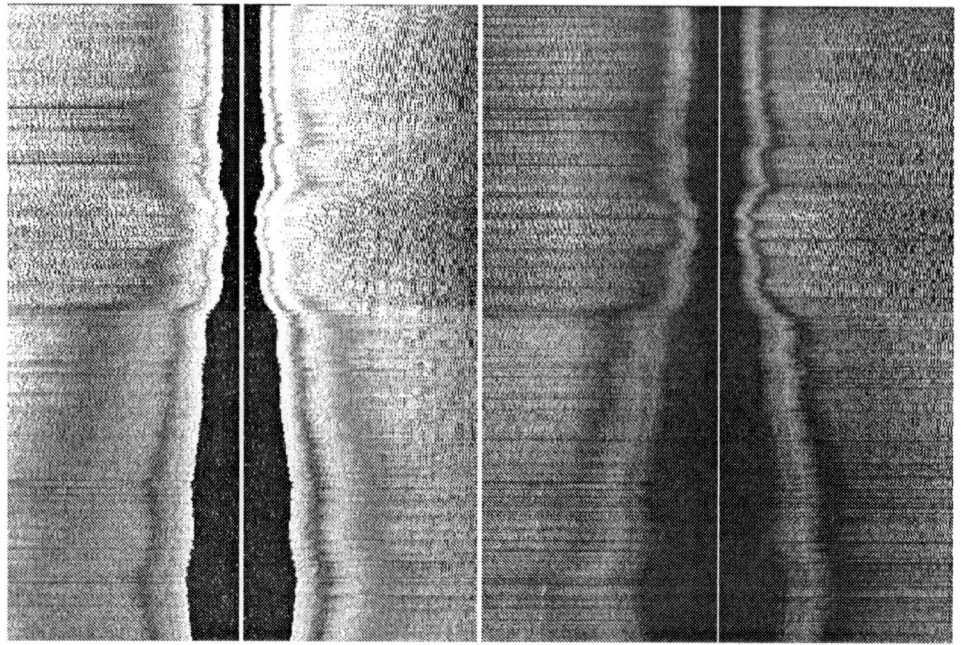

FIGURE 4. Sonar images L0710002.mst and H0710002.mst.

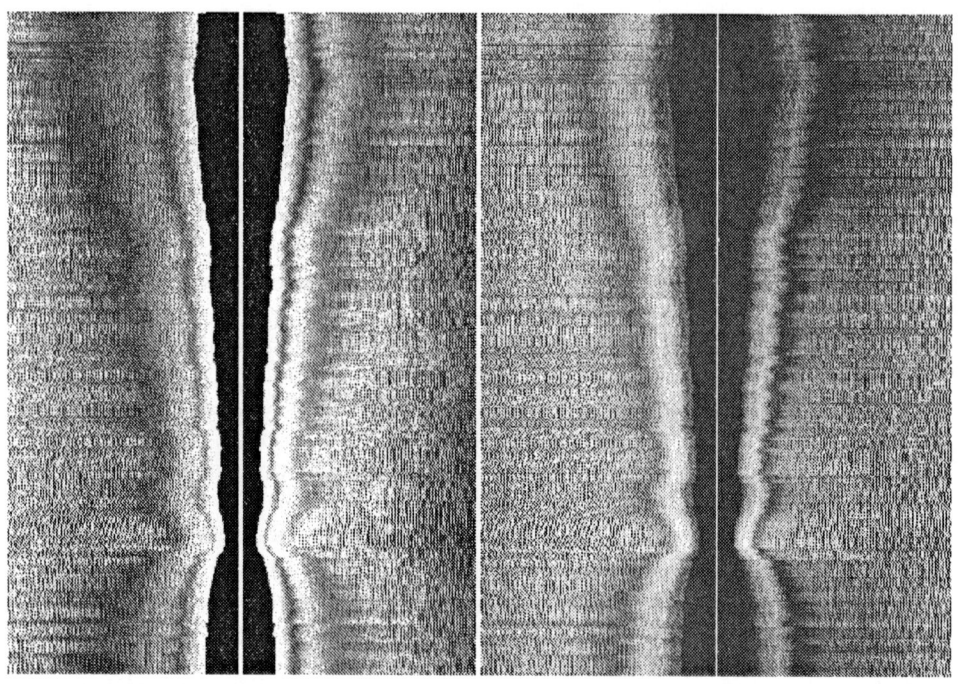

FIGURE 5. Sonar images L0710010.mst and H0710010.mst.

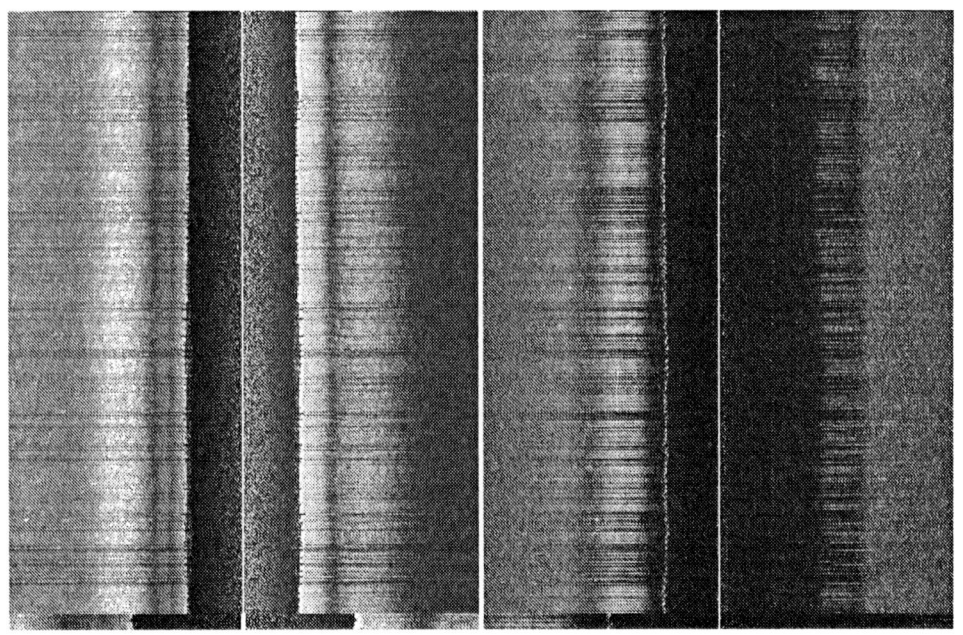

FIGURE 6. Sonar images L0710021.mst and H0710021.mst.

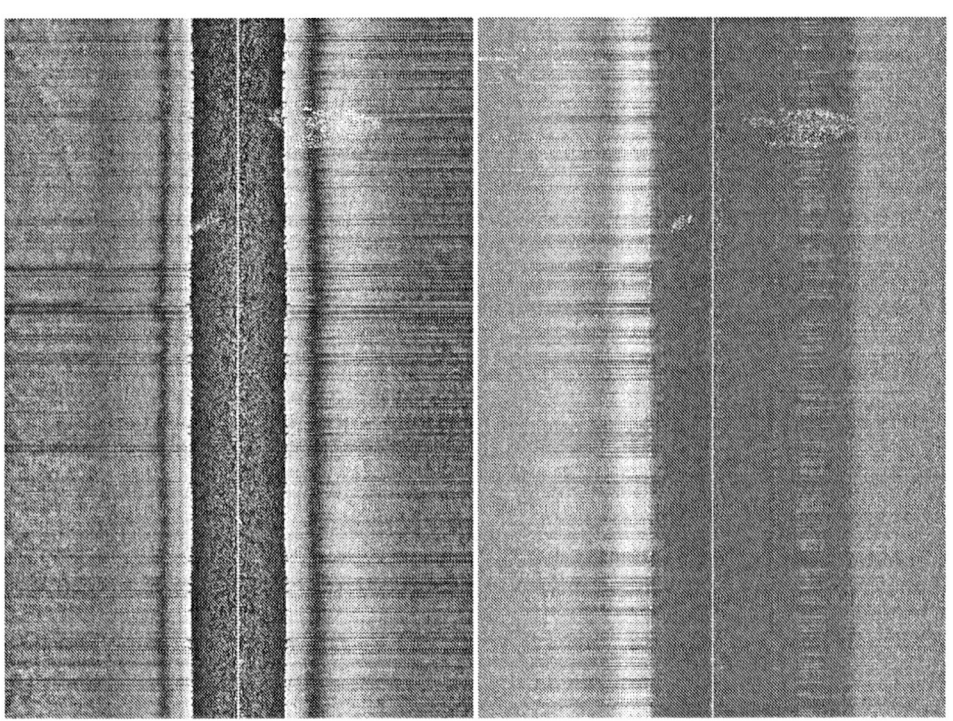

FIGURE 7. Sonar images L0710027.mst and H0710027.mst.

position more steadily than the run to the north (see Figs. 6 and 7). (For these images, the range scale was set to 200 m.) Again, possible ridge-like structures on the shoreward (a and c) side could be seen. These features appear to be similar structures seen in the Kauai HFX Experiment Multibeam Bathymetry.[5]

SUMMARY AND CONCLUSIONS

Significant problems in the collection of these data were experienced: the sea state was the major factor affecting the survey with the required operating depth a close second in terms of handling the activity. Excellent data in the shallower, more benign region south of the HFX range were collected. Although of very little value to the HFX experiment itself, these data did demonstrate that the SSS functioned well under less restrictive conditions, and the survey provided useful SSS data in their own right. For the HFX range, the conclusions indicate that the bottom is generally uniform with probable sand ripples of 1-m wavelength in the downslope direction and some globular inhomogeneities at a scale of about 3 m. These are important considerations for the HFX experiment because they represent the more straight-line paths between some of the more separated HFX sources and receivers. There were some large irregularities, perhaps ridges, shoreward of the 100-m contour, particularly at the southeastern end of the range.

REFERENCES

1. High-Frequency Channel Characterization Experiment (HFX), http://ososd.saic.com/KauaiEx/.
2. J.W. Caruthers, R.G. Goodman, M. Wilson, S. Stanic, "High-Frequency Acoustic Seafloor Scattering Data with Emphasis on Development of a New Side-Scan Sonar," *MARINE FRONTIERS* MTS/IEEE Proceedings of OCEANS'02, pp. 363-368, Oct. 2002.
3. The FTP site contains the following directory providing the Kauai SSS survey data (subdirectory 'FieldWorksData/') and this report and its figures (subdirectory 'InitialReport/').
4. Discussions with Dr. Charlotte A. Brunner, Professor, The University of Southern Mississippi.
5. The Joint Hydrographic Center, University of New Hampshire, HFX Experiment Multibeam Bathymetry, Kauai, HI, 25-30 June 2003. (Can be found under the link "Bathymetry" at the web site given in ref. 1)

High Frequency Tomography Using Bottom-Mounted Transducers

James K. Lewis, Peter J. Stein, Subramaniam Rajan, Jason Rudzinsky*, Amy Vandiver, The KauaiEx Group

Scientific Solutions, Inc., P.O. Box 1029, Kalaheo, HI 96741
*Applied Physical Sciences, Corp., 2 State Street, Suite 300, New London, CT 06320

Abstract. The sources and receivers at the Pacific Missile Range Facility (PMRF) provide for performing acoustic tomography. A primary problem is that acoustic signals of interest interact with the ocean surface, and surface wave fields result in considerable variability in arrival times. Arrival times of rays that have interacted with the moving ocean surface are obtained by averaging over a number of pings to eliminate errors due to Doppler shifts. We also present transforms using differences between observed travel time anomalies and those calculated using an ocean circulation model to make adjustments to model-predicted water temperatures, salinities, and currents (model-oriented acoustic tomography).

INTRODUCTION

The Pacific Missile Range Facility (PMRF) off Kauai, Hawaii, has 15 bottom-mounted sources (8-15 kHz) and 178 bottom-mounted receivers. These assets provide for the possibility of performing acoustic tomography throughout the range. The distances between source-receiver pairs in the shallow water range are relatively small (≤10 km), meaning arrival times are readily detectable. However, the acoustic signals of most interest interact with the ocean surface. As a result, the surface wave field results in considerable variability in arrival times at 8-15 kHz.

Arrival time anomalies are determined relative to a monthly sound velocity structure based on a three-dimensional grid of a hydrodynamic model of the PMRF region. The model provides its own estimate of the four-dimensional sound speed structure. We present transforms for determining a model-related travel time anomaly along the path of a transmitted acoustic ray. The differences between the two travel time anomalies can be transformed back to adjustments of model-predicted water temperatures, salinities, and currents using the Physical-space Statistical Analysis System (PSAS) data assimilation scheme[1].

THE OCEAN MODELS FOR PMRF

An adaptation of the hydrodynamic model of the Blumberg and Mellor[2] has been implemented for Kauai, Hawaii (Fig. 1). Bathymetry is from the Smith-Sandwell topography[3] augmented with data from NOAA and PMRF.

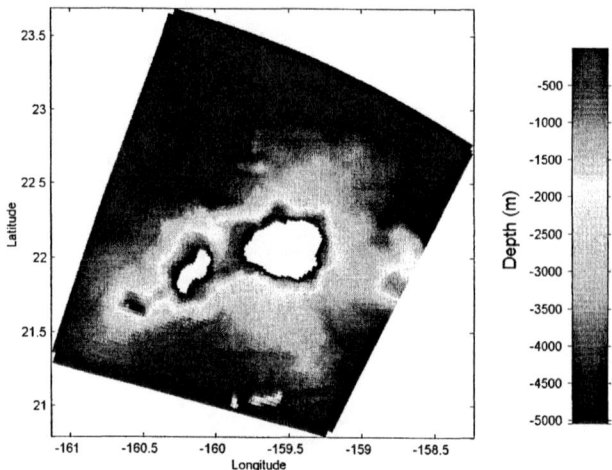

FIGURE 1. Domain and depth field for the Kauai ocean model.

Observed temperature and salinity (T-S) characteristics were used in specifying the vertical resolution of the model. The vertical grid structure has 28 active levels with higher resolution within the top 100 m and at depths at which salinity extremes exist.

We utilized the open boundary condition presented in Lewis et al.[4], specifying the M_2, S_2, N_2, O_1, K_1, and P_1 tidal sea level elevations and phases along the open boundaries of the model domains. The boundary values were from the TPXO.3[5] tidal model but "tuned" to match observed amplitudes and phases for Kauai.

Surface waves were modeled using Delft University of Technology's SWAN (Simulating WAves Nearshore[6]). SWAN is a two-dimensional wave spectra model that can use a curvilinear-orthogonal grid. This allows SWAN to use the same computational grid (and associated depths) as the ocean circulation model.

Initialization and Forcing Fields

The model uses the Navy's daily Modular Ocean Data Assimilation System (MODAS) to estimate the three-dimensional T-S structure. This is used to introduce the mesoscale circulation field into the model. Atmospheric forcing from the National Centers for Environmental Prediction (NCEP) includes momentum, heat, and mass fluxes at the air-sea interface. The SWAN model is forced by the NCEP surface wind velocity and, wave spectra along the open boundaries from NOAA's WaveWatch III wave model is used to account for wave energy that propagates into the region.

Since waves can have a significant impact on ocean circulation[7], the circulation model utilized the surface wave model results to calculate 1) wave-enhanced bottom friction, 2) Stokes drift and the Coriolis wave stress, 3) radiation stresses, 4) wave-related mixing length at the ocean surface, and 5) the virtual tangential surface stress.

CONSTRUCTS RELATED TO THE PSAS DATA ASSIMILATION

Simulating Acoustic Paths and Travel Times

We used the locations of the PMRF sources and receivers, model-predicted sound speed profiles (SSP), and acoustic propagation models to calculate paths of acoustic rays between the sources and receivers. In many cases, direct-path rays and single-surface bounce rays had arrival times very close to one another. This was verified with actual field data.

Our simulations indicated that multiple surface bounce acoustic rays were stable in the paths they took from source to receiver, and their arrival times were well separated from the earlier arrivals. Moreover, multiple surface bounce rays provide a better sampling of the water column. Due to the drop in signal-to-noise for higher multiple surface bounce acoustic paths, we concentrated on analyzing paths that bounced off the ocean surface only twice. Our analyses require a reference sound speed structure c_R. This was determined using monthly climatological T-S fields for the region. The PSAS assimilation scheme also requires reference fields for T, S, and current velocities. Again, the monthly climatological T-S fields were used, while a reference velocity of 0 m/s was used throughout space and time.

Observed and Model-Predicted Travel Time Anomalies

The observed travel time t_o for an acoustic path is combined with a reference ocean arrival time t_R to determine a travel time anomaly: $\Delta\tau_R = t_o - t_R$. Knowing the path an acoustic ray would take through the model domain (individual grid cells denoted by i = 1,2,3,...,N), we can calculate an estimated arrival time: $t = \Sigma [\Delta L_i / (c_i + U_i)]$, where c_i is the sound speed in the ith grid cell, ΔL_i is the distance that the ray travels through the ith grid cell, and U_i is the component of the three-dimensional current along a particular direction of interest responsible for effectively increasing or decreasing the sound speed.

The reference travel time t_R uses the reference T-S vs. depth to give individual values for $c_{R,i}$, and $U_{R,i}$ is always zero:

$$t_R = \Sigma [\Delta L_i / c_{R,i}]. \qquad (1)$$

We used a different expression when dealing with the model-predicted travel time anomalies. A model-predicted travel time is

$$t_m = \Sigma [\Delta L_i / (c_{m,i} + U_{m,i})] \qquad (2)$$

where the m denotes model-predicted values. The model sound speed anomaly is

$$\Delta c_i = (c_{m,i} + U_{m,i}) - c_{R,i}. \qquad (3)$$

Rearranging (3), substituting into (2), linearizing using $c_{Ri}^2 \gg \Delta c_i^2$, and rearranging the result in terms of a travel time anomaly give

$$\Delta \tau_m(t) = -\sum_{i=1}^{N} \frac{\Delta L_i \Delta c_i(t)}{c_{R,i}^2} + e_{LIN}(t) + e_{DIS}(t) = \mathbf{b}\,\Delta \mathbf{c}^T + e_{FM}(t). \quad (4)$$

where N is the number of model grid cells through which the ray travels. Errors associated with the linearization approximation are in $e_{LIN}(t)$, and the discretization errors are in $e_{DIS}(t)$. The $e_{LIN}(t)$ term also includes those errors resulting from the assumption that the ray path does not vary with time.

On the right hand side of (4), we have represented the summation as the multiplication of the **b** vector of the constant $-\Delta L_i/c^2_{R,i}$ terms and the $\Delta \mathbf{c}$ vector of the time-varying terms $\Delta c_i(t)$. All errors have been combined into $e_{FM}(t)$.

We can group all of the model-related acoustic travel time anomaly measurements into a *measurement matrix equation*:

$$\mathbf{z}(t) = \mathbf{H}\mathbf{x}(t) + \mathbf{e}_{FM}(t) + \mathbf{e}_z(t). \quad (5)$$

The M travel time anomaly measurements ($\Delta\tau$'s) of P source-receiver transects made at a given time t are in the column vector $\mathbf{z}(t)$. The rows of the **H** matrix are the **b** vectors. The number of columns in **H** will be the maximum of the N's (N_{max}), and there can be a number of zero entries in **H**. The $\Delta \mathbf{c}$ for each grid cell through which a ray path travels is the N_{max} column vector **x**.

Tomographic Transformation of Equation (5)

Equation (5) must be transformed to relate travel time anomalies to ocean model variables. We express the sound speed as a reference c_R and a perturbation:

$$c = c_R + \frac{\partial c}{\partial T}\Delta T + \frac{\partial c}{\partial S}\Delta S + U = c_R + \delta c. \quad (6)$$

For small T and S variations, we can approximate the two partial derivatives as:

$$\frac{\partial c}{\partial T}\Delta T \approx 4.947\,\Delta T = \alpha\,\Delta T \qquad \frac{\partial c}{\partial S}\Delta S \approx 1.34\,\Delta S = \beta\,\Delta S$$

where α has units of m/s/°C and β has units of m/s/ppt. We can rearrange (6) to give

$$\Delta c_i = \alpha \Delta T + \beta \Delta S + U \quad (7)$$

We use (7) to transform (5) or, in its form in (4), the expression for the model-related travel time anomaly along a given path of N grid cells:

$$\Delta \tau_m = \mathbf{b}\,\Delta \mathbf{v}^T \quad (8)$$

where now

$$b = [-\Delta L_1/(\Delta c_{R,1})^2 \ -\Delta L_1 \alpha /(\Delta c_{R,1})^2 \ -\Delta L_1 \beta /(\Delta c_{R,1})^2 \ \ldots$$
$$-\Delta L_N/(\Delta c_{R,N})^2 \ -\Delta L_N \alpha /(\Delta c_{R,N})^2 \ -\Delta L_N \beta /(\Delta c_{R,N})^2]$$

and the Δv vector is

$$\Delta v = [U_1 \ \Delta T_1 \ \Delta S_1 \ \ldots \ U_N \ \Delta T_N \ \Delta S_N].$$

Each of the parameters in Δv is the known model-predicted variable relative to the reference value. We will use the above b for the rows in H and Δv for the column vector x. Our tomographic relationship is (8), relating acoustics to T, S, and U.

Assimilation of Tomographic Information Into The Ocean Model

We define ocean parameters as x, x^F and x^A, vectors representing the *true state*, the *forecasted estimate*, and the *analysis estimate*, respectively. The vectors x, x^F and x^A are time dependent, and the three-dimensional T, S, and U fields form our state vector.

The basic expression to determine the analysis (updated) field combines the forecast estimate x^F with the acoustic-related measurements z_R (the $\Delta \tau_R$'s) using the model-related measurement matrix H as follows (the PSAS formulation)[1]:

$$x^A = x^F + K(z_R - Hx^F) \qquad (9)$$

Here, $z_R - Hx^F$ is the *measurement residual*. K is the residual (Kalman) gain matrix:

$$K = P_F H^T (HP_F H^T + R)^{-1}.$$

P_F is the forecast error covariance

$$P_F = E\left[(x - x^F)(x - x^F)^T\right]$$

and R is the observation error covariance

$$R = E\left[(e_{FM} + e_z)(e_{FM} + e_z)^T\right] = R_{FM} + R_z.$$

Here, the total observation error covariance is expressed as a sum of the model error covariance and the measurement error covariance. The former can be estimated using archived ocean model output to evaluate the expected travel time differences between a linear discrete acoustic model (such as that embodied here in the measurement matrix H) and an acoustic propagation model based on a solution to the wave equation. The latter term can be estimated from the second order statistics of an ensemble of arrival time observations. The Kalman gain distributes the measurement residual throughout the forecast model domain. Corrections are assigned mostly to regions closest to the observations and where the forecast model uncertainty is highest.

Use of Equation (9) For This Study

For this study, (9) was simplified assuming that travel time anomalies were primarily a result of the differences between predicted and ocean water temperatures. In this case, $\mathbf{x}^F$ and $\mathbf{x}^A$ are vectors of forecasted and analysis temperatures relative to monthly reference temperatures: $\mathbf{x}^F = \mathbf{T}_{model} - \mathbf{T}_{reference}$ and $\mathbf{x}^A = \mathbf{T}_{analysis} - \mathbf{T}_{reference}$. Since $\mathbf{T}_{reference}$ appears on both sides of (9), that the expression can be simplified to

$$\mathbf{T}_{analysis} = \mathbf{T}_{model} + \mathbf{K}(\mathbf{z}_R - \mathbf{H}\mathbf{x}^F).$$

The rows of the $\mathbf{H}$ matrix now consist of

$$\mathbf{b} = [-\Delta L_1 \alpha /(\Delta c_{R,1})^2 \ \ -\Delta L_2 \alpha /(\Delta c_{R,1})^2 \ \ \ldots \ \ -\Delta L_N \alpha /(\Delta c_{R,N})^2]$$

for each ray path. $\mathbf{z}_R$ is the column vector of "observed" travel time anomalies. Thus, all the terms on the RHS of (9) are defined, and we can solve for $\mathbf{T}_{analysis}$.

Calculating Covariance Functions

In this study, the error covariance matrices $\mathbf{P}_F$ and $\mathbf{R}$ were derived from the estimates of the spatial covariance of the model-predicted temperatures:

$$P_F = E\left[(T_{MODEL} - E(T_{MODEL}))(T_{MODEL} - E(T_{MODEL}))^T\right] \qquad (10)$$

$\mathbf{P}_F$ can be estimated using ensembles of model temperature time series. This always results in over-estimating the error covariance. As a result, the rate of spatial decorrelation of a model variable is underestimated. Thus we would expect the observations to be more spatially limited in their impact on the analysis fields.

Implementation

In implementing our data assimilation scheme, we limited the model grid cells impacted by acoustic observations to those within 10 km of any ray path being considered. If L_{max} is the number of grid cells within the 10 km range, then $\mathbf{P}_F\mathbf{H}^T$ for the p ray is an L_{max} column vector. There are 12 monthly column vectors for any ray path. All terms in the $\mathbf{HP}^F\mathbf{H}^T$ matrix are known, and each monthly matrix was calculated. $\mathbf{HP}^F\mathbf{H}^T$ was used to represent the $\mathbf{R}$ array until some future time when we can make an accurate estimate of $\mathbf{R}$. The monthly $\mathbf{HP}^F\mathbf{H}^T$ matrices were inverted and multiplied by $\mathbf{P}_F\mathbf{H}^T$ to give 12 L_{max} x P_{max} arrays, where P_{max} is the number of ray paths being considered.

Thus, the analysis and assimilation software only requires the monthly databases of 1) the $\mathbf{P}_F\mathbf{H}^T(\mathbf{HP}^F\mathbf{H}^T)^{-1}$ elements, 2) the reference temperatures along the acoustic paths being considered (for Δv in (8)), and 3) the $-\alpha \Delta L_N / c_{R,N}^2$ values (for b in (8)).

ACOUSTIC DATA ACQUISITION SYSTEM (ADAS)

To allow near autonomous repetitive collection of acoustic data using the PMRF hydrophone network, we implemented an Acoustic Data Acquisition System (ADAS). Transmissions can be made from any the PMRF acoustic projectors. Any combination of the PMRF receivers can be specified for recording acoustic signals.

Each transmitted signal is replica-correlated with the received signal. The received signal is modeled as a sum of ray arrivals given by

$$r(t) = \sum_n a_n s(t - \tau_n)$$

where a_n is the weight associated with each arrival, s(t) is the transmitted signal and τ_n is the delay associated with each arrival. This model is not strictly correct when dealing with propagation in a shallow water environment. However, at source frequencies of 10 kHz, the effect of dispersion is negligible, and the errors in the model can be ignored.

Two choices for the transmit signal were a chirp signal and a phase-coded sequence. In areas where the acoustic signal interacts with waves on the ocean surface, there can be a Doppler shift that compresses or elongates the signal envelop. Matched filter output of this Doppler-shifted signal gives rise to errors in a) estimating the arrival time and b) the amplitude of the matched filter output of each arrival. Analysis of errors in estimating arrival times showed that the errors in the phase-coded sequence were less than that of the chirp signal. However, the amplitude of the matched filter output remained practically unaltered in the case of chirp signal, while in the phase-coded sequence it is reduced substantially. Since averaging can eliminate the error in the arrival time due to surface motion, a chirp signal was considered a better choice for the transmit signal. The chirp signals had a center frequency of 9.5 kHz and a bandwidth of 3 kHz.

In order to improve the signal-to-noise ratio and to reduce the impact of surface motion, we averaged over a number of acoustic transmissions. Under normal circumstances, it would have been appropriate to send a large train of acoustic pulses and perform an average over this train of pulses. However, this was not possible because of "cross talk" between the transmitter and receivers. Instead, we transmitted a sequence of 12 acoustic transmissions that consisted of 4 groups separated by 8 sec. Each acoustic transmission had a duration of 0.1 sec, with an interval of 0.4 sec between each transmission. The number of pulses in a group was restricted to 3 transmissions to avoid interference due to cross talk. The distance to the nearest receiver set this limitation. The maximum distance between the source and selected receivers dictated the 8 sec separation between groups of transmissions. The string of 12 transmissions was repeated 3 times with an interval of about 30 sec.

The matched filter output for a varying number of transmissions is shown in Fig. 2. We see that a considerable enhancement of the signal-to-noise ratio is achieved by averaging over 36 transmissions. The arrival structure in Fig. 2 consists of one stronger group of arrivals followed by three weaker groups of arrivals. An eigenray[8] analysis for this particular source/receiver pair was performed using a sound

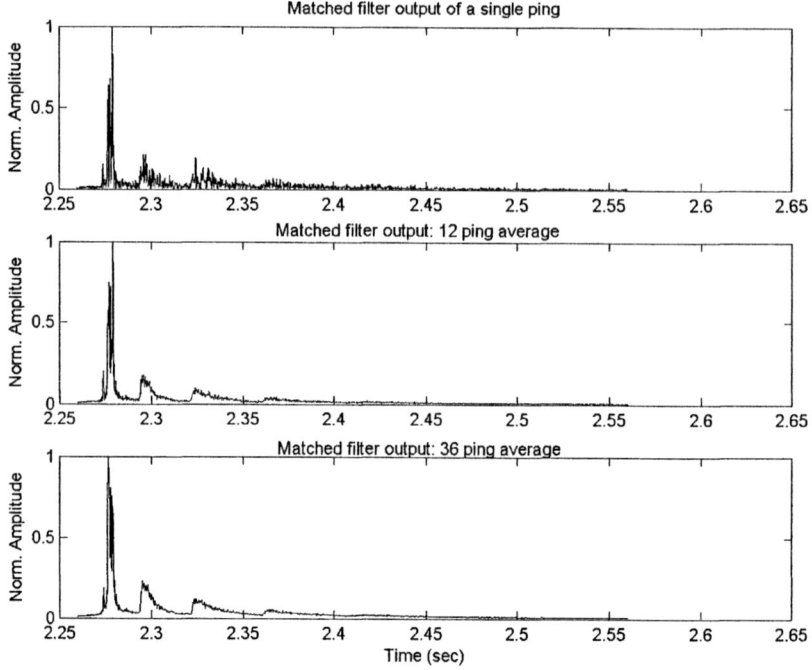

FIGURE 2. Matched filter output for a single transmission (top), an average of 12 transmissions (middle), and an average of 36 transmissions (bottom).

speed field for the region obtained from the ocean model. The earliest group of arrivals consists of rays that travel from source to receiver without interacting with either the ocean surface or bottom and others that include a ray that interacts with the surface only once and rays that hug the bottom and have repeated interactions with the bottom. The arrival time of rays that have only interactions with the bottom carry little information about the bulk of the water column. This plus the problem of delineating individual ray arrival times within the first group of arrivals lead us to neglect these arrivals in our tomography analysis.

The subsequent three arrivals in Fig. 2 correspond to rays that have 2, 3, and 4 surface bounces, respectively. We concentrated on analyzing the path that bounced off the ocean surface only twice due to the drop in signal-to-noise for higher multiple surface bounce acoustic paths (the third and fourth groups of arrivals).

The ADAS collected travel time observations for a double-surface bounce ray path just offshore of the 90 m isobath. The ray path was between a source (D9) some 3.5 km from a receiver (D12). The arrival time of a ray with two surface bounces is about 2.3 seconds. This is used to readily identify the arrival time of the two-surface bounce ray in the ADAS data. An enhanced view of the matched filter output corresponding to the arrival time for this ray (not shown) indicates a distinct peak, with other peaks that are likely the result fluctuations in arrival time caused by motion of the ocean surface and the roughness of the ocean bottom. In order to better determine an arrival time, we low-pass filtered the matched filter output.

DATA ASSIMILATION TEST CASE, JULY 2003

During June-July, 2003, a series of thermistor strings were placed along the 90 m isobath just shoreward of the D9-D12 source-receiver pair. The distance between the D9-D12 ray path and the thermistor arrays was ~0.5 km. The thermistor data provide a means of assessing the impact of assimilating the D9-D12 arrival time anomaly data into the ocean circulation model.

The ocean model was first executed without any acoustic data assimilation, and the model-predicted water temperatures were compared to the thermistor data for July 3, 2003. The rms differences between the model results and the thermistor data are shown in Fig. 3. These range from 0.4-1.5°C, with maximum differences occurring in the lower part of the water column. The bottom of the surface mixed layer is approximately 70-90 m, and tidal forcing can result in considerable semi-diurnal temperature oscillations at this depth.

A comparison of the model and observed water temperatures at one thermistor string in shown in Fig. 4. There is a distinct bias near the ocean surface where the model temperatures are too warm. For the cooler waters at depth, the scatter is considerable, with the model predictions being as much as 2°C too warm or too cool.

When dealing with only one ray path, there is a single element of $\mathbf{K}$ that is multiplied times the corresponding element of $\mathbf{z}_R - \mathbf{Hx}^F$ for each model grid cell falling within the 10 km volume around the ray path. Thus, a larger value of the element of $\mathbf{K}$ for a grid cell results in a greater modification in the model-predicted water temperature in that grid cell during the PSAS assimilation process. An analysis of $\mathbf{K}$ showed that all the larger values were at water depths from 80-300 m. Thus, the assimilation of travel time anomalies will have the greatest impact on the lower levels

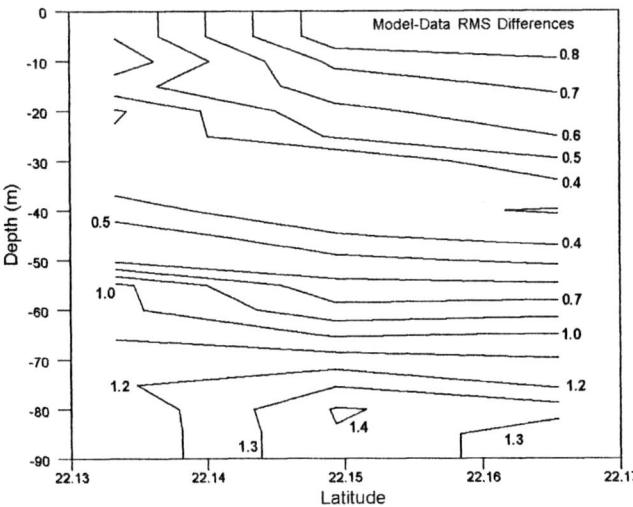

FIGURE 3. Contours (°C) of the rms differences between model-predicted water temperatures and thermistor data for July 3, 2003, without any assimilation of arrival time anomalies.

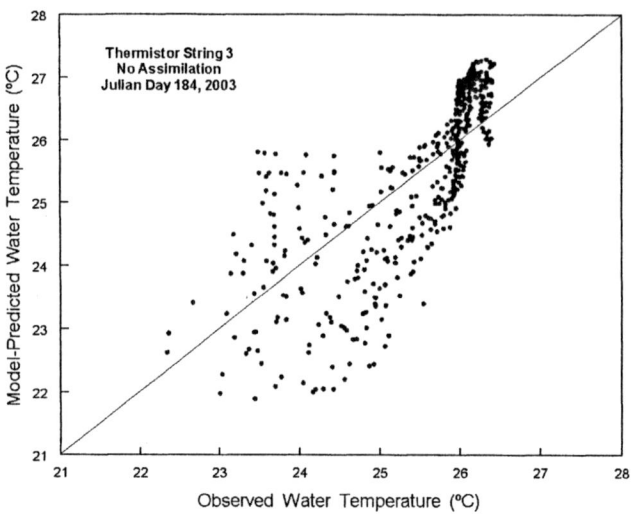

FIGURE 4. A scatter plot of observed and predicted water temperatures at one of the thermistor strings for July 3, 2003 without assimilation.

of the model grid cells, precisely where the rms differences in Fig. 3 are the largest.

During June 30-July 3, 2003, 8-11 kHz chirps (linear FM sweep) with durations of 100 ms were transmitted from D9 36 times over a 2.5-minute period every half hour. Output of each transmission was run through the matched filter process, and then the 36-transmission average was calculated. After determining the arrival times of the double surface bounce path from D9 to D12, travel time anomalies were calculated.

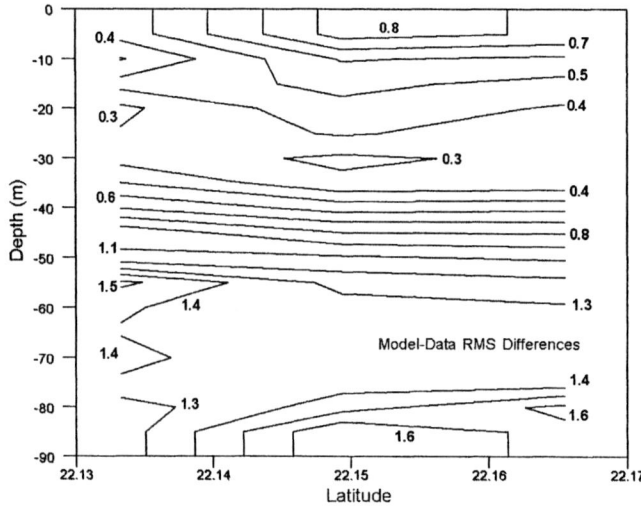

FIGURE 5. Contours (°C) of the rms differences between model-predicted water temperatures and thermistor data for July 3, 2003, with the assimilation of arrival time anomalies.

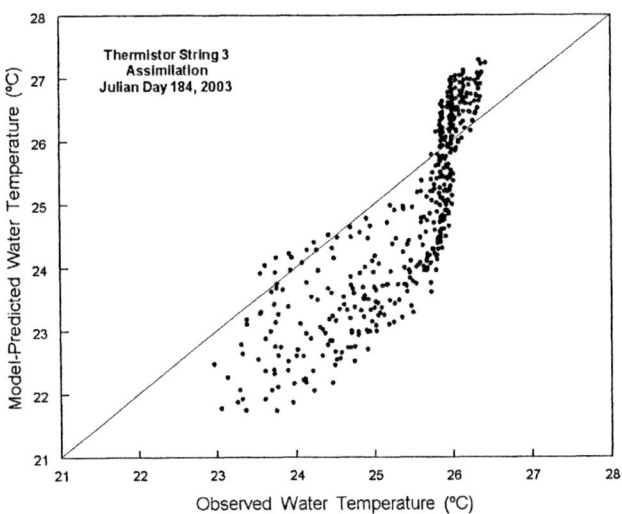

FIGURE 6. A scatter plot of observed and predicted water temperatures at one of the thermistor strings for July 3, 2003 with assimilation.

The model simulations were repeated with the assimilation of the acoustic data. The rms differences are shown in Fig. 5. We see that the rms differences have been reduced slightly in the top 30 m but increased somewhat in the lower layers of the water column.

In Fig. 6 we again show a comparison between the observed and model temperatures at the same thermistor string used in Fig. 4. We see that, although the assimilation process has reduced the scatter of the model predictions, the warmer bias near the surface still exists, and there is now a distinct cooler bias in the lower part of the water column. Comparing Figs. 4 and 6, we would conclude that assimilating the one ray path has had a positive impact on the model predictions in that it has reduced the scatter of the model water temperatures relative to the observed water temperatures. However, the assimilation has failed to eliminate the bias of the model predictions (e.g., making the surface temperatures cooler and the lower temperatures warmer in Fig. 6).

CONCLUSIONS AND RECOMMENDATIONS

We have put forward a model-oriented acoustic inversion and assimilation technique for arrival time anomalies from bottom-mounted sources and receivers. A test of this technology used arrival time anomalies from only one acoustic path. The acoustic information did not have a significant impact on reducing the rms differences between the model and observed water temperatures (Figs. 3 and 5). However, it did help in reducing the scatter of the model temperatures relative to the observed temperatures (Figs. 4 and 6).

To explain the character of the scatter plot in Fig. 6, we note that the **K** elements with the larger values were associated with water depths from 80-300 m. Thus, the assimilation of travel time anomalies had a greater impact on the lower level model grid cells. The acoustic information resulted in a general cooling of the model temperatures (compare Figs. 4 and 6). The cooling would be applied more at the lower parts of the water column and less at the upper parts due to the values of the corresponding elements of the **K** matrix. The result is the "cold" bias in Fig. 6.

The PSAS scheme is critically dependent on the model and observation error covariance matrices. Here we approximated both error covariance matrices by P_F. We recommend that **R** be estimated using archived model sound speed structure. Travel times could be computed using both a sophisticated acoustic propagation model and the simplified model of equation (4). Comparison of the two model results will allow us to characterize the travel time error statistics and error correlation structure.

We also need to improve the estimate of the model error covariance P_F. If we use our most sophisticated simulations as a representation of the true ocean, we could subtract this from corresponding "degraded" model simulations. Degradations could include the exclusion of certain forcing or the use of climatological T-S fields. After subtracting the "true" ocean forecast temperatures from the degraded ocean forecast temperatures, we could then calculate the spatial covariances to obtain P_F.

ACKNOWLEDGMENTS

This work was supported by the Office of Naval Research and the National Defense Center of Excellence for Research in Ocean Sciences. The KauaiEx Group is: M. Porter, P. Hursky, M. Siderius (SAIC), M. Badiey (Univ. Delaware), J. Caruthers (Univ. Southern Miss.), W. Hodgkiss, K. Raghukumar (Scripps Inst. of Oceanography), D. Rouseff, W. Fox (Univ. Washington), C. de Moustier, B. Calder, B. J. Kraft (Univ. New Hampshire), K. McDonald (SPAWARSYSCEN), P. Stein, J. K. Lewis, and S. Rajan (Scientific Solutions, Inc.).

REFERENCES

1. Cohn, S. E., da Silva, A., Jing Guo, Sienkiewicz, M., and Lamich, D., "Assessing the effects of data selection with the DAO Physical-space Statistical Analysis System". *Mon. Weather Rev.*, 126 (11), 1998, pp. 2913-2926.
2. Blumberg, A. F., and Mellor, G. L., "A description of a three-dimensional coastal ocean circulation model," in *Three Dimensional Coastal Models, Coastal and Estuarine Sciences, 4*, edited by N. S. Heaps, Amer. Geophys. Union Geophysical Monograph Board, 1987, pp. 1-16.
3. Smith, W. H. F., and Sandwell, D. T., "Global sea floor topography from satellite altimetry and ship depth soundings". *Science*, v. 277, 1997, pp. 1956-1962.
4. Lewis, J. K., Shulman, I., and Blumberg, A. F., "Assimilation of Doppler radar current data into ocean circulation models". *Cont. Shelf Res.*, 18, 1998, pp. 541-559.
5. Egbert, G. D., "Tidal data inversion: interpolation and inference". *Prog. Oceaongr.*, 40, 1997, pp. 53-80.
6. Booij, N., Ris, R. C., and Holthuijsen, L. H., "A third-generation wave model for coastal regions, Part 1, model description and validation. *J. Geophys. Res.*, 104 (C4), 1999, pp. 7649-7666.
7. Lewis, J. K., "A three-dimensional ocean circulation model with wave effects", in *Estuarine and Coastal Modeling V*, ed. M. Spaulding and A. F. Blumberg. Am. Soc. Civil Eng., 1998, pp. 584-600.
8. Porter, M. B., and Bucker, H. P., "Gaussian beam ray tracing for computing ocean acoustic field", *J. Acoust. Soc. Am.*, 82, 1987, pp. 1349-1359.-

Results from the Elba HF-2003 experiment

Finn Jensen, Lucie Pautet*, Michael Porter, Martin Siderius[†],
Vincent McDonald**, Mohsen Badiey[‡], Dan Kilfoyle[§] and Lee Freitag[¶]

*NATO Undersea Research Centre, La Spezia, Italy
[†]Center for Ocean Research, SAIC, 10260 Campus Point Drive, San Diego, CA
**Space and Naval Warfare Systems Center, San Diego, CA
[‡]University of Delaware, Newark, DE
[§]Science Applications International Corp., Woods Hole, MA
[¶]Woods Hole Oceanographic Institute, Woods Hole, MA

Abstract. In October of 2003, a high-frequency propagation and acoustic communications experiment was conducted off the Italian island of Elba. The experiment followed closely a previous experiment off Kauai (Hawaii Islands) [1]. In particular, a 5 km propagation path along the 100-m isobath was selected. Relative to the Kauai Experiment, the Elba test was significant both in terms of what was similar and what was different. The experiment geometry was identical and a similar mixed layer structure was expected. However, since NURC has worked extensively in this area in past tests we were able to confidently select two sites, one with a very soft bottom and one with a very hard bottom. The comparison between measurements at the two sites in Elba and in Kauai is very illuminating in terms of the propagation conditions and the performance of the acoustic communications scheme. A final significant change was the inclusion of multiple input/multiple output (i.e. using source/receive arrays) communications schemes. We summarize preliminary results from this experiment.

INTRODUCTION

The ELBA HF-2003 trial was a pre-cursor to a planned 3-year (2004–06) collaborative program on High Frequency Acoustics, involving the following institutions: NURC, APL-UW, NRL, SAIC, SPAWAR, UDEL, WHOI and Univ. of Algarve (PO). This joint research effort seeks to significantly improve our understanding of the propagation and scattering of high frequency (5–50 kHz) acoustic waves in the presence of oceanographic variability in shallow water. Yearly field tests are planned to characterize the propagation as a function of 1) source/receiver geometry, 2) arrival angle, 3) carrier (center) frequency, 4) ocean volume structure, 5) bottom type and roughness, and 6) boundary dynamics, including effects of surface waves, bubbles, and noise. Much of the characterization will depend on accurate propagation modeling. The modeling effort seeks to explain, and ultimately predict, the factors that significantly alter operational effectiveness of acoustic communications for applications such as AUV-based MCM detection and classification systems.

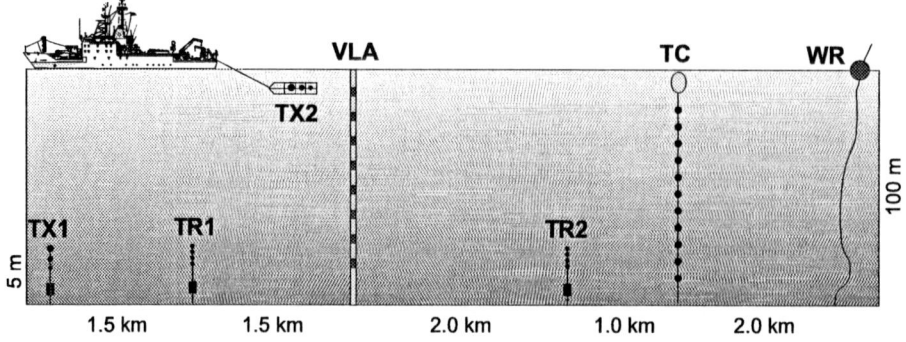

FIGURE 1. Schematic of the experimental setup.

EXPERIMENTAL SETUP

Two experimental sites in approximately 100 m water were selected: One north of Elba with a hard sand bottom and one south of Elba with a soft silt bottom. A schematic of the equipment deployed at each site in shown in Fig. 1. Two telesonar testbed sources and two receivers were supplied by SPAWAR. These are autonomous bottom-moored systems deployed as shown in the figure. One source (TX1) was moored, with the three transducers placed around 5 m off the bottom. The second source (TX2) was towed up and down the acoustic track by *R/V Alliance*. These sources cover a frequency range of 8–50 kHz. The sources were programmed to transmit a sequence of different signals, including LFM's and a variety of communication encodings, to be repeated every 5 min.

The two telesonar receivers (4 hydrophones each, 5 m off the bottom) were placed at ranges of 1.5 (TR1) and 5.0 km (TR2) from the moored source. In addition, the NURC vertical line array (VLA) with 8 hydrophones covering most of the water column was moored at a range of 3.0 km from TX1. The VLA data, with an upper frequency limit of 16 kHz, were received directly on board *Alliance* via a radio link.

A second acoustic experiment was done with the SAIC/WHOI equipment to test the MIMO (Multi-Input/Multi-Output) concept for acoustic communications. Here two drifting ships were employed, one with the source array suspended over board, and one with the receiver array. Transmissions were done at different ranges both north and south of Elba.

Environmental monitoring was done with an 11-element thermistor chain (TC) and a waverider buoy (WR). In addition, several CTD's and XBT's were taken throughout the trial period. Whereas sediment properties for the southern site are well-known from previous experiments, there is no historical information for the northern site. Hence sediment grab samples were collected and seismic profiling carried out for the full track. Finally, ambient noise measurements for geoacoustic inversions were done on a vertical array suspended from *R/V Alliance*.

PROPAGATION ENVIRONMENT

The thermistor chain result north of Elba over a period of 42 h is shown in Fig. 2. Note that the upper 60 m of the water column is quite stable and well mixed. Below there is a sharp thermocline where the temperature drops around 4°C within a few meters. Near the bottom there is colder water, which generates a sound channnel limited below by the bottom and above by the thermocline. We also see that the there is strong internal wave activity causing the thermocline to move up and down by around ±5 m. Despite this strong internal wave activity, the Elba site presented much less variability in the lower part of the water column than Kauai. In Kauai, it was not unusual for the thermocline to complete disappear and reappear over the course of a day.

Figure 3 shows a single sound-speed profile measured in the northern site with the associated ray trace (results are from the BELLHOP Gaussian beam-tracing model). Note the so-called bouncing ball paths refracted in the lower duct.

The BELLHOP model can also predict the accordion pattern of arrivals that would be seen by a vertical array in the water column. In particular, we simulate the field due to an impulse transmitted from the telesonar testbed to the VLA which was placed at a range of about 3 km. This result is seen in the left in Fig. 4 where each fold of the accordion represents a successive surface or bottom reflection. The arrivals come in pairs with the first path directed up into the water column and the second being simply the downgoing path, which is almost immediately reflected from the bottom. Since the source is close to the bottom the upgoing and bottom-reflected paths follow almost the same trajectory from then on to the receiver. One can also see the energy of the 'bouncing ball' paths in the lower part of the water column. Since the sound speed is lower near the bottom, those paths are not the first arrivals.

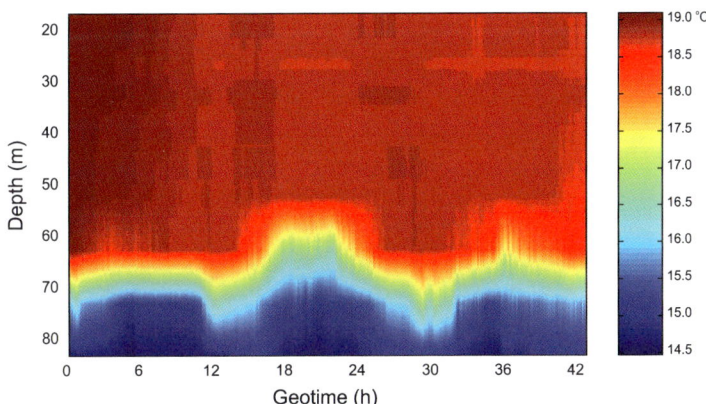

FIGURE 2. Moored thermistor chain data recorded over a 42-h period North of Elba.

PRELIMINARY RESULTS

Measured and modeled impulse responses

An important part of this experiment is to gain confidence in our ability to predict acoustic propagation at these high frequencies. Figure 4 shows a comparison between the modeled and measured impulse responses on the VLA. This result is obtained by using pulse compression techniques (matched-filter or replica correlation) to convert the LFM's or chirps into an equivalent impulse. Note the precise correspondence between the model and the data.

The experimental sites were carefully selected to present very different bottom types. Figure 5 shows a comparison between the impulse responses at the two sites. Note that as expected, the northern site with the very reflective bottom shows extensive multipath spread, i.e. lots of echoes, whereas the southern site shows almost no multipath. Multipath spread, as discussed later, is a critical parameter in prediction acoustic modem performance.

Hybrid (LF/HF) bottom characterization

An additional goal of the experiment was to test a proposed hybrid (HF/LF) scheme to derive bottom (geoacoustic) properties. The scenario envisioned is one where a system uses a compact array of HF sensors to measure the directionality of the ambient noise. As shown by Harrison and Simons [2] there is a simple relationship between the ambient noise that appears to come from the surface and that coming from the bottom. In essence, the surface is considered as a nearly perfect mirror and the bottom as a somewhat murky one. Noise in the HF band is predominantly due to breaking waves and therefore widely distributed. Noise seen on a vertical line array looking towards surface or bottom is then really a sum due to the 'barbershop mirrors' formed by surface and bottom. However, it can be shown that if the bottom is murkier, then the ratio of energy between looking up to the surface and down to the bottom is a direct measurement of the bottom reflection coefficient. This sort of measurement is easily made with small HF arrays and can then

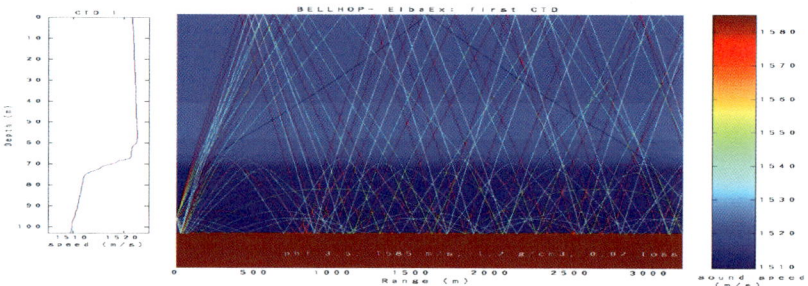

FIGURE 3. Representative sound-speed profile and ray trace for source near the bottom.

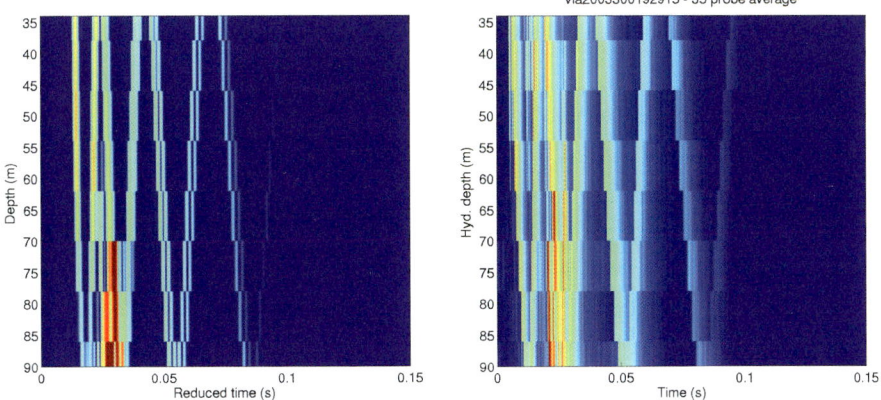

FIGURE 4. Comparison of modeled (left) and measured (right) impulse responses along track N. The data is derived from the VLA positioned about 3 km from the source.

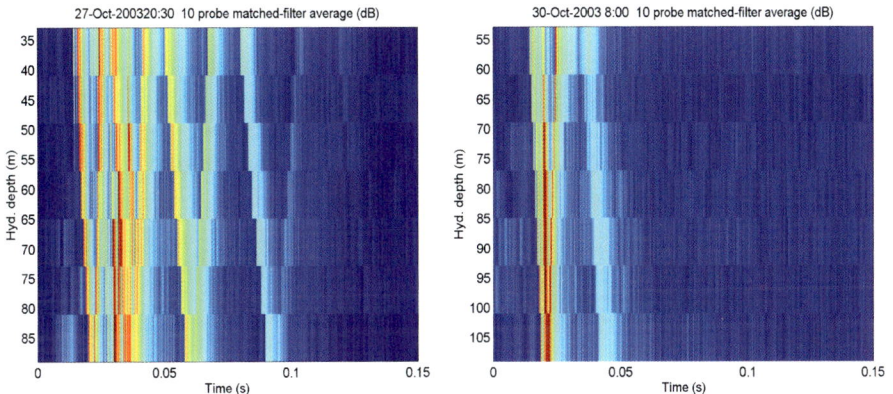

FIGURE 5. Comparison of measured impulse responses along tracks N and S.

be extrapolated down to lower frequencies of interest for other applications.

An example of the technique is shown in Fig. 6. The left panel shows the directionality of ambient noise as a function of frequency as measured north of Elba. Note that the most noise comes from shallow angles. Ambient noise at steeper angles is absorbed in the bottom. Taking the ratio of the up- and down-going energy yields the reflection coefficient as a function of angle of incidence and frequency as shown in the right panel of Fig. 6.

This technique is potentially of great importance in providing immediate knowledge of the bottom properties. Future work in this program will examine the ability to extract both surface and bottom losses.

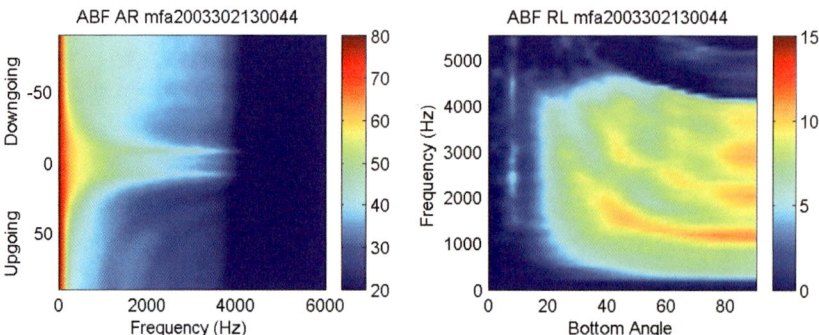

FIGURE 6. Directionality of ambient noise (left) for the northern site. Bottom reflectivity derived by processing the ambient noise (right).

Effects on acoustic communications

The variation in bottom conditions and multipath spread is of critical importance for acoustic communications. For instance, one very common approach to communications involves frequency-shift keying (FSK). This may be compared to playing a piano where each chord is used to encode information. The decoder is nothing more than a spectrum analyzer, which, like the human ear, detects which tones have been played. In some ways, the ocean may be compared to a badly designed concert hall with excess reverberation. To decode the pattern of notes, time must be allowed between each chord so that the reverberation can die down. Thus the multipath spread can limit the transmission rate. Therefore one consideration for optimal data rates is to seek a channel with low multipath spread, i.e. low reverberation.

Interestingly, conditions of low multipath spread are essentially the opposite of what we normally consider to be 'good' propagation conditions. The direct path (neglecting refractive focusing) is a spherical wave expanding from the source and losing energy in a spherical manner. In essence, it is the multipath that gives us the much-improved cylindrical spreading law because the surface and bottom continually reflect the energy that otherwise would be lost. However, that multipath is often simply clutter for the simple FSK communication approach.

These effects of multipath are illustrated in Fig. 7. Note first of all that the northern site with its reflective bottom shows significantly higher SNR. However, the bit error rates are actually much higher. These results are derived from our MFSK algorithm at 2400 bps as recorded on the VLA. It should be noted that in a practical implementation we also include channel coding to dramatically reduce the errors. We generally prefer to study environmental effects on the uncoded waveforms to minimize the number of transmissions required.

Numerous other modulation schemes, including Direct Sequence Spread Spectrum and various coherent Phase-Shift Keying methods were also tested in the experiment and are currently being processed.

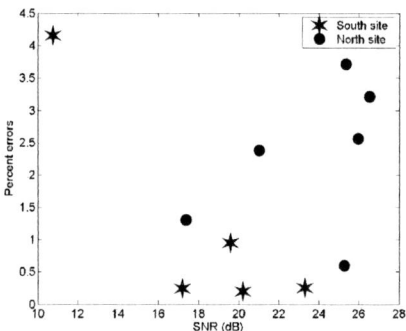

FIGURE 7. Comparison of measured bit error rates along tracks N and S.

High-data rate acomms (MIMO)

An additional focus in this experiment was to test Multiple Input, Multiple Output communications (MIMO). MIMO is a general term applied to systems that use multiple sources and receivers to increase the data rates and is currently an area of very active research in wireless communications (cell phones). In the ocean and with single sources and receivers we can today obtain bandwidth efficiencies of typically 3 bits/Hz, yielding perhaps 16 kbits/sec at distances of 5 km in shallow water. Obviously quality video images from AUV's or raw time-series data from off board horizontal line arrays require much higher data rates, motivating the development of MIMO systems. (Other terms used for MIMO are 'Spatial Modulation', and BLAST, which refers to a particular approach developed at Bell Labs.) Bandwidth efficiencies of 28 bits/Hz have been demonstrated in electromagnetic applications.

Roughly speaking, MIMO systems work by transmitting independent data streams along independent propagation paths. For instance one can imagine a source array that sends energy along different eigenrays and a receive array that separates the data streams by beamforming in the direction of those same independent eigenrays connecting source and receiver. However, such MIMO systems are limited by their ability to separate the arrivals. The various data streams are interfering with each other and the receive array has the difficult task of suppressing the interference to extract each data stream. For this reason, the promise of high-data rates with MIMO systems is generally accompanied by a requirement of high SNR.

Two groups fielded MIMO systems in the Elba experiment. The SPAWAR Systems Center, together with co-investigators from Northeastern University and Arizona State fielded a variety of STAP (space-time adaptive processing) schemes based on current work in wireless electromagnetic communications. The SAIC/WHOI team fielded a scheme previously developed by Kilfoyle [3] and referred to as 'spatial modulation'.

Tables 1 and 2 compare the performance of the latter in the northern and southern Elba sites. These are really extrapolations or estimates of performance derived from the measured data rates. Interestingly, the two sites again showed radically different performance. In both cases the bandwidth efficiency tends to peak at about 3 'channels', i.e. processing 3 independent data streams. With additional data streams the mutual inter-

TABLE 1. MIMO results along tracks N.

	1 MIMO channel	2 MIMO channels	3 MIMO channels	4 MIMO channels
Average SNR$_{output}$(dB)	19.0	15.0	13.8	11.3
	–	13.2	11.6	9.3
	–	–	9.6	4.9
	–	–	–	3.2
Capacity (bits/use)	6.3	9.5	**11.9**	10.7

TABLE 2. MIMO results along tracks S.

	1 MIMO channel	2 MIMO channels	3 MIMO channels	4 MIMO channels	5 MIMO channels	6 MIMO channels
Average SNR$_{output}$(dB)	17.8	15.9	15.1	13.4	9.4	7.9
	–	15.0	14.0	12.2	7.9	6.3
	–	–	12.7	11.3	7.2	5.7
	–	–	–	10.4	5.7	5.3
	–	–	–	–	4.7	4.2
	–	–	–	–	–	3.1
Capacity (bits/use)	5.4	10.1	13.9	**15.9**	12.8	12.2

ference starts to degrade the performance. In the northern site a bandwidth efficiency of 11.9 bits/Hz is attained which is a significant improvement over standard single channel systems. In the southern site with much less multipath, an even better bandwidth efficiency of 15.9 bits/Hz is attained.

These results clearly point to the importance of the environment in determining modem performance. Unfortunately our ability today to accurately predict multipath spread as well as the dynamics (fluctuations) of the multipath is very limited in the 8-50 kHz band which is currently the focus of underwater acoustic modems. Indeed, unexplained modem failures and associated network outages are a fairly common experience. The current research initiative in high-frequency acoustics promises to greatly improve our capabilities in this area.

ACKNOWLEDGMENTS

Appreciation is expressed to the Captain and Crew of the *R/V Alliance*, to Scientific and Technical personnel from NURC, in particular to Mark Stevenson for performing the duty of Chief Scientist on the second part of the trial. Portions of this work were supported by ONR under contract N00014-00-D-0115.

REFERENCES

1. Porter, M.B. and the KauaiEx Group, "The Kauai Experiment," this volume.
2. Harrison, C.H. and Simons, D.G., Geoacoustic inversion of ambient noise: A simple method, *J. Acoust. Soc. Am.* **112**, 1377–1389 (2002).
3. Kilfoyle, D.B., Spatial demodulation in the underwater acoustic communication channel. Doctoral Dissertation, Massachusetts Institute of Technology (2002).

Panama City 2003 Broadband Shallow-water Acoustic Coherence Experiments

Steve Stanic[*], Edgar Kennedy[*], Dexter Malley[*], Bob Brown[*], Roger Meredith[*], Robert Fisher[*], Howard Chandler[*], Richard Ray[**], and Ralph Goodman[***]

*Code 7184, Naval Research Laboratory, Stennis Space Center, Ms. 39529
** Code 7330, Naval Research Laboratory, Stennis Space Center, Ms. 39529
*** Department of Marine Sciences, University of Southern Mississippi
Stennis Space Center, Ms 39529

Abstract. In June 2003 a series of acoustic propagation experiments were conducted off the coast of Panama City, Florida. The experiments were designed to measure and provide an understand of signal phase and amplitude fluctuations, and signal spatial and temporal coherence over several large horizontal and vertical arrays. The propagation measurements were conducted in a water depth of 8.8m and at ranges of 70 m and 150 m. The acoustic measurements cover frequencies from 1 to 140 kHz. The propagation measurements were supported by data obtained by wave rider buoys, CTD's, thermister chains and current meters. Bottom penetration data was also obtained using a buried hydrophone array. The experiments will be outlined and the data sets described.

INTRODUCTION

Measurements of acoustic signal fluctuations caused by random medium inhomogeneities have been made by numerous investigators and several papers have extensive bibliographies [1-7]. However, only a few measurements have been made in very-shallow-waters [6-11]. Since 1998, The Naval Research Laboratory's (NRL) high-frequency program has focused on understanding very-shallow-water high-frequency acoustic signal fluctuations caused by random medium inhomogeneities. The results of these measurements were shown to correlate acoustic signal phase variability with small-scale water column thermal variability and ocean swell conditions [12]. In June 2003, another series of shallow-water broadband propagation measurements were conducted. The objectives of these experiments were to measure and understand broadband horizontal and vertical signal coherence over large vertical and horizontal array apertures, and to measure and understand acoustic penetration into the sediment at sub-critical grazing angles. These measurements were conducted over a broad range of frequencies and changing oceanographic conditions. In addition, to minimize the effects of the ocean's swell, the experiment was configured so that the acoustic propagation path was always parallel to the shore line.

This paper will describe the experimental layout and types of measurements taken. We will seek to understand the acoustic measurements in terms of the environmental conditions. The analysis of these data is just beginning, and at this time only a limited attempt is made to model the acoustic results. In the papers that follow, only a few preliminary acoustic measurement results will be presented. Detailed analysis and modeling of the data will be the subject of future publications. For these proceedings, several papers will be presented describing the data acquisition and environmental measurement systems. A description of the oceanographic conditions during the experiments and initial coherence and buried hydrophone results will be presented.

EXPERIMENTAL CONFIGURATION

The experiments took place in a 300 m by 300 m area just off the beach in Panama City, Florida (Fig. 1). The area has several offshore sand bars, one very close the beach and the other about 100 m from the shore line. The measurements took place 545 m from the shore line in a water depth of 8.8 m. The ocean sediments in this area are generally a mixture of medium sand and shell hash with the occasional large shell fragment.

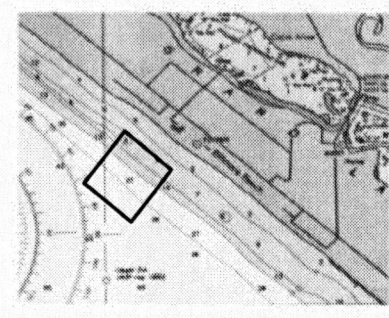
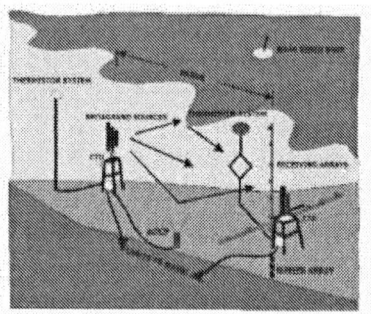

FIGURE 1. Experimental location   **FIGURE 2.** Schematic of experimental setup

A schematic of the experimental layout is shown in Fig. 2. Four broadband source arrays were mounted on a 3-axis positioning system that was mounted on one of NRL's shallow-water towers. The high-frequency source arrays covered a frequency range from 18 kHz to 200 kHz. The arrays were designed to have very narrow vertical beamwidths that ranged from 2^0 (-3dB) at 18 kHz to less than 1^0 (-3dB) at 200 kHz. Horizontal beamwidths ranged from 58^0 (-3dB) at 18 kHz to 44^0 (-3dB) at 200 kHz. The maximum response axis (MRA) of these sources was 4.1 m above the bottom. These narrow vertical beamwidths assured that there would be no boundary interactions over the propagation ranges used (70 m and 150 m). The low-frequency source array (1 kHz to 10 kHz) was a G34 supplied by the Naval Undersea Warfare Center Newport RI. This source was mounted 2.7 m above the bottom and had almost omni-directional beam patterns over the frequency range used.

The receiving arrays consisted of a 6 m vertical array and a 12 m horizontal array. The vertical array had ten equally spaced hydrophones while the horizontal array had seven unequally spaced hydrophones and was mounted 1.75 m above the ocean sediment. Both arrays were rigidly mounted to one of the small towers (Fig. 2). The data from these arrays were used to measure both the vertical and horizontal coherence of the low-frequency multi path signals. These array configurations and hydrophone spacings are shown in Fig. 3.

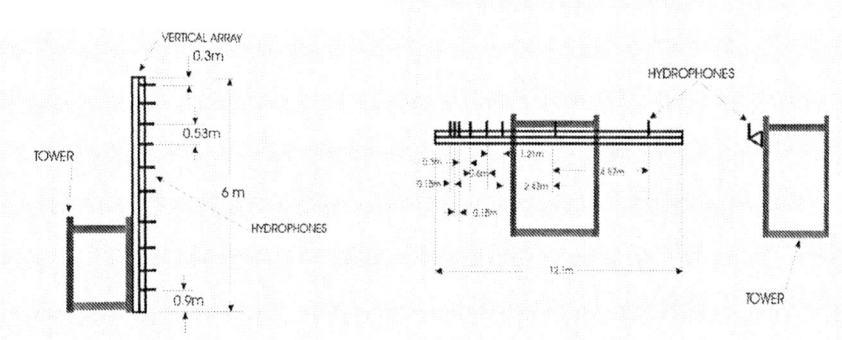

FIGURE 3. Low-frequency vertical and horizontal array configurations

Also mounted on the receive tower were several vertical high-resolution high-frequency receiving arrays with beam patterns similar to those of the high-frequency source arrays. These arrays were used the measure high-frequency temporal phase variability caused by only the random medium inhomogeneities in the ocean's water column. A small 1 m long by 1 m high high-frequency receiving array was also mounted on the receive tower. This array had 6 hydrophones mounted on both the vertical and horizontal arms. These high-frequency hydrophone array configurations are shown in Fig. 4. These arrays are also mounted on a three axis positioning system. Fig. 5 are photographs of the assembled transmit and receive system.

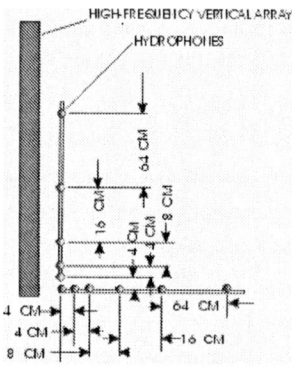

FIGURE 4. High-frequency receive arrays

FIGURE 5. Shallow-water array systems

A six-element, 0.5 m long hydrophone array was water jetted into the sediment at the foot of the vertical array and was used to measure the energy propagating into the sandy bottom at sub critical grazing angles.

Data from all the receiving array systems were wired into electronic canisters located at the base of each tower. The data was then multiplexed on to fiber optic cables and send to an instrumentation van located on the beach. Here the signals were filtered, amplified, digitized at 1 MHz and recorded. Each acoustic measurement sequence consisted of 500 to 1,000 pings with a repetition rate of 1 Hz.

ENVIRONMENTAL MEASUREMENTS

Obtaining an accurate characterization of an oceanographic environment is a difficult task. In order to understand acoustic propagation and signal coherence measurements a number of environmental measurements systems were designed and deployed to measure the important water column properties and their fluctuations. A complete description of the sensor systems and descriptions of the general environmental conditions are given in a following paper.

Figure 6 shows the placement of the environmental sensor systems. Two high-resolution thermistor systems (FRTS and TMMS) were deployed along the propagation path. The vertical system (FRTS) which was connected to the source tower canister, had 9 thermistors that covered the entire water column. The second thermistor system (TMMS) was deployed at the height of the MRA of the high-frequency source arrays and measured the small scale spatial and temporal thermal variations over a 1 m square area. The output from both systems was cabled back to the instrumentation van and displayed in real time.

An RDI ADCP system was deployed close to the receive tower in an upward looking configuration. The output of the ADCP was also cabled back to the shore station and displayed in real time. To measure the near bottom currents, an S4 current meter was mounted on a 1 m high frame and deployed near the ADCP.

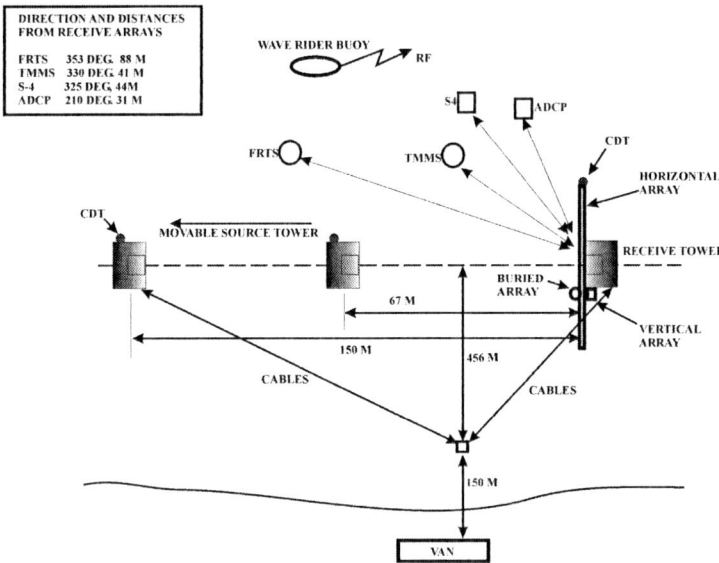

FIGURE 6. Placement of acoustic and environmental sensor systems

A Sea bird model 49 FastCAT CTD was mounted on each tower. These systems measured the temperature and salinity time series. The outputs of each were also displayed on monitors in the shore station.

A Neptune Sciences wave rider buoy was deployed south of the center of the propagation path. It provided a 17-minute average of the wave directional spectra each hour. A weather station measured wind vectors, atmospheric temperature, and humidity.

ACOUSTIC MEASUREMENTS

The acoustic propagation measurements were taken first at a range of 70 m. The source tower was then moved out to a range of 150 m and additional propagation measurements taken. At each range, propagation measurements were taken over a wide range of environmental conditions and frequencies. These ranged from days with very light winds, days with stable water columns, and days when the wind velocities reached 35 to 40 knots. Measurements were also taken during times when the water column thermal variabilities approached $1.0\,^{0}$ C. Table I is a listing of the transmitted signals and their characteristics.

TABLE 1. Transmitted signal characteristics

Range (m)	Source	Pulse type	Frequencies (kHz)	Pulse Length (ms)
70	G34	Broadband	1 to 10 kHz	4
	HF1	CW	18,20,22,24	1
150	HF2	CW	40,60,80	1
	HF3	CW	100,120,140	1

Low-frequency propagation measurements were taken using the G34 at both the 70 m and 150 m ranges. A broadband pulse (Fig. 7) was transmitted and the signal forward scattering from the ocean surface and bottom received on the large aperture horizontal and vertical arrays.

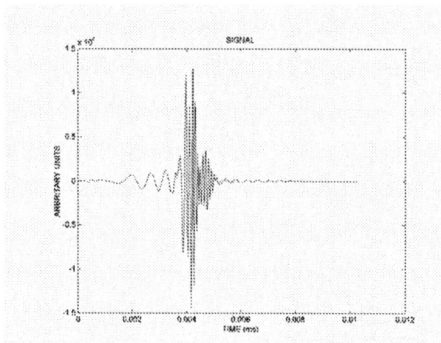 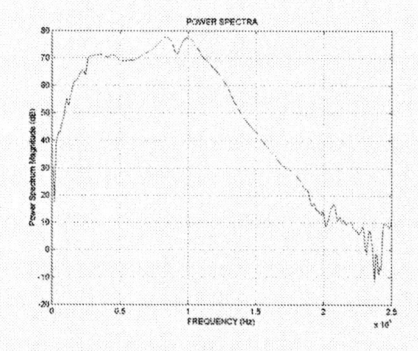

FIGURE 7. Low-frequency broadband pulse and spectra

Figure 8 shows a time history of several signals envelopes received on one of the horizontal array channels on a very calm day. It is clear that the beginning signal is stable for the first direct path arrival. The rest of the signal is dominated by the multi path arrivals from the surface and bottom. These data are being analyzed to calculate horizontal and vertical coherence as a function of frequency, hydrophone separation, range, and time.

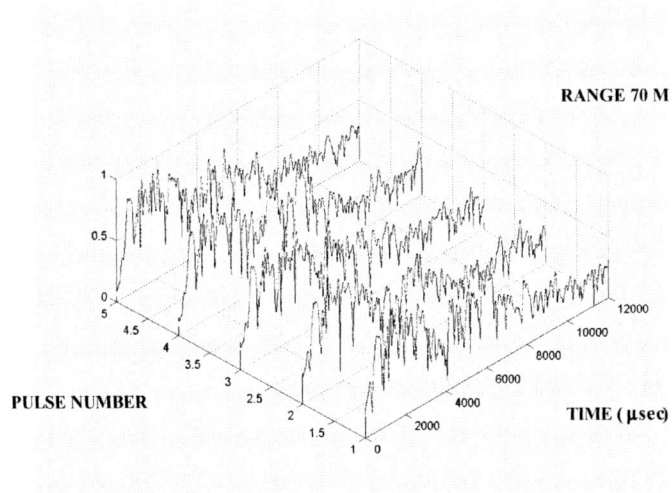

FIGURE 8. Broadband signals received on one of the horizontal array channels

The buried hydrophone array measured the levels of these low-frequency signals as they propagated into the sediment at angles below critical. The data will be used to determine acoustic attenuation, and the penetration ratios. OASES [13] will be used to model the penetration and identify the penetration mechanisms.

High-frequency propagation measurements were taken using the narrow beam width vertical source and receiving arrays. Figure 9 is an example of the high-frequency signal envelopes from one of the array channels. A 1 ms long CW was transmitted. Because of these very narrow vertical beamwidths, the measured signal fluctuations were due only to the randomness in the water column. The received signals clearly show no evidence of multipath interference.

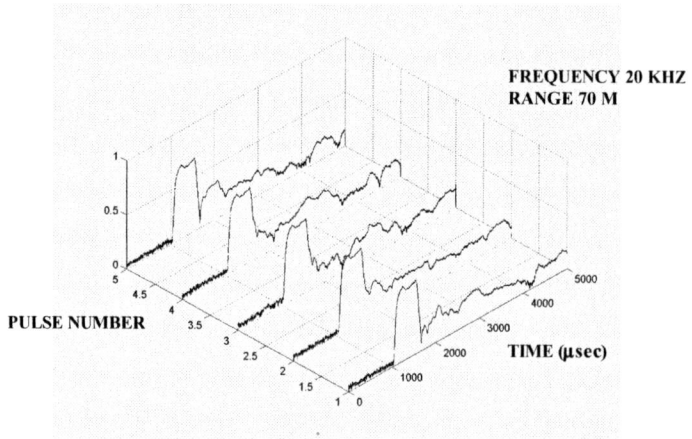

FIGURE 9. Received direct path high-frequency CW pulse

Signal fluctuation statistics will be calculated and correlated with thermal variabilities measured by the thermistor systems. At the same time the small vertical and horizontal high-frequency arrays will measure the high-frequency signal correlations as a function of frequency, range, and hydrophone spacing.

SUMMARY

A shallow-water coherence experiment was conducted during the summer of 2003. The experiments were designed to measure and understand broadband spatial and temporal coherence as a function of propagation range, frequencies, hydrophone separation, and changing oceanographic parameters. The important environmental parameters and their fluctuations which are important to modeling and understanding the coherence measurements were also measured.

ACKNOWLEDGMENTS

This work was supported by the Office of Naval Research, technical management by the Naval Research Laboratory under program element 62435N.

REFERENCES

1. R. M. Kennedy, "Phase and Amplitude Fluctuations in Propagating Through a Layered Ocean," J. Acoust. Soc. Am., vo.1 46, pp-737-745, 1969.

2. D.C. Whitmarsh, E. Skudrzyk, and R. J. Urick, "Forward Scattering of Sound in the Sea and its Correlation with Temperature Microstructure", J. Acoust. Soc. Amer. Vol. 29, p. 1124, (1957).
3. R. F. Shvachko, "Sound Fluctuations in the Upper Layer of the Ocean and their Relation to the Random Inhomogeneities of the Medium", Soviet Physics-Acoustics, Vol. 9, p. 280, (1964).
4. 2 . H. Medwin, "Sound Phase and Amplitude Fluctuations due to Temperature Microstructure in the Upper Ocean," J. Acoust. Soc. Am., vol 56, pp 1105-1110, 1974.
5. 3 . T. Ewart, "Acoustic Fluctuations in the Open Ocean-A Measurement using a fixed Refracted Path," J. Acoust. Soc. Am., vol 60, pp 46-59, 1976.
6. 4 . D. M. Farmer, S. F. Clifford, and J. A. Verrall, "Scintillation of a Turbulent Tidal Flow," J. Geophysical Research, vol 92 pp 5369-5382, 1987.
7. 5. B. J. Uscinski, "Broadband Acoustic Transmission Intensity Fluctuations in the Tyrrhenian Sea," J. Acoustic. Soc. Am., vol 100, pp 784-796, 1996.
8. 6 . P. T. Gough and M. P. Hayes, "Measurements of Acoustic Phase Stability in Loch Linnhe, Scotland," J. Acoustic. Soc. Am., vol 86, pp 837-839, 1989.
9. 7 . J. T. Christoff, C. D. Loggins, and E. L. Pipkin, "Measurements of the Temporal Phase Stability of the Medium," J. Acoust. Soc. Am., vol 71, pp 1606-1607, 1982.
10. 8 . S. Guyonic, " Experiments on a Sonar with a Synthetic Aperture Array Moving on a Rail," Proceedings of the IEEE Ocean 94, vol III, pp 571-576, 13-16 Sept., 1994.
11. 9 . O. Bergem, N. G. Pace, and D Di Iorio, "Surface Wave Influence on Acoustic Propagation in very Shallow-Water," Proceeding of the MTS/IEEE Oceans 99, vol 1, 13-6 Sept., 1999.
12. 10. S. J. Stanic, R. R. Goodman, R. W. Meredith, and E. Kenedy, ' Measurement of High-Frequency Shallow-Water Acoustic Phase Fluctuations," IEEE J. of Oceanic Engineering, vol 25, No. 4, pp 507-515, 2000.
13. 11. H. Schmit, "OASES Version 2.2 Users Guide and Reference Manual," Massachusetts Instsitite of Technology, Cambridge, MA. 1999.

Panama City 2003 Acoustic Coherence Experiments: Environmental Characterization

Roger Meredith, Robert Fisher, Steve Stanic, Edgar Kennedy, Dexter Malley, and Bob Brown

Code 7184, Naval Research Laboratory, Stennis Space Center, Ms. 39529

Abstract. During June 2003, the Naval Research Laboratory conducted a series of acoustic propagation experiments to measure both high (20 to 150 kHz) and low (1 to 10 kHz) frequency spatial and temporal coherence in very shallow water. Environmental data collected to support the acoustic measurements included water column current, bottom current, sea-surface wave height, tide height, CTD water column profiles and mid-water time series, two-dimensional micro-scale seawater temperature, and weather parameters. Wave periods varied from 3 to 7 seconds and wind speeds ranged from 4 to 35 knots throughout the experiment. Temperature and salinity profiles characterized periods when the water column was isovelocity and periods when the water column was stratified with a strong depth dependence of temperature and salinity. Current magnitudes were always less than 25 cm/s. Experimental geometry and methods of environmental data collection are briefly described and environmental conditions and their impact on the propagation environment are emphasized.

INTRODUCTION

In June 2003 the Naval Research Laboratory (NRL), conducted a series of acoustic propagation experiments to measure both high and low-frequency spatial and temporal coherence. The high-frequency propagation measurements spanned 10 to 150 kHz while the low-frequency measurements spanned a frequency band from 1 to 10 kHz. Acoustic measurements were made as a function of frequency, source to receiver range, and receiver separations [1]. Environmental measurements made in support of the acoustic exercise included near-bottom and water column currents, sea-surface wave spectra, conductivity-temperature-depth (CTD) profiles, and time-series, micro-scale seawater temperature, weather, and tides. The objective of the environmental measurements was to first generally characterize the propagation environment in terms of the physical oceanography and second, attempt to quantify the variability in the environment that affects high-frequency propagation. All measurements took place in an area just off the coast of Panama City Florida in a water depth of approximately 8.8m.

ENVIRONMENTAL SENSOR CONFIGURATION

Environmental sensors were mounted on each acoustic tower and deployed at locations near the acoustical path of the experiment (Fig. 1)[1]. The two towers, one configured with acoustic sources and one configured with acoustic receivers, were deployed parallel to the shoreline approximately 545 m seaward of the beach in 8.8m of water. During the experiment, the receive tower's position was fixed while the source tower was repositioned to obtain data sets at two ranges (70 and 150 m).

Figure 1 shows the location of the environmental sensors while Table 1 gives a synopsis of the environmental data collected by each. An ADCP, two tower-mounted CTDs, and two temperature arrays (TMMS and FRTS) were each powered from shore and their data transmitted to shore with the acoustic data stream. As a result, these data were recorded in real-time, but only during acoustic data runs. A wave and tide gauge, Trident wave buoy, and S4 current meter were autonomous instruments that were powered and logged data internally. They recorded data on a programmed schedule, were later recovered and downloaded. A Davis weather system recorded barometric pressure, temperature, UV radiation, solar radiation, humidity, wind speed and direction, and rainfall from an integrated suite of sensors. This sensor suite was mounted on a 10m tower, approximately 100m from the shoreline. In addition to mounted and moored sensors, CTD profiles were collected from an inflatable boat at times permitted by the experiment schedule and weather conditions. Acoustic properties of the bottom sediments were measured during previous NRL experiments at this site.

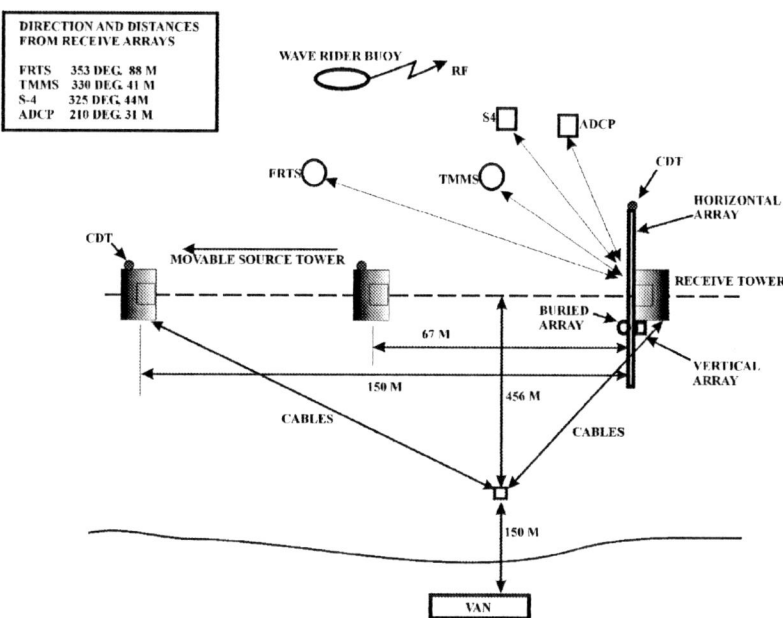

FIGURE 1. Experimental configuration schematic diagram.

TABLE. Synopsis of environmental date collected during Panama City 2003 experiment.

Measurement	Instrument	Recording Period
Tide and Wave height	InterOcean Model WTG 904	Sampled Time-series (4min. wave, 10min. tide); Internal power and data recording
Sea-surface wave height and period	Neptune Sciences Trident Wave Buoy	Processed time-series (17min each hour); internal power and data recording
Currents, Water Col.	RDI 1200 kHz ADCP	Real-time during LF and HF acoustic runs; Shore power and data recording
Currents, Near-btm	InterOcean Model S4 Current Meter	Sampled Time-series (4 min) Internal power and data recording
Cond-Temp-Depth		
Time Series (Receive Tower)	SeaBird SBE49 CTD	Real-time during LF and HF acoustic runs; Shore power and data recording
Time Series (Source Tower)	SeaBird SBE49 CTD	Real-time during LF and HF acoustic runs; Shore power and data recording
Profiles	Ocean Sensors OS 200 CTD	Multiple casts on acoustic data collection days; Internal power and data recording
Temp. vert. structure	NRL thermistor array (FRTS)	Real-time during LF and HF acoustic runs; Shore power and data recording
Temp. microstructure.	NRL TMMS system	Real-time during LF and HF acoustic runs; Shore power and data recording
Weather, Local	Davis Health Enviromonitor	Sampled time-series (30 min); Shore power and data recording

ENVIRONMENTAL CONDITIONS

A thorough description of the general area is available in Salsman and Ciesluk [2]. Data collected during the Panama City 2003 experiment characterize the local environmental conditions and quantify the variability in the environment. Weather during the experiment was typical of the northeastern Gulf Coast during summer. Most days were sunny with high temperatures in the mid-80° F. Generally, wind speeds were less than 18 knots with the direction depending on the existence of an afternoon sea breeze. Periods of higher winds occurred and acoustic data were collected during both periods of high and low winds (Fig. 2).

Tide height measured by the WTG904 wave and tide gauge compared favorably to data from NOAA's tide station 8729210 located in Panama City Beach, FL and indicated a maximum tidal range of ~1.0m. When overlaid with collection times of the acoustic data runs, these data illustrate the variety of tidal conditions under which acoustic experiments were conducted (Fig. 3).

Wave height, period, and direction data, collected by Neptune Science's Trident wave buoy, characterized the sea surface during the acoustic data runs. The time series in Fig. 4 shows that wave heights were typically less than 1m with a general decrease in wave height over the course of the experiment that correlates well with wind speed (See Fig. 2). Each datum from the wave buoy is the result of a 17-minute average, which is a time period comparable to a single acoustic run and does not address the degree of variability that may exist within sea-surface wave trains. An indication of this variability in sea-surface waves may be seen in directional wave spectra computed from wave buoy data (Fig. 5). Examples from two separate days, approximately 1 hour apart are plotted with intensity as a function of the wave direction and wave period. Most spectra show a number of wave periods ranging from approximately 2 to 7 seconds.

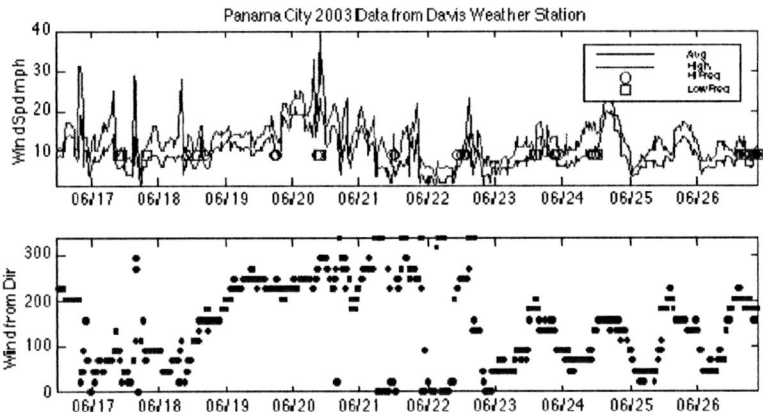

FIGURE 2. Local wind speed and direction during Panama City 2003 experiment with times of acoustic data runs.

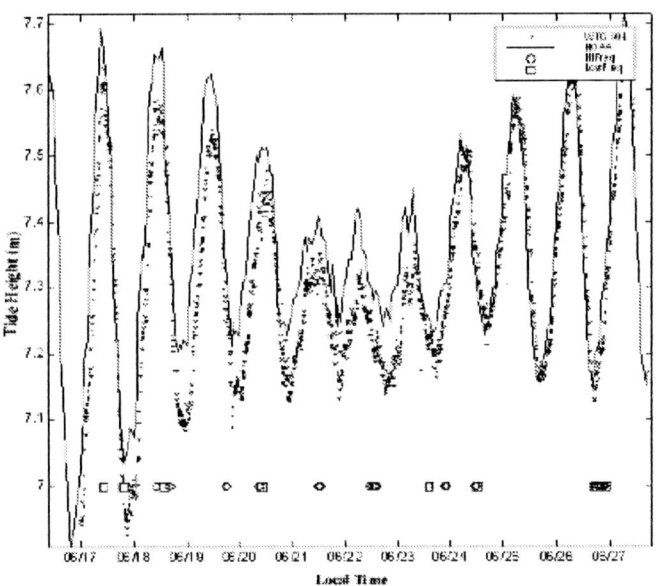

FIGURE 3. Measured tides and acoustic data run times.

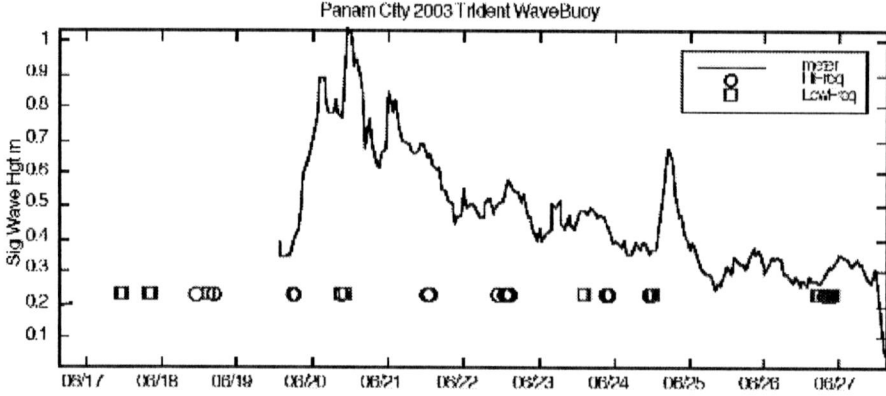

FIGURE 4. Significant wave height measured by the Trident wave buoy.

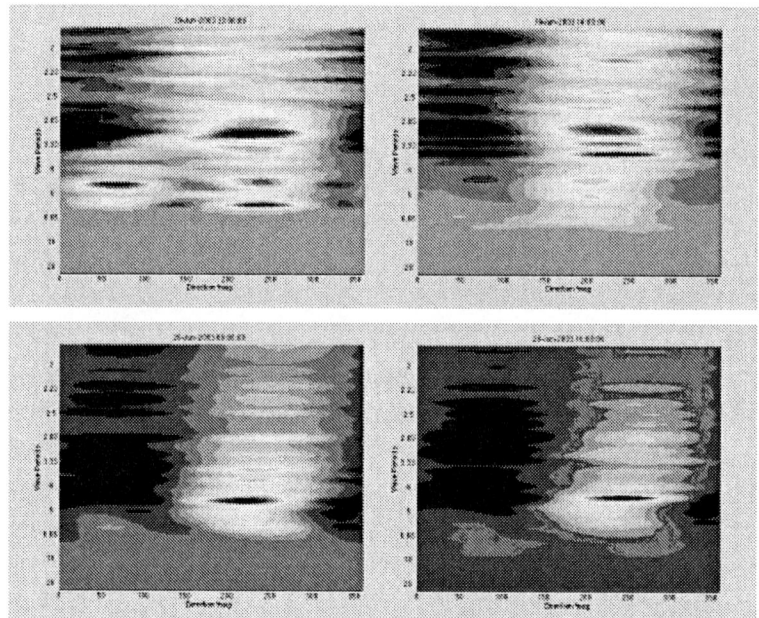

FIGURE 5. Computed directional wave spectra from 2 separate days, ~1 hour apart; 19 June (top), 20 June (bottom).

Two prevalent current directions are indicated from ADCP water column current data collected during each acoustic run (Fig. 6). When currents are low (less than 7 cm/s) the direction tends to be more variable, but when currents are higher (greater than 12 to 15 cm/s) the direction is almost always either approximately 125° Mag. (long-shore current towards the east) or ~320° Mag. (long-shore current towards the west). Figure 6 also shows current standard deviation increases as current magnitude increases, indicating that higher currents introduce greater current fluctuations.

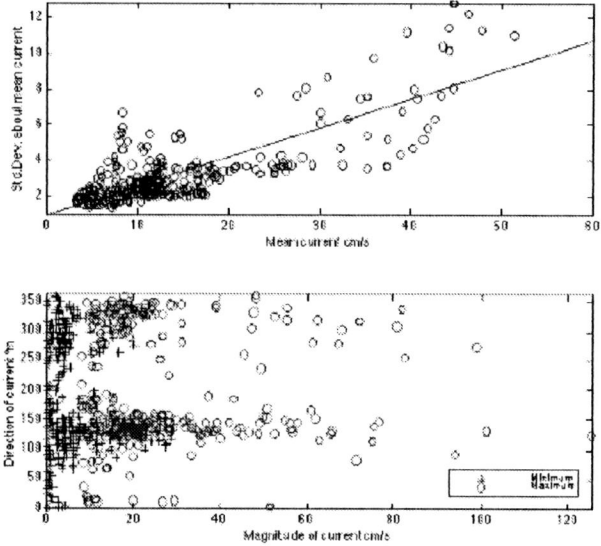

FIGURE 6. Typical water column current magnitude and direction (ADCP).

Near bottom currents were measured by a S4 current meter moored to the bottom and generally agreed well with ADCP water column results in magnitude and direction. The S4 provided a time-series from the entire 12-day measurement period. Figure 7 shows this current spectrum for both north and east components. It is notable that the lower frequency (corresponding to periods of ~1.5 - 30 minutes) components agree quite well with the predictions of isotropic turbulence for frequency dependence (straight line).

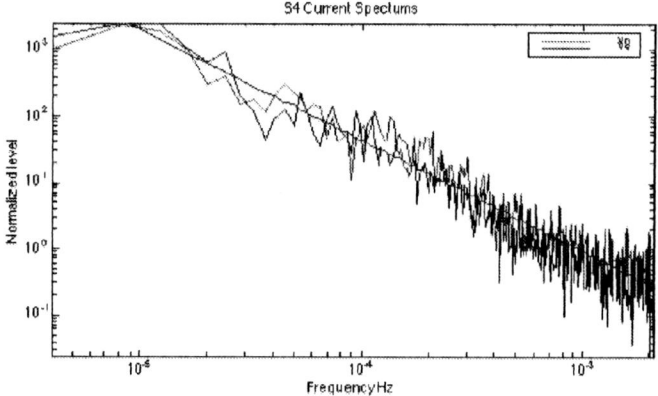

FIGURE 7. Bottom current spectra.

Profiles of temperature, salinity, and computed sound speed were used to characterize the propagation environment. CTD casts were collected from a small boat along an offshore transect at stations that can be typically described as (1) seaward of the buoy field, (2) within the propagation path, and (3), shoreward of the buoy field. Over the course of the experiment, water column properties ranged from profiles with isovelocity conditions to profiles with distinct layers. Figure 8 shows examples of computed sound speed versus depth for two days of sampling plus a temperature vs. salinity diagram for each day to identify density structure. In this figure, 18 June is an example of a well mixed, nearly isovelocity water column. A low salinity filament, running diagonally through the acoustic measurement path, was indicated in salinity profiles, is easily identified in the Temperature/Salinity diagram. This lower salinity filament produces a sound speed change of approximately 0.7 m/s, but since it extends to less than 3m depth, it should have minimal influence on mid-water high-frequency propagation. The mid-water sound speeds appear stable in the profiles collected within the acoustic measurement path. 21 June is an example of layered water where the Temperature/Salinity diagram illustrates the similarity of these profiles and the density stratification in the water column. This stratification of density produces a mid-water sound channel axis near 4 m depth.

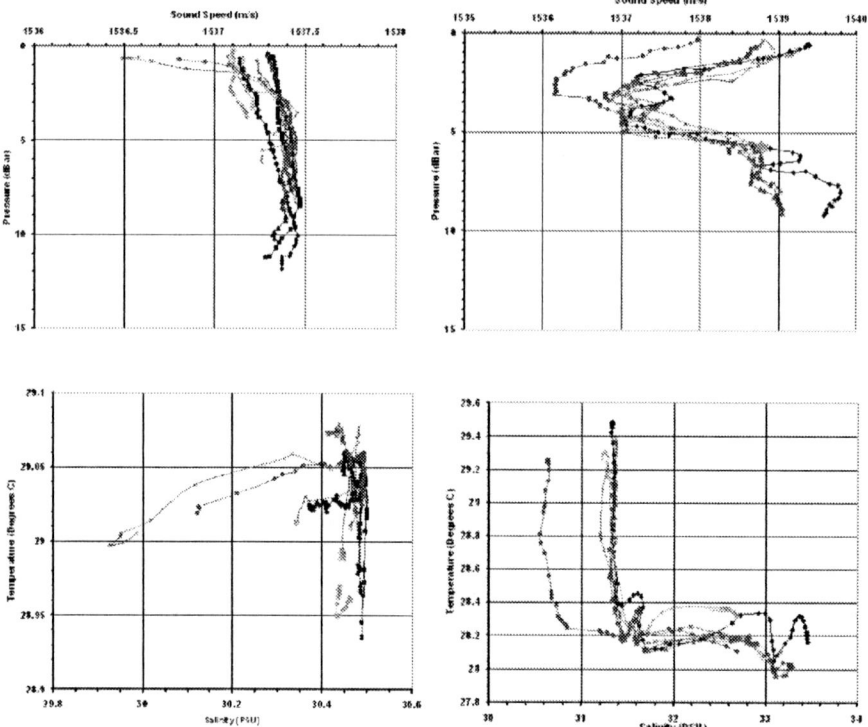

FIGURE 8. CTD profiles indicating isovelocity conditions on 18 June (left column) and a layered water column on 21 June (right column).

Tower mounted CTD's measured temperature, salinity, and pressure time series at approximately a 5 Hz sampling rate during all acoustic runs, but only for Runs 1 through 5 at the source location. Examples of time-series in Fig. 9 illustrate the fluctuations measured.

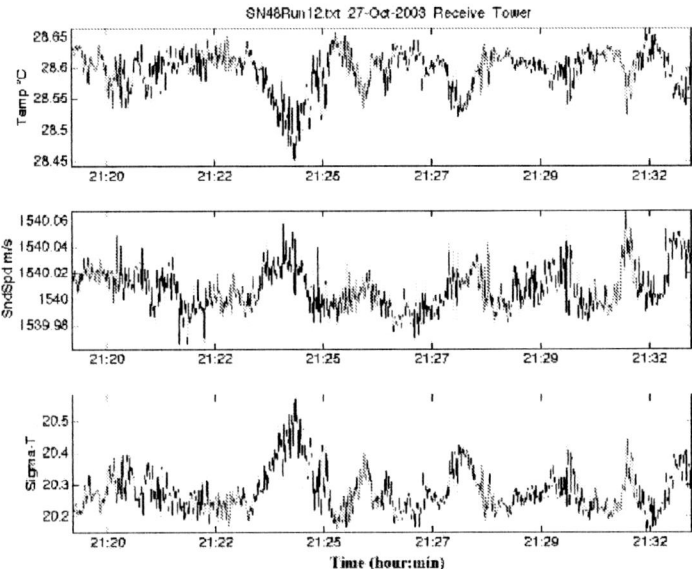

FIGURE 9. Temperature, computed sound speed, and sigma-t time-series from the receive tower CTD during acoustic Run 12 (short range).

High resolution, small scale temporal, and two-dimensional spatial changes in ocean temperature were measured by TMMS which can distinguish temperature changes as small as 0.007°C with absolute temperature accuracy of 0.05°C. As an example of TMMS data, Fig. 10 gives the average temperature spectra during Run 11. In this figure, spectra from TMMS vertical and horizontal arrays are nearly identical except in the periods ranging 8-10 seconds. The spectral slope generally agrees well with that predicted by isotropic turbulence; represented by the straight line. A two-dimensional index of refraction can be estimated from the temperature field as can the two-dimensional spatial scale.

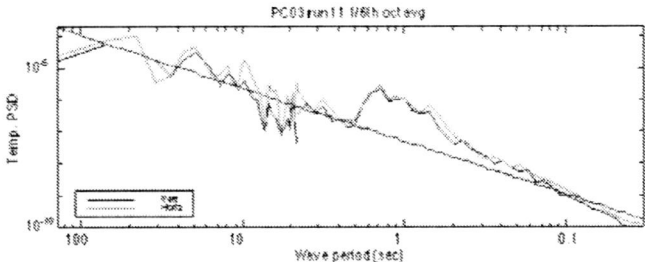

FIGURE 10. Temperature array spectra computed from TMMS data.

Water column temperature structure was also characterized during acoustic runs by the Fast Response Temperature System (FRTS) vertical thermistor array. Examples over the course of the experiment include a nearly uniform temperature profile and a profile showing colder water near the bottom and at the surface with warmer water near the acoustic source depth (Fig. 11).

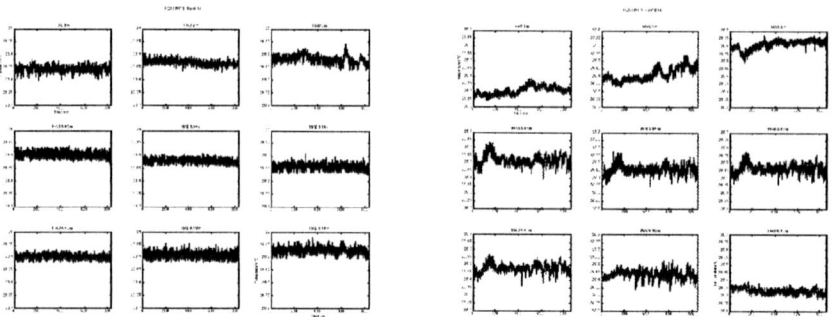

FIGURE 11. FRTS temperature time-series spanning the water column showing a nearly uniform temperature profile (Run 4 - Left) and a stratified profile with colder water near the bottom and at the surface and warmer water near the acoustic source depth (Run 12 - Right).

Table 2 summarizes the general environmental conditions during seven sample acoustic runs and presents several two-dimensional micro-scale temperature fluctuation parameters calculated from the TMMS data

Table 2. Summary of environmental conditions during Panama City 2003 Experiment

Parameters	Short Range Acoustic Runs				Long Range Acoustic	
	Run 2 (6/17/2003) (10:05:00 AM)	Run 4 (6/17/2003) (7:42:00 PM)	Run 6 (6/18/2003) (2:13:00 PM)	Run 12 (6/20/2003) (9:20:00 AM)	Run 26 (6/23/2003) (2:19:00 PM)	Run 35 (6/24/2003) (11:57:00 AM)
Wind						
Wind Speed (mph)	8-10	5-10	8-15	20-40	8-12	12-15
Wind Direction (°Mag)	100	75	150	250	170	150
Currents - Water Column						
Current Magnitude @ srce dpth (cm/s)	4	7.4	9-12	20-30	4-6	7-9
Current Direction @ srce dpth (°mag)	180°	110°	300°	140°	300	20°
Bottom Current Magnitude (cm/s)	< 5	< 5	< 5	< 5	< 5	10
Conductivity - Temperature						
CTD Profiles	-		isovel.>2m		2 layers TS spikes	2 layers up refract
CTD Computed Sound Speed (m/s)	-	±0.5	±0.5 grad increase	0.05	0.3 TS spikes	±0.05 oscillatory
Tide Height (Water height)						
Water Depth (m); Wave-Tide Gauge	8.6	8.6	8.6	8.5	8.3	8.3
Wave Height and Direction						
Signif. Wave Ht.(m); Wave-Tide Gauge	0.6	0.5	0.8	1.6	0.7	0.6
Signif. Wave Height (m); Wave Buoy	-			0.8	0.5	0.4
Wave Direction (°Mag); Wave Buoy				240	190	200
Wave Period						
Wave Period (sec); Wave-Tide Gauge	4.4	4.4	4.4	3.9	4.5	4.2
Wave Period (sec); Wave Buoy	-			5.5	6.5	5.5
Microscale Temperature (TMMS)						
Brunt-Väisälä Frequency (cy/hr)	unstable	4	2	unstable	19	9
Vertical AutoCorrelation Time (Sec)	0.57	0.54	0.62	0.69	0.87	0.62
Vertical Spatial Coherence Length (m)	0.46	0.39	0.48	0.6	0.29	0.24
Horizontal AutoCorrelation Time (Sec)	0.62	0.53	0.69	0.6	0.87	0.65
Horiz. Spatial Coherence Length (m)	0.51	0.56	0.57	0.55	0.51	0.51

For example, the Brunt - Väisälä frequency (or natural buoyancy frequency) indicates that both stable and unstable water conditions were encountered during the experiment. Vertical and horizontal autocorrelation times were computed for consecutive five-second windows for each thermistor. An average over the ensemble of thermistors for each array (vertical or horizontal) is reported in the table. Similarly, the cross-correlation as a function of separation length (array element spacing) for each five-second window was calculated to compute the thermal spatial coherence length. Five seconds was chosen as a reasonable compromise between the acoustic pulse length and the nominal sea-surface swell period.

SUMMARY

Environmental data were collected that characterized the propagation environment in terms of the physical oceanography and quantified variability in the environment that may affect propagation. During the experiment, environmental conditions were encountered that included different tidal stages, different sea-surface wave heights and periods, high and low winds, high and low currents, water column conditions that were both stable and unstable, and sound speed profiles that ranged from isovelocity to stratified.

ACKNOWLEDGMENTS

This work was supported by the Office of Naval Research, technical management by the Naval Research Laboratory under program element 62435N.

REFERENCES

1. Stanic, S., Kennedy, E., Malley, D., Brown, B., Meredith, R., Fisher, R. A., Chandler, H., Ray, R., and Goodman, R., "Panama City 2003 Broadband Shallow-water Acoustic Coherence Experiments," in *Proceedings of the High-Frequency Ocean Acoustics Conference*, 2004.
2. Salsman, G.G. and Ciesluk, A. J. "Environmental Conditions in Coastal Waters Near Panama City Florida" Naval Coastal Systems Center, NCSC TR-337-78, Aug 1978.

Broadband Horizontal and Vertical Spatial Coherence Measurements

Timothy H. Ruppel*, Steve Stanic*, Guy V, Norton†, Roger W. Meredith*, Edgar T. Kennedy*, Ralph R. Goodman** and Marcia A. Wilson*

*Code 7184, Naval Research Laboratory, Stennis Space Center, MS 39529
†Code 7181, Naval Research Laboratory, Stennis Space Center, MS 39529
**Department of Marine Sciences, University of Southern Mississippi, Stennis Space Center, MS 39529

Abstract. Initial results of broadband (1 to 10 kHz) spatial coherence measurements taken during the June 2003 shallow-water (8 m) propagation experiments will be presented. The results will show spatial coherence estimates over a 12 m long horizontal array and over a 6 m vertical array. The data was taken over a range of sea states and at ranges of approximately 70 and 150 m.

INTRODUCTION

Shallow-water horizontal and vertical coherence measurements were taken during a series of experiments in June, 2003 off the coast of Panama City, FL. This paper reports initial results of these measurements at 70.4 m range. One set of measurements was taken during calm surface conditions, and one during the rough surface conditions that accompanied numerous thunderstorms. A number of other measurements were made over several days. Analysis of that data is ongoing. A brief summary of the important environmental parameters will be presented, followed by the horizontal and vertical coherence measurement results. Finally, conclusions will be drawn and our continuing analysis outlined.

ENVIRONMENT

A detailed description of the experimental procedure and the measured environment are found in other papers presented at this conference.[1, 2] In particular, Ref. [1] includes a brief survey of the previous work in this field. To summarize the procedure, the experiment took place in 8.8 m of water with a sandy bottom. A broadband source was mounted to a tower such that the source was 5.8 m below the surface. A 12 m horizontal array of eight unequally-spaced hydrophones was mounted on a separate tower 70 m away from the source tower. The horizontal array was mounted 7.1 m below the surface. A vertical array of ten hydrophones with a 0.53 m spacing was also mounted on the same tower as the horizontal array. Acoustic pings were transmitted once per second. The calm sea run was conducted on June 17 with winds at 5 knots under clear

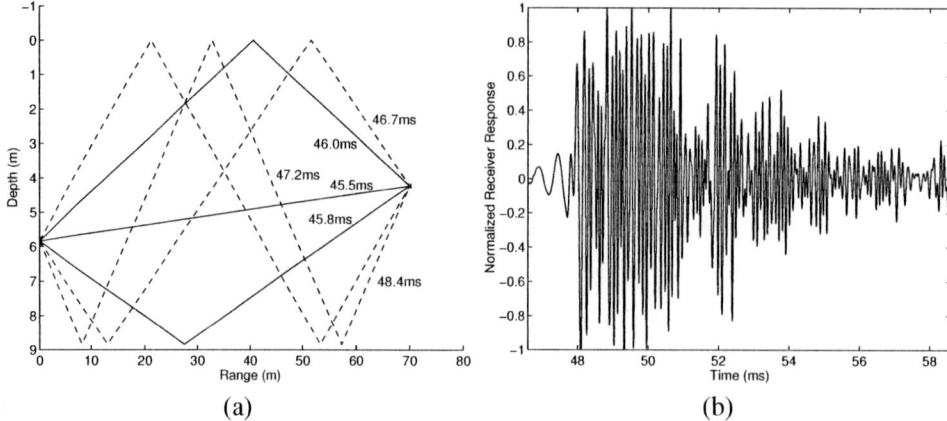

FIGURE 1. Ray traces from the source to one of the transducers on the vertical array assuming a constant 1539 m/s sound speed in the water channel (a), and a sample receiver response for the calm sea run (b). The values shown on the ray trace are the arrival times for each ray.

conditions. The rough sea run was conducted on June 20 with winds at 13–17 knots during numerous thunderstorms.

Ray traces for an iso-velocity sound profile (measured at 1539 m/s), a sandy bottom, and smooth surfaces are shown in Fig. 1(a). Only the first few rays are shown. Plotted with each ray is its calculated arrival time. The shallow water and close range result in arrivals which are difficult to separate with only fractions of a millisecond separating them. The bottom-reflected ray arrives only 0.3 ms later than the direct ray, and the surface-reflected ray arrives only 0.2 ms after that. Fig. 1(b) shows the receiver response as a function of time for one of the hydrophones on the horizontal array during the calm sea run. A very complex multi-path arrival structure is evident in the signal.

HORIZONTAL COHERENCE MEASUREMENTS

Fig. 2 shows a time history and histogram of the horizontal coherence between two hydrophones 2.41 m apart during the calm sea run. Note the trend toward decreasing coherence (and increasing variation) with frequency. Fig. 3 shows the results of similar calculations for the rough sea run. As one might expect, the horizontal coherence measured is smaller, and the variation in those measurements is larger in the rough sea case. Fig. 4 shows the variation of the mean and standard deviation of the coherence between these hydrophones with each frequency calculated. The general downward trend in coherence is seen here as well, and is somewhat more pronounced in the rough sea data. Also seen in Fig. 4 are the large coherence variations. Fig. 5 summarizes the values for all the hydrophones in the array. Plotted are the mean values and standard deviations of coherence measured throughout the run for the frequencies shown in Fig. 2. Note the more rapid reduction of coherence with distance in the rough sea case, but the variations are quite large even for the calm sea run.

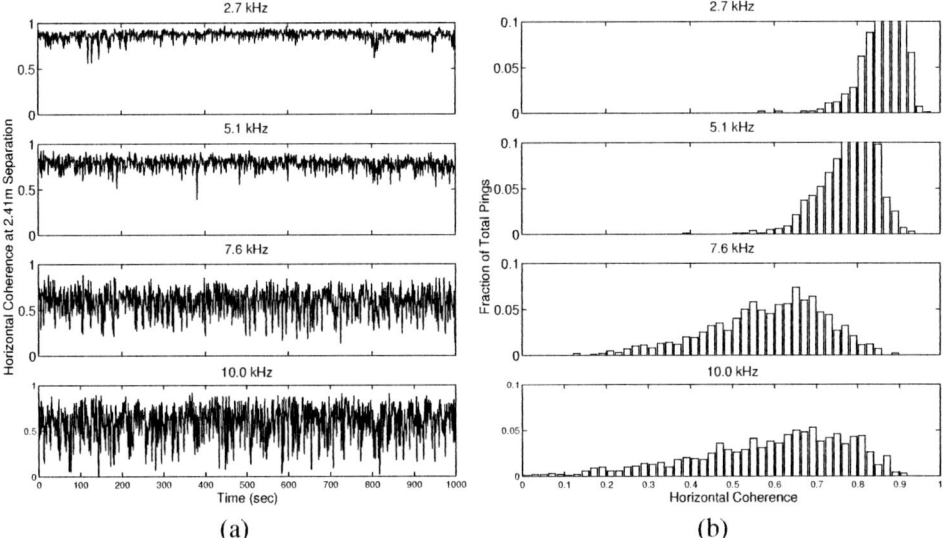

FIGURE 2. The measured horizontal coherence vs. time for four frequencies during the calm sea run (a), and a histogram of that data (b).

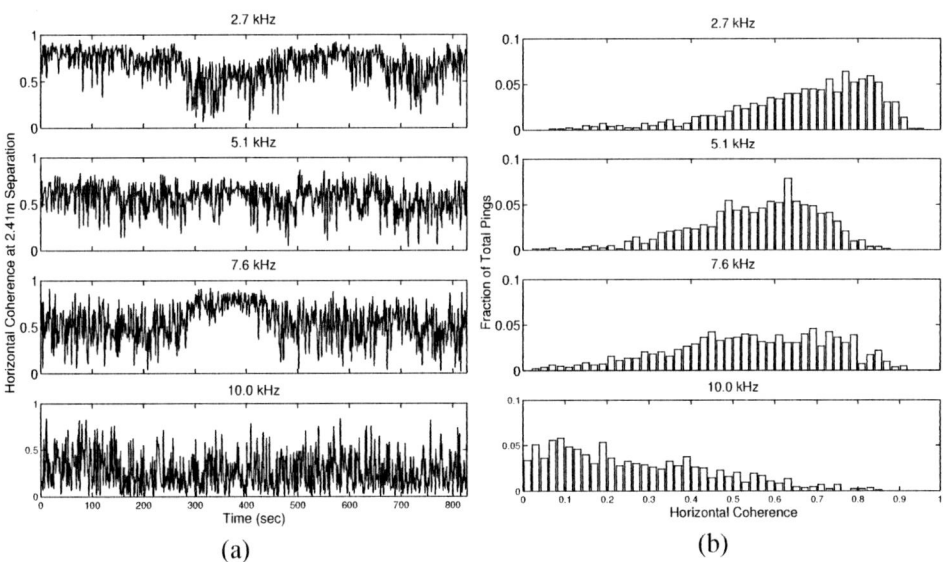

FIGURE 3. The horizontal coherence vs. time for four frequencies during the rough sea run (a), and a histogram of that data (b).

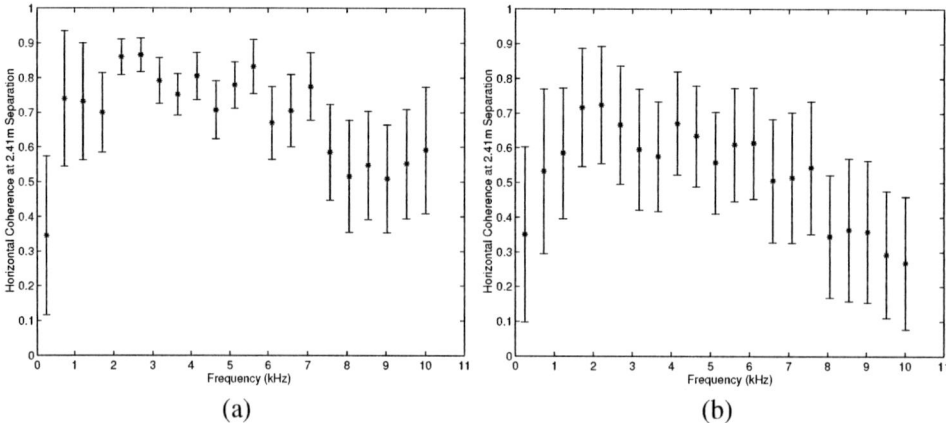

FIGURE 4. Frequency variation of mean and standard deviation of horizontal coherence measurements between two hydrophones 2.41 m apart during the calm sea (a) and rough sea (b) runs.

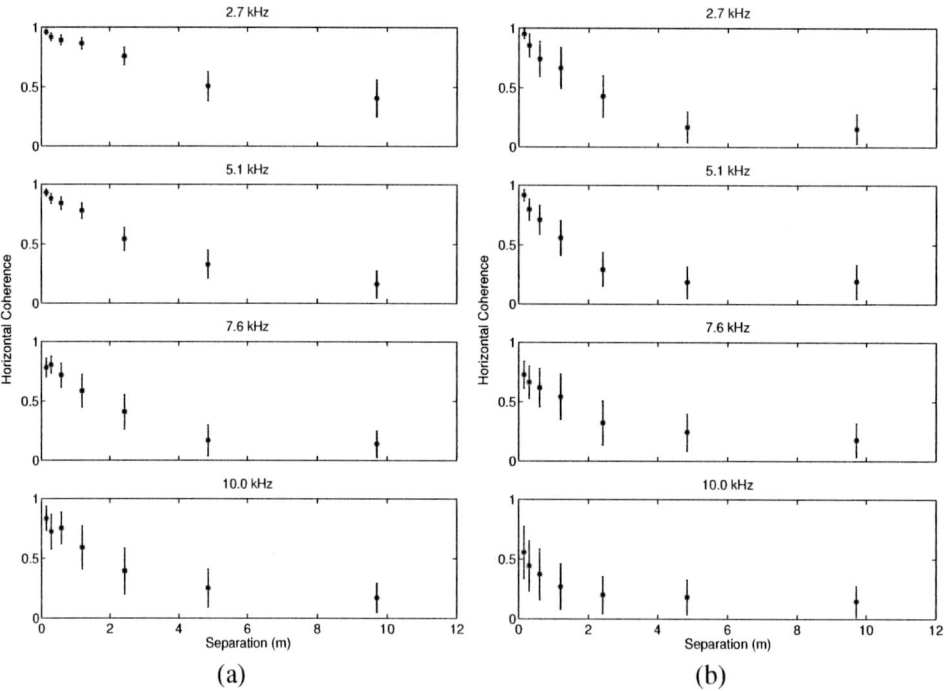

FIGURE 5. Variation of mean and standard deviation of horizontal coherence measurements with distance during the calm sea (a) and rough sea (b) runs.

VERTICAL COHERENCE MEASUREMENTS

Fig. 6 shows a time history and histogram of the vertical coherence between two hydrophones 2.12 m apart during the calm sea run. Here the large variation appears in all frequency bins, though coherence seems to increase at 10 kHz. Also plotted in Fig. 6 are the results of a numerical model for an ideal flat sea with an iso-velocity sound speed profile. While the numerical model also predicts some increase in coherence at this frequency, it greatly underestimates the observed values. Fig. 7 shows analogous results for the rough sea run, again showing small vertical coherence, and large variation in that coherence. Note that the increased correlation at 10 kHz is not as evident here. Fig. 8 shows the variation of the mean and standard deviation of the coherence between these hydrophones with each frequency calculated. Again, the ideal flat sea numerical predictions are plotted with the calm sea data in Fig. 8(a). Note that the mean values for the coherence do not vary as much with frequency in the rough sea data. Also note that the numerical model predicts a peak in coherence at about 9 kHz, so that the higher measured value at 10 kHz may be due to the slightly rough surface. Fig. 9 summarize the values for all the hydrophones in the array. It can be seen that the variation in vertical coherence is quite large for both the calm sea and rough sea runs, and for each hydrophone.

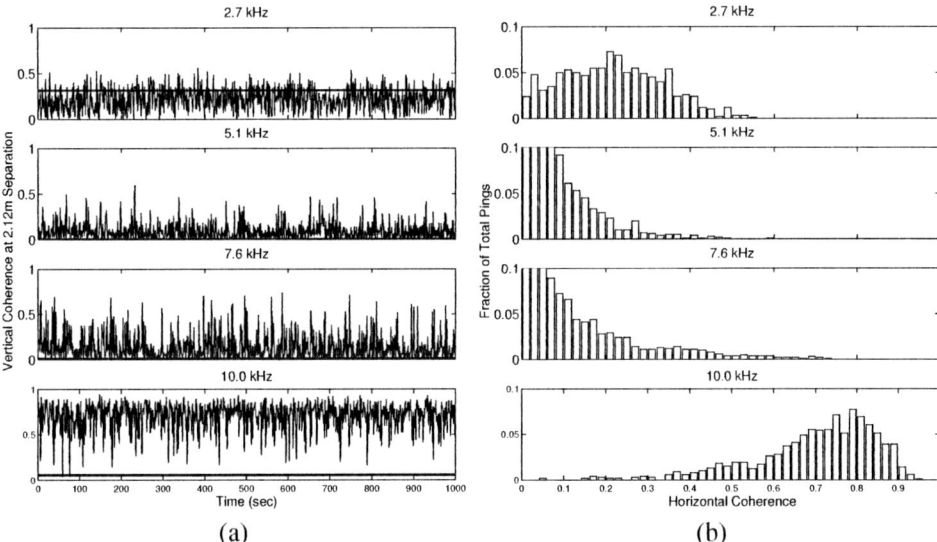

FIGURE 6. Vertical coherence vs. time for four frequencies during the calm sea run (a), and a histogram of that data (b). Also plotted are the results from an numerical model of an ideal flat surface, shown as a solid line.

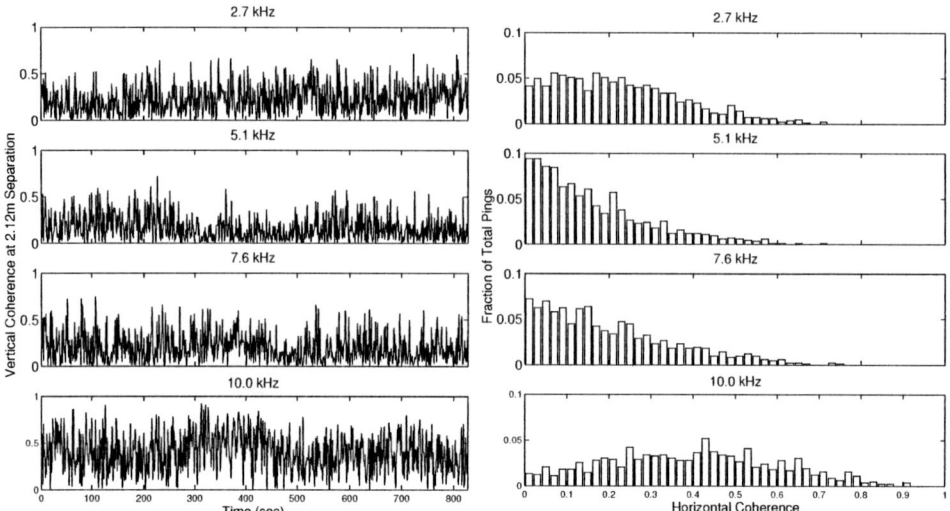

FIGURE 7. Vertical coherence vs. time for four frequencies during the rough sea run (a), and a histogram of that data (b).

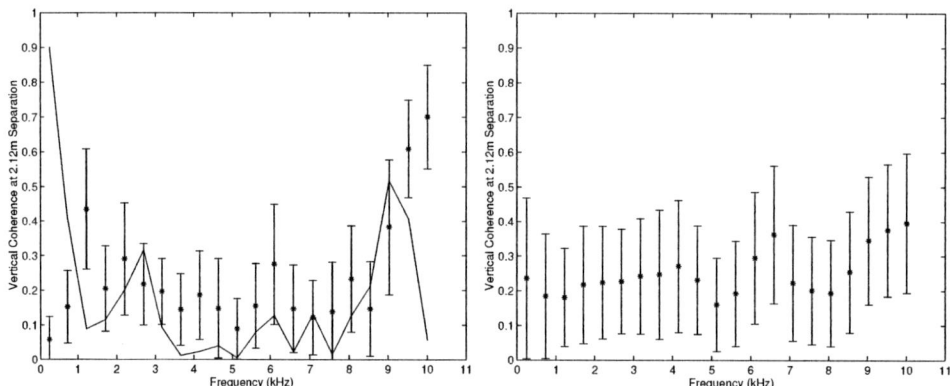

FIGURE 8. Frequency variation of mean and standard deviation of vertical coherence measurements between two hydrophones 2.12 m apart during the calm sea (a) and rough sea (b) runs. Numerical model results for an ideal flat-surface environment are plotted as a solid line with the calm sea data (a).

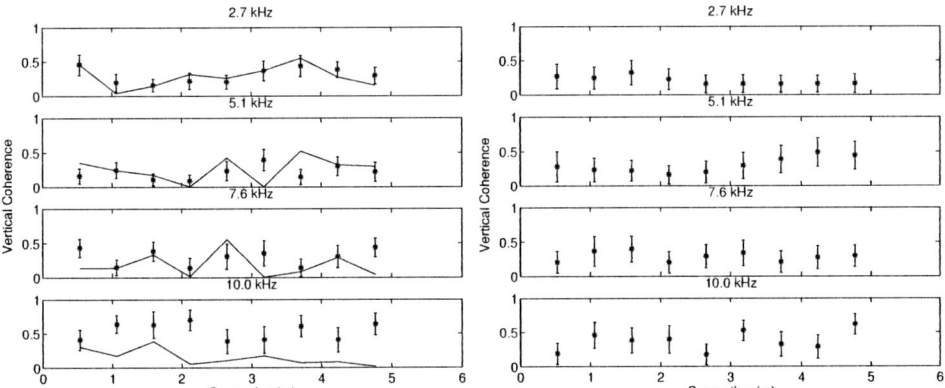

FIGURE 9. Variation of mean and standard deviation of vertical coherence measurements with distance during the calm sea (a) and rough sea (b) runs.

CONCLUSIONS

Preliminary results concerning the measured horizontal and vertical coherence from the June 2003 Panama City experiment have been presented. It has been shown that for 1–10 kHz significant variations in coherence have been observed, even for fairly calms seas. Larger variations, particularly in horizontal coherence, are evident in rough seas. Measurements from different surface conditions and at the more distant source range have yet to be analyzed, and further analysis of all the data is planned.

ACKNOWLEDGMENTS

This work was supported by the Office of Naval Research, technical management by the Naval Research Laboratory under program element 62435N.

REFERENCES

1. Stanic, S., et al., "Panama City 2003 Broadband Shallow-Water Acoustic Coherence Experiments," in *Proceedings of the High-Frequency Ocean Acoustics Conference*, 2004.
2. Meredith, R., et al., "Panama City 2003 Acoustic Coherence Experiments: Environmental Characterization," in *Proceedings of the High-Frequency Ocean Acoustics Conference*, 2004.

Broadband Temporal Coherence Results From the June 2003 Panama City Coherence Experiments

H. Chandler*, E. Kennedy*, R. Meredith*, R. Goodman**, S. Stanic*

*Code 7184, Naval Research Laboratory Stennis Space Center, Ms. 39529.
** Department of Marine Sciences, University of Southern Mississippi
Stennis Space Center, Ms. 39529

Abstract. During the month of June, 2003, the Naval Research Laboratory conducted a series of coherence experiments in shallow water (approximately 9 meters) off Panama City Beach, Florida. Examined here are preliminary mid frequency (1 - 10 kHz) results of analyzed temporal coherence data. For this experiment, a G34 omnidirectional source, mounted approximately 2.7 meters from the bottom, ensonified a vertical and horizontal array of hydrophones mounted on a submerged tower 70 and 150 meters down range in the along shore direction. Results will be shown for both macro (5 minutes), and micro events (<20 seconds).

INTRODUCTION

Temporal and spatial coherence of acoustic signals propagating in shallow water have a direct impact on the performance of Synthetic Aperture Sonar (SAS) and underwater communications systems. Causes of temporal coherence variability are numerous and are driven by oceanographic and meteorological dynamics which can include microstructure dynamics of the water column, sound speed variability, and boundary scattering. Even on relatively calm days, movement of the sea surface can present a randomly changing scattering surface. Volume scattering, either from biological scatterers, or, from entrained bubble plumes on rough days, can have a large effect on received signal amplitude and phase. In shallow waters, multipath interference compounds the effects of the dynamically changing waveguide. At present time, there is little experimental data available to adequately model these effects.

The objective of this paper is to present preliminary results of broadband temporal signal coherence from data obtained during the Panama City 2003 experiments. These measurements were conducted over a broad range of frequencies and changing oceanographic conditions over a relatively smooth, sandy bottom. There was no evidence of biological scatterers present during the examined runs. This paper will focus primarily on mid-frequency broadband results of 20 June 2003. These results represent the effects seen during a 10-minute period on a relatively calm day with an isovelocity water column, [1]. Coherence for the entire received signal is calculated i.e. no attempt has been made to resolve individual multipath arrivals; so the data have not been match filtered. We also present some preliminary results of time of arrival fluctuations for these runs, and contrast them with results obtained on a stormy day.

EXPERIMENTAL CONFIGURATION

Figure 1 is a schematic representation of the experimental geometry. Broadband pulses were generated by a G34 projector 2.7 m above the ocean bottom. The G34 is a mid-frequency omnidirectional source having a bandwidth of 1 to 10 kHz and is capable of generating peak source levels of about 170 dB at 3.5 kHz. The source tower was deployed 545 m from the shoreline in a depth of approximately 8.8 m.

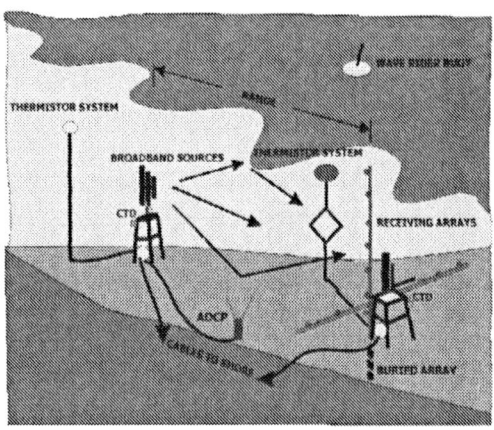

FIGURE 1. Experimental Layout.

Downrange, in a direction nearly parallel with the shoreline, a similar tower was deployed upon which were mounted various acoustic receiving arrays. These included a 12 m horizontal array with eight unequally spaced broadband hydrophones. This horizontal array was mounted on the tower 1.75 m above the bottom, and was orientated nearly perpendicular with the line of propagation from the source tower. Data were also collected on a 6 m vertical array of 10 equally spaced (0.53 m) low-frequency hydrophones but these data have not yet been examined. A complete description of the measurements, program objectives and experimental configuration is given in [2].

Figure 2 shows the low-frequency transmitted waveform and its spectrum. It is a 4 ms 1-10 kHz linear FM pulse equalized to compensate for the varying transmitting voltage response of the source transducer.

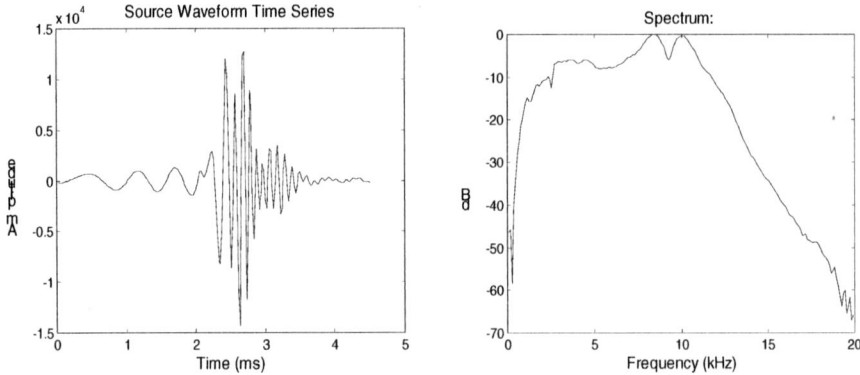
FIGURE 2. Time series and spectrum of the low-frequency source waveform

CALCULATING BROADBAND COHERENCE

For this experiment, pulses were transmitted at a rate of one per second, giving a temporal separation period (ΔT_p) of 1 second between pings. This allowed the comparison of signals separated in time by an integral number of time periods between pings (the temporal delay):

$$T_d = n\Delta T_p. \tag{1}$$

Source triggering and data collection were synchronized to a common clock so that each received pulse was digitally sampled within a 12 ms frame starting 46.6 ms after the commencement of acoustic transmission, with a sampling period. This yielded 800 to 1000 time aligned frames of received pings of 12 bit quantized hydrophone voltage,

$$E_i(m\Delta T_s), \; m = 1, M \tag{2}$$

where $M = 1200$ samples, and i is the pulse number.

Broadband coherence in the time domain is defined in terms of the cross correlation function

$$C_{i,j}(\tau) = \sum_{m=-M}^{M} E_i(\tau) E_j(m\Delta T_s + \tau). \tag{3}$$

This term is then normalized to give a value of 1 for identical signals.

$$\rho_{i,j}(\tau) = \frac{C_{i,j}(\tau)}{\sqrt{C_{i,i}(0)C_{j,j}(0)}}. \tag{4}$$

Temporal coherence is given by the maximum value of the cross correlation function

$$\varphi_{i,j} = \left|\rho_{i,j}(\tau)\right|_{max}. \tag{5}$$

For this study coherence as a function of time and temporal delay was examined for n=1, 2, 3…N, for N=20 delays.

$$\varphi(t_i, T_d) = \left| \rho_{i,i-n}(\tau) \right|_{max}, \quad n = 1,2,3,\ldots 20. \tag{6}$$

Figure 3 is a graphic illustration of how pulse pairs are selected for coherence processing. The dots represent the received pings on a particular hydrophone and are numbered in order of reception. The rows represent how pairs are selected for increasing temporal delay, T_d. For each n, a time series is formed. For example, on the first row, the first two dots are surrounded by a box to illustrate that adjacent pings are compared. This gives the first term in a time series. The box is then slid to the right one ping, and another comparison is made of adjacent pings. This process is repeated until the last ping is processed, and a time series is formed. The second row represents the case where the temporal delay between pings ($n\Delta T_p$) is 2 seconds and illustrates that the first comparison is made between the first ping and the ping $2\Delta T_p$ later. Again the box is slid to the right one ping and the second and 4th ping are compared, and so on to give a second time series.

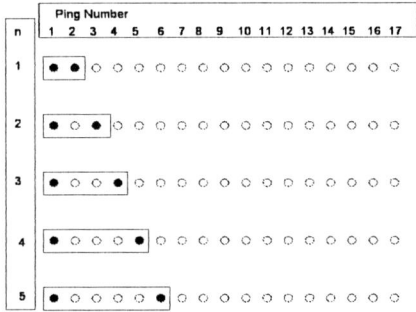

FIGURE 3. Coherence processing for temporal delay of 1 through 5 seconds

Figure 4 shows values of the ping-to-adjacent-ping (n=1) coherence as a function of time for all 8 hydrophones on the horizontal array for a calm day. During this 6.5 minute period, the temporal coherence seems to affect each channel in the array simultaneously. Though not illustrated, this apparent correlation of coherence

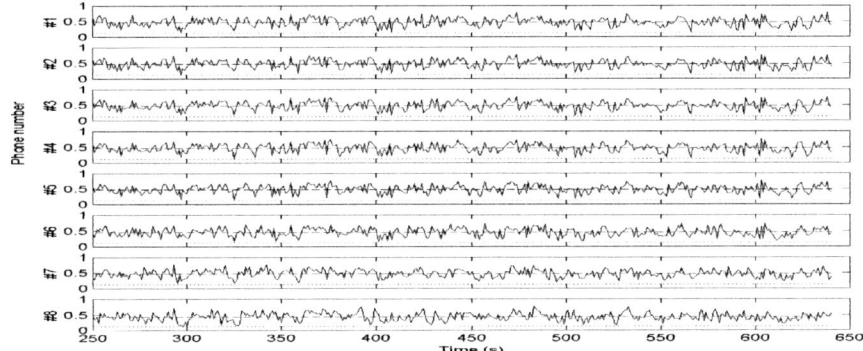

Figure 4. Temporal coherence vs. time for 1 second temporal delay

between hydrophones is evident for all delays processed. Figure 5 shows the mean coherence over the entire 6.5 minute period as a function of hydrophone. We find that there was little spatial dependence. Only hydrophone number 8 gave a somewhat lower coherence than the other hydrophones. Standard deviation is nearly constant at 0.14 across all phones.

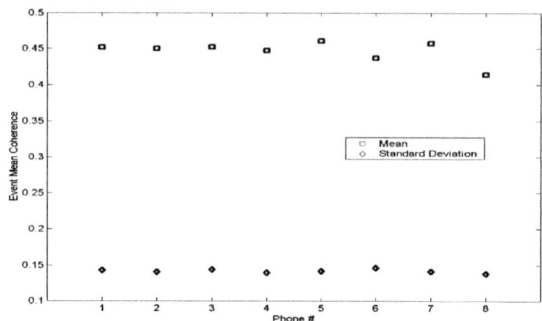

FIGURE 5. Mean coherence vs. hydrophone

Figure 6 shows the mean ping-to ping coherence of the entire 6.5 minute data set. each line represents a different hydrophone with the eighth hydrophone measuring somewhat lower the rest. In this figure we see that the coherence over the 6.5 minutes varies only slightly about the mean of around 0.45.

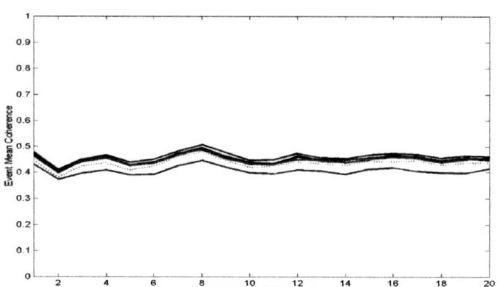

FIGURE 6. Mean coherence vs. temporal delay

Figure 7 shows the measured coherence, for N = 20 calculated temporal delays, as a function of time on a calm day (winds less than 10 knots) at a range of 70 meters. Twenty successive pings are shown, with each of the 20 lines representing one of the a different temporal delay interval, $T_d = n\Delta T$, for n = 1,2,3…20. We observe that coherence as a function of time and for all 20 T_d varies randomly about a central mean as shown by the solid line in Fig 8.

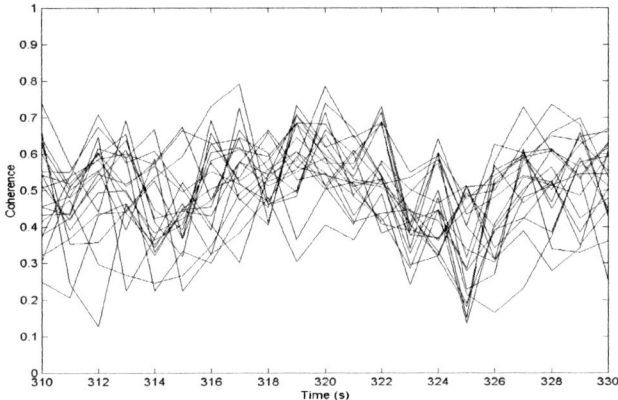

FIGURE 7. Coherence vs. time for all 20 computed temporal delays

The fluctuation of the mean as a function of time seems to be consistent with the swell period. Furthermore, Fig 8 shows that the standard deviation (dashed line below) of the individually computed temporal delays is nearly stationary for this 20-second period, with standard deviation of 0.11.

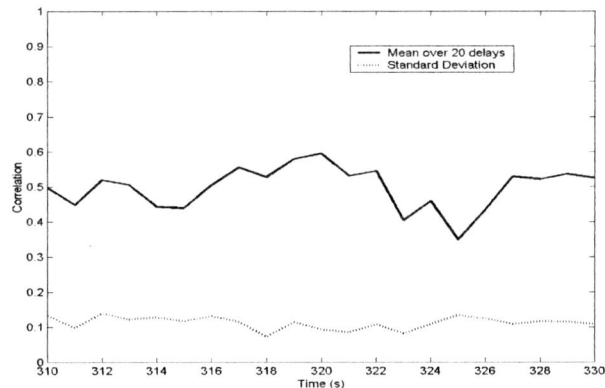

FIGURE 8. Mean and standard deviation of coherence over 20 pulses

AMPLITUDE AND TIME OF ARRIVAL FLUCTUATIONS

Table 1 summarizes the time of arrival for the first significant source-receiver propagation paths as calculated from a ray-trace model. The first bottom bounce arrives 256 μsec after the direct path. The surface multipath does not contaminate the received signal until 456 μsec after receiving the leading edge of the direct path arrival. Since both source and receiver are fixed, the first bottom bounce has no appreciable effect on temporal coherence. Therefore, measuring the fluctuations in the time of arrival of the source waveform using the leading 450 μsec of the arriving pulse will give some indication of the effects of water column dynamics independent of surface scattering dynamics, since the surface bounce will not have arrived until after this time.

TABLE 1. Ray Travel Time

Ray path	arrival angle (deg)	arrival time (ms)
direct	-1.318	45.496
first bottom bounce	-6.202	45.752
first surface bounce	8.186	45.952

Figure 9 shows waterfall plots of the leading 500 μsec (3 cycles) of the time signal for 400 pulses received on a single hydrophone near the center of the horizontal array at a range of 70 m. These pulses have been rectified to show only the positive peaks of the linear FM received wave train. Fig. 4(a). was from a calm day, and Fig. 4(b) a windy day (maximum recorded winds 20 m/s). On the calm day, very little fluctuation of arrival time is evident. Contrast this with the windy day, where there is a great deal of both temporal and amplitude fluctuation in the arriving pulses.

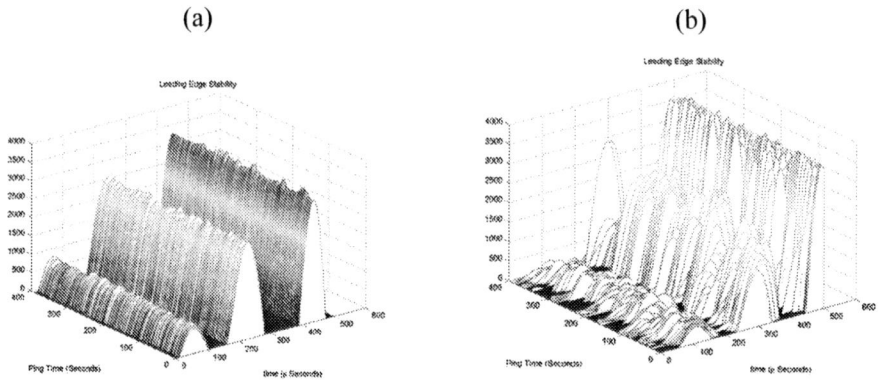

FIGURE 9. Waterfall plot of the leading edge of received pulses Vs time

These results are summarized in Fig 10, which shows the probability density functions for the arrival times for the calm and windy days. On the calm day, the time of arrival was most consistently measured to be 47.06 ms, with a standard deviation of 1 μsec (the sampling period). On the rough day, a spread of nearly 150 μsec was measured, suggesting a nearly 4.7 m/s fluctuation in average sound speed. The nearly 0.04 ms offset in the mean arrival time for each run is believed to be caused by variation in soundspeed. It is conjectured that this was due to bubble effects caused by the breaking waves and whitecaps which occurred during the rough day; a result that is in agreement with [3].

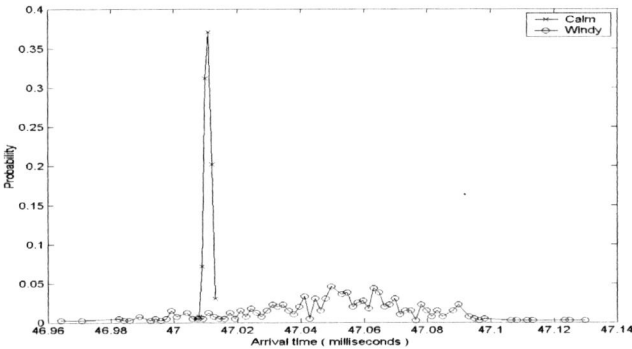

FIGURE 10. Distribution of arrival times for a calm and a windy day.

SUMMARY

During these experiments the mean of the measured broadband temporal coherence for the time scales examined was seen to be 0.47 and was relatively stationary as a function of temporal delay. This result is in agreement with [1] for the Rayleigh noise case. There was a fluctuation about the mean of approximately +/-10%. The time of arrivals showed a significant difference between the calm day and the stormy day. While there was very little arrival fluctuation evident on the calm day, there was still significant de-correlation. This would suggest that on calm days, the effect of water column sound speed microstructure dynamics is small compared to surface scattering multipath effects, a result consistent with [4]. Earlier work [5] has reported higher values of coherence for a similar environment. However, comparison of these results with those presented here is not possible since we have not matched filtered our data prior to calculating coherence. Subsequent work will address these issues

ACKNOWLEDGMENTS

This work was supported by the Office of Naval Research, technical management by the Naval Research Laboratory, under program element 0602435N.

REFERENCES

1. Meredith, R., et al., "Panama City 2003 Acoustic Coherence Experiments: Environmental Characterization," in *Proceedings of the High-Frequency Ocean Acoustics Conference, 2004.*
2. Stanic, S., et al., "Panama City 2003 Broadband Shallow-Water Acoustic Coherence Experiments," in *Proceedings of the High-Frequency Ocean Acoustics Conference, 2004.*
3. Chotiros, N. P., et al, "Acoustic backscattering at low grazing angles from the ocean bottom. Part II. Statistical characteristics of bottom backscatter at a shallow water site," in J. Acoust. Soc. Am., 77(3), March 1985.
4. Goodman, R. R., et al, "Observations of High-Frequency Sound Propagation in Shallow Water with Bubbles Due to Storm and Surf.," *in IEEE Journal of Oceanic Engineering, Vol 25, No.4, October, 2000.*
5. Badiey, M., et al, "Signal Variability in Shallow-Water Sound Channels.," *in IEEE Journal of Oceanic Engineering, Vol 25, No.4, October, 2000.*

Panama City 2003 Acoustic Coherence Experiments: Low Frequency Bottom Penetration Fluctuation Measurements in a Multipath Environment

Roger W. Meredith, E. Ted Kennedy, Dexter Malley, Robert A. Fisher, Robert Brown, and Steve Stanic

Naval Research Laboratory, Code 7184, SSC, MS 39529

Abstract. This paper is part of a series of papers describing acoustic coherence and fluctuations measurements made by the Naval Research Laboratory in the Gulf of Mexico near Panama City Beach, FL during June 2003. This paper presents low frequency (1-10 kHz) buried hydrophone measurements and preliminary results for two source-receiver ranges with grazing angles less than two degrees (realtive to the direct-path to the seafloor at the receiver location). Results focus on fluctuations after acoustic penetration into the sediment. These fluctuations are correlated with environmental influences.

INTRODUCTION

As given in the earlier paper [1], the omnidirectional G34 transmitted a 4.5 second pulse using a LFM with a low-frequency bandwidth of 500-5000 Hz and a high-frequency bandwidth of 5-12 kHz. Corresponding beamwidths varied from omnidirectional at 1 kHz to approximately 35 degrees at 10 kHz. The G34 has source levels of about 170 dB at 3.5 kHz and was mounted 2.7 m above the sea bottom in approximately 8.8 m of water. The burried receiver was a 6-element, linearly-spaced hydrophone array with a 50 cm aperture. It has a free-field voltage sensitivity (FFVS) of -180 dB re 1V/μPa at 2 kHz and -168 dB at 10 kHz. It was vertically buried at the foot of the vertical low-frequency array. The signals from each hydrophone were cabled back to an instrument canister mounted on the tower. Signals are multiplexed, sent via fiber optic cable to shore, digitized at 100 kHz, and recorded. Data is presented for two source-to-receiver ranges, nominally 70 and 150 m. These correspond to direct path grazing angles of less than two degrees.

An example of the received signals for a single ping on each buried hydrophone is shown in Figure 1. For each acoustic run, five hundred pings were acquired with greater than 20 dB signal-to-noise ratio, and these pings form the basis for the bottom-penetration fluctuation processing. Channel 1 is the topmost phone and was positioned in the water column on the ocean-sea bottom interface and Channel 6 is the deepest buried phone at 50 cm.

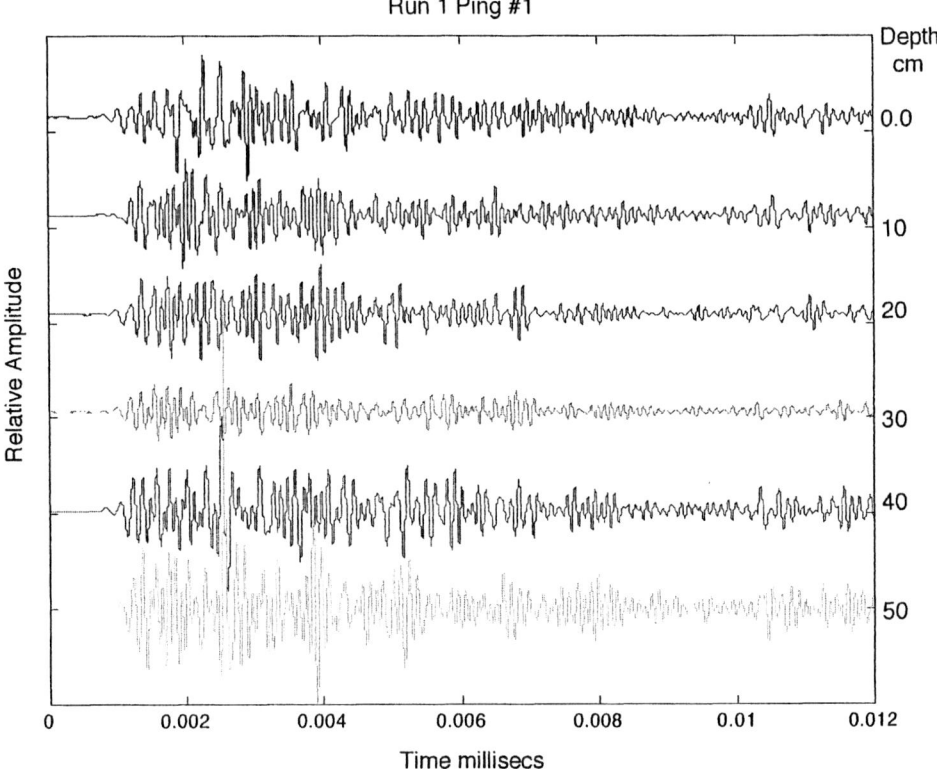

Figure 1. Relative amplitudes for a single ping for each array hydrophone at the short range (~70 m).

Based on the source-receiver geometry for Channel 1 and measured sound speed profiles, an ideal normal mode intensity plot can be generated of the sound field in the water column. Examples are shown in Figure 2 to provide a sense of structure in the sound field. Ray tracing gives more than six arrivals with times less than a pulse length for the short-range geometry.

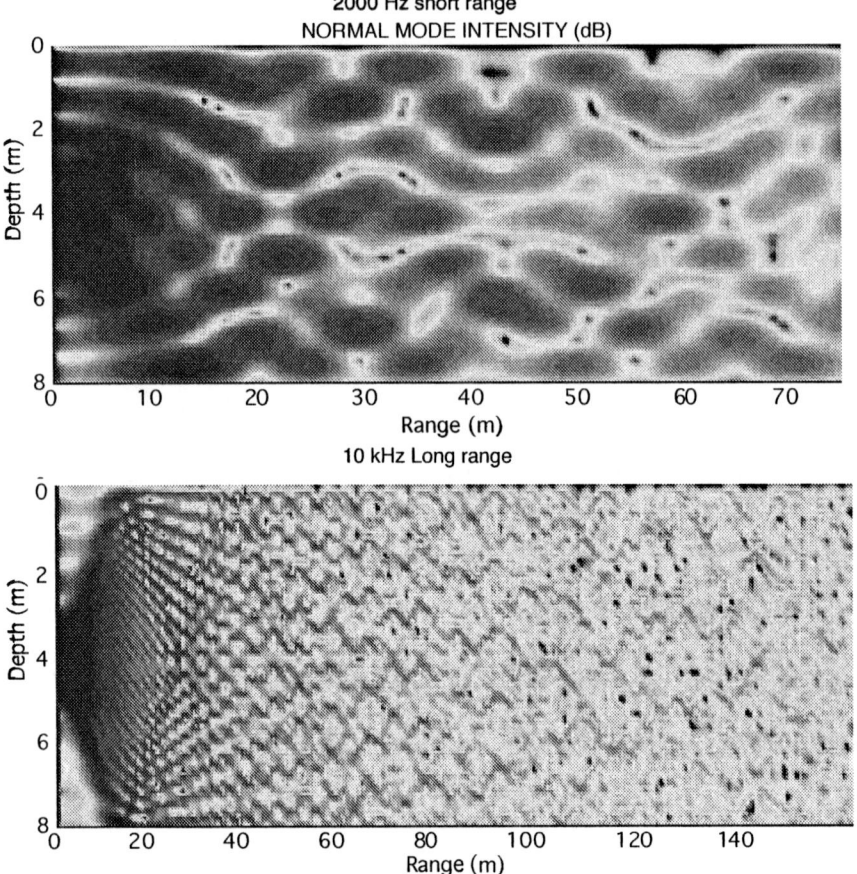

Figure 2. Normal mode intensity plots for the source receiver geometry's using measured sound speed profile averages.

BOTTOM PENETRATION RATIO

This paper focuses on fluctuations in the sediment after acoustic penetration into the bottom sediment. The penetration ratio is a ratio of spectral densities [2, 3] where P is the Fourier transform of the pressure time series from a hydrophone. This ratio represents the pressure produced by unit amplitude of the incident wave and is defined by:

$$P_r(f) \equiv \left| \frac{P_i(f)}{P_{ref}(f)} \right|^2 \qquad (1)$$

where f is frequency Hz, and i is an index to the hydrophone array (i.e., burial depth). This frequency-dependent penetration ratio is the ratio of the spectral density from a hydrophone buried in the sediment to the spectral density of a reference hydrophone.

Typically, the reference hydrophone is located in the water column at or slightly above the sea bottom. A negative dB value of the penetration ratio indicates an attenuation of energy (absorption and scattering) relative to the energy at the reference phone. Scattering and reflections may give rise to positive dB values. To minimize the effects from shadowing and reflections from the tower structure, the reference hydrophone for this work was chosen to be the shallowest buried phone (0.10 cm depth) rather than the phone in the water column. Therefore, this ratio represents the acoustic pressure produced by an incident acoustic wave originating inside the sediment, below the water-seabed interface, and is called the post-penetration ratio in Figure 3. Effects such as seafloor scattering, sea-surface movement, thermal microstructure fluctuations, and grazing angle fluctuations are also reduced since these effects occur in the propagation path prior to arriving at the reference phone location.

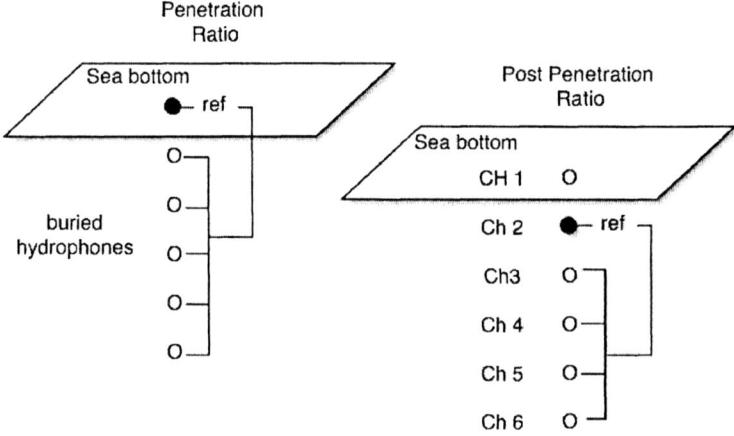

Figure 3. Definition of penetration ratios.

The ping-to-ping amplitude-squared fluctuations of this post-penetration ratio is pursued here to (1) determine the minimum ping-to-ping fluctuations one could expect in the penetration ratio for sandy bottoms, and (2) to determine correlation with environmental influences.

EXAMPLE DATA

A random example of the post-penetration ratio for a single ping is given in Figure 4. The two prominent features of this spectrum are consistent with the penetration ratio and were first explained by Maguer et al. [2, 3]. Their experiments and modeling showed that the linear falloff in the 1-6 kHz band was associated with the evanescent wave propagation. The scalloping in the 6-12 kHz band was associated with Bragg scattering, and sediment-volume scattering. Another noticeable feature of the post-penetration ratio spectrum is that the relative magnitude (dB) does not decay linearly with depth. This holds for both ranges and is in part attributed to changes in the physical properties and composition of the sediment in both depth and range. Bottom cores from previous experiments indicate that water content and porosity vary actively with depth but are more uniform with range. The quantity of gravel is depth dependent, while sand and clay constituents are more persistent in both depth and range.

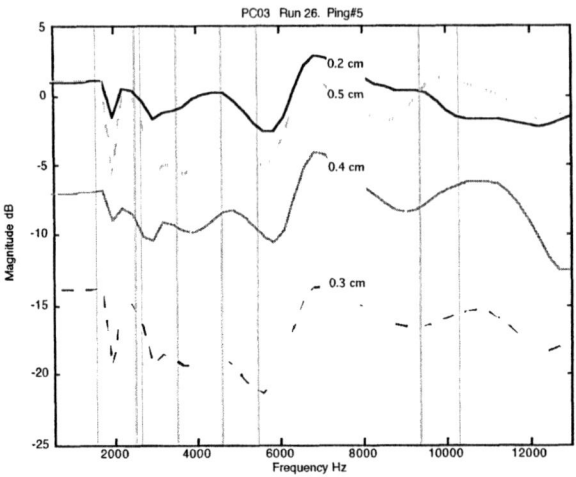

Figure 4. Example of post-penetration ratio for four burial depths for the longer-range geometry.

For this paper, it was necessary to limit the number of frequencies at which ping-to-ping fluctuations would be examined. Four bands were chosen, the beginning, middle, and end of the evanescent wave frequency band and the middle of the frequency band associated with scattering. The mean over a 500 Hz band about each center frequency is computed for each ping to obtain a time series of the post-penetration ratio at each depth. It is with this time series that we begin our analysis of the ping-to-ping relative magnitude fluctuations. An example is shown in Figure 5.

Two metrics chosen to characterize the temporal fluctuations shown in Figure 5, are the standard deviation, (STDEV) and the Gaussian fit error, (GFE). The standard deviation represents the estimate of the error in the mean of the relative magnitude and the Gaussian fit error gives a measure of the deviation from a Gaussian distribution. GFE is the least-square error between the cumulative distribution function of each ping-to-ping time series in Figure 5 and the cumulative distribution function of a Gaussian distribution with the same mean and standard deviation as the ping-to-ping time series. A graphical comparison is given in Figure 6. The symbols are the data probabilities and the line the Gaussian distribution probabilities. Smaller GFE values indicate a more Gaussian distribution.

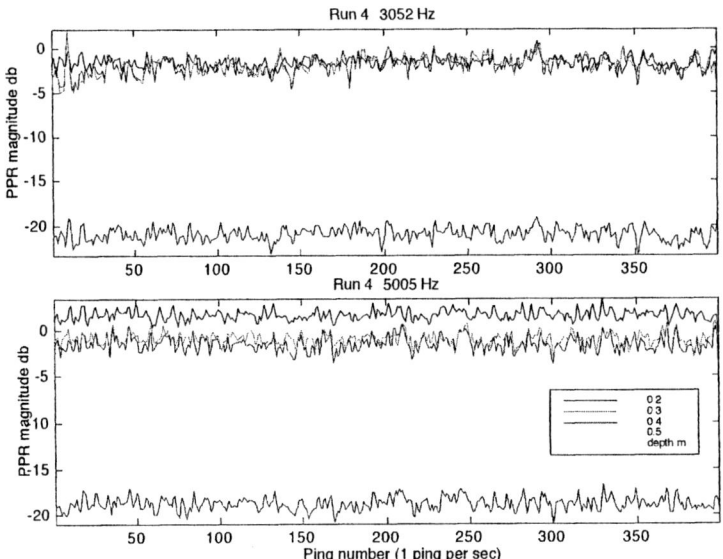

Figure 5. Ping-to-ping time series of post-penetration ratio (PPR) magnitude for two frequency bands for the longer range geometry.

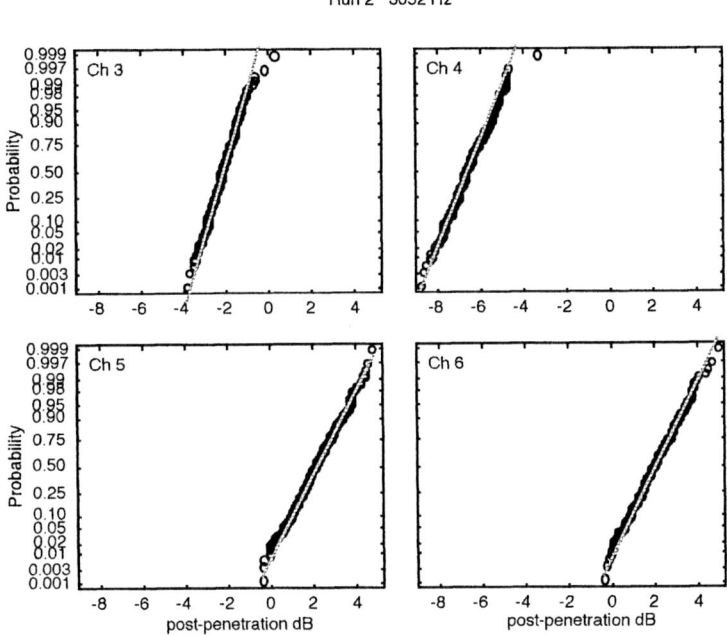

Figure 6. GFE example for single frequency for the shorter range geometry.

RESULTS

The STDEV and GFE of the post-penetration ratio for each buried phone were computed for multiple runs in both the short-range and long-range tower configurations. Each run was approximately 10 minutes in duration. This allowed a sufficient number of pings for comparison with oceanographic conditions to determine the environmental influences on fluctuations. The runs were were separated by at least 8 hours and span a total of nine days in June 2003.

Run Number Analysis

Figure 7 shows the STDEV of the post-penetration ratio ping-to-ping fluctuations plotted with run number. Results from all four frequency bands (Figure 7) for each run are included.

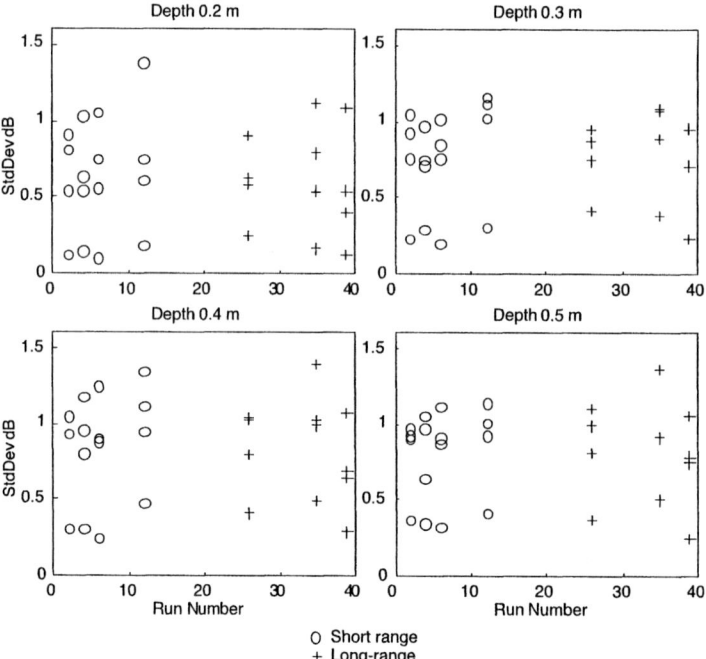

Figure 7. Post-penetration ratio ping-to-ping fluctuations STDEV results.

There is no single run that has the highest or lowest STDEV values for all four depths, but the highest STDEV occurs for runs 12 and 35 for the shorter and longer range respectively. In general, the STDEV is about equal for both ranges. For a threshold of 0.8 dB, the number of occurrences in which the STDEV is greater than the threshold increases with depth to 0.4 m and then levels off. The highest STDEV occurred for Run 35 that correlates with a higher anisotropy index in the TMMS temperatures.

Although not included here, a similar plot was created for GFE vs run number. Overall, the GFE is small. However, run 26 shows a consistently higher GFE for all depths. Otherwise, the GFE is about equal for both ranges. The equality of the results for the two ranges are not unexpected because (1) the direct-path grazing angles are less than two degrees for both ranges, and (2) the number of multipaths are large, and (3) there are multiple interactions with both boundaries for each multipath (Figure 5).

Frequency Band Analysis

Figure 8 shows the STDEV vs frequency band (indicated in Fig. 5) for the post-penetration ratio ping-to-ping fluctuations. Results from all runs at both ranges are included. At the smallest separation (0.2 m depth in Figure 8) from the reference hydrophone, the STDEV increases non linearly with frequency band. For the other three separations, the STDEV is almost constant above 3 kHz. Below 2 kHz, the STDEV is < 0.5 dB.

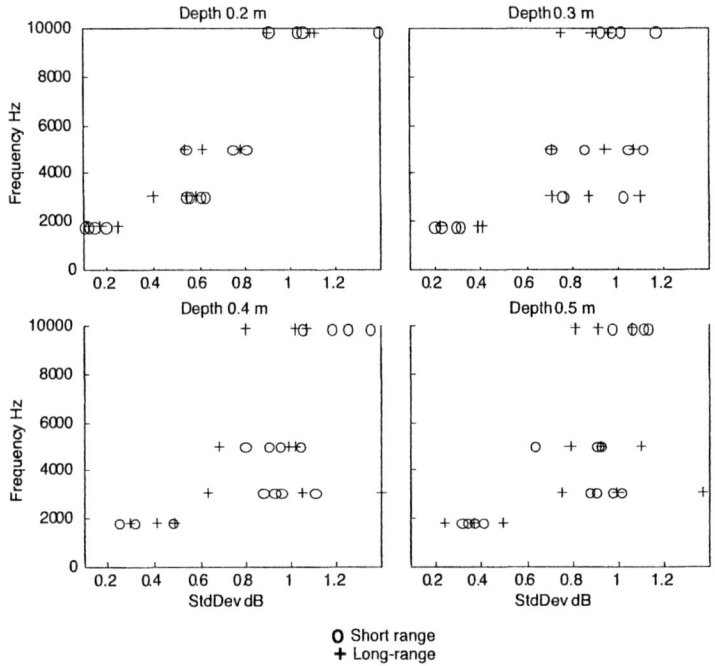

Figure 8. Post-penetration ratio ping-to-ping fluctuations STDEV results with frequency band.

The GFE was found to be small, averaging 0.14, and was insensitive to frequency band for all runs and all depths. This analysis has revealed runs 12, 26, and 35 to warrant further ping-to-ping investigation.

Fluctuation Spectrum Analysis

The Fourier transform of the ping-to-ping time series of post-penetration ratio magnitudes (an example was shown in Figure 5) gives rise to a fluctuation spectrum. Although these spectrums appear to have some structure, they are noisy. A fifth-order low-pass elliptical filter was applied to each spectrum and followed with an ensemble average over all the runs for each range separately. These results are shown in Figure 9 for each of the four frequency bands but is limited to hydrophone CH 3 (0.2 m depth) only, which is the hydrophone closest to the reference phone (see Figure 3).

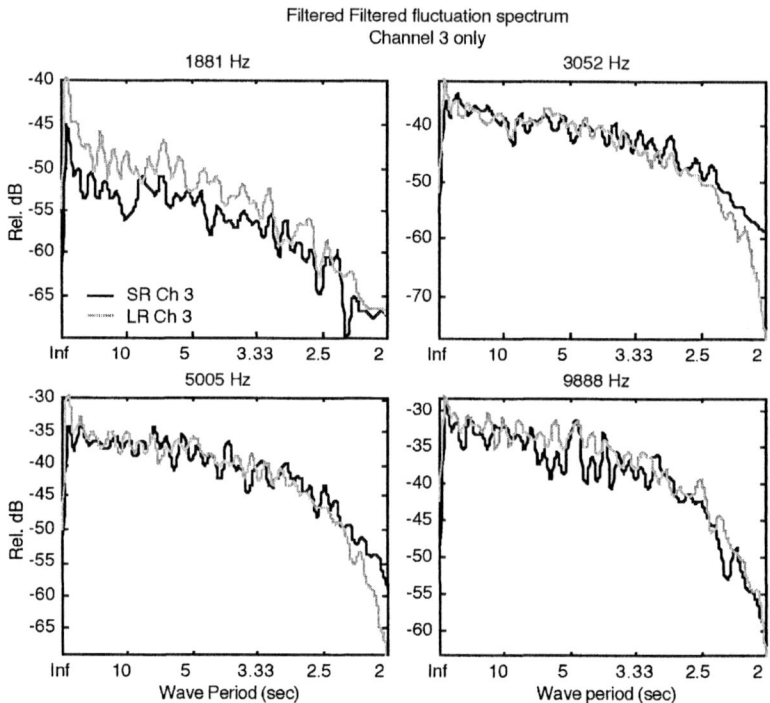

Figure 9. Channel 3 (0.2 m) post-penetration ratio ping-to-ping fluctuation spectrums after low-pass filtering and averaging over runs for each range.

The filtering and ensemble averaging (over runs) reveals some interesting features in the fluctuation spectrum. Only the 1881 Hz band appears to be a linear power-law spectrum. The other frequencies are curvilinear. What stands out in all four frequency bands and at both ranges, are peaks and a slightly raised relative spectral level over the wave periods from 3.5 - 7 seconds, the same range of measured periods for the sea-surface wave spectrum. Even with large number of multipaths, and a large number of surface wave interactions (as indicated in Figure 2), the effects of the moving surface do not average out. Surface effects are more noticeable in the short-range average and perhaps more well defined, although the definition may be filtering effect. Also, the multiple peaks in the 3.5 - 7 second portion of the post-penetration fluctuation spectrum are consistent with the observation that the measured sea-surface spectrum (computed from a 17 minute time frame) is composed of multiple instances of a sinusoids with different periods not a single periodic wavetrain with multiple periods.

SUMMARY

Among the short range runs, the STDEV of the post-penetration ratio was largest for Run 12, which recorded higher wind speeds, stronger currents, and larger sea-surface wave-heights. The GFE was constant for the short-range runs, and about 10-15% higher for the long-range runs. This correlates with water-column stability. The fluctuations were generally less Gaussian (higher GFE) for stable water conditions (which also coincides with the longer range). The highest STDEV for the long-range runs occurred for Run 35 that recorded the most measured anisotropy index in the TMMS temperature arrays and the smallest sea-surface wave height.

This data has great potential for analyzing environmental effects on ultra-low grazing angle multipath propagation. During the coming year, this data will be modeled using the OASES [4] seismo-acoustic model for comparison with post-penetration ratio measurements. Future data analysis will include cross-correlation as a function of sensor spacing and hydrophone depth for the six-element buried array. This analysis has revealed runs 12, 26, and 35 to warrant further ping-to-ping investigation. The low-level fluctuations in the post-penetration ratio are a result of sea-surface motion.

ACKNOWLEDGMENTS

This research was supported by the Office of Naval Research with technical management by the Naval Research Laboratory under program element 62435N.

REFERENCES

1. Steve Stanic, et.al, "Panama City 2003 broadband shallow-water acoustic coherence experiments," in *Proceedings of the High Frequency Ocean Acoustics Conference*, La Jolla, CA, March 2004.
2. Alain Maguer, Warren L.J. Fox, Henrick Schmidt, Eric Pouliquen, and Edward Bovio, "Mechanisms for subcritical penetration into a sandy bottom: Experimental and modeling results," J. Acoust. Soc. AM., 107 (3), 1215-1225, March 2000.
3. Alain Maguer, Edward Bovio, Warren L.J. Fox, Henrick Schmidt, "In situ estimation of sediment sound speed and critical angle," J. Acoust. Soc. AM., 108 (3), Pt. 1., 987-996, Sept. 2000.
4. Henrik Schmidt, "OASES Version 2.2 User Guide and Reference Manual," Massachusetts Institute of Technology. Cambridge, MA 1999.

A High-Speed, Multi-Channel Data Acquisition System

Dexter Malley, Bob Brown, Edgar Kennedy, Roger Meredith, and Steve Stanic

Naval Research Laboratory, Code 7184, Stennis Space Center, Ms. 39529

Abstract. The Naval Research Lab is currently conducting research programs in MCM detections and classifications using both low and high frequency acoustics. These include target detections, target imaging, proud and buried target detections and classifications using structural clues. To determine the limitations that a fluctuating environment places on these target detection methods, a data acquisition system was developed. The data acquisition system consists of multi-channel, high-speed A/D's with remote, variable gain control, and FPGA technology. Each A/D is synchronously sampled at a rate of 1 MHz and using time-division multiplexing techniques, is sent down an optical fiber at 1.3 Gbps. The sampled data is then separated back to its original channel and recovered back to an analog signal along with the original clock. Precision filters and high speed transient recorders utilizing fast CAMAC crate controllers are then employed to sample, simultaneously, all data channels with sample rates up to 3Msps. Acoustic and environmental real-time software were developed using National Instruments Labview to generate the CW source signals that went from 10 kHz to 200 kHz, monitor acquired data, and control sample and repetition rates.

INTRODUCTION

In June 2003, the Naval Research Laboratory conducted a series of vary-shallow-water, broadband coherence experiments. These measurements used large aperture vertical and horizontal low-frequency receiving arrays, a pair of multi-channel, high frequency receiving arrays, and a multi channel buried hydrophone array. Since it was necessary to record data simultaneously from many of these channels, a new digital multi-channel data acquisition system was designed and built. This system acquired data from 44 channels and multiplexed this data onto one optical fiber. This data was then sent to an instrumentation van located on shore. A complete description of the measurements, program objectives, and at-sea experimental configuration is given in [1]. This paper will describe the signal generation and the multi-channel data acquisition system, and control functions.

System Description

An overview of the complete system is shown in Figure 1. The system is divided

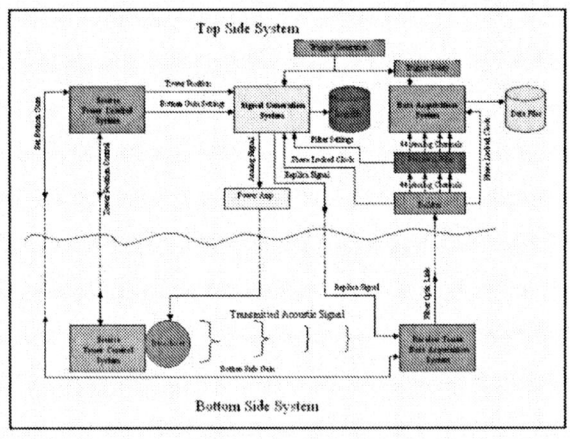

FIGURE 1. System Overview

into two distinct sections: acoustic signal generation / data acquisition, and tower control (position and amplifier gain). Each of these sections include both a topside and a bottomside component. The topside signal generation is illustrated in Figure 2.

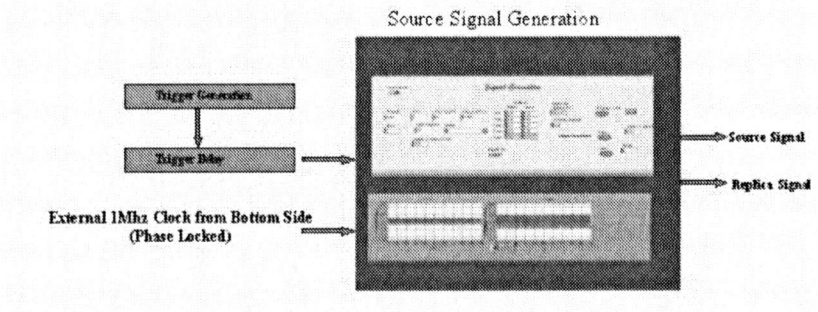

FIGURE 2. Topside Signal Generation

A 10 MHz master clock signal is generated in the bottomside receive electronics and transmitted to the topside signal generation system via a fiber optic link. This clock is divided down to generate a 1 Hz trigger and the 1 MHz sample clock. Acoustic signal generation is accomplished using a National Instruments PC-MIO 16 Multifunction I/O Board controlled by a PC with a 1 GHz processor. User Interface software is written in National Instruments LabView, and provides the capability of generating high-frequency CW signals with a user specified amplitude and pulse length. It was also used to generate a broadband, low-frequency, digital signal from a user supplied data file. During these measurements, two signals are generated: the source signal which is transmitted via the source transducer, and a replica signal which is transmitted through all of the receive electronics to quantify the system electronic noise and phase stability.

Bottomside Electronics

The bottomside system was designed to acquire 44 channels of real-time data. Each individual data channel was synchronously digitized at 1 MHz with addressable, programmable gain up to 58 dB. The channels are multiplexed down to 11 channels and then serialized into a single bit stream. This bit stream is then transmitted via optical fiber to a receive station located onshore. Once received, the bit stream is then deserialized back to 11 channels and demultiplexed back to 44 channels. The original sampling clock is also retrieved from the incoming bit stream. Finally, each of the 44 individual digital channels is passed to digital-to-analog (D/A) converters, which restores the analog signals to their original form. A simple system block is shown in Figure 3.

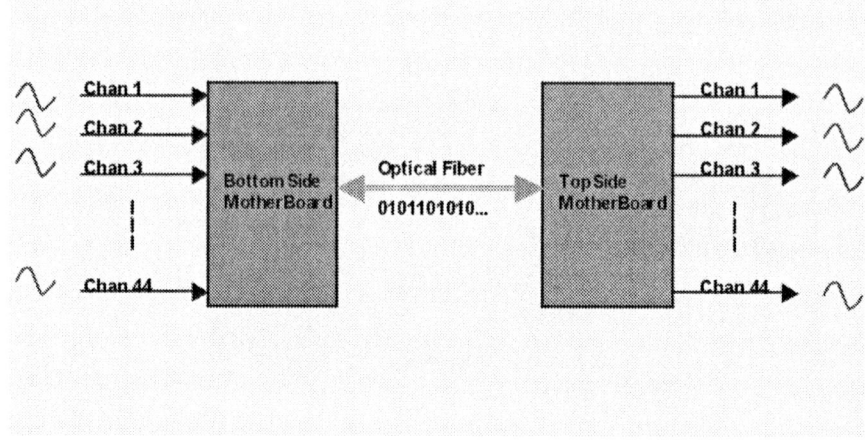

Figure 3. System Block Diagram

The electronics for the bottomside consists of 2 major parts, the A/D boards and the bottomside motherboard. A block diagram of the A/D board is seen in Fig. 4.

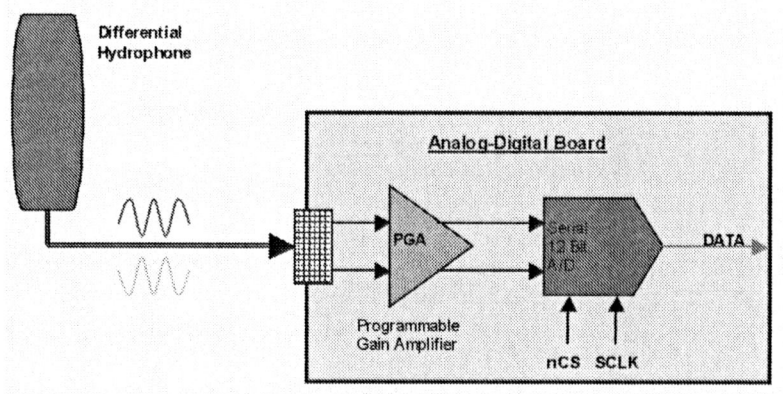

FIGURE 4. A/D Board

The output of the hydrophone goes to a Programmable Gain Amplifier (PGA) which has selectable gain from 0 to 58 dB in 6 dB increments. Each module has an 8-bit address and a 4-bit gain register. A PC computer topside writes down individual gain settings for each A/D board. The differential signal from the PGA is then passed to an anti-alias filter, and converted to a digital stream using a 12-bit A/D converter. The conversion and timing are controlled using chip select, nCS, commonly known as a framing bit, and a serial clock, SCLK.

The final part of the bottomside electronics is the motherboard. As shown in Figure 5, the bottomside motherboard is made up of 3 major components: a Field-Programmable Gate Array (FPGA), a serializer, and a fiber optic transmitter. The first component, the FPGA, was used mainly because they are reprogrammable, flexible, low cost, and low power. Because FPGA's are reprogrammable, it removes the non-recurring engineering cost from prototyping and testing new designs. Once a design is created, simulation routines can be done on a PC to debug any hardware or timing issues. Lastly, a FPGA allows the platform to be reconfigured for future designs.

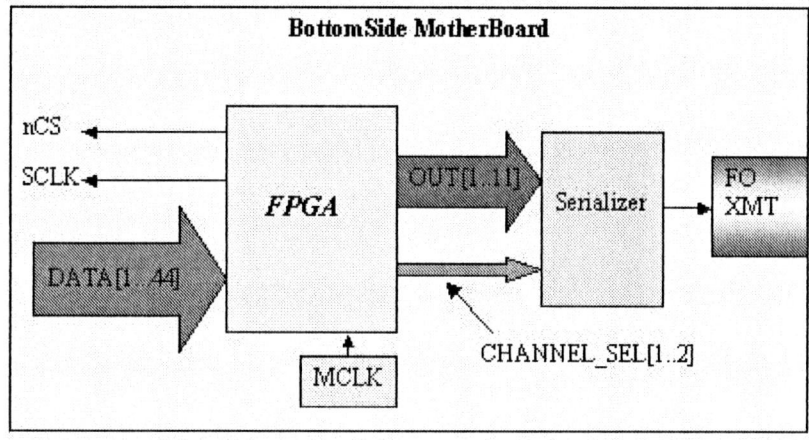

FIGURE 5. Bottomside Motherboard

The FPGA for this project provides three major functions which are clock dividing, multiplexing and counting. As stated above, the A/D modules need two timing and control signals, SCLK and nCS. These two signals are created in and fanout from the FPGA. In this case, SCLK is 18 MHz. This signal is generated by taking the free-running master clock, MCLK, which is 72 MHz and dividing it by four. To get the framing or sync signal, nCS, the SCLK is fed into a counter. For every 16th clock pulse of SCLK, nCS will go high for 1 clock pulse. This clock pulse initiates the synchronous sampling for all of the A/D's.

The digital data from the 44 A/D modules are routed to individual input pins on the FPGA. Inside the FPGA, the digital data from each A/D channel is input into a 4 channel, 11-bit wide multiplexer. These 44 data channels are separated into 4 groups of eleven as seen in Figure 6. Each group is sampled or switched into the output of the multiplexer at a rate of MCLK (72 MHz) or 4 times the frequency of SCLK.

The CHANNEL_SEL[1..2] is a binary 2-bit counter that repeatedly counts from 1 to 4 to sequentially switch the groups.

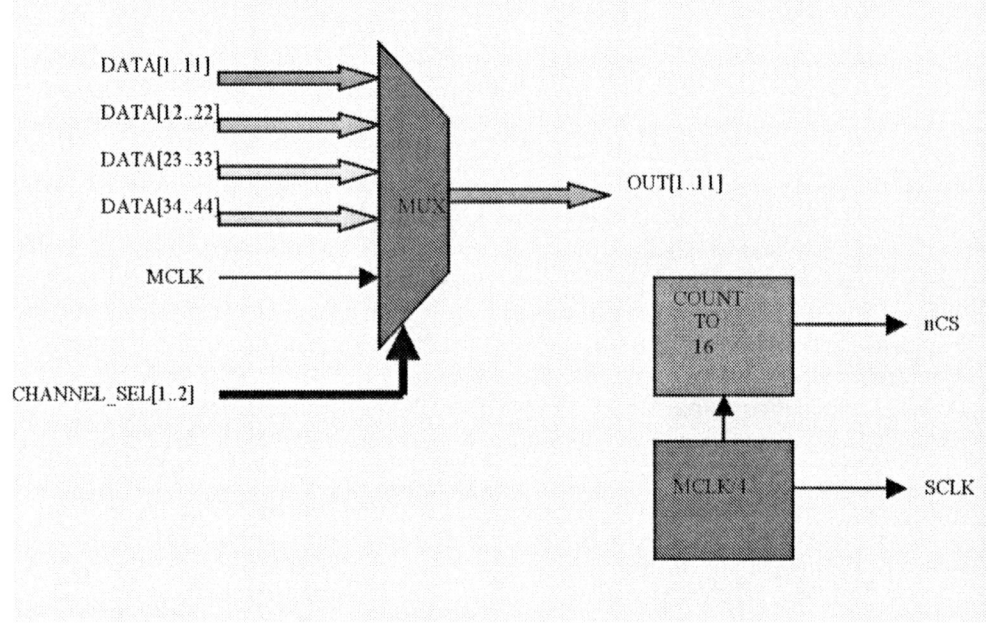

FIGURE 6. FPGA Multiplexer and Clocking

The next major component of the bottomside motherboard is the serializer. The serializer that was used is a 16-to-1, meaning it serializes 16 inputs into 1 output. For this application, only 13 inputs will be used; 11 from the output of the multiplexer and 2 from CHANNEL_SEL. The 3 remaining unused inputs are tied low. Using the serializer has several advantages. One major advantage is that the serializer uses 8B/10B encoding which guarantees at least one transition for every character which maximizes a synchronization sequence [2]. Next, the serializer's counterpart, the deserializer, generates the original master clock utilizing an internal phase-lock loop, PLL. Because of the high-speed signal, the serializer outputs a low-voltage, differential signal, LVDS. This helps minimize crosstalk and noise throughout the circuitry. The serializer transmits the data and clock bits (16+2 bits) at 18 times the MCLK frequency. This is calculated by the following:

MCLK * (16 inputs + 1 start bit + 1 stop bit) = 72 MHz * 18 = 1.296 Gbps. [3]

The final component is the fiber optic transmitter. The transmitter accepts the differential output of the serializer, converts it to light and transmits the data through a single-mode fiber to the receiving topside motherboard located onshore.

Topside Electronics

Like the bottomside, the topside motherboard is made up of similar components: an FPGA, deserializer, and a fiber optic receiver. The fiber optic receiver takes the optical data where is it converted again to a low-voltage, differential signal. It is passed to the deserializer, which separates it into the original 11 data signals and the 2 CHANNEL_SEL lines that were input into the serializer on the bottomside. Another important output the serializer generates is the regenerated MCLK signal, which is the phase-locked replica of the bottomside clock.

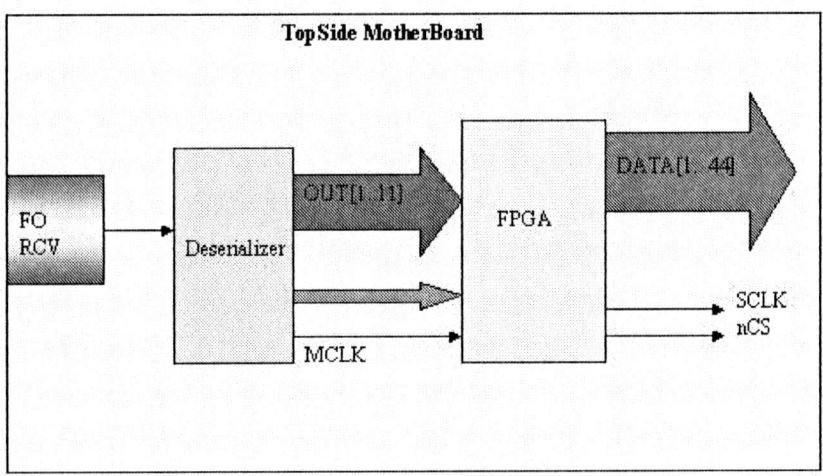

FIGURE 7. Topside Motherboard

The above outputs of the deserializer are passed to the FPGA. To retrieve the data, the process in the topside FPGA must be the reverse of the bottomside FPGA. To begin this process, four shift registers are created. Each shift register is 11 bits wide. The channel select lines that were passed from the deserializer are used to select the corresponding shift register which corresponds to the same channel from the bottomside multiplexer. For example, in Fig. 8, when CHANNEL 1 goes high, it allows the contents of the data bus to load only Shift Reg #1 which is clocked in by the synchronous MCLK. This data corresponds to the original first 11 A/D channels, DATA[1..11].

Now that the original 44 digital data channels have been reconstructed, the digital data channels can be recorded in their present form or converted back to their original analog form using D/A converters. The D/A converter accepts 3 inputs which are the serial data, the sync bit, nCS, and the serial clock, SCLK. Both the nCS and SCLK are created inside the FPGA by the phase-locked replica of MCLK. Lastly, MCLK is divided down to 10 MHz to provide signal synchronization for topside acquisition and control.

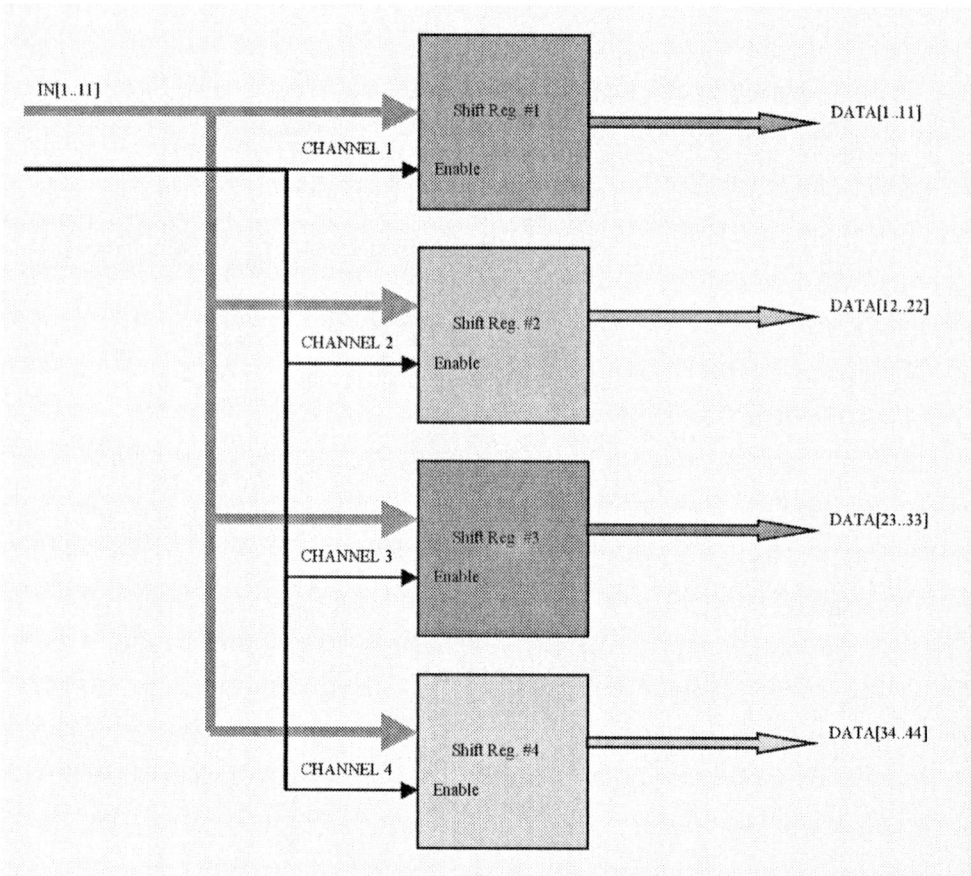

FIGURE 8. FPGA Demultiplexer

Topside Data Acquisition

The topside data acquisition system is shown in Figure 9. The signal received on each of the low and high-frequency acoustic arrays is bandpass filtered. For the high frequency data sets, the passband was 0.5 kHz – 200 kHz., and for the low-frequency data sets the passband was 0.5 kHz – 12 kHz. To insure that the system is phase locked to the topside data acquisition system, the sample clock and trigger are derived from the 10 MHz clock transmitted from the topside electronics which is phase-locked to the bottomside MCLK. Analog to digital conversion is accomplished using a CAMAC Crate configured with seven Joeger TR612/3 Transient recorders and a Fast CAMAC SCSI Crate controller. Each transient recorder is capable of simultaneous sampling six analog channels at a maximum sample rate of 3 MHz per channel. Each channel has 512 Mbytes of on board memory. Trigger and sample clocks are daisy chained across all seven recorders to insure that the conversion process is phase-locked. The digital data acquisition system is controlled using a PC with 3 GHz processor running Windows XP, and National Instruments LabView. The data was recorded on hard drive in real-time, and later archived to DVD's. Topside user interface software was designed to allow the operator to select which transient recorders are active and to monitor any of the active recorders. The capability to monitor the individual channels in real-time, allows the data acquisition trigger to be delayed until the received signal is completely contained within the digitized window.

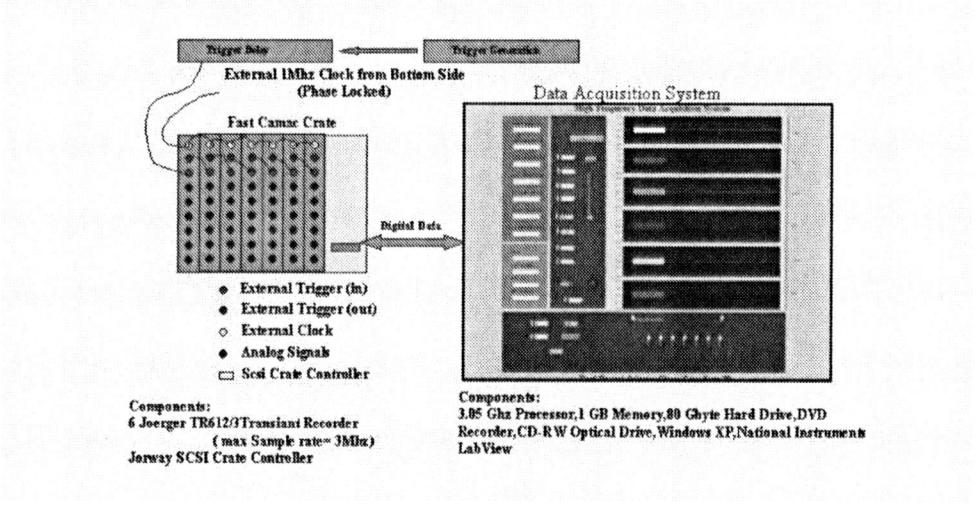

FIGURE 9. Analog Data Acquisition System

Summary

A multi-channel acquisition system was developed to support broadband coherence experiments that were performed in June 2003. Because of the technology used for the project, the system can be easily adapted with a minimal effort to support other future experiments and requirements.

ACKNOWLEDGMENTS

This work was supported by the Office of Naval Research, technical management by the Naval Research Laboratory under program element 62435N.

REFERENCES

1. Stanic, S.,et al., "Panama City 2003 broadband shallow-water acoustic coherence experiments," in *Proceedings of the High-Fregquency Ocean Acoustics Conference*, 2004.
2. F. Xiong, *Digital Modulation Techniques* . Norwood, MA: Artech House Inc., 2000.
3. "DS92LV16 16-Bit Bus LVDS Serializer/Deserializer," National Semiconductor Corporation, Santa Clara, Ca. 2001.

TARGET MODELING, SYSTEMS, AND APPLICATIONS

Navy Applications of High-Frequency Acoustics

Henry Cox

Lockeed Martin Orincon Defense, 4350 North Fairfax Drive, Suite 470, Arlington, VA 22202

Abstract. Although the emphasis in underwater acoustics for the last few decades has been in low-frequency acoustics, motivated by long range detection of submarines, there has been a continuing use of high-frequency acoustics in traditional specialized applications such as bottom mapping, mine hunting, torpedo homing and under ice navigation. The attractive characteristics of high-frequency sonar, high spatial resolution, wide bandwidth, small size and relatively low cost must be balanced against the severe range limitation imposed by attenuation that increases approximately as frequency-squared. Many commercial applications of acoustics are ideally served by high-frequency active systems. The small size and low cost, coupled with the revolution in small powerful signal processing hardware has led to the consideration of more sophisticated systems. Driven by commercial applications, there are currently available several commercial-off-the-shelf products including acoustic modems for underwater communication, multi-beam fathometers, side scan sonars for bottom mapping, and even synthetic aperture side scan sonar. Much of the work in high frequency sonar today continues to be focused on specialized applications in which the application is emphasized over the underlying acoustics. Today's vision for the Navy of the future involves Autonomous Undersea Vehicles (AUVs) and off-board ASW sensors. High-frequency acoustics will play a central role in the fulfillment of this vision as a means of communication and as a sensor. The acoustic communication problems for moving AUVs and deep sensors are discussed. Explicit relationships are derived between the communication theoretic description of channel parameters in terms of time and Doppler spreads and ocean acoustic parameters, group velocities, phase velocities and horizontal wavenumbers. Finally the application of synthetic aperture sonar to the mine hunting problems is described.

INTRODUCTION

Most World War II sonars operated in the 15 to 30 kHz frequency range and would be considered "high frequency" today. Long range detection of submarines has been the primary concern that has motivated Navy sponsored research and development for the last several decades. This has led to an emphasis on low-frequency underwater acoustics. During this period there has been a continuing interest in high-frequency acoustics for special applications. The attenuation of sound in sea water is approximately 1dB/km at a frequency of 10 kHz and increases as frequency-squared so that attenuation is about 4 dB/km at 20 kHz. This severe attenuation limits the domain of applicability of high-frequency underwater acoustics to relatively short range. There are, however, many advantages to high frequencies for short range applications including high spatial resolution, wide bandwidth, small size and low cost. These characteristics are attractive for many commercial applications of underwater acoustics. The availability of low cost powerful computing has led to commercial systems of increased sophistication. The result is that the Navy is not the only major user of high-frequency underwater systems. Much of the work involving high-frequency underwater systems is focused on the particular application rather than basic research in the underlying acoustics. Navy applications include fathometers, mine hunting and avoidance, under ice navigation, torpedo homing, bottom mapping and communication. Fathometers, bottom-mapping sonars, and communication systems have a sufficient commercial market that they are advertised on the world wide web where specifications and pictures are presented.

Today, there is a new vision for the Navy that involves use of off-board sensors for Anti-Submarine Warfare (ASW) and Autonomous Undersea Vehicles (AUVs) for a number of missions. In these emerging systems, high-frequency acoustics will play a significant role for sensors and communications. Again the application interest predominates over the underlying acoustics.

ACOUSTIC COMMUNICATIONS

High-frequency acoustics is a key enabling technology for both AUV communication and AUV sensors. Today, unmanned Remotely Operated Vehicles (ROVs) are commonly used for inspection of oil rigs and similar commercial applications with very severe range restrictions. These vehicles are tethered to a mother platform with communications of nearly unlimited bandwidth via an optical fiber in the tether. With a submerged AUV there is no tether and no practical alternative to acoustic communications, except possibly at extremely short ranges. AUVs are envisioned as having multiple missions, involving different ranges and data rate requirements. Requirements for range and data rate drive the selection of frequency, bandwidth and power for an acoustic communications system. Desired data rates and ranges may or may not be compatible with a practical implementation. Thus, there will be a need to compromise between what is (really) required and what is feasible. In addition, in military applications there will be a premium placed on communication reliability, with an associated reduction in data rate to increase reliability. It is reasonable to consider multiple communication modes using different powers and different frequencies to achieve different data rates at different ranges and perhaps different channel conditions.

A complicated communication channel with multi-path and temporal variability is frequently characterized by two parameters: the time–spread τ and the Doppler-spread B or their reciprocals: the fading bandwidth W and the coherence time T [1]. Uncompensated time-spread limits the coherent processing bandwidth W to a value $1/\tau$. Similarly temporal variability limits the coherent processing time T to 1/B. The amount of coherent processing gain that the channel can support is TW or $1/B\tau$. For underwater communication applications with moving sources and receivers, such as when the AUV must be free to move with respect to the mother platform during communications, the temporal instability may depend more on source-receiver motion that on the temporal variability of the ocean. The motion converts spatial variability to temporal variability. For the situation of motion induced temporal variability; we can express the communication characteristics, B and τ in terms of ocean acoustic parameters.

Consider the situation of two multi-paths of comparable strength so that they interfere. The time spread between the two components is simply the difference in travel time from source to receiver. This can be expressed in terms of range R and the difference in the reciprocals of the average group velocities $\delta(1/u)$ as follows:

$$\tau = R \, \delta(1/u) \tag{1}$$

An isospeed approximation is sometimes useful in shallow water when the range is much greater than the water depth. For horizontal propagation in an isospeed channel, the group velocity of a ray can be expressed in terms of the sound speed c and the grazing angle or ray angle θ.

$$u = c \cos\theta \tag{2}$$

Then

$$\tau = R \, \delta(\cos\theta)/(c \cos\theta_1 \cos\theta_2) \approx R \, [\theta_2^2 - \theta_1^2]/2c \tag{3}$$

The small angle approximation is used to express results directly in terms of ray angles. For example, two rays with ray angles of 0 and 10 degrees (0.174 radians) produce a time-spread of about 1.5% of the propagation delay or 10ms/km. If the angles were 0 and 20 degrees, the time spread would increase by a factor of four to 6% or 40ms/km.

The radial coherence length or spatial scale in range ΔR of the resulting interference pattern can be expressed in terms of the difference in horizontal wavenumbers Δk or equivalently the difference in the reciprocals of the phase velocities $\delta(1/c_p)$ as follows:

$$\Delta R = 2\pi/\Delta k = 1/[f \, \delta(1/c_p)] = \lambda / \delta(\cos\theta) \approx 2\lambda / [\theta_2^2 - \theta_1^2] \tag{4}$$

At longer ranges in shallow water, higher modes or equivalently steeper angles may be stripped away due to multiple lossy reflections off the ocean bottom. Then, the steepest propagating angle and $\delta(\cos\theta)$ decrease slowly with range. This change with range is much slower than the interference induced change so that (4) remains valid as long as the correct interfering components are used. The coherence time is obtained by dividing this distance by the range rate v.

$$T = 1/[f \, \delta(1/c_p) \, v] = \lambda / [\delta(\cos\theta) \, v] \tag{5}$$

For the example of 0 and 10 degree rays, the coherence length is equal to 67 wavelengths. If the angles were 0 and 20 degrees this would reduce to about 17 wavelengths. At a frequency of 10 kHz, these coherence lengths become 10 m and 2.5 m respectively. At a 3 knot range rate, the coherence times become 6.7 s and 1.7 s respectively.

We can use (1) and (5) to express the product of Doppler-spread and time-spread in terms of ocean acoustic parameters.

$$B\tau = R\,f\,v\,\delta(1/c_p)\,\delta(1/u) \approx R\,f\,v\,[\theta_2^2 - \theta_1^2]^2/4\,c^2 \tag{6}$$

Small values of $B\tau$ are good. For the previous examples, at a range of 10km, the values of $B\tau$ are 0.015 and 0.24. Notice that doubling the steepest angle from 10 to 20 degrees caused a factor of 16 increase in the $B\tau$ product. The $B\tau$ product is directly proportional to range, range-rate and frequency and for the isospeed case approximately proportional to the square of $\delta(\cos\theta)$ or the fourth power of the steepest angle. These dependencies plus the dependence of attenuation on frequency-squared frame the problem of system design for AUV acoustic communications.

If the channel impulse response were known, it could be taken into account in the signal design and the processing. Even if it were known, it would change after one coherence time interval. A number of communication techniques use probe signals and adaptive equalization algorithms that involve measuring and compensating for the channel properties in an attempt to overcome these limitations. These algorithms are an active area of research for the application to the acoustic channel. For the case in which the temporal variability of the channel is due to known radial motion, it may be possible to take the properties of the acoustic propagation into account in order to extend the domain of applicability of a channel measurement. Let $h(\tau; R)$ be the measured impulse response at range R. What is it at range $R + \delta R$? A first order approximation is

$$h(t; R + \delta R) = h(t[1 - \beta\,\delta R/R]; R) \tag{7}$$

The parameter β is the wave guide invariant [2, 3]. It is approximately unity for shallow water. The first approximation to the effect of a range increase is a stretch in the impulse response. This can be used to extend the useful life of a measurement when range and range rate are known [3, 4]. The inclusion of the physics of propagation into the techniques of adaptive channel equalization may give rise to improved practical approaches to acoustic communications between moving platforms.

For nearly stationary sources, motions that are on the order of a wavelength can cause loss of coherence. At 10 kHz the wavelength is only six inches so that small motions are significant. Motion measurement and compensation can be used to improve performance. One approach is to use the phase of a known reference signal for compensation.

While the investigation of sophisticated processing techniques to maximize the efficiency of a communication system is a popular area of research, the designer of a practical system may opt for less sophistication and lower data rate in order to achieve robustness. Thus, an incoherent processing approach such as Multiple Frequency Shift Keying (MFSK) with redundancy coding may be chosen to provide reliable low data-rate communications in difficult environments. In AUV applications there may well be a need for multiple modes to accommodate the different range, data-rate and maneuvering requirements. For example, there might be a combination of a high-frequency, high data-rate, short range, coherent communication system and a long range, lower frequency, low data rate, incoherent MFSK system.

While the requirements of some Navy systems may be special and appropriate for custom design, for other applications it would be desirable to have a "handbook", catalog, or MATLAB program that would allow tradeoffs in power, frequency, bandwidth, range and data-rate. The processing requirement for an acoustic communication system can be handled on a modern PC so that system hardware is not much more than a laptop computer, a transducer and a power amplifier. The cost can be kept reasonable for routine applications. The power amplifier and the computer can be common across a variety of applications so that a family of systems with different transducers at different frequencies can provide capability at different ranges and data-rates. A step in this direction is available in the commercial sector as can be seen by visiting the internet where a number of commercial acoustic modems are advertised.

The application of acoustic communications to off-board sensors is also of current Navy interest. For sensors the major tradeoff is between more processing in the sensor to reduce the communication data-rate and simpler in-sensor processing with more data being sent in the sensor reports. The sonobuoy has been a widely used off-board sensor for several decades. It uses radio communications. For some other sensors radio communications is also an attractive option.

An application in which acoustic communication is the only practical alternative is a Reliable Acoustic Path (RAP) acoustic receiver for either passive or active ASW. The concept is simple and serves as a specific example of a well posed design problem. The solution will be left to the reader. The acoustic receiver is deployed from an aircraft or ship and then sinks to the bottom in deep water with a depth of several thousand meters. The deep receiver location provides direct path acoustic coverage to a range of several water depths. The sensor is battery powered so that the power budget for communication is a concern. It reports via an acoustic link to a ship or sonobuoy that relays the data by radio the ultimate user. The acoustic link must function reliably over a path-length of up to 18 km. The acoustic propagation is simple with the only multi-path arising from a bottom reflection at the reporting sensor and a surface reflection at the receiver. Modest directivity can be used to minimize the effects of these multi-paths. The grazing angle of the propagation is steep so that refraction is not significant. The ambient noise falls off with frequency at about 6 dB per octave. Some array gain against ambient noise can be provided in the receiving sonobuoy to reduce the power requirement. The sonobuoy relay is drifting and experiencing motion due to surface wave action. The acoustic propagation loss at of 18 km is about 85 dB plus frequency dependent absorption. The problem is to provide the trade-offs concerning frequency, power and data-rate that can feed into other total system design issues such a processing approach, battery and system life.

SYNTHETIC APERTURE SONAR (SAS)

While the use of Synthetic Aperture in Radar (SAR) is very mature, the sonar counterpart is in its early stage of development and application. It is an exiting application area for high-frequency acoustics. The basic idea is that coherently adding data from successive pings with a moving real side-looking aperture looking at a fixed scene is equivalent to observing the scene with a larger aperture.

The spatial sampling requirement is that the motion between pings should be less than one-have the length of the real aperture. The maximum unambiguous range is one-half the distance sound travels between pings. These two conditions restrict the search rate. One can either ping less frequently, move more slowly and work to longer range, or move more rapidly, ping more often and work to shorter range. The result is that the search rate is proportional to the real aperture-length not the speed of advance. The shape of the searched area depends on speed but the search rate does not. The speed of sound is relatively slow compared to the speed of light so that vehicles speeds are slow and it takes minutes to build a synthetic sonar image. An issue has been whether or not the sonar environment would be sufficiently stationary over the time it takes to build an SAS image. The answer seems to be yes. A set of practical difficulties are associated with motion compensation for deviations from constant linear motion by the array. Algorithms for motion compensation, similar to the ones used in radar have now been successfully applied in sonar. The signal processing for SAS has been made practical by the advent of small, cheap, powerful computers. Research systems have been built in several countries and in the US under sponsorship of DARPA and ONR. They have demonstrated impressive improvement in range and resolution over conventional side looking sonar.

The Navy application of most current interest is mine hunting. Increased swath width increases search-rate but more importantly the large increase in resolution provides finer shape information that greatly simplifies the problem of distinguishing mines from other objects on the bottom. Because in mine countermeasures no false alarm can go uninvestigated, a reduction in false alarms greatly reduces the time it takes to clear an area. An AUV with a SAS sensor is a particularly attractive concept for mine hunting in which high-frequency acoustics provides the fundamental enabling technology for both sensing and communications.

CONCLUSION

High-frequency acoustics has many applications to Navy problems. Two of the most technologically advanced are acoustic communications and SAS. These come together in the vision of the future Navy that involves AUVs and off-board sensors. The current interest seems to be more in the application than in the underlying acoustics and the research has a heavy signal processing flavor. The discussion of the communication problem presented for the first time explicit relationships between the signal processing quantities parameters of time and Doppler spread and the underlying ocean acoustic and physical quantities such as horizontal wavenumbers, group velocities, and phase velocities, range and range rate.

REFERENCES

1. A.B. Baggeroer, "Acoustic Telemetry – An Overview", *IEEE J. Oc. Acoustics*, 9 (4), 1984
2. G.L. D'Spain and W.A. Kuperman, "Application of waveguide invariants to analysis of spectrograms from shallow water environments that vary in range and azimuth", *JASA*, 106 (5), 1999
3. H. Cox and K. Heaney, "Interference Patterns in Range Dependent Environments", 144th Meeting of the Acoustical Society of America, December 2002. *JASA* 112(5), 2002
4. H. Cox, H. Lai and M. Hirano, "Reciprocity Based Channel Compensation for Wideband Communication in a Multipath Environment", *Proceedings of the 27th Asilomar Conference on Signals, Systems and Computers*, Pacific Grove California Nov. 1-3, 1993.

Virtual Source Approach to Scattering from Partially Buried Elastic Targets

Henrik Schmidt

Department of Ocean Engineering, Massachusetts Institute of Technology, Cambridge, MA 02139

Abstract. A hybrid modeling framework for scattering from general 3D elastic targets in a stratified ocean waveguide is presented, incorporating multiple scattering between the target and the stratification, and allowing for targets to be completely or partially buried in a stratified seabed. The approach uses a generalized wavefield superposition, or virtual source approach, together with a Fourier-Bessel spectral representation of the Green's function in a stratified ocean, to model scattering from targets described solely by their arbitrary surface geometry and dynamic stiffness. The hybrid approach is here implemented within the OASES-3D modeling framework, with the target surface stiffness determined either analytically, or numerically using Finite Elements.

INTRODUCTION

A modeling framework for scattering from proud or completely buried targets has been developed earlier, based on a single scatter approximation[1]. It uses OASES to compute the incident field at the target position, anywhere in a stratified fluid-elastic waveguide. This field is then convolved with the free-field scattering function for elastic targets such as spherical shells, effectively replacing the target by a virtual, multipole source at the target position, the 3D radiation from which is directly computed using OASES-3D [4]. It has been used extensively in the analysis of experimental data collected during the GOATS'98, 2000, and 2002 experiments carried out jointly by MIT and SACLANTCEN[2, 3]. However, this single-scattering approach ignores all multiple interactions between the target and the seabed, and is therefore incapable of treating partially buried targets.

To more realistically represent the shallow water mine-countermeasures problem, the OASES-3D target modeling framework has been combined with a high-fidelity Finite Element modeling framework, FESTA [5, 6], developed at SACLANTCEN, to form a new hybrid modeling capability for completely or partially buried targets, incorporating seabed interference. The coupling is here achieved by representing the scattered field as a superposition of fields produced by a distribution of virtual sources, of unknown strengths and phases, within the target surface, and including the seabed interaction in the Green's function.

The virtual source field is superimposed with the incident field, and the virtual source strengths are determined from the known boundary conditions on the surface of the target. The boundary conditions for any elastic target may be expressed in terms of the dynamic stiffness matrix, expressing the unique relation between the surface pressure and the normal displacements. The stiffness matrices, which are independent of the surrounding medium, may be computed by any independent, suitable method, e.g. using finite elements for complex targets, or Green's theorem for homogeneous targets.

The scattered field in the waveguide is then computed using the 'exact' multipole expansion of point source distributions inherent in OASES-3D. As opposed to other coupling approaches such as the 'scattering chamber' approach, this virtual source approach does not require the treatment of the outer medium by the target model. Thus, once the dynamic stiffness matrix for the target is determined, e.g. in-vacuo or submerged in a homogeneous medium, it can be used for arbitrary orientation and burial of the target. This approach is, therefore, convenient for investigating sensitivity of the scattering to seabed properties, burial dept, insonification geometry, etc.

The present virtual source approach may be considered a full 3D generalization of the so-called internal source density method which has been applied in the past to model the scattering from targets of simple geometry or boundary conditions, For example, it has been used by Stepanishen [7] to model 3D scattering from objects of revolution with ideal Dirichlet or Neumann boundary conditions. However, in addition to the homogeneous boundary conditions and the axisymmetric target geometry, it was assumed that the outer medium be an infinite, homogeneous fluid, as has been the case in most other applications as well. As an exception, Kessel used a similar internal multipole expansion method in combination with a modal Green's function to model the scattering from objects in horizontally stratified waveguides, again assuming ideal, homogeneous boundary conditions [8]. Also, that approach does not allow the target to penetrate the waveguide interfaces, and does not incorporate multiple scattering between the target and the adjacent boundaries.

In contrast to such earlier work, the present approach applies to general elastic objects with full 3D geometry, with all required being a frequency-dependent stiffness matrix, uniquely associated with the internal structure and composition. Also, the present approach allows the target to be penetrating any interface in a horizontally stratified ocean environment, thus providing a versatile numerical method for analysis of scattering from partially and completely buried targets, incorporating multiple scattering effects, within the targets as well as between the target and the environmental stratification.

The dynamic stiffness matrix for the target may be modeled using any applicable approach. Thus, for a homogeneous fluid object it may be determined using a 'reverse' virtual source approach, while for a spherical shell it may be computed using an exact spherical harmonics representation, or using a more general numerical method such as Finite Elements.

THEORY

Virtual Source Approach

The virtual source approach is fundamentally a wavefield superposition approach, replacing the target by a distribution of acoustic sources placed in the background medium, inside the surface of the target, and of unknown magnitude and phase. These virtual source strengths are then found from the condition that the superposition of their

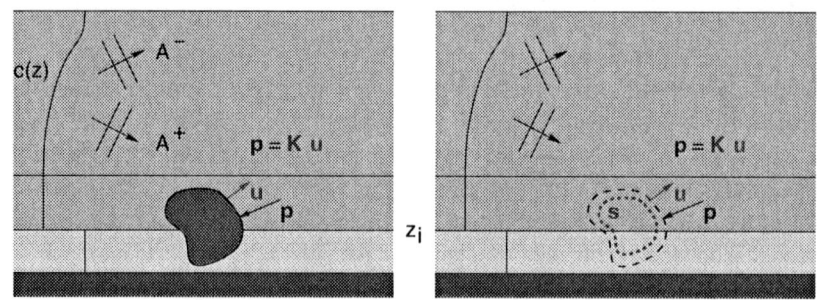

FIGURE 1. Virtual source approach to scattering from partially buried targets in stratified ocean waveguides. The target is replaced by an internal, virtual source distribution generating a field in the background environment which superimposed with incident field satisfies the boundary condition $\mathbf{p} = \mathbf{Ku}$, representing the target's dynamic stiffness properties.

generated field with the incident field on the surface of the volume occupied by the target must satisfy the boundary conditions associated with the true target.

This simple superposition principle is illustrated schematically in Fig. 1. The plot to the left shows an arbitrarily shaped object in a stratified ocean, here partially buried in the seabed at depth z_i. The stratification can include fluid as well as elastic layers, but it is here for simplicity assumed that the layers containing the target are isovelocity fluid media. In the plot to the right, the target is removed and replaced by a continuously stratified medium with a discrete distribution of N simple point sources, the unknown, complex strengths of which are represented by the vector $\mathbf{s}$. This source distribution is assumed to generate a field which is identical to the scattering produced by the target.

If the surface of the target is discretized in N nodes, the total pressure $\mathbf{p}$ and normal displacement $\mathbf{u}$ at the nodes are decomposed into the known incident field contribution $\mathbf{p}_i, \mathbf{u}_i$, and the scattered field, $\mathbf{p}_s, \mathbf{u}_s$,

$$\mathbf{p} = \mathbf{p}_i + \mathbf{p}_s \tag{1}$$
$$\mathbf{u} = \mathbf{u}_i + \mathbf{u}_s, \tag{2}$$

The scattered field is generated by the virtual source distribution $\mathbf{s}$,

$$\mathbf{p}_s = \mathbf{Ps} \tag{3}$$
$$\mathbf{u}_s = \mathbf{Us} \tag{4}$$

with $\mathbf{P}$ and $\mathbf{U}$ being $N \times N$ matrices containing the the pressure- and normal displacement Green's functions, respectively, between the N virtual sources and the N surface nodes.

The total field on the virtual target surface must now satisfy the boundary conditions associated with the real target. Thus, the field inside the true target must satisfy Green's theorem, providing a unique relation between the pressure and normal displacements on the surface. In a discrete representation with N surface nodes, this relation can be expressed in terms of a frequency-dependent, dynamic stiffness matrix $\mathbf{K}$,

$$\mathbf{p} = \mathbf{Ku} \tag{5}$$

Combining Eqs. (1)-(5) then leads to the following matrix representation for the virtual source strengths.

$$\mathbf{s} = [\mathbf{P} - \mathbf{KU}]^{-1}[\mathbf{Ku_i} - \mathbf{p_i}] \qquad (6)$$

The scattered field now follows anywhere in the external medium by superposition, using the Green's function for the continuous medium, in this case the stratified ocean waveguide.

Fourier-Bessel Green's Function

The Green's function for a stratified ocean needed for the scattered waveguide field may be computed using any of the established approaches, wavenumber integration, normal modes, or the parabolic equation [9]. However, the Green's functions in eq. (6) are to be evaluated in the near field, ignored by the standard approaches, which also assume the source to be at the origin. However, the Fourier-Bessel wavenumber integration formulation [4] for stratified waveguides overcomes both of these complications. Thus, the field produced by a horizontal distribution of sources can be expressed in an azimuthal Fourier series of the displacement potential $\phi(r, \theta, z)$,

$$\phi(r,\theta,z) = \phi_S + \phi_H = \sum_{m=0}^{\infty} [\phi_S^m(r,z) + \phi_H^m(r,z)] \left\{ \begin{array}{c} \cos m\theta \\ \sin m\theta \end{array} \right\}, \qquad (7)$$

where $\phi_S^m(r,z)$ and $\phi_H^m(r,z)$ are the Fourier coefficients for the direct source contribution and the field produced by the boundary interactions, respectively. Both components are represented in terms of horizontal wavenumber integrals,

$$\phi_S^m(r,\theta,z) = \frac{\varepsilon_m}{4\pi} \int_0^\infty \left[\sum_{j=1}^N S_j \left\{ \begin{array}{c} \cos m\theta_j \\ \sin m\theta_j \end{array} \right\} J_m(k_r r_j) \frac{\exp jk_r|z-z_j|}{jk_z} \right] k_r J_m(k_r r) dk_r \qquad (8)$$

$$\phi_H^m(r,\theta,z) = \int_0^\infty \left[A_m^+(k_r) e^{jk_z z} + A_m^-(k_r) e^{-jk_z z} \right] k_r J_m(k_r r) dk_r \qquad (9)$$

where k_r, k_z are the horizontal and vertical wavenumbers, S_j is the complex source strength of source j at (r_j, θ_j, z_j), and $A_m^+(k_r)$ and $A_m^-(k_r)$ are the complex azimuthal Fourier coefficients of the up-and downgoing wavefield amplitudes produced by the multiple boundary interactions. They are found by matching the boundary conditions at all horizontal interfaces. ε_m is a factor which is 1 for $m = 0$, and 2 otherwise.

In the present implementation, we apply the full Direct Global Matrix [9] implementation of OASES-3D for evaluating eq. (8)- (9). However, for the Green's functions needed for eq. (6) we will only consider interactions with the target interface z_i, allowing the amplitudes A_m^+ and A_m^- in eq. (9) to be expressed explicitly in terms of the plane wave reflection and transmission coefficients associated with this interface. This approximation therefore ignores multiple interactions with all other interfaces in the stratification. This is done for efficiency, though it is justified physically for most practical applications, but it is no fundamental limitation, since the full Green's function formulation in eq. (8)-(9) may be used also locally.

Numerical Issues

The efficiency of the virtual source approach hinges on a number of numerical issues, primarily associated with the details of the distribution of surface nodes and virtual sources, but also the implementation of the spectral integral representations of the waveguide Green's function.

Virtual Source Distribution. Work is ongoing in terms of developing a systematic, adaptive approach to the distribution of the virtual sources inside the targets. However, it has been found empirically that a consistent convergence is achieved by distributing the surface nodes with a separation which is proportional to the local radii of curvature, and by placing a virtual source along the inward normal at each node, at a depth of approximately 0.6 times the node separation. This seems to provide the optimal compromise between diagonal dominance of the matrix to be inverted in eq. (6) and efficient use of the dynamic range.

Fourier-Bessel Green's Function Representation. The computationally most intensive component of the present approach is the evaluation of the *NxN* pressure and displacement Green's function matrices **P** and **U** in eq. (6) through the Fourier-Bessel representations in Eqs. (8)-(9). Here it is extremely important to take advantage of any target symmetries. Thus for example, for targets with vertical axisymmetry, the virtual sources and surface nodes are naturally placed in 'rings' at constant depth, thus reducing the number of required values of the depth-separated Green's function. Also, this will reduce the number of required values of the Bessel functions. In that regard, the fact that the Green's functions are only needed in the near field with $k_r r$ small, makes it convenient to pre-compute a table of Bessel functions with $\Delta k_r r \simeq \pi/20$ from which the required values are extracted by simple interpolation. Further, the wavenumber integration interval may be reduced significantly by replacing eq. (8) with the exact free field Green's function $exp(jkR)/4\pi R$ for all source-receiver pairs in the same layer.

Finally, for computing the scattered field, significant computational gains are achieved by performing the inner summation in eq. (8) for a reference depth, e.g. z_i, within each layer before solving for the homogeneous solution in eq. (9), essentially collapsing the virtual source distribution into a multipole at $(r,z) = (0, z_i)$. It should be pointed out that this procedure yields incorrect field solution within the depth-interval occupied by the target due to incorrect representation of the evanescent contribution. However, as long as the field is not needed at these depths, it yields significant computational savings.

NUMERICAL RESULTS

Spherical Shell in Free Space

The first example concerns the free field scattering of a plane wave incident on a spherical, elastic shell in an infinite medium. The sphere considered is the 1.06 m outer diameter air-filled steel sphere used as a target in the GOATS experiments, and

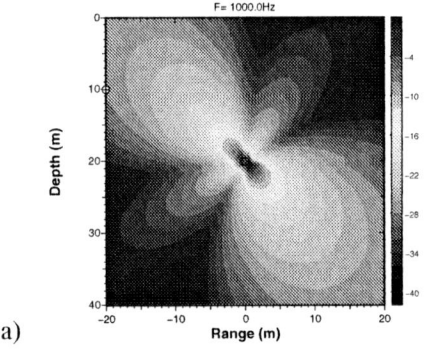 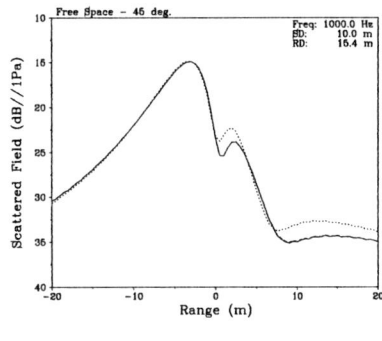

FIGURE 2. In-plane scattered field for spherical shell in a homogeneous fluid medium insonified by a unit amplitude plane wave incident at 45°. a) Contours of scattered pressure in dB in vertical plane. Spherical target at origin as indicated. (b) Scattered pressure 4.6 m above target center in dB Solid curve: Reference solution using spherical harmonics; Dashed curve: Virtual source solution with 1146 nodes; Dotted curve: Virtual source solution with 732 nodes.

extensively analyzed in terms of its scattering characteristics [10]. The shell thickness is 3 cm and the sound speed in the surrounding fluid is assumed to be 1500 m/s.

Figure 2(a) shows contours in dB of the in-plane scattered field for a 1 kHz plane wave incident at 45°, as computed using an exact multipole expansion obtained by spherical harmonics, but with the field evaluated using the Fourier-Bessel spectral representation in eq. (8). Note the small errors at the depth of the target due to the point-multipole representation of the vertically extended target as discussed above. Figure 2(b) shows the scattered field in negative dB along a horizontal line 4.6 m above the target. The solid curve shows the spherical harmonics solution, while the dotted curve shows the result obtained by the virtual source approach using 732 virtual sources (24 horizontal rings). Almost indistinguishable from the 'exact' spherical harmonics result, a dashed curve shows the virtual source result obtained with 1,146 sources (30 rings). The target stiffness matrix was in both cases computed using both spherical harmonics, and FESTA, with virtually identical results.

In the absence of interfaces penetrating the target both virtual source results in Fig.2 were obtained by solving eq. (6) using the exact free field Green's functions in

To illustrate the accuracy of the Fourier-Bessel spectral representation of the Green's function, Fig. 3 compares the in-plane scattered field 1m from the target center, in terms of (a) pressure, and (b) particle velocity, using the Fourier-Bessel representation with a dummy interface at the depth of the target center (blue, solid curve), and using the exact spherical Green's function for the same virtual source distribution (red, dashed curve).

Flush-buried Spherical Shell

The final example demonstrates the significance of properly incorporating multiple scattering when modeling the response of targets close to the interfaces in the stratification. Here the spherical shell treated above is assumed to be flush-buried in a seabed

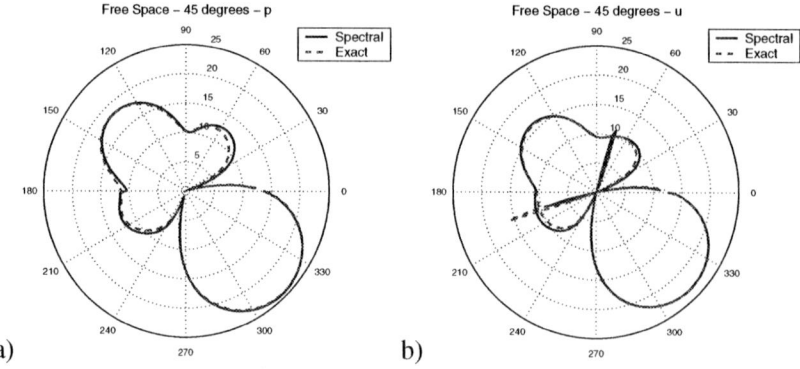

FIGURE 3. In-plane scattered field from spherical shell in infinite, homogeneous medium, 1 m from target center, using multipole, spectral representation of Green's functions (blue, solid curves) and free field exact spherical Green's function $exp(ikR)/R$ (red, dashed curves). (a) Pressure; (b) Normal Velocity.

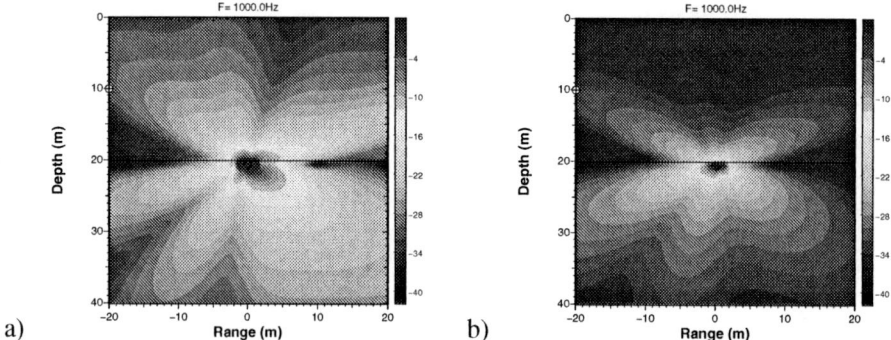

FIGURE 4. Pressure contours in dB of in-plane scattering from flush-buried spherical shell for plane wave incident at 10° grazing onto seabed. (a) Multiple scattering included using spectral representation of dual-halfspace Green's function. (b) Single scatter approximation using free field Green's function for virtual source contributions.

with sound speed 1700 m/s and density 1.8 g/cm^3, and insonified by a 1 kHz plane wave from the water column, at sub-critical 10° grazing angle, thus yielding an evanescent incident field in the seabed. The 1146 node target stiffness matrix of the previous example is used here unchanged. Figure 4(a) shows the in-plane pressure contours in dB using the Fourier-Bessel integral representation of the Green's functions in eq. (6), while (b) shows the corresponding result using the free-field Green's functions for the sediment. In both cases the resulting scattered field is evaluated using the full stratified Green's function, and the significant difference observed in the scattered field illustrates the importance of incorporating multiple scattering for these problems. In contrast Fawcett [11] found single scattering to be adequate, but for deeper buried targets.

SUMMARY

A hybrid modeling framework for scattering from general 3D elastic targets in a stratified ocean waveguide has been presented, incorporating multiple scattering between the target and the stratification, and allowing for targets to be completely or partially buried in an elastic seabed. The main components of the hybrid framework are (i) a stiffness matrix uniquely describing the dynamic properties of the target in vacuo, (ii) the spectral Fourier-Bessel representation of the Green's function in a horizontally stratified ocean. The two components are coupled through a wavefield superposition, or virtual source approach, where the scattered field on the surface of the target is 'generated' by a distribution of virtual sources inside the volume occupied by the target, and which superimposed with the incident field must be consistent with the surface stiffness properties of the target. The hybrid approach is here implemented within the OASES-3D wavenumber integration modeling framework, with the target response determined using spherical harmonics [12] for spherical shells, or FESTA [5] for more general targets.

ACKNOWLEDGMENTS

The author appreciates the collaboration with Dr. Mario Zampolli and David Burnett of SACLANTCEN on the integration of FESTA and OASES. The SACLANTCEN support of the collaborate modeling effort through the GOATS and the Hybrid Modeling JRPs is highly appreciated, as is the support by the US Office of Naval Research.

REFERENCES

1. Schmidt, H., and Lee, J., *J. Acoust. Soc. Am.*, **105**, 1605–1617 (1999).
2. Maguer, A., Fox, W., Schmidt, H., Pouliquen, E., and Bovio, E., *J. Acoust. Soc. Am.*, **107(3)**, 1215–1225 (2000).
3. Schmidt, H., "Bistatic Scattering from Buried Targets in Shallow Water," in *Proceedings, GOATS 2000 Conference*, edited by E. Bovio and H. Schmidt, SACLANTCEN Conference Proceedings Series CP-46, NATO SACLANT Undersea Research Centre, La Spezia, Italy, 2001.
4. Schmidt, H., and Glattetre, J., *J. Acoust. Soc. Am.*, **78**, 2105–2114 (1985).
5. Burnett, D., and Zampolli, M., Development of a Steady-State, 3-d Acoustics Code for Target Scattering, SR 379, NATO Undersea Research Centre, La Spezia, Italy (2003).
6. Burnett, D., "Finite-Element Methods for Structural Acoustics: Physics, Mathematics and Modeling," in *Proceedings*, 10th Internat. Congress on Sound and Vibration, Stockholm, Sweden, 2003.
7. Stepanishen, P., *J. Acoust. Soc. Am.*, **101**, 3270–3277 (1997).
8. Kessel, R. T., *Scattering of Elastic Waves in Layered Media: A Boundary Integral-Normal Mode Method*, Ph.D. thesis, University of Victoria (1996).
9. Jensen, F., Kuperman, W., Porter, M., and Schmidt, H., *Computational Ocean Acoustics*, AIP Press, New York, 1994.
10. Tesei, A., Lim, R., Maguer, A., Fox, W., and Schmidt, H., *J. Acoust. Soc. Am.*, **112(5)**, 1817–1830 (2002).
11. Fawcett, J., "Scattering from an Elastic Cylinder Buried Beneath a Rough Water/Basement Interface," in *High Frequency Acoustics in Shallow Water*, edited by N. Pace, E. Pouliquen, O. Bergem, and A. Lyons, SACLANTCEN Conference proceedings Series CP-45, 1997.
12. Schmidt, H., *J. Acoust. Soc. Am.*, **94(4)**, 2420–2430 (1993).

A Finite-Element Tool for Scattering from Localized Inhomogeneities and Submerged Elastic Structures

Mario Zampolli, David S. Burnett, Finn B. Jensen, Alessandra Tesei*,
Henrik Schmidt[†] and John B. Blottman III[**]

NATO Undersea Research Centre, 19138 La Spezia, Italy
[†]*Department of Ocean Engineering, Massachusetts Institute of Technology, Cambridge, MA 02139*
[**]*Naval Undersea Warfare Center, Division Newport, Code 2131, Newport, RI 02841-1708*

Abstract. A steady-state 3-D finite-element tool called FESTA (Finite-Element STructural Acoustics), is being developed at the NATO Undersea Research Centre. The code is geared towards a variety of applications in underwater acoustics, such as multistatic scattering from localized inhomogeneities, scattering across interfaces between fluids and/or solids, and multistatic scattering from single and multiple fluid-loaded elastic targets. The hp-adaptive finite-element technology used to develop FESTA allows the user to optimize the convergence as well as the demand on computing resources by selectively changing the element size (h-refinement) and/or by increasing the order of the polynomial finite-element shape functions (p-enrichment). An efficient hybrid tool for the computation of multistatic scattering from targets buried, partially buried or proud inside shallow water waveguides is being developed in conjunction with MIT. In this hybrid tool FESTA is used to perform target computations in the near field of the scatterer, while the waveguide propagation model OASES computes the far field propagation of the incident and scattered acoustic fields. The presentation focuses on the most relevant details of the formulations implemented in FESTA, as well as on application examples for frequencies ranging from 1 kHz to a few tens of kHz.

INTRODUCTION

The steady-state **F**inite **E**lement **ST**ructural **A**coustics code called **FESTA**, which is under development at the NATO Undersea Research Centre, is based on a fully 3-D continuum mechanics description of the acoustics in both fluid and elastic media. The basic building block for the tool is a state-of-the-art commercial finite element kernel and library called ProPHLEX, which is developed at COMCO/Altair Engineering (see [1]). FESTA is capable of computing efficiently the multistatic scattered field in the vicinity of localized inhomogeneities and/or of elastic objects with internal structure, such as mines. Because of the computational cost associated with 3-D finite element computations, one is usually limited to finite element domains which are not larger than at most a few tens of wavelengths. Common problems of scattering from objects in underwater waveguides, on the other hand, require the computation of the acoustic field in much larger domains. To overcome this limitation, FESTA has been interfaced with the MIT wavenumber integration propagation tool OASES [2, 3]. The resulting hybrid FESTA/OASES tool is capable of treating 3-D elastic objects of arbitrary shape inside ocean waveguides. This paper focuses on the technique developed to couple FESTA with OASES by giving a brief description of the relevant mathematical formulations. Several applications of FESTA and of the hybrid tool are discussed in the oral conference presentation.

THEORY

This section describes briefly the steady-state structural acoustics equations used in FESTA, and the boundary conditions implemented in the current version of the tool. A special boundary condition which is used to couple FESTA with underwater propagation tools is also presented. For a more detailed discussion of the derivations and theory underlying FESTA, the reader is referred to [4, 5].

Assuming $\exp(+i\omega t)$ time-dependence, where ω is the angular frequency and t represents time, the linear wave equations for an anisotropic elastic solid in a cartesian coordinate system can be written as:

$$\left(c_{\alpha l \beta m} u_{\beta,m}\right)_{,l} + \omega^2 \rho^s u_\alpha = 0, \qquad (1)$$

with $\alpha, \beta = 1,\ldots,6$, and with the latin indices $l, m = 1, 2, 3$ representing the cartesian coordinates x, y and z. The displacement vector in the solid is:

$$(u_1,\ldots,u_6) = \left(u_x^R, u_y^R, u_z^R, u_x^I, u_y^I, u_z^I\right), \qquad (2)$$

where (u_x, u_y, u_z) is the complex displacement vector in cartesian coordinates, and the superscripts R and I denote the real and imaginary parts, respectively. According to the definitions of tensor calculus, repeated indices in a term imply summation, and a comma preceding a subscript index denotes differentiation with respect to the coordinate associated with the index. The material properties in equation (1) are given by the tensor of elastic moduli $c_{\alpha l \beta m}$ and by the density ρ^s. In its most general form, $c_{\alpha l \beta m}$ can describe fully anisotropic elastic materials.

The complex pressure $p = (p^R, p^I)$ in the fluid is described by the Helmholtz equation

$$\left(\frac{1}{\omega^2 \rho^f} p_{\alpha,m}\right)_{,m} + K_{\alpha\beta} p_\beta = 0. \qquad (3)$$

The material properties of the fluid are the density ρ^f and the compressibility matrix

$$[K] = \frac{1}{B^R(1+\delta^2)} \begin{bmatrix} 1 & -\delta \\ \delta & 1 \end{bmatrix}, \qquad (4)$$

where B^R is the dynamic bulk modulus, and δ is the damping factor. Although the writing of complex equations as systems of purely real equations may seem somewhat awkward to the reader, the reason motivating this choice in the present paper is that the ProPHLEX FE development library and kernel used by the authors requires all quantities to be purely real (see [1]).

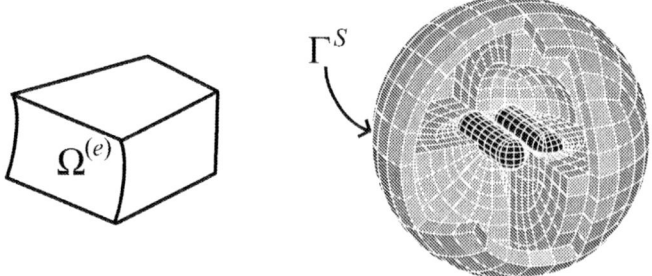

FIGURE 1. A generic finite element $\Omega^{(e)}$, and a view into a 3-D finite element mesh of two hollow steel cylinders (black elements) surrounded by a ball of fluid elements. The ball of fluid is bounded by the spherical surface Γ^S, on which radiation conditions are applied to ensure the outward radiation of the scattered field.

Finite element equations

Instead of deriving separate finite element equations for a solid element and for a fluid element, one can construct a single type of finite element which represents either a solid or a fluid, depending on the choice adopted for the constants in the equations. Figure 1 shows an example of a single element and of a full finite element mesh consisting of approximately 2000 elements, some of which are solid elements (two cylinders in the center of the mesh), while the others are fluid elements. The first step in writing a single FE equation for the two media is to group equations (1) and (3) into one single expression:

$$\left(a_{\alpha l \beta m} q_{\beta,m}\right)_{,l} + c_{\alpha\beta} q_\beta = 0, \quad (5)$$

with $\alpha, \beta = 1, \ldots, 8$. The unknown vector quantity $\{q\}$ is defined by

$$\{q\} = \{u_x^R, u_y^R, u_z^R, u_x^I, u_y^I, u_z^I, p^R, p^I\}^T, \quad (6)$$

where the superscript T denotes the transpose. The wave equation for a solid is obtained from equation (5) by choosing

$$a_{\alpha l \beta m} = \begin{cases} c_{\alpha l \beta m} & \alpha, \beta = 1, \ldots, 6 \\ 0 & \text{otherwise} \end{cases} \quad (7)$$

$$c_{\alpha \beta} = \begin{cases} \omega^2 \rho^s \delta_{\alpha\beta} & \alpha, \beta = 1, \ldots, 6 \\ 0 & \text{otherwise}. \end{cases} \quad (8)$$

The symbol $\delta_{\alpha\beta}$ represents the Kronecker Delta, which is equal to unity if and only if $\alpha = \beta$, while it is zero for $\alpha \neq \beta$. The equations for a fluid are obtained easily from (5) by setting

$$a_{\alpha l \beta m} = \begin{cases} \frac{\delta_{\alpha\beta} \delta_{lm}}{\omega^2 \rho^f} & \alpha, \beta = 7, 8 \\ 0 & \text{otherwise} \end{cases} \quad (9)$$

$$c_{\alpha\beta} = \begin{cases} K_{\alpha\beta} & \alpha, \beta = 7, 8 \\ 0 & \text{otherwise}. \end{cases} \quad (10)$$

The starting point for the derivation of the Bubnov-Galerkin integral equations for a generic fluid/solid element is the expansion of each of the 8 field variables q_β in a series of v basis functions,

$$q_\beta = \Phi_{\beta j} Q_j, \quad \beta = 1,\ldots,8; \, j = 1,\ldots,8\,v. \tag{11}$$

The basis functions, which in the present case are the hierarchic polynomials defined in [1], are grouped into the matrix

$$[\Phi] = [\phi_1[I],\ldots,\phi_v[I]], \tag{12}$$

where $[I]$ is an 8x8 identity matrix. The vector Q_j contains the unknowns, which are the coefficients of each individual basis function.

Equation (5) is premultiplied by $[\Phi]^T$ and integrated over the element volume $\Omega^{(e)}$. Application of the divergence theorem and the substitution of Eq. (11) into the $a_{\alpha l \beta m}$ and $c_{\alpha\beta}$ integrals yields the finite element linear matrix equation

$$\left[\int_{\Omega^{(e)}} \Phi_{i\alpha,l} a_{\alpha l \beta m} \Phi_{\beta j,m} d\Omega - \int_{\Omega^{(e)}} \Phi_{i\alpha} c_{\alpha\beta} \Phi_{\beta j} d\Omega \right] Q_j =$$

$$\underbrace{\int_{\partial\Omega^{(e)}} \Phi_{i\alpha} a_{\alpha l \beta m} q_{\beta,m} n_l \, d\Gamma}_{I_{\partial\Omega^{(e)}}}, \quad i = 1,\ldots,8v, \tag{13}$$

where $\partial\Omega^{(e)}$ represents the boundary of the element (e). The vector n_l is the unit normal on $\partial\Omega^{(e)}$ pointing outward. The boundary conditions on the faces of an element are applied via the integral $I_{\partial\Omega^{(e)}}$.

Once the computational domain has been divided into a mesh of elements, the linear equations (13) corresponding to each element are assembled into a global sparse linear system of equations. Assembly enforces the inter-element continuity of normal particle displacement within the solid domain and the inter-element continuity of acoustic pressure inside the fluid domain.

For any pair of adjacent fluid elements, the assembly process also produces a difference of two $I_{\partial\Omega^{(e)}}$ integrals on the interface shared by each of the two elements. Since $a_{\alpha l \beta m} q_{\beta,m} n_l$ represents the normal particle displacement, which is required to be continuous across the interface between two adjacent elements, the difference of two $I_{\partial\Omega^{(e)}}$ integrals is set to zero. Similarly, the difference of two $I_{\partial\Omega^{(e)}}$ integrals on the interface between two adjacent solid elements is set to zero because $a_{\alpha l \beta m} q_{\beta,m} n_l$ represents the normal component of the stress tensor, which also has to be continuous. The continuity of normal stress and displacement cannot be imposed naturally by the assembly process across the interface between a fluid and a solid element. To join the solid and the fluid domains, the continuity of normal stress and particle displacement on the common interfaces between adjacent solid and fluid elements is guaranteed by solid/fluid coupling surface integrals, which convert the solid degrees of freedom to compatible fluid degrees of freedom and vice-versa. Other boundary conditions which can be applied to the finite element equation are the Dirichlet, Neumann, and Mixed conditions on the free faces of an element. To apply the Sommerfeld radiation condition for unbounded domains in an approximate form, the Bayliss-Turkel first order radiation conditions (see [6]) are used on the free faces of a spherical surface enclosing the finite element computational domain. The reader is referred to [4, 5] for a more detailed discussion of the boundary conditions implemented in FESTA.

Hybrid finite element/propagation tool modeling

In many practical applications it is necessary to compute the acoustic field inside a shallow water waveguide containing one or more targets. For such problems, the size of the overall computational domain is usually around a few hundreds of wavelengths in range and depth, while the size of the target is on the order of one to a few tens of wavelengths. Experience shows that the complete solution of such problems by the finite element technique alone is not feasible because of the computational cost associated with 3-D FE codes. On the other hand, many underwater propagation tools, such as OASES/SAFARI [2, 3], compute the propagation inside ocean waveguides efficiently, but lack the capability of treating arbitrarily shaped 3-D elastic targets. These different characteristics of the FE technique and the propagation tools suggest that the 3-D scattering problems in underwater ducts can be solved by a technique in which the finite element tool and a propagation code are interfaced to form a hybrid tool. The hybrid technique presented here is based on the decomposition of the problem into a long-range propagation sub-problem, which can be solved by the propagation tool, and a nearfield scattering sub-problem to be solved by the finite element tool. The boundary data required for the definition of the local finite element scattering problem is assigned via the boundary condition presented below.

The communication between the finite element model and the wavenumber integration tool occurs through the three step procedure outlined in Fig. 2. In the first step, the propagation tool computes the acoustic field inside the shallow water waveguide in absence of the target. In the second step, the incident field computed in step one and its normal derivative are assigned on Γ^S (see figure 1) as incident field data for the local scattering computation. An approximate form of the Sommerfeld radiation condition to ensure the outward radiation of the scattered field is also needed on Γ^S. At this point, the finite element scattering problem is solved by FESTA, and the total field resulting from the computation is sampled on the boundary of the sphere. Subtraction of the incident field from the FESTA result yields the scattered field on Γ^S. In the third and last step, the scattered field and its normal derivative are passed back to the propagation tool, which constructs a multipole having the same radiation pattern as the scattered field. The multipole is then used as a source for the propagation of the scattered field through the underwater channel.

The boundary conditions for coupling FESTA to the propagation tool in step two are obtained from the boundary integral $I_{\partial\Omega^{(e)}}$ in Eq. (13). To achieve this, the boundary integral for a fluid is expressed explicitly in terms of the gradient of the acoustic pressure as

$$I_{\partial\Omega^{(e)}} = \int_{\Gamma^S} \frac{1}{\omega^2 \rho^f} \Phi_{i\beta} p_{\beta,l} n_l d\Gamma, \tag{14}$$

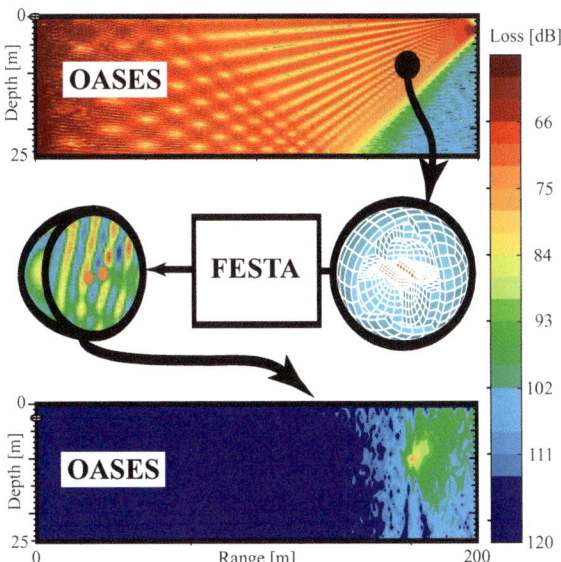

FIGURE 2. Example of a coupled FESTA/OASES computation. Step 1: incident field computation with OASES. Step 2: scattered field computation with FESTA, using the incident field from step 1 as a BC on the spherical surface Γ^S. Step 3: the scattered field is propagated in the waveguide by OASES.

with $\beta = 1, 2$. Denoting the incident field by p_β^{inc} and the scattered field by p_β^{scat}, the total pressure p_β can be written as

$$p_\beta = p_\beta^{\text{inc}} + p_\beta^{\text{scat}}. \tag{15}$$

The approximate radiation condition for the scattered field used here is the one developed by Bayliss, Turkel and co-workers ([6]):

$$\frac{1}{\rho^f} p_{\beta,l}^{\text{scat}} n_l = R_{\beta\alpha} p_\alpha^{\text{scat}}, \quad \alpha = 1, 2, \tag{16}$$

where the matrix $R_{\beta\alpha}$ represents the first order linear Bayliss-Turkel radiation operator. Substitution of equation (15) into equation (14) and application of the radiation condition (16) yield after some simple manipulation:

$$I_{\partial\Omega^{(e)}} = \int_{\Gamma^S} \frac{\Phi_{i\beta}}{\omega^2} R_{\beta\alpha} p_\alpha \, d\Gamma + \int_{\Gamma^S} \frac{\Phi_{i\beta}}{\omega^2} \left[\frac{1}{\rho^f} p_{\beta,l}^{\text{inc}} n_l - R_{\beta\alpha} p_\alpha^{\text{inc}} \right] d\Gamma. \tag{17}$$

The first integral in equation (17) contains the unknown p_α and thus it is moved to the left-hand side of the FE linear equations. The other integral remains on the right-hand side of the FE system because it contains the incident field p_α^{inc} and its gradient, which are both computed by the propagation tool in the first step of the coupling procedure.

NUMERICAL RESULTS

To illustrate a possible use of the hybrid FESTA/OASES tool, an acoustic barrier for the detection of objects in shallow water is studied (Fig. 2). In the envisioned application, a focused field is generated in an area across which one wants to detect intruding foreign objects. The detection is based on the measurement of perturbations in the quiescent region below the focus, which are caused by the forward scattering from an object crossing the barrier.

The environment considered here is a Pekeris waveguide consisting of a 25m deep water layer with an infinite layer of sand below it. A vertical time-reversal mirror array, located at range 0m, focuses the emitted sound at range 200m, and at 2.5m depth. At the frequency of 10 kHz used in the computations, the area shown in Fig. 2 spans over 1300 wavelengths in range, and approximately 170 wavelengths in depth. The scatterers penetrating the focused acoustic field are two closely spaced void steel cylindrical shells with hemispherical endcaps, modeled with the mesh depicted in Fig. 1. Each cylinder has a diameter of approximately 18 cm (roughly 1 wavelength) and a length of 60 cm (4 wavelengths). A contour plot of the transmission loss against range and depth computed by OASES in the first step of the hybrid procedure, showing the unperturbed focused acoustic field inside the waveguide, is displayed in the top panel of Fig. 2. The acoustic field in the sediment is not plotted in the figure. In the second step of the computation, the targets are located at range 170m and at depth 20m (large dot in the figure), with the axes of the cylinders lying perpendicular to the range-depth plane shown in the figure, and with the centers of the cylinders aligned parallel to the bottom. The middle panel of the figure shows a representation of the FESTA computation, and the bottom panel shows the scattered field in the waveguide computed by OASES. The top arrow connecting OASES to FESTA and the bottom arrow connecting FESTA to OASES represent respectively the sampling of the incident field and of the scattered field on the surface Γ^S. On a high-end Unix RISC workstation, the typical CPU times for computations like the one shown here are on the order of one to a few hours for the FESTA step, and a few minutes for each of the two OASES steps.

CONCLUSIONS

The hybrid finite element scattering/wavenumber integration propagation technique presented makes it possible to address the problem of scattering from 3-D objects in shallow water environments. In the example shown above, the multiple scattering between the cylinders is treated properly by one single finite element computation because both cylinders belong to the same finite element mesh, on which the full acoustic wave equation is solved. The multiple scattering between the targets, the sea surface and the bottom, on the other hand, can be computed only by approximating the Born series of the scattering via multiple iterations of the described three-step procedure. Although the importance of multiple scattering effects is of lower order for situations like the one discussed above, the interactions between the targets and the sediment can dominate the scattering from partially buried or flush buried objects. To take such effects properly into account, an environment-independent coupling technique based on the characterization of the target via in-vacuo response matrices is being investigated by the authors at the present time. The approach is based on the finite element analysis of the elastic responses of the target to a series of independent localized forces applied to its surface in the absence of a surrounding fluid. As a result of such computations, one obtains an admittance matrix in which the displacements and the forces on the surface of the target are related to each other. Using such a matrix as an input, OASES can rapidly compute the field scattered by the target for a given scenario, including the case of a partially or flush buried target. An alternate environment-independent coupling technique is obtained by considering an independent set of spherical harmonic fields incident on the target immersed in a fluid. For each different spherical harmonic case, a scattered field is computed, and the result is stored in a database. The scattering for a given incident field, like for example the focused field considered above, can be easily computed by decomposing the incident field into spherical harmonics, and by successively superimposing each spherical harmonic scattered field stored in the database according to the coefficients of the expansion. The common advantage associated with the environment-independent coupling techniques, is the capability to synthesize efficiently the scattered field for a given scenario by using a set of FE results which are calculated a priori, thus eliminating the need to repeat the computationally expensive finite element analyses for every single case considered.

ACKNOWLEDGMENTS

The collaboration of John Fawcett from DRDC Canada in the verification of cylinder scattering results with his thin shell FEM/BEM scattering tool is gratefully acknowledged. The authors are also grateful to Mark Stevenson and Reginald Hollett of the NATO Undersea Research Centre for the stimulating discussions on the focused field scattering problem.

REFERENCES

1. Liszka, T., Tworzydlo, W., Bass, J., Sharma, S., Westermann, T., and Yavari, B., *Comput. Methods in Appl. Mech. and Engrg.*, **150**, 251–271 (1997).
2. Schmidt, H., OASES: User Guide and Reference Manual , Tech. rep., Dept. Ocean Engrg., MIT (1999), available via ftp: *//keel.mit.edu/pub/Oases/oases.pdf*.
3. Schmidt, H., SAFARI: Seismo-acoustic fast field algorithm for range independent environments , Tech. Rep. SR-113, NATO Undersea Res. Ctr., La Spezia, Italy (1987).
4. Burnett, D. S., and Zampolli, M., Development of a finite-element, steady-state, 3-D acoustics code for target scattering , Tech. Rep. SR-379, NATO Undersea Res. Ctr., La Spezia, Italy (2003).
5. Burnett, D. S., "Finite-element methods for structural acoustics: physics, mathematics and modeling," in *Proc. 10th Internat. Cong. on Sound and Vibration*, Stockholm, Sweden, 2003.
6. Bayliss, A., Gunzberger, M., and Turkel, E., *SIAM J. Appl. Math.*, **42**, 430–451 (1982).

High-Frequency Material-Dependent Scattering Processes for Tilted Truncated Cylindrical and Disk-Shaped Targets

Philip L. Marston

Department of Physics, Washington State University, Pullman, WA 99164-2814

Abstract. When evaluating potential applications of high-frequency scattering by targets, it is important to be able to anticipate which features of the scattering may be attributed to material properties of the target and which features would also be present for rigid targets. These concerns may be especially important for the classification of completely exposed and partially buried targets. The overview here emphasizes previously published relevant experimental and theoretical results for backscattering by bluntly-truncated targets in water.

INTRODUCTION

This survey is concerned with the scattering of sound in water by truncated objects when the size significantly exceeds the acoustic wavelength. Diffraction from the edges of the object similar to the diffraction from a rigid object can be important for some orientations of the target. For certain viewing angles, however, elastic contributions to the scattering, can be much larger in magnitude than simple edge diffraction. Such elastic contributions can greatly enhance the visibility of truncated tilted cylinders imaged with a high frequency sonar [1]. There has been significant progress in understanding such enhancements with quantitative ray theory. In most of the cases reviewed here, distinct signatures are present which depend on the material properties of the target. In most of the cases noted the signatures depend on guided wave reflection properties at truncations. Examples of such truncations include the end of a cylinder or the rim of a tilted circular disk. In most of the cases noted, at the optimum tilt condition the enhancements have only a weak dependence on frequency and the enhancements are not dependent on the presence of a global normal-mode or "resonance" of the target. That is because the high-frequency target normal-modes are often significantly damped and the strong coupling of elastic guided waves with the acoustic field is such that local coupling processes can affect the scattering. Those local processes are often describable by ray theory.

To describe the general magnitude of the scattering, in the case of finite circular cylinders it is convenient to introduce the following normalization for a dimensionless form function f of which is related to the far-field scattered pressure as follows:

$$p_{scattered} = p_{incident} (a/2r) f \exp(ikr), \qquad (1)$$

where a is the radius of the cylinder and r is the distance from the cylinder and the time convention exp(-iωt) is used. With this normalization, if the cylinder is replaced by a fixed-rigid sphere having the same radius, then when ka >> 1, |p$_{scattered}$| is such that |f| is close to unity except in the case of near-forward scattering.

Bluntly-Truncated Rigid Cylinders

Prior to examining elastic scattering enhancements it is appropriate to consider the scattering by a perfectly rigid circular cylinder having flat ends. At high frequencies, there are significant enhancements in the directions of specular reflections from the sides and ends of such a target. For the case of backscattering, these enhancements are for broadside and end-on illumination. Elsewhere, however, at high frequencies the scattering is dominated by the interference pattern of the diffraction effects of the ends. In terms of the normalized form function used in Eq. (1), the magnitude of the end diffraction is of order (ka)$^{-1/2}$ and |f| is much less than unity when ka is large. See for example [2] and Section 2.12 of [3]. If ka is large and the ends of the cylinder are rounded, as is the case of a cylinder having hemispherical end-caps, then for backscattering by a tilted rigid cylinder, |f| can be relatively close to unity.

MERIDIONAL AND HELICAL RAY BACKSCATTERING ENHANCEMENTS

An important feature of the backscattering enhancements visible in the sonar images in [1] is that the enhancements were limited to specific ranges of tilt angles of the cylinder. It was correctly hypothesized in [1] that the observed enhancements were associated with elastic excitations of the cylinder (supersonic or "leaky" Lamb waves) described by rays that travel down the meridian of the cylinder and reflect off the end. The relevant meridian is in the plane containing the incident wave vector and the cylinder's axis. Since this type of scattering contribution was outside the scope of previous quantitative ray theories for scattering by shells (as well outside the scope of computational studies available at that time) Marston *et al.* [4-7] extended quantitative ray theory for cylinders to include this type of backscattering enhancement. During the development of the ray theory, it turned out to be useful to test ray theory for the related case of scattering by tilted infinitely-long solid cylinders [4] and cylindrical shells [8].

Solid Cylinders: Meridional, Helical and Face-Crossing Rays

The first quantitative experimental test of the aforementioned theory concerned the backscattering by a bluntly truncated solid stainless-steel cylinder where (in the frequency range of interest) the relevant leaky wave was analogous to a Rayleigh wave on a flat surface [5]. The magnitude of the meridional peak was |f| $\approx$ 2.5, but elsewhere typically |f| $\approx$ 0.2. The theory (see [4] and Appendix A of [5]) described the general magnitude and width of the peak, though the observed peak was offset from the predicted location for far-field scattering. That offset was caused by the limited size of the facility used in the measurements. In modeling the scattering, it was necessary to approximate the reflection coefficient of a Rayleigh wave from the end of the cylinder. In related work, backscattering enhancements were also observed that were associated with helical rays [5,9] and rays crossing the flat face of the ends of the cylinder [9]. The interference pattern associated with diffraction from the edges at the ends of the cylinder was also visible but was generally weaker than the aforementioned enhancements [9].

Meridional Ray Enhancements for Tilted Cylindrical Shells

Computations [6-8] show that at high frequencies, the relevant meridional rays for shells are typically associated with the lowest antisymmetric and symmetric generalizations of Lamb waves commonly designated respectively as a_0 and s_0 waves. Of these waves, the a_0 wave has a phase velocity that depends significantly on frequency so that the tilt condition for the enhancement depends on frequency. This was confirmed by measuring the backscattering as a function of frequency over a wide range of scattering angles [6,10]. The a_0 wave studied was on the branch that is important above the coincidence frequency of the shell. The coupling conditions were also confirmed by extending Rumerman's method [11] of approximating the backscattering by truncated thin shells (based on partial-waves) to the situation here where thin-shell assumptions were no longer applicable [5]. Subsequently qualitative [10] as well as quantitative [7,12] experiments demonstrated that in certain frequency ranges the simply-supported boundary condition in Rumerman's partial-wave approximation needed to be modified. These end corrections are especially significant for the a_0 wave when the threshold frequency of the a_1 mode is approached [7,10,12]. The cut-off a_1 mode is associated with a nonpropagating flexure of the shell localized near the ends of the shell. The non-propagating mode is analogous to the "nonpropagating near-field deflection" of an elastic beam discussed by Junger and Feit [13]. This can be demonstrated by noting that for plates in a vacuum, the complex dispersion relation of the cut-off branch of the a_1 mode of a plate wraps around (with an imaginary wavenumber) and at low frequencies becomes the nonpropagating flexural branch of the fourth-order plate dispersion relation. See for example Fig. 19 of Mindlin [14]. In the present case of a fluid-loaded shell the cut-off a_1 mode causes an extra radiation of sound near the ends of the shell which (from energy conservation) reduces the magnitude of the reflection coefficient of the propagating a_0 wave [7]. (This reduction of the reflection coefficient associated with fluid-loading at a truncation could at least in principle be described by hybrid FEM-BEM computational methods.) Quantitative ray theory described the general magnitude and tilt-conditions of the backscattering enhancements. With the normalization used in Eq. (1), the backscattering peak typically had $|f| \approx 6$ and an angular width given by theory [7,12]. The transient responses in the time-domain and time-frequency domains were also investigated [6,12,15].

Helical Ray Backscattering Enhancements for Tilted Cylindrical Shells

Away from the meridional ray coupling condition, qualitative measurements demonstrated that helical ray contributions can be significant [6,10,12,15]. A ray model for these enhancements was developed and confirmed with quantitative measurements of the backscattering of tone bursts [16] and the case of a water-filled shell was investigated [17]. Helical waves were also computed and modeled in the infinite shell case [18].

SCATTERING ENHANCEMENTS FOR FLAT ELASTIC AND PLASTIC CIRULAR DISKS

For certain tilt conditions, experiments by Hefner and Marston [19,20] demonstrated enhanced backscattering caused by leaky Lamb wave launched on an elastic disk. Acoustic holography was used to verify the identity of the waves associated with the enhancements [19-21] and a quantitative ray theory was developed that gives a useful approximation of the magnitude [21]. These enhancements are superposed on the edge diffraction contributions for the disks. The specular and edge diffraction contributions were usually weaker except when the incident wave vector was nearly perpendicular to the face of the disk. For plastic disks the shear wave velocity was less than the speed of sound in water and different types of scattering enhancements were observed [21,22].

ENHANCED BACKSCATTERING BY A TILTED ELASTIC CUBE

A leaky wave mechanism exists for producing a locally flat backscattered wavefront from a tilted elastic cube [23]. Only one of the cube's three Euler angles is constrained to lie in a narrow range. Consequently, for a randomly oriented cube (or certain other square-cornered objects) this mechanism becomes the most likely cause of large high frequency backscattering. Calculations based on ray theory predict enhanced backscattering. Measurements confirmed the general magnitude of the prediction in the case of an appropriately tilted stainless steel cube.

CYLINDERS FILLED WITH LOW SHEAR VELOCITY MATERIALS

For some cases of interest, it is anticipated that thin-walled cylindrical shells may be filled with materials having shear wave velocities less than the speed of sound in the surrounding water or sediment. To gain insight, the scattering was measured and modeled for bluntly truncated solid plastic and rubber cylinders in water [24]. For a range of tilt angles a significant backscattering enhancement was anticipated and was observed. That enhancement is associated with a caustic that is partially related to the acoustic "rainbow" of the tilted cylinder. Typical peak values of $|f|$ exceeded 2.5 and the enhancement was evident for ka as small as 10. A ray theory for this enhancement was confirmed [24] and aspects of the theory were also confirmed with a light scattering experiment [25].

Leaky Waves and Internal Loading of Shells by Plastics

For some cylinders of interest, the attenuation rate of meridional leaky or helical waves on cylinders will be affected by the presence of internal loading by plastic-like materials. Recent calculations by Marston suggest, however, that conditions can be found where the enhanced attenuation caused by placing a solid plastic cylinder inside a metal shell will not be sufficient to eliminate the aforementioned meridional ray enhancements.

Low-Frequency Modes of Water-Filled Cylinders Having Open Ends

Experiments with low-frequency modes of water-filled shells suggest that scattering enhancements associated with low-frequency internal length-wise modes can be significant [26]. An approximate theory was developed [27] based on the application of the generalized optical theorem [28]. One potential application in underwater acoustics is that such open-ended water-filled cylinders could be used as test targets having a simple spectrum of resonances.

ACKNOWLEDGMENTS

The Office of Naval Research supported this research.

REFERENCES

1. Kaduchak, G., Wassmuth, C. M., and Loeffler, C. M., "Elastic wave contributions in high resolution acoustic images of fluid filled, finite cylindrical shells in water," *J. Acoust. Soc. Am.* **100**, 64-71 (1996).
2. Ufimtsev, P. Y., "Theory of acoustical edge waves," *J. Acoust. Soc. Am.* **86**, 463-474 (1989).
3. Marston, P. L., "Geometrical and catastrophe optics methods in scattering," in *Physical Acoustics*, edited by R. N. Thurston and A. D. Pierce (Academic, Boston, 1992), Vol. 21, pp. 1-234.
4. Marston, P. L., "Approximate meridional leaky ray amplitudes for tilted cylinders: End-backscattering enhancements and comparisons with exact theory for infinite solid cylinders," *J. Acoust. Soc. Am.* **102**, 358-369 (1997).
5. Gipson, K., and Marston, P. L., "Backscattering enhancements due to reflection of meridional leaky Rayleigh waves at the blunt truncation of a tilted solid cylinder in water: Observations and theory," *J. Acoust. Soc. Am.* **106**, 1673-1680 (1999).
6. Morse, S. F., Marston, P. L., and Kaduchak, G., "High frequency backscattering enhancements by thick finite cylindrical shells in water at oblique incidence: experiments, interpretation and calculations," *J. Acoust. Soc. Am.* **103**, 785-794 (1998).
7. Morse, S. F., and Marston, P. L., "Meridional ray backscattering enhancements for empty truncated tilted cylindrical shells: Measurements, ray model and effects of a mode threshold," *J. Acoust. Soc. Am.* **112**, 1318-1326 (2002).
8. Morse, S. F., and Marston, P. L., "Meridional ray contributions to scattering by tilted cylindrical shells above the coincidence frequency: Ray theory and computations," *J. Acoust. Soc. Am.* **106**, 2595-2600 (1999).
9. Gipson, K. and Marston, P. L. , "Backscattering enhancements from Rayleigh waves on the flat face of a tilted solid cylinder in water," *J. Acoust. Soc. Am.* **107**, 112-117 (2000).
10. Morse, S. F., and Marston, P. L., "Degradation of meridional ray backscattering enhancements for tilted cylinders by mode conversion: Wide-band observations using a chirped PVDF sheet source," *IEEE J. Ocean. Eng.* **26**, 152-155 (2001).
11. Rumerman, M. L., "Contribution of membrane wave reradiation to scattering from finite cylindrical steel shells in water," *J. Acoust. Soc. Am.* **93**, 55-65 (1993).
12. Morse, S. F., "High Frequency Acoustic Backscattering Enhancements for Finite Cylindrical Shells in Water at Oblique Incidence," (Ph.D. diss., Washington State University, Pullman, WA, 1998).
13. Junger, M., and Feit, D., *Sound, Structures, and Their Interaction*, 2nd ed. (American Institute of Physics, Woodbury, 1993) p. 206.

14. Mindlin, R. D., "Waves and vibrations in isotropic, elastic plates," in *Structural Mechanics, Proceedings of the First Symposium on Naval Structural Mechanics* (Pergamon, Oxford, 1960).
15. Morse, S. F., and Marston, P. L., "Backscattering of transients by tilted truncated cylindrical shells: time-frequency identification of ray contributions from measurements," *J. Acoust. Soc. Am.* **111**, 1289-1294 (2002).
16. Blonigen, F. J. and Marston, P. L., "Leaky helical flexural wave backscattering contributions from tilted cylindrical shells in water: Observations and modeling," *J. Acoust. Soc. Am.* **112**, 528-536 (2002).
17. Blonigen, F. J. and Marston, P. L., "Leaky helical flexural wave backscattering contributions from tilted water-filled cylindrical shells," *J. Acoust. Soc. Am.* **113**, 309-312 (2003).
18. Blonigen, F. J. and Marston, P. L., "Leaky helical flexural wave scattering contributions from tilted cylindrical shells: Ray theory and wave-vector anisotropy," *J. Acoust. Soc. Am.* **110**, 1764-1769 (2001).
19. Hefner, B. T. and Marston, P. L., "Backscattering enhancements associated with the excitation of symmetric Lamb waves on a circular plate: direct and holographic observations," *Acoustics Research Letters Online* **2**, 55-60 (2001) {http://ojps.aip.org/ARLO/top.html}.
20. Hefner, B. T., and Marston, P. L., "Backscattering enhancements associated with antisymmetric Lamb waves confined to the edge of a circular plate: direct and holographic observations," *Acoustics Research Letters Online* **3**, 101-106 (2002) {http://ojps.aip.org/ARLO/top.html}.
21. Hefner, B. T., "Acoustic Backscattering Enhancements for Circular Elastic Plates and Acrylic Targets, the Application of Acoustic Holography to the Study of Scattering from Planar Elastic Objects, and Other Research on the Radiation of Sound" (Ph.D. diss., Washington State University, Pullman, WA, 2000).
22. Hefner, B. T., and Marston, P. L., "Acoustic backscattering and coupling processes for elastic and plastic circular disks in water: Direct and holographic observations," *J. Acoust. Soc. Am.* **109**, 2489 (A) (2001).
23. Gipson, K., and Marston, P. L., "Backscattering enhancements due to retroreflection of ultrasonic leaky Rayleigh waves at corners of solid elastic cubes in water," *J. Acoust. Soc. Am.* **105**, 700-710 (1999).
24. Blonigen, F. J., and Marston, P. L., "Backscattering enhancements for tilted solid plastic cylinders in water due to the caustic merging transition: Observations and theory," *J. Acoust. Soc. Am.* **107**, 689-698 (2000).
25. Marston, P. L., Zhang, Y. B., and Thiessen, D. B., "Observation of the enhanced backscattering of light by the end of a tilted dielectric cylinder owing to the caustic merging transition," *Appl. Opt.* **42**, 412-417 (2003).
26. Osterhoudt, C. F., and Marston, P. L., "Tilt angle dependence of backscattering enhancements from organ pipe modes of open water-filled cylinders: Measurements and models," *J. Acoust. Soc. Am.* **113**, 2334 (A) (2003)
27. Marston, P. L., and Osterhoudt, C. F., "Modeling scattering enhancements at isolated resonances using energy conservation, reciprocity, symmetry, and the optical theorem," *J. Acoust. Soc. Am.* **113**, 2284-2285 (A) (2003)
28. Marston, P. L., "Generalized optical theorem for scatterers having inversion symmetry: Applications to acoustic backscattering," *J. Acoust. Soc. Am.* **109**, 1291-1295 (2001).

Detection of Direct-path Arrivals for Multi-Narrowband Sequences (3-30 kHz) In Shallow Water

A. Zoksimovski, C. de Moustier

Center for Coastal and Ocean Mapping, University of New Hampshire
24 Colovos Road, Durham, NH 03824

Abstract. In an effort to measure underwater acoustic transmission loss over direct-path lengths ranging from a few hundred meters to ten kilometers in shallow water, a sequence of 16 gated pure tones (3-30 kHz) was transmitted every 10 s from a towed source and received at moored sonobuoys. The magnitude of multipath arrivals often exceeded that of direct-path arrivals, resulting in variable detection performance of simple matched filtering techniques. More reliable signal recognition was obtained via iterative least square time constraints on the arrival times across all frequencies in a sequence, based on the known time intervals between transmitted tones. Signal detection improvement was obtained also by searching for the direct-path arrival near the global maximum of the sum of the rectified correlograms of the received sequences. These methods allowed detection in environments characterized by multipath interferences, as well as low signal-to-noise ratio and fading, and in the presence of other unrelated sonar signals that cause large detection errors. It also improved the direct-path signal strength estimation, and associated transmission loss computation, by bounding the time interval over which to compute the signals' autocorrelations and estimate their power. These algorithms were tested on a limited data set recorded in the Southern California Offshore Range, confirming that frequencies below 6 kHz suffered less direct-path transmission losses than higher frequencies (7-30 kHz).

INTRODUCTION

Efficient underwater acoustic communication depends on the selection of a frequency band that is well adapted to the propagation channel. In shallow water environments, this channel is characterized by rapid changes, with range and depth, in the physical properties of the ocean volume and its boundaries. A measure of the frequency response of the channel can be obtained by estimating the transmission loss over a specified path at a range of frequencies. In the following, we shall focus on acoustic frequencies from 3 kHz to 30 kHz and report on direct-path transmission loss measurements over ranges extending to a few kilometers.

The experimental data were collected at three sites in the Southern California Offshore Range, West of San Clemente Island, in December 2000, and in February and June 2001. Two sites were in water depths ranging from 100 m to 300 m: one was between Cortes and Tanner Banks (32°41'N/119°11.07'W), and the other was South of San Clemente Island's China Point (32°45.17'N/118°25'W). The third site was in deep water (1600 m) in San Nicholas Basin (32°55'N/118°54'W). The environmental part of the experiment included conductivity/temperature/depth measurements, as well as shipboard and moored acoustic Doppler current profiles.

The acoustic signals were transmitted from an omni-directional transducer (ITC-1007) [1] towed by a ship, and were recorded at two identical SSQ-57B sonobuoy hydrophones [2] deployed on a mooring line at 27.4 m and 61 m depths. The intent was to place one hydrophone in, and one below the mixed layer, which is the upper nearly uniform portion of the sound speed vs. depth profile. Likewise, the source was towed in and below the mixed layer. In the Southern California Bight, the depth of the mixed layer decreases from about 50 m in January, to around 20 m in April [3]. During the experiment, the base of the mixed layer was typically found between 10 m and 40 m below the sea surface. However, the mixed layer was not always strongly defined and the depth of its base varied along the transmission path.

A transmission sequence consisted of 16 continuous wave (CW) pulses at discrete frequencies ranging from 3 kHz to 12 kHz in 1 kHz steps, and from 12 kHz to 30 kHz in 3 kHz steps. Each tone lasted 10 ms, with a 90 ms gap between pulses. The complete sequence lasted 1.6 s, followed by a 10 s quiescent period before the next transmission sequence, yielding a repetition period of 11.6 s for single frequency tones.

As shown in Fig. 1, the transmitter source level ranged from 173 to 195dB re 1 μPa @ 1 m when driven at constant power. The transmitted signals were recorded also at an omni-directional reference hydrophone (ITC-1042) [1], attached on the towing line 3 m away from the source in order to monitor the source levels. To avoid saturation of the highly sensitive sonobuoy receivers (Fig. 1) when approaching them, source levels were reduced in 10 dB steps from the maximum values shown in Fig. 1.

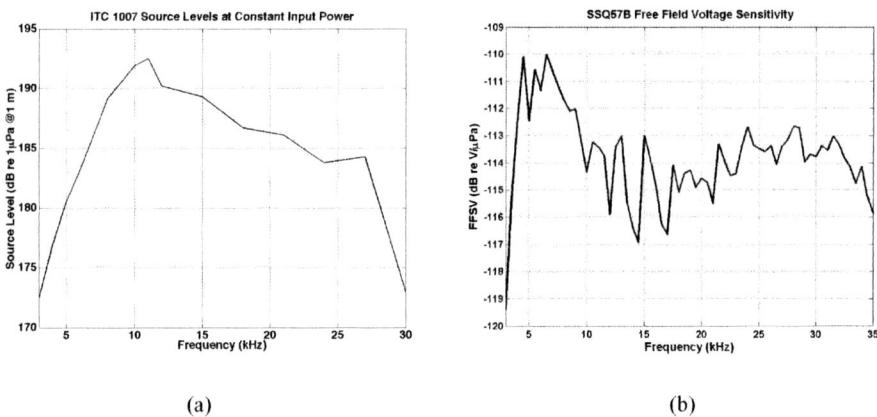

FIGURE 1. Source Level calibration of the ITC-1007 transducer (a) and Free Field Voltage Sensitivity calibration of the SSQ-57B receiver (b).

The ship towed the source at about 2.5 m/s on orthogonal traverses roughly 18 km long, and centered on the hydrophones mooring. At each ping, the direct-path transmission loss, TL, across the distance that separated the reference hydrophone and the two moored hydrophones was estimated by taking the ratio of their respective received sound intensities I_0 and $I_{1,2}$, according to [4, Chapter 5]:

$$TL = 10\log_{10}\left(\frac{I_0}{I_{1,2}}\right) \quad dB \qquad (1)$$

To first order, underwater acoustic transmission loss includes spreading and absorption losses. In shallow water, spreading losses at range R can be spherical ($20\log_{10}R$) or cylindrical ($10\log_{10}R$), whereas absorption losses vary with local temperature and increase in proportion to the square of the acoustic frequency [4]. Other factors of acoustic field attenuation are refraction, reflections from the seafloor and the sea surface, the various kinds of sound channels existing in the sea, and the multipath effects causing fluctuations in sound reception.

In this experiment, direct arrival detection could not be achieved with conventional detection algorithms, such as matched filter or maximum likelihood methods [5-7], because of the high correlation between the multipath and direct-path signals, and because the multipath arrivals often were stronger than the direct-path arrivals. To overcome this problem, we have devised two different detection methods based on combined frequency analysis. One method relies on a least square time constraint applied to all 16 signals in a sequence. The other uses the sum of the rectified autocorrelograms of the transmitted pulses as a detection template to search for the direct-path arrival near the global maximum of the sum of the rectified correlograms of the received time series. We describe the characteristics of the received time series in Section II, and present the algorithms for our direct-path arrival detection methods in Section III. Examples of the resulting estimates of transmission loss versus range and frequency are given in Section IV.

SIGNAL PROPERTIES

A typical spectrogram of the received signal is shown in Fig. 2. The received signals stand out in intensity at the corresponding frequencies. Acoustic intensity exceeding 120 dB re 1 µPa was detected at each transmitted frequency with longer duration than the transmission time of each pulse (10 ms). The extended time of each received signal is mostly the result of reflections from the seafloor and sea surface. Because of the limitations of the acoustic source used (Fig. 1), the transmitted power was stronger at the mid frequencies (8-12 kHz) than at the others, with a maximum at 11 kHz. Therefore, it takes longer for the acoustic field to settle back to the ambient noise level after the transmission period for pulses near 11 kHz. Higher harmonics of some frequencies, e.g. 7, 8, 9, 10 kHz, due to clipping of the multipath signals, are observable as well in Fig. 2. Most notably, the strong third harmonics of these frequencies (21, 24, 27, and 30 kHz) fall in transmission bands and can cause detection errors at those frequencies.

Interference from signal replicas that arrive at the receiver as reflections from objects or surfaces, e.g. schools of fish, water surface or bottom, presents another critical obstacle for smooth processing of the received signal. In shallow water, this phenomenon is not as uniformly distributed in range as noise, but is range and location dependent due to relatively close boundaries. Fig. 3 shows a sequence of consecutive pulses at the same frequency, for both moored hydrophones. Assuming an average speed of the ship around 2.5 m/s, 6 pings yield a travel path of 6×1.6×2.5 = 174 m.

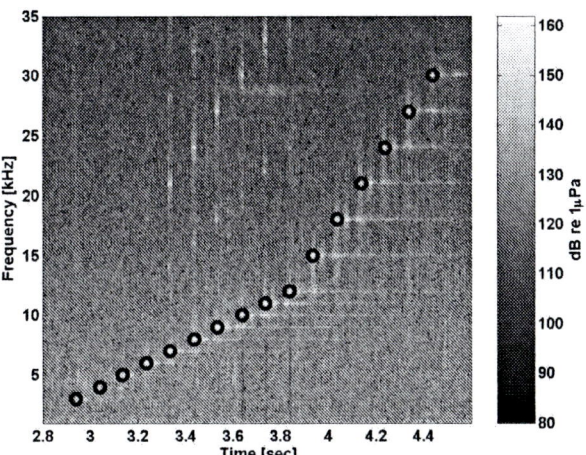

FIGURE 2. Spectrogram of the time series received at one hydrophone for one transmission sequence (range ≅ 2 km). Circles identify direct-path arrivals, and trailing signals represent multipath arrivals.

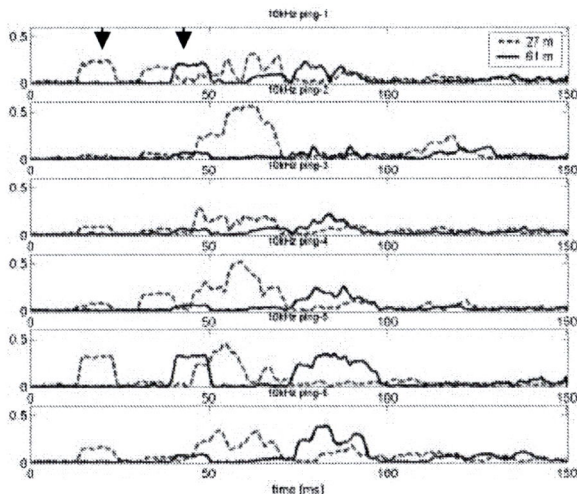

FIGURE 3. Pressure envelope (units of Pa) of 6 consecutive pings from measurements made at the Cortes-Tanner Banks site. Depths for the source and the receiving hydrophones were respectively about 40 m, 27 m, and 61 m. The arrows point to the first arrivals of the signals at the two different receiver depths.

We can see that within a 174 m difference in range, the direct-path pressure magnitude fluctuates between 0 and 0.5 Pa. In addition, the multipaths have different patterns at 61 m and 27 m, thus interferences vary with depth. Nonetheless, the magnitude of the direct-path signal is very similar at the two depths. This should not be the case if the source and the shallower of the receivers were inside the mixed layer, because the wave front would be mostly trapped inside the layer [4].

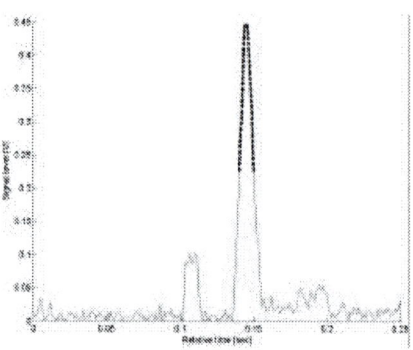

FIGURE 4. Envelope of a received signal with a strong multipath component. The direct-path arrival starts at 0.1 s and the taller peak starting at 0.13 s is due to multipath. The dashed line is the envelope of the signal detected by matched filtering.

SIGNAL DETECTION METHODS

In Fig. 4, the direct-path signal is followed by a few reflections. Because of the additional loss at the boundaries, the reflected acoustic waves are expected to have lower magnitude than the original [8]. Besides dynamic change in shape and magnitude from location to location (Fig. 3), this type of undesired signal, called acoustic echo [5] or multipath [6][7], is highly correlated to the direct-path signal. If a matched filter [5] were applied to this kind of pattern, it would almost definitely recognize the reflection with maximum magnitude as the most probable direct-path signal, which is incorrect in this example. In order to overcome this problem, we have devised and implemented two different algorithms: one based on a least square time constraint applied to the arrival times in a sequence, and the other based on the sum of the rectified correlograms of the received signals.

Least Square Time Constraint

Observing that multipath is a relatively random phenomenon (Fig. 3), a regular pattern would not be expected in the time difference between the direct-path arrival and reflections at N different frequencies. Therefore, we devised a method that detects each CW pulse independently, and then applies a tone-to-tone time constraint to get a more accurate estimate of the direct-path arrival times of the pulses, and to help distinguish direct arrivals from reflections. The algorithm contains the following steps applied to each received transmission sequence:

1) Band-pass filtering, centered at each of the transmitted frequencies. The bandwidth of each band-pass filter should be at least as wide as the associated Doppler shift Δf (Hz), which for a ship speed $v_s = 2.5$ m/s, sound speed $c = 1500$ m/s, and acoustic frequency f_m (Hz), is given by:

$$\Delta f = v_s f_m / c \approx f_m / 600 \ (Hz) \tag{2}$$

2) Match-filter each of the N outputs from step 1 and retain the time of the maximum amplitude.

3) Find the largest subset of the N time points from step 2 whose least-square fit to a straight line has a 100 ms gradient (time difference between the tones) with less than Δt uncertainty. The uncertainty factor should be chosen not to be smaller than the biggest pulse-to-pulse time offset. With 1.6 s between the first and the last transmitted pulse, the biggest time offset will be $\Delta t = 1.6 \times v_s / c = 2.7$ ms to first approximation.

4) Set a confidence interval around the least square fit values from step 3 for every point. Check whether the points belong to the designated interval. Every point found outside the boundaries is deemed an outlier.

5) Repeat step 2 for each outlier in step 4, restricting the filtering to later or earlier times toward the least square estimate for that tone.

6) Repeat steps 3 to 5 until all the points are within the region of confidence specified by Δt.

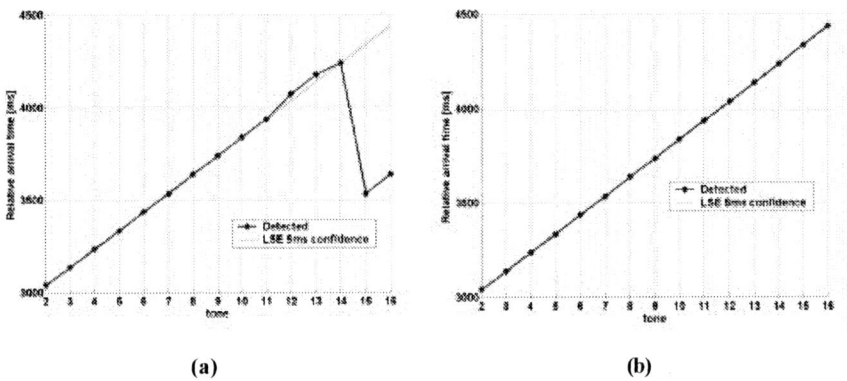

(a) (b)

FIGURE 5. An example of a successful signal detection completed after two iterations. The line with symbols represents the times. The parallel straight lines delimit the 5ms confidence region of the least square fit.

Figure 5 shows an example of the iterations of this algorithm applied to the transmission sequence whose spectrogram is displayed in Fig. 2. After the initial iteration (Fig. 5a), 11 points of detected arrival times are in good agreement with the 100 ms tone-to-tone time difference. There is another group of two points about 100 ms apart, tones 15 (27 kHz) and 16 (30 kHz). They are the detected third harmonics of tones 7 (9 kHz) and 8 (10 kHz). Tones 12 and 13 are also on a line whose gradient is around 100ms. These two points probably belong to multipath arrivals. In step 3 of the algorithm, the largest group is chosen for the direct-path arrivals. At the second iteration, step 4 proceeds forward from the current estimated arrival times for tones 15 and 16, and backward for tones 12 and 13. A solution is reached when all the pulses are detected within a 5 ms tolerance interval (Fig. 5b). In these operations over the limited data set, we did not use the 3 kHz signal, because its signal-to-noise ratio was often lower than the detection threshold. A repetitious false detection of one CW pulse would introduce high variance in the least square estimate, which would lead to slower convergence, and require redundant iterations. The potential direct-path arrival time at this frequency can be extrapolated from the measurements at the other frequencies. The output of the described example was used in plotting the circles in Fig. 2.

Coherent Summation of Correlograms

Since the data show that multipath arrivals were often much stronger than the direct-path arrivals, we developed an algorithm to search the correlograms for the global maximum, and for preceding local maxima that might correspond to the direct-path arrivals. Our rationale was that (1) detection results from one of the applied matched filters may help signal detection at other frequencies, for which the direct-path signal might have largely faded away, and/or the signal-to-noise ratio is below the detection threshold because of the bandwidth limitations of both the source transducer and the hydrophone (Fig. 1). Hence, by combining the correlograms for the 16 frequencies, one could take advantage of stronger signals at some frequencies to compensate for the limitations at others. And (2), the complexity of the algorithm should decrease by combining the correlograms into one generic sequence, and then searching one instead of N different sequences.

The algorithm starts with the broadband sequence y(m), which is digitized and recorded at the receiver, then band-pass filtered (impulse responses h_n) at each of the N transmitted frequencies:

$$x_n(m) = conv(y(m), h_n(m)), \quad n = 1,..N \quad (3)$$

Next, outputs from the N filters are matched with a time-inverted replica h_M of the test signal, containing M samples, at the corresponding frequency [5]:

$$r_n(m) = conv(x_n(m), h_M^n(m)), \quad n = 1,...N \quad (4)$$

$$\text{with } h_M^n(n) = x_n(M-1-m), \quad m = 0,...M-1 \quad (5)$$

The resulting N correlograms (r_n) are then delayed (N-n)T time units, where T is the transmission time difference between two consecutive CW pulses (100 ms). To increase the accuracy, the time delay should be consistent with the ship's speed and the geometry of the problem. This operation aligns in time all the correlograms of one transmission sequence, before rectifying and summing them into a global correlogram r_s:

$$r_s(l) = \sum_{n=1}^{N} abs(r_n(l - (N-n)T \times f_s)), \quad l = 1,...L \quad (6)$$

with sampling frequency f_s, and for a received sequence of length L. An example of such a global correlogram is shown in Fig. 6.

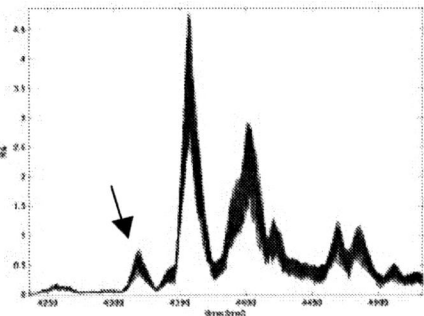

FIGURE 6. Sum of the 16 rectified correlograms received for one transmission sequence. The arrow points to the detected direct path arrival.

The algorithm searches the samples of the envelope of r_s, in the vicinity of the peak value, for a local maximum that might belong to the direct-path arrival.

Matches between segments of the sequence r_s, spanning ±10 ms around each detected local maximum, and an auto-correlation template are evaluated through a cost function (7). The template is generated by the same set of operations leading to r_s, but they are applied to the known transmitted signal, weighted by the source function (Fig. 1). The template is therefore the sum the rectified auto-correlation of each transmitted CW pulse. We expect more distortions in the multipath than the direct-path signals; hence the direct-path should retain more coherence with the signal auto-correlation in the correlograms. Since each transmitted CW pulse is 10ms long, the auto-correlograms ideally should be 20 ms long [5].

If I local maxima higher than the detection threshold are found, a cost function $c(i)$ is used to select one of them:

$$c(i) = \sum_{k=-K}^{K} \left| \frac{r_s^{template}(k) - r_s^{measured}(p(i)-k)}{r_s^{template}(k)} \right| \quad i = 0,\ldots I-1 \qquad (7)$$

where K is the number of samples equivalent to the duration of each CW pulse (10 ms). A sample p will be selected as the end of the direct-path signal arrival, if it is the earliest of all the samples for which the cost function is below the threshold:

$$p_{dpa} = \min\left(p(i) | c(i) < threshold\right) \qquad (8)$$

For example, p_{dpa} was about 4320 ms in Fig. 6. If none of the local maxima satisfies the above criteria (7,8), the global maximum is retained as the direct-path arrival. The time of arrival for each CW pulse is then estimated by subtracting the corresponding time shift from (6):

$$p_n = p_{dpa} - (N-n)T \times f_s, \quad n = 1,\ldots N \qquad (9)$$

TRANSMISSION LOSS ESTIMATION

The coherent sum of correlograms was used to estimate the propagation loss over one track at the Cortes-Tanner Banks site. The results are displayed in Figs. 7-8. Negative ranges correspond to the source moving towards the receiver. Detection threshold was set to 6 dB above the average noise power computed in each frequency band over a 100 ms time window, starting 2 s prior to each transmission sequence.

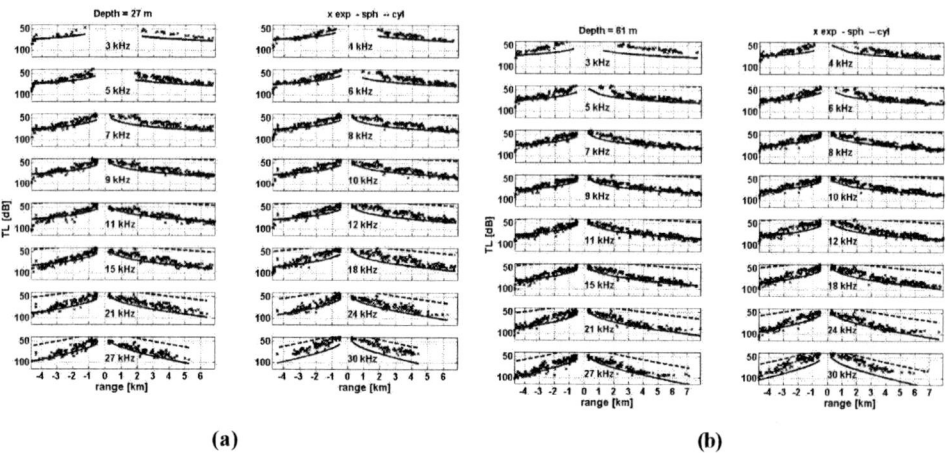

FIGURE 7. Transmission loss vs. range at the 16 frequencies, measured at Cortes-Tanner Banks, for hydrophone depths of 27 m (a), and 61 m (b). The experimental results are marked with 'x', while solid and dashed line correspond to the theoretical transmission loss for absorption with spherical and cylindrical spreading, respectively.

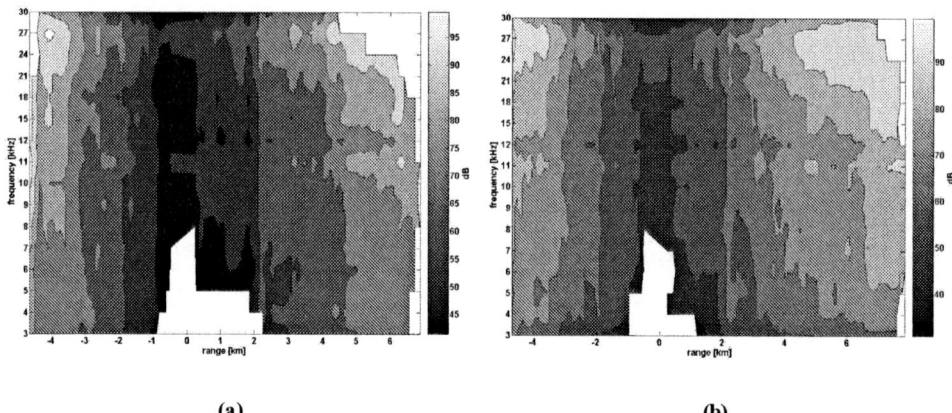

FIGURE 8. Contours of transmission loss vs. range and frequency, measured at Cortes-Tanner Banks, for hydrophone depths of 27 m (a), and 61 m (b). The white regions correspond to SNR < 6 dB.

The experimental transmission loss values follow the theoretical spherical spreading and absorption [9-10] curves for all frequencies except for 3 kHz and 30 kHz (Fig. 7). Results shown in Fig. 8, excluding the regions with SNR lower than 6 dB, indicate that within the limited range (< 10 km) of the experiment, frequencies below 6 kHz are more appropriate for signal transmission at either receiver depth (27 m and 61 m) for a source depth of about 40 m. It must be emphasized that these results correspond to direct-path attenuation in the 3-30 kHz band. In addition, low frequency (< 5 kHz) masking noise due to the towing vessel at short ranges, and low transmit response at 3 kHz and 30 kHz, as well as low receive sensitivity at 3 kHz (Fig. 1) were reasons for poor signal to noise ratio in the white range-frequency space in Fig. 8. Pending a more detailed range propagation study, we have assumed that no boundary interactions are involved, and that the frequency dependence is strictly due to absorption and scattering in the water column, as well as ambient noise. By contrast, results obtained at lower frequencies (e.g. 50 Hz - 3.2 kHz [11-13]) have shown that in waters shallower than 100 m, the optimum underwater acoustic communication frequency is around a few hundred Hz.

CONCLUSION

Direct-path transmission loss measurements were made with 16 CW pulses at frequencies from 3 kHz to 30 kHz. Two different methods of direct-path signal detection were implemented and tested on a limited data set. The first method detects each CW pulse independently and then applies a tone-to-tone time constraint to obtain a more accurate estimate of the direct-path arrival times of the pulses. The second algorithm is based on searching the recent history of the global maximum of the sum of the rectified correlograms received in a given transmission sequence.

The first method is robust against the inevitable timing offsets due to a moving transmitter, or nonlinear effects in the communication channel. The strength of the second method is its ability to merge the information from the CW pulses into a single function, and make a more reliable detection of the direct-path arrival for each tone in one step, whereas the first method requires a few iterations.

ACKNOWLEDGMENTS

For the experimental work that produced the data presented here, we wish to thank the Captain and crew of R/V New Horizon, Sean Wiggins who led the Feb. 2001 data collection, as well as Howard McManus, Seth Mogk, Carl Mattson, Scott Hiller, Earl Heckman, Erin Oleson and Melissa Hock who made all the pieces work together. P. Taylor and G. Wilkes of the US Naval Oceanographic Office funded the data collection work. A. Zoksimovski's MS thesis work is supported in part by the Center for Coastal and Ocean Mapping.

REFERENCES

1. http://www.itc-transducers.com
2. http://www.fas.org/man/dod-101/sys/ship/weaps/an-ssq-57.htm
3. U. S. Department of the Interior, POCS Technical Paper No. 82-2, *Physical Oceanography and Meteorology of the California Outer Continental Shelf*, August 1982.
4. Urick, R.J., *Principles of Underwater Sound*, McGraw-Hill, 1983.

5. Vaseghi, S.V., *Advanced Digital Signal Processing and Noise Reduction*, John Wiley, June 2000.
6. Ehrenberg, J.E., Ewart, T.E., and Morris, D.R., "Signal Processing techniques for resolving individual pulses in a multipath signal", *J. Acoust. Soc. Am.* **63(6)**, 1861-1865, (1978).
7. Moghaddam, P. P., Amindavar H., and Kirlin, R.L., "A new time-delay estimation in multipath", *IEEE Transactions on Signal Processing* **65**, 2503-2504 (1994).
8. Officer, C.B., *Introduction to the Theory of Sound Transmission – with application to the ocean*, McGraw-Hill, 1958.
9. François, R.E., and Garrison, G.R., "Sound absorption based on ocean measurements. I Pure water and magnesium sulfate contributions", *J. Acoust. Soc. Am.* **72(3)**, 896-907, (1982a).
10. François, R.E., and Garrison, G.R., "Sound absorption based on ocean measurements. II Boric acid contribution and equation for total absorption", *J. Acoust. Soc. Am.* **72(6)**, 1879-1890, (1982b).
11. Jensen, F.B., "Wave Theory Modeling: A Convenient Approach and Pulse Propagation Modeling in Low-Frequency Acoustics", *IEEE Journal of Oceanic Engineering*, vol. **13**, no. **4**, 186-197 (1988).
12. Jensen, F.B., and Kuperman, W.A., "Optimum frequency of propagation in shallow water environments", *J. Acoust. Soc. Am.* **73(3)**, 813-819 (1983).
13. Akal, T., *Effects of environmental variability on acoustic propagation loss in shallow water*, Impact of Littoral Environmental Variability on Acoustic Predictions and Sonar Performance – conference book, pp. 229-236, Kluwer Academic Publishers, 2002.

A New Synthetic Aperture Sonar Design with Multipath Mitigation

Marc Pinto*, Andrea Bellettini*, Lian Sheng Wang*, Peter Munk*, Vincent Myers* and Lucie Pautet*

*NATO Undersea Research Centre, La Spezia 19138, Italy

Abstract. Sonar performance in shallow water is severely degraded by multipath which reduces image contrast and degrades the performance of interferometric processing. This is an important limitation for high resolution applications such as minehunting, where target recognition exploits chiefly the shape and size of the target shadow. Experimental data showing the nature and importance of the multipath is presented together with a new sonar design, optimized to achieve a high level of multipath rejection at large range to water depth ratio.

INTRODUCTION

As is well known, synthetic aperture sonar (SAS) has the potential to provide very high cross-track resolution at long ranges. In practice, however, multipath interference can be a dominant cause of performance degradation, especially in shallow water. Multipath, besides the well known effect of ghost targets, leads to loss of image contrast (with consequent filling in of shadows) and degrades the quality of bathymetric estimates when interferometry is used. These effects, which are not SAS-specific, have nonetheless enhanced relevance in synthetic aperture imaging, since SAS aims naturally to extend the range of a sonar to fully exploit the gain in cross range resolution.

In addition, multipath affects specifically SAS performance because of the influence on the data-driven methods, such as the Displaced Phase Centre Array (DPCA) micronavigation, used to estimate the platform trajectory. The DPCA technique makes use of the correlation of the sea bottom direct backscatter to estimate the displacement of the SAS between pings, and depends critically [1] on a generalized Signal to Noise Ratio (SNR), where the signal is the seafloor backscatter coming from the direct path, while the noise consists of background noise of the sea, system noise, surface and volume reverberation and, last but not least, multipath interference of various orders.

We will adopt the convention of naming a multipath by the a combination of letters 'b' (for bottom) and 's' (for surface), with a lower letter indicating a specular bouncing and a capital letter indicating a non-specular scattering. In Fig. 1 first and second order multipaths from bottom scattering are shown.

Note that these plots show the trajectories for the same arrival time, and not, as it is more usual, for multiple returns from the same target. The focus will be, unlike in [2], on the multipath effects on the sea bottom direct backscatter, i.e., on the generalized SNR at a given range, which has direct implications on the DPCA technique and, at certain conditions, on the shadow contrast. In other words, this paper will investigate how multipath affects SAS even in the absence of targets.

It will be argued that the second order multipath 'bsB' and its reciprocal 'Bsb' constitute a major obstacle for obtaining high generalized SNR at large range to waterdepth ratios, both for physical and synthetic aperture imaging. Note that because of the different spatial correlation properties, no SNR gain due to synthetic aperture processing of the kind described in [2] for targets is expected in the case of sea bottom backscatter, except when the SAS is oversampled (i.e., it moves less than half of the sonar length between pings).

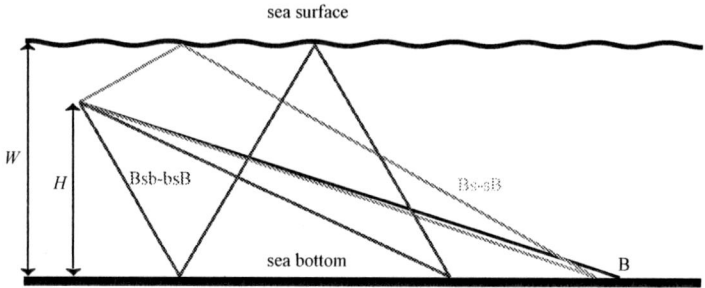

FIGURE 1. Multipath from bottom backscatter.

EXPERIMENTAL SETTING

To investigate the importance of higher order multipath for SAS performance, two experiments were conducted in June 2002 and November 2003.

In the first experiment, a 100 kHz sonar was deployed vertically on a fixed tower at a height H of 10.7 m and in a water depth W of about 20 m, in the vicinity of La Spezia. The seafloor was hard mud and the sea was calm during the experiment. No targets

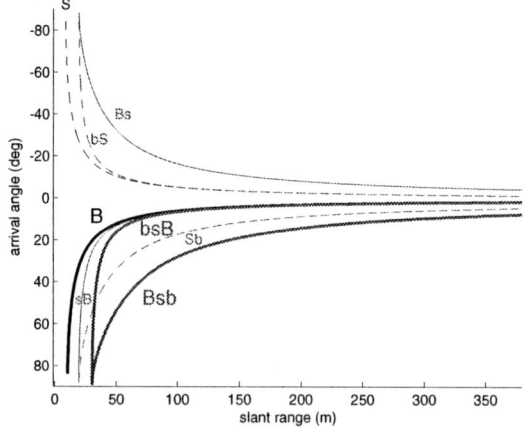

FIGURE 2. Direct and multipath returns as a function of range and arrival angle at the sonar, for the geometry of the experiment.

were deployed. In Fig. 2 the arrival angle in function of arrival time (expressed in terms of slant range of the direct bottom return) is plotted for this geometry, assuming a flat bottom.

The sonar array consisted of 256 receiver elements spaced at 7.5 mm to form an aperture of 1.92 m. When the sonar is mounted in vertical configuration, the vertical and horizontal beamwidths of the elements are about 40 degrees and 100 degrees respectively. The 64 channels at the center of the array formed a fully programmable transmitter, allowed different vertical transmission beampatterns to be synthesized (Fig. 3). The waveform used was a 95-105 kHz, 10 ms chirp. A previous experiment [3] with the same sonar deployed vertically had been conducted in 2001, but unlike this experiment, no tower was used, the sonar being lowered with ropes from the R/V Alliance. Therefore, the sonar moved horizontally too much to allow studying the ping-to-ping correlation of the data, since the horizontal beamwidth of 100 degrees gives, in fact, a spatial correlation length of only 0.8 cm. To begin, a broad transmission beam, shaped

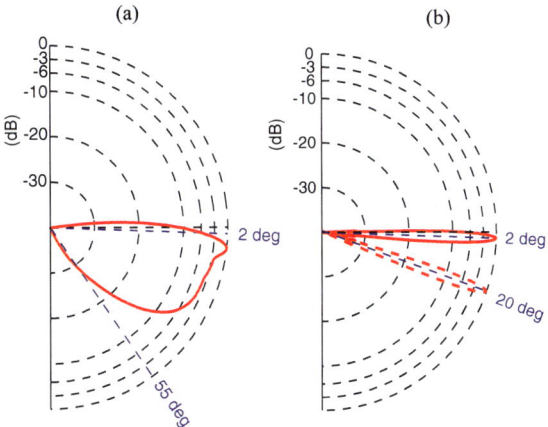

FIGURE 3. Vertical transmission beampatterns used in the experiment: (a) Wide beampattern (b) Narrow beampatterns.

to ensonify a wide swath of the seafloor while avoiding surface direct ('S') and first order multipaths ('Sb', 'sB'), was synthesized using the flexibility of the programmable transmitter. The beam, shown in figure Fig. 3(a), captures the main features of a typical conventional sonar design. In Fig. 4(a) the beamformed data are presented as a function of slant range and arrival angle. A time-varying gain has been applied to the data. A comparison with Fig. 2 indicates that it is impossible at long range to separate the arrival direction of the direct return 'B' from the second order multipath 'bsB'.

A 7 degree vertical beam with -20 dB sidelobes was synthesized in reception. The SNR, derived from the ping to ping correlation, is plotted in Fig. 5(a) as a function of time (expressed as equivalent slant range distance) and the receive depression angle. The white line represents the direct bottom return arrival.

The SNR is seen to fall off with range, well before the range where noise is dominant, indicating that there are other contributions than the direct seafloor return 'B'.

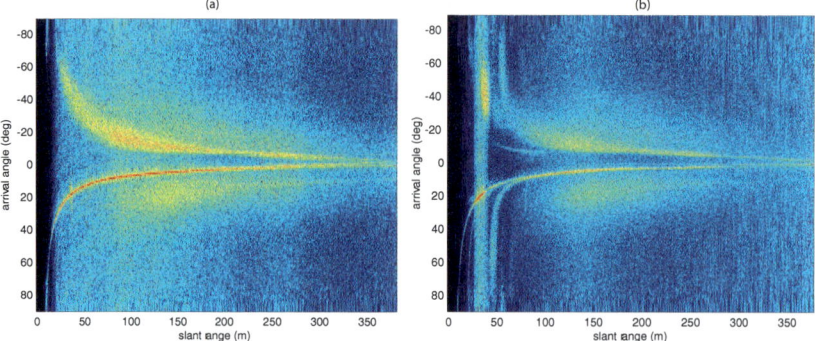

FIGURE 4. Direct and multipath returns as a function of slant range and arrival angle. Figures (a) and (b) correspond to transmission beampatterns as in Fig. 3(a) and (b), dashed line, respectively. The total dynamic range is 60 dB.

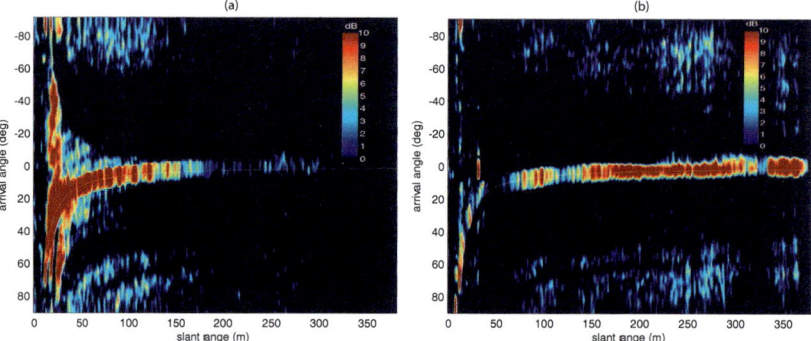

FIGURE 5. Generalized SNR as a function of range and the depression angle of the 7 degree receive beam. Figures (a) and (b) correspond to transmission beampatterns as in Fig. 3(a) and (b), continuous line, respectively.

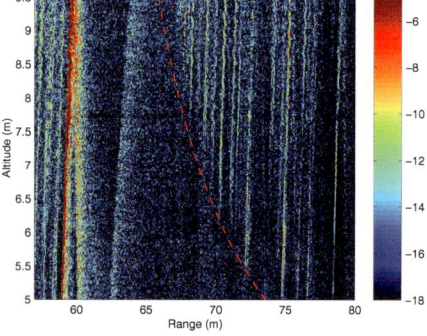

FIGURE 6. Direct and multipath returns from a sphere at 59 m range (Klein 5500 data), plotted for several sonar altidudes in the water column. The dashed red line shows the maximum shadow range, according to the geometry of the experiment.

The assumption is that the drop in SNR is due to high order multipath, excited at short range. To validate this assumption, a narrow transmission beam (3 degrees at 3 dB) steered at close range (32 m) was synthesized (Fig. 3(b), dashed line) and the corresponding data are plotted in Fig. 4(b). The 'bsB' multipath whose bottom specular reflection is at 32 m is clearly visible in the region around 145 m. Other multipath returns of first, second, third and fourth order are also visible, but the 'bsB' return is by far the most important, because its reception angle is nearly the same as that of the direct returns at far range. This explains the drop in correlation shown in Fig. 5(a) for the broad sector ensonification.

Thus, to achieve high SNR at very large relative ranges $r = R/W$, where R is the slant range, it is necessary not to ensonify the seafloor at short ranges, to avoid 'bsB' multipath whose arrival angle is impossible to separate in reception (similarly, a narrow receive beam is required to rule out the reciprocal 'Bsb' multipath).

To validate this assumption, the 3 degree transmission beam was steered at far range, as in the continuous line of Fig. 3(b). The corresponding SNR, obtained as above, is plotted in Fig. 5(b). The increase in SNR at long ranges over Fig. 5(a) is evident.

In the second experiment, conducted jointly with Defense Research & Development Canada (DRDC), Atlantic, a Klein 5500 sidescan sonar was deployed in 12 m water depth on a telescopic tower. The 455 kHz sonar was then moved vertically from 5 m to 9.5 m height while pinging in the direction of a 1 m diameter sphere placed at 59 m range. Figure 6 shows the sonar data with the strong direct return from the sphere followed by the multipath returns 'Bs' and 'Bsb'. The multipath intensity is shown to be approximately constant with the sonar altitude and comparable to the reverberation level after the end of the sphere shadow, indicated by the red dashed line. In this case, the first order multipath is stronger than second order one because of the wide vertical beampattern both in transmission and in reception.

MUTIPATH-REJECTING SONAR DESIGN

To design a sonar with high generalized SNR at long range, a key parameter is, therefore, the difference between the angle of the direct signal and the second order multipath signals ('Bsb', 'bsB') arriving at the same time.

By geometric considerations, we have

$$u_{Bsb} = \frac{H(H+2W)}{H^2 + (H+W)W u_B^2} u_B, \tag{1}$$

where u_B and u_{Bsb} are the sines of the receive angles of the bottom direct and the second order 'Bsb' multipath. Defining the relative height $h = H/W$ and the In terms of relative range r, we have,

$$u_{Bsb} = \frac{(2+h)r}{1+h+r^2}. \tag{2}$$

Therefore for relative range $r \gg 1$, $u_{Bsb} \cong (2+h)/r$ and

$$u_B - u_{Bsb} \cong 2/r. \tag{3}$$

This formula gives an important design criterion for the beamwidth necessary to achieve high SNR at long range. For example, at a relative range $r = 10$, it gives an angular separation of less than 8 degrees. Given the narrowness of the beams required at large r and the need to maintain a full swath imaging, a sonar design allowing different beamwidths at different ranges seems necessary.

To address these issues, a novel sonar design has been devised by Nato Undersea Research Centre. Besides incorporating all the experience in synthetic aperture sonar the Centre has developed, it provides multipath mitigation in order to fully exploit the gain in cross range resolution even in very shallow water. The aim is to achieve good shadow contrast up to a range of more than 10 times the water depth. The multipath suppression is achieved by transmitting two beams, a wide one steered at short range and a narrow one steered at long range, with two disjoint 30 kHz frequency bands to be transmitted simultaneously. Similarly, two different receive beams with null-to-null beamwidth of 28 and 14 degrees, respectively, and a 4 degree difference in depression angle, are used (Fig. 7). In other words, the sonar design is similar to a 'two sonar in one'

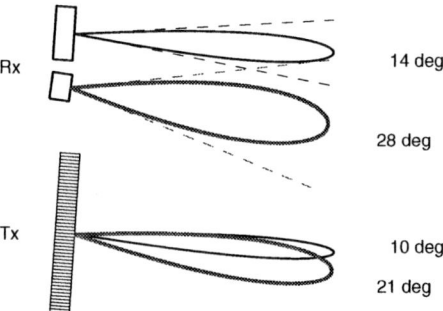

FIGURE 7. Vertical receive and transmit beampatterns for the NATO Undersea Research Centre. The wider beam is for short range while the narrow beam is for long range, up to more than 10 times the waterdepth. The two beams use disjoint sub-bandwiths of the signal.

design, with at least two important differences. The first is that only one transmitter is used, albeit a quite flexible one. The second is that a time-switched acquisition between the two staves of the receiver is implemented, reducing the incoming data rate.

The performance of this type of sonar was assessed using the ESPRESSO sonar performance prediction tool [4]. An SNR in excess of 10 dB is obtained out of at least one of the two beams up to 220 m range in 20 m water depth (Fig. 8). The sonar altitude in this example is 15 m above the sea bottom. The SNR which corresponds to a vertical beampattern similar to Fig. 3(a) is shown for comparison and indicated as 'conventional', and gives results with a good qualitative match with Fig. 5(a). It is worth mentioning that this agreement is obtained although ESPRESSO uses a model of multipath for which each reflection is purely specular. Figure 9 gives the direct and multipath returns modeled by ESPRESSO for a narrow beam steered at 20 degrees. It is evident from the comparison with the experimental data of Fig. 4(b) that the model

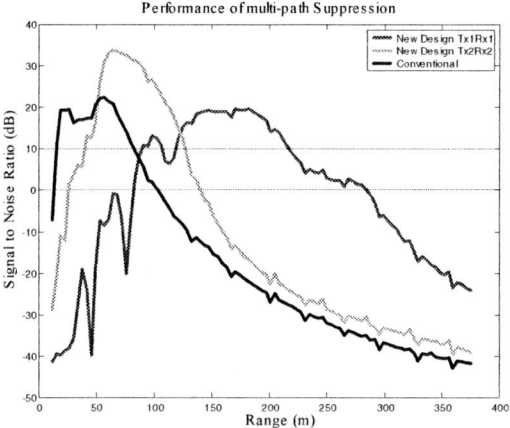

FIGURE 8. Performance prediction of the sonar design.

is only a first order approximation and that considerable spreading in angle due to non-specular reflections on the sea and bottom interfaces.

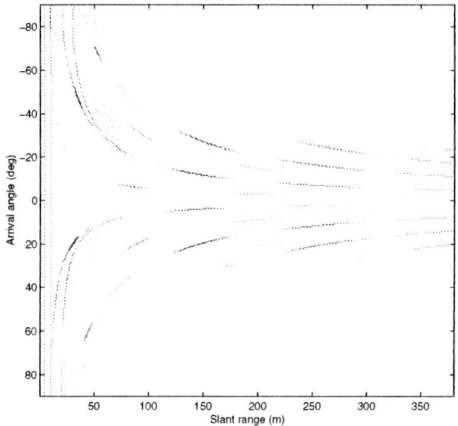

FIGURE 9. Direct and multipath returns for the transmission beam as in Fig. 3(b), steered at 20 degrees, according to the sonar performance prediction tool ESPRESSO.

CONCLUSIONS

Experimental evidence for the importance of (second order) multipath in degrading the SNR at large range to water depth ratios has been provided. This suggests that in shallow water, it is not enough to shape the beams in such a way to reject signal coming from the surface, and that a narrow beam pointing at long range can provide significant improvement. A synthetic aperture sonar design which achieves full swath imaging transmitting two beams, a wide one steered at short range and a narrow one steered at long range, with two different frequency bands to be transmitted simultaneously is discussed.

ACKNOWLEDGMENTS

The authors are grateful to John A. Fawcett and Terrance L. Miller, DRDC Atlantic, Canada, for the use of the Klein 5500 sonar.

REFERENCES

1. Bellettini, A. and Pinto, M. A., "Theoretical Accuracy of Synthetic Aperture Sonar Micronavigation using a Displaced Phase Centre Antenna," *IEEE J. Oceanic Eng.*, **27**, 780–789 (2002).
2. Davis, B., Gough, P. and Hunt, B., "Sea Surface Simulator for Testing a Synthetic Aperture Sonar," in *Impact of Littoral Environmental Variability on Acoustic Predictions and Sonar Performance*, edited by N.G.Pace and F.B.Jensen, Kluwer Acad. Publ., Lerici, Italy, 2002, pp. 473-480.
3. L. Wang, G. Davies, A. Bellettini and M. Pinto, "Multipath Effect on DPCA Micronavigation of a Synthetic Aperture Sonar," in *Impact of Littoral Environmental Variability on Acoustic Predictions and Sonar Performance*, edited by N.G.Pace and F.B.Jensen, Kluwer Acad. Publ., Lerici, Italy, 2002, pp. 465-472.
4. G. Davies, ESPRESSO Sonar Prediction Tool Software, under development at NATO Undersea Research Centre, La Spezia, Italy, version 0.6.

Mid- to High-Frequency Ambient Noise Anisotropy and Notch-Filling Mechanisms

Patrick Ferat & Juan Arvelo

Johns Hopkins University Applied Physics Laboratory, 11100 Johns Hopkins Rd., Laurel, MD 20723-6099

Abstract. Mid to high-frequency (1-20 kHz) noise is generally dominated by wind-driven wave activity but under certain conditions potentially exploitable ambient noise fields can be severely degraded by nearby ships, especially in the lower end of the band. The use of directive elements and adaptive methods are shown as possible ways to mitigate this problem. For a submerged receiver in a downward-refracting environment without nearby ships, a vertical noise notch that can offer increased array gain over the directivity index can be filled in by scattering from volume inhomogeneities. Just outside the notch, the ambient noise vertical directivity is sensitive to the assumed surface source directionality. A ray-based model is used to assess the sensitivity of the performance of a vertical line array and a volumetric array to these mechanisms in a tactically relevant environment.

INTRODUCTION

An understanding of the physics of ambient noise generation in the mid to high-frequencies is required in order to assess the capabilities of underwater acoustic systems that operate in that band. Some typical applicable systems are underwater communications, tomography and both active and passive sonar. For example, one way to mitigate passive (or active) sonar clutter due to distant shipping in bearing time recorder (BTR) displays is to operate at higher frequencies. In addition, higher frequency systems are attractive because of the reduced size and greater portability promised by high frequency arrays. A brief review of past ambient noise work is given first, followed by a description of current model developments.

Underwater acoustic ambient noise contains a surface-generated anisotropic component that is typically dominant over the isotropic thermal noise [1]. The ubiquitous thermal noise is due to molecular agitation and represents an absolute minimum level independent of sea conditions. At high frequencies, the surface generated component consists mainly of wind-driven (collapsing bubbles) noise with occasional shipping and biologic contributions. The theoretical study in [1] assumed straight-line propagation of rays and estimated the spatial correlation for monopole and dipole point sources.

A few years later Liggett and Jacobson [2] used a ray-based approach under idealized environmental conditions to compare source element directionality on estimated correlation function between vertically spaced point receivers. Two decades later Kuperman and Ingenito [3] developed a normal-mode approach for a stratified ocean, but their solution implies the use of a single layer of monopole point sources. Hamson [4] modified Kuperman and Ingenito's approach to model point sources with general vertical directivity to determine the effects of environmental parameters and source directionality on the noise level and the array response.

An accurate simulation of the performance of a passive sonar system requires an equally accurate estimation of the ambient noise. Since ambient noise is not isotropic, arrays with vertical aperture may be capable of harnessing additional gain (when steering towards the horizontal or below) over what it would otherwise have under isotropic conditions.

Wind-generated ambient noise is generally highest at elevation angles looking towards the sea surface and lower at angles towards the ocean floor. The difference in levels between the positive and negative elevation angles is a manifestation of the bottom loss at dominant grazing angles (when including corrections to account for refraction and attenuation). The oceanic waveguide's multipath effect and downward-refractive sound speed profile cause near horizontal ambient noise to be very low. This noise notch may be filled by the isotropic component of the ambient noise. Aredov [5], for example, hypothesizes an isotropic floor mechanism due to volume scattering. However, there is no known experiment that has been able to confirm this possible notch-filling mechanism.

Kennedy and Szlyk [6] collected ambient noise on two ten-wavelengths (16 & 32 kHz) vertical arrays off the Bahamas for a one-year interval. They were able to collect samples at various wind speeds, but only the broadside beam data were available. Despite the limitation of their measurement, they were able to show a distinct difference in the noise statistics that appear to correlate with the presence or absence of whitecaps. They found that their noise model requires dipole sources when whitecaps (high sea states) are present and a layer of monopole sources in their absence (low sea states). In this paper, the reference to the Kennedy model applies to the former.

Harrison [7] justified the use of a simple ray-based approach for high-frequency wind-driven noise modeling by demonstrating similar coherence function in a range-independent waveguide to that obtained with a normal-mode approach. He later extended the ray-based approach to account for range-dependent environments [8]. An analytic model similar to that of Harrison is used model wind-driven noise limited by a volume scattering component. An efficient ray-based computational model is described to include a shipping noise component. The resulting simulation tool is used here to estimate mid to high-frequency passive array performance in the presence of nearby shipping noise in addition to the wind-driven component.

THEORY

The noise model developed by the Johns Hopkins University Applied Physics Laboratory (JHU/APL) is a hybrid model that has made use of published models and theories obtained in the literature. The calculation of the vertical noise directivity is based on [7], but modified to allow for arbitrary surface source directionality [6] and also includes a volume scattering mechanism [5]. Once the ambient noise directionality is obtained, we calculate a Cross Spectral Density (CSD) matrix for an array of elements, as described by [7,9]. The CSD for shipping is separately calculated by incoherently summing the CSD matrices of individual ships. For a particular environment, ship locations and speeds (i.e., source level) are described by a realization of the Historical Temporal Shipping (HITS). In the present model, ships are stationary and the bathymetry is range-independent. The following sections provide additional detail on each of these sub models.

Wind-Generated Ambient Vertical Noise Directivity

Wind-generated ambient noise is assumed to be azimuthally isotropic but anisotropic in vertical. The anisotropy in the vertical is due primarily by the sound speed profile, the bottom loss and the depth of the receiver. We ignore any azimuthal or range dependence in bathymetry, wind speed, bottom composition and sound speed in the water column. Therefore, the model is applicable for predicting the noise field from noise sources that are relatively close to the receiver. In addition, the model does not account for any temporal fluctuations of the sound speed typical of internal waves which can have a substantial effect on propagation in certain environments.

The following ambient noise spatial coherence equation was derived in [7]:

$$\rho(d,\gamma) = 2\pi \int_0^{\pi/2} \frac{e^{ikd\sin(\phi_r)\sin(\gamma)} e^{-as_p} + R_b e^{-ikd\sin(\phi_r)\sin()} e^{-a(s_c - s_p)}}{[1 - R_s R_b e^{-as_c}]} \times \quad (1)$$

$$J_o(kd\cos\phi_r \cos\gamma) \sin^{2m-1}\phi_s \cos\phi_r d\phi_r$$

where k is the wavenumber, d is the element separation, γ is the element pair orientation, ϕ_r is the vertical angle at the receiver measured from the horizontal, ϕ_s the vertical angle of the surface source measured from the horizontal, $2m-1$ is the surface source directivity exponent of the sine function, R_s and R_b are the surface and bottom plane-wave reflection coefficients, respectively. The exponential terms containing a are absorption losses for the ray paths s_c and s_p.

When the receiver separation $d=0$ (and $\gamma=0$), the spatial coherence is equivalent to the omni noise power. Assuming dipole sources ($m=1$), negligible absorption and surface losses, Eq. (1) can be simplified to give the element omni power N:

$$N = 2\pi \int_0^{\pi/2} \frac{1+R_b}{[1-R_b]} \sin(\phi_s) \cos(\phi_r) d\phi_r$$

$$= 2\pi \int_0^{\pi/2} \frac{1}{[1-R_b]} \sin(\phi_s) \cos(\phi_r) d\phi_r + 2\pi \int_0^{\pi/2} \frac{R_b}{[1-R_b]} \sin(\phi_s) \cos(\phi_r) d\phi_r \quad (2)$$

The first integral term represents the noise contribution coming from above the receiver where the integrand contains the Vertical Noise Directionality (VND) function $\frac{\sin(\phi_s)}{1-R_b}$ and the solid angle component $\cos(\phi)d\phi$. The second integral represents the noise contribution from below the receiver, where the integrand VND component is $R_b \frac{\sin(\phi_s)}{1-R_b}$, simply the upward directivity scaled by the bottom loss.

In a hypothetical deep water iso-speed environment $\theta_s \sim \theta_r$ and the second integral term vanishes when $R_b \ll 1$ (e.g., large bottom loss). When applying the dipole surface source strength A [12] Eq. (2) simplifies to:

$$N_{deep\,water} = 2\pi A \int_0^{\pi/2} \sin(\phi_r)\cos(\phi_r)d\phi_r = \pi A \qquad (3)$$

The evaluation of Eq. (3) represents the omni power in a deep water, high bottom loss and iso-speed environment. This result is identical to Eq. (55) in [12] which was derived from equations based on a different noise model than shown here.

The derivation shown in Eq. (3) could be re-derived for different values of m, but the integral may not be as simple to evaluate analytically as it is for the case of dipole surface sources. Although convenient, the surface directionality function need not be described via simple trigonometric functions like $\sin^n(\phi_s)$, but rather it may be represented by the general function $D(\phi_s)$. Since A is applicable to dipole surface sources, a scaling factor is applied to insure that the deep water, iso-speed omni power due to any arbitrary source directivity function $D(\phi_s)$ is identical to that obtained by Eq.(3). In the following expressions, A_s represents the effective arbitrary surface source level which also includes the 2π term.

The JHU/APL model can also describe the surface noise source directionality proposed by Kennedy [6], which is most sensitive to whether or not white caps are present. In the absence of white caps, an effective source directivity function is modeled by incoherently summing the contribution of monopole source layers below the surface. With white caps present, the source directionality is described by dipoles. Kennedy acknowledges that under whitecap conditions, sub-surface monopole sources are present but suggests that the wave crashing and spray events are more dominant.

Therefore, the general expression for VND becomes (negative angles looking up):

$$VND(\phi_r^-) = \frac{A_s \cdot l_D \cdot D(\phi_s)}{\sin\phi_s \cdot (1-R_s(\phi)) \cdot R_b(\phi) \cdot l_C)}$$

and,

$$VND(\phi_r^+) = R_b(\phi) \cdot l_{DC} \cdot VND(\phi_r^-) \qquad (4)$$

where A_s is the scaled surface source strength applicable to an arbitrary surface source directivity D. The terms l_D, l_C and l_{DC} represent the absorptive losses of the surface to receiver ray, the ray path cycle distance and the difference path, respectively.

A volume scattering mechanism is implemented to limit the level of the noise at low elevation angles [5]. The model assumes that each elemental volume in the water column scatters noise omni-directionally. The incident power is the omni noise power (assumed to be depth independent) and the volume scattering strength is assumed to be depth-independent. The contribution of distant volume scatterers is limited by absorption. Currently, the volume scattering strength can be user input or selected from average values listed in [12].

An example of the sensitivity of the VND model to surface directionality is shown in Fig. 1. The sound speed profile (GDEM, [13]), bottom loss [12], volume scattering strength of -72 dB and wind speed (5 m/s) are typical of the summer conditions in the Strait of Hormuz (25° 48 N, 56° 48 E). The VND assuming dipole surface sources are shown in (a) and VND using Kennedy's monopole surface directionality model for low wind speeds (e.g., model of a depth distribution of monopole sources when white caps are absent at low wind speed) are shown in (b), negative elevations look up to the surface. Sub-plots (c) and (d) show the sound speed and bottom loss. Of note are the significant differences in the shape of VND for angles outside the notch.

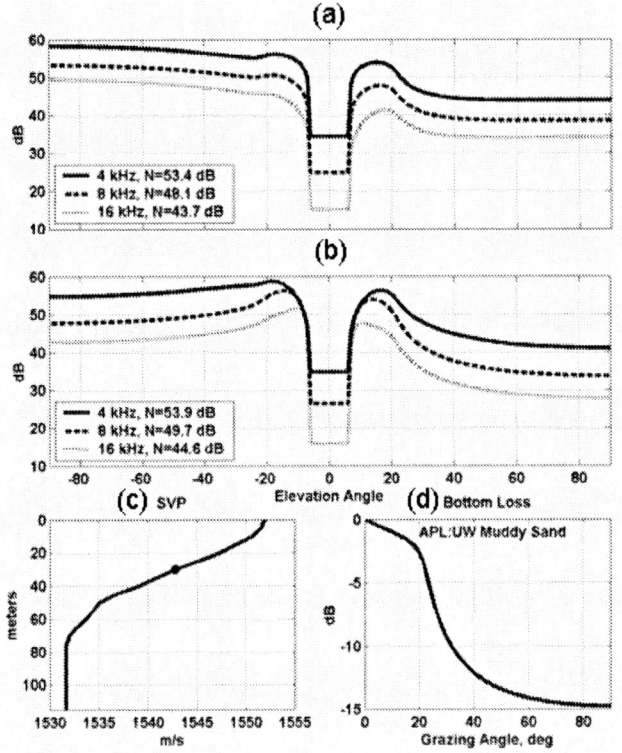

FIGURE 1. Ambient vertical noise directionality in Strait of Hormuz for receiver at 30 m 5 m/s wind speed and VSS=-72 dB, (a) dipole surface source directionality, (b) Kennedy surface source directionality, (c) sound speed profile and (d) bottom loss versus grazing angle

Via [9,10], VND can be mapped to a CSD matrix for an arbitrary array of elements. The implementation is straightforward and is not repeated here.

Cross Spectral Density of Shipping Noise

The CSD of shipping noise is determined by the incoherent sum of the CSD of all ships. The ray-based model CASS/GRAB [13] provides the eigenray launch, arrival angles and transmission loss versus range for a source at 5 meters. The same environmental inputs used for the noise (SVP, interface losses, etc) are used to maintain consistency between the ambient noise and shipping models. A separate module generates a shipping realization of the HITS database [13], for a given season/month and location. This realization provides the location (azimuth and range) of each ship and also specifies the ship type and speed. A source level is associated with ship type and speed based on [14]. A shipping realization in the Straits of Hormuz overlays unclassified bathymetry, Fig. 2.

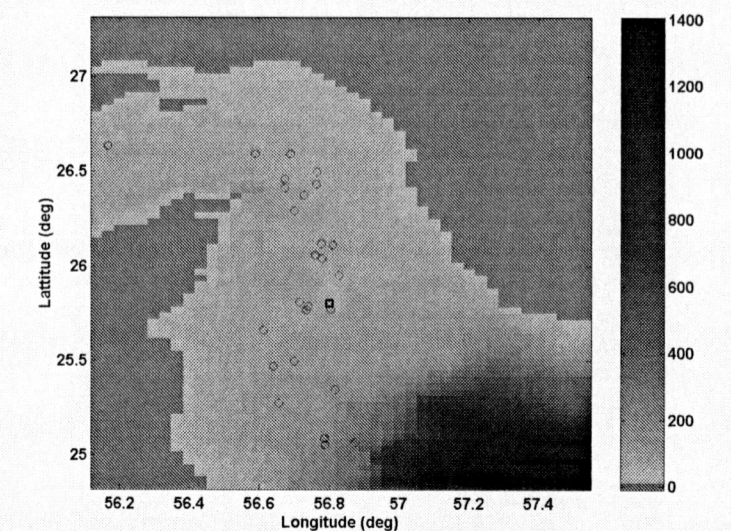

FIGURE 2. Bathymetry (in meters) and a shipping realization in the Straits of Hormuz. The circles represent ships and the square represents the location of the receiver arrays.

For a ship at a given range, consider that there are M eigenrays to the center element of an array of N elements. Letting p be the complex pressure matrix of size NxM, each element of this matrix p is evaluated for element n and ray m as:

$$p_n^m = a_m \cdot \exp(i(k_x^m x_n + k_y^m y_n + k_z^m z_n)) \quad (5)$$

where a_m represents the received level pressure amplitude, and the terms in the exponential are the plane-wave arrival vectors for each element position for the given ray m. Ignoring any multi-path interference, the CSD matrix for a single ship s is simply:

$$R_{SHIP(s)} = p\,p' \quad (6)$$

and the total CSD matrix for all ships is:

$$R_S = \sum_{s=1}^{\infty} R_{SHIP(s)} \quad (7)$$

RESULTS

Consider two co-located receiver arrays located at a position represented by the square in Fig. 2. The first array is a 3-plane 72 element volumetric array with 10 kHz design frequency in a plane, but with a 0.8λ plane separation. The second array is a 16 element Vertical Line Array (VLA) with a design frequency of 10 kHz. The element positions and corresponding beam patterns at 8 kHz are shown in Fig. 3.

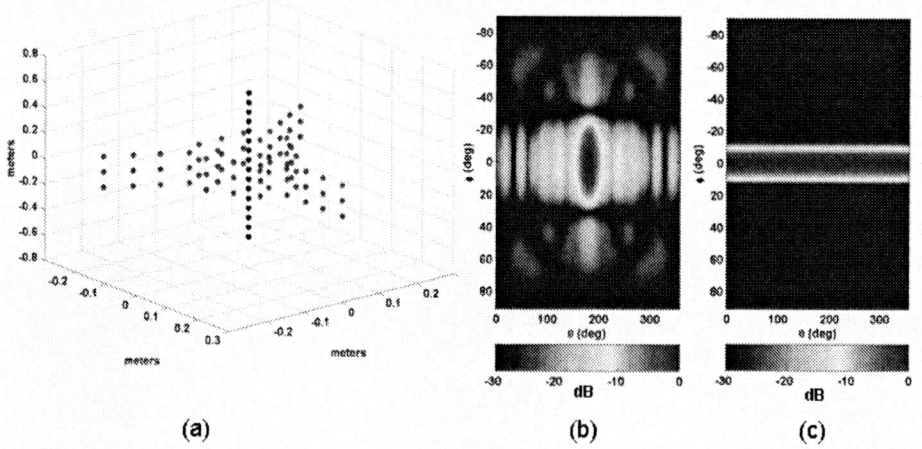

FIGURE 3. (a) Element positions of VLA and volumetric arrays, 8 kHz beam pattern steered horizontally for the volumetric array (b) and VLA (b).

Unity spatial weights are used for the volumetric array and Chebychev (-30 dB sidelobes) for the VLA. The volumetric array ambient noise response at 8 kHz versus vertical steering angle (negative looks up) and one azimuth direction is shown in Figure 4, given the noise directivities shown in Figure 1. [The response of all azimuthally steered beams will be essentially identical; however slight variations are expected due to the azimuth-dependent beam patterns]. The left panel shows the response using the dipole directionality model, and the right panel shows the response using the Kennedy monopole directionality model. The solid lines represent conventional beamforming (CBF) output and the light dotted lines modeled adaptive beamforming (MABF) assuming unconstrained minimum variance distortionless rejection (MVDR).

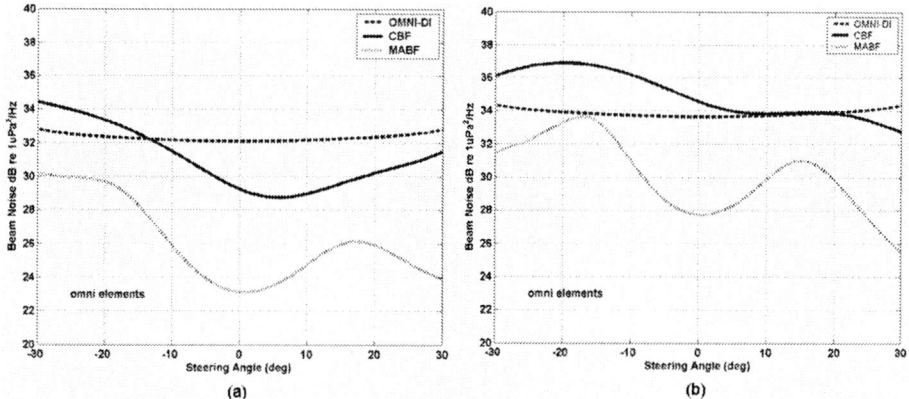

FIGURE 4. Ambient noise array response versus elevation angle at 8 kHz of the volumetric array assuming in the Straits of Hormuz summer (a) dipole source directionality and (b) Kennedy monopole source directionality.

The beam levels assuming isotropic noise are shown by the solid dashes in Fig. 4. The elevated response of the Kennedy model compared to the dipole model is expected based on the slight increase in omni level (Fig.1). Note the ability of adaptive methods to better resolve the ambient noise field structure in the notch region at the expense of possible poor white noise gain given the unconstrained adaptive processor. In recent work, not available at the time of writing of this paper, we have implemented a white noise gain constraint which provides robustness to uncorrelated noise passing through the beamformer.

Now consider that shipping noise generated via Eq. (9) is incoherently added to the CSD of the ambient. At 8 kHz, only ships near the receivers shown in Figure 2 are expected to generate sufficient noise to contaminate the ambient background. At 172°T, a tanker is located 4 km away, and a cluster of two tankers and one supertanker at ~270°T are located ~7 km away and another tanker is observed at ~0°T 17 km away. The response of the volumetric array (unity shading) to ships and ambient (Kennedy monopole source directionality model) at 8 kHz are shown in Fig. 5. The left panel shows the CBF response over all azimuthal steering directions (5° increments), as well as the level of the shipping omni level (44 dB). The right panel shows the result of MABF. Note the severe influence of the nearby ships on a significant number of azimuthal beams using CBF; considerable sidelobe leakage is apparent due to the poor sidelobe control in vertical and azimuth as shown in Fig. 3 (the remaining beams reach the ambient, similar to the solid curve in Fig. 4 (b)). This leakage is virtually eliminated using MABF; four distinctly elevated azimuthal beams are observed (individual ships), all other azimuthal beams reach the ambient levels, similar to the dashed curve in Fig. 4(b).

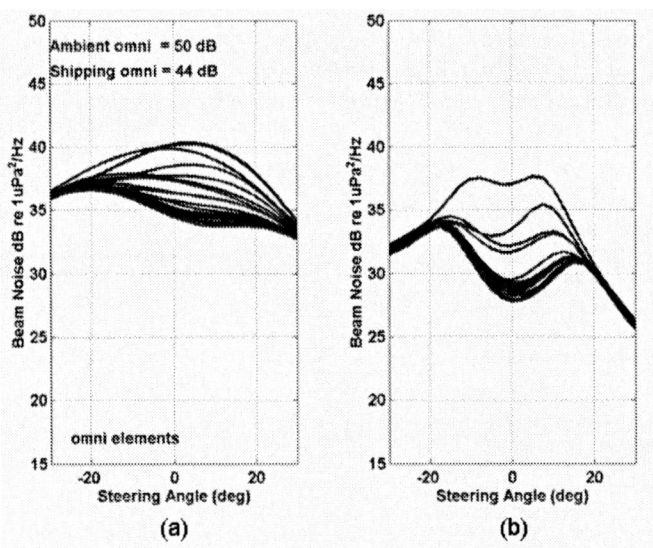

FIGURE 5. Ambient and shipping noise array response versus elevation angle at 8 kHz of the volumetric array assuming Kennedy monopole source directionality for ambient, (a) CBF processing and (b) MABF processing.

For sake of comparison, Figs. 6 and 7 show the response of the VLA in the same shipping field, using omni-directional and cardioid elements, respectively. For both cases, the Chebychev spatial window was used and the cardioid null was steered towards the nearest ship at 172°T. The dashed curves show the response of the arrays to ambient alone.

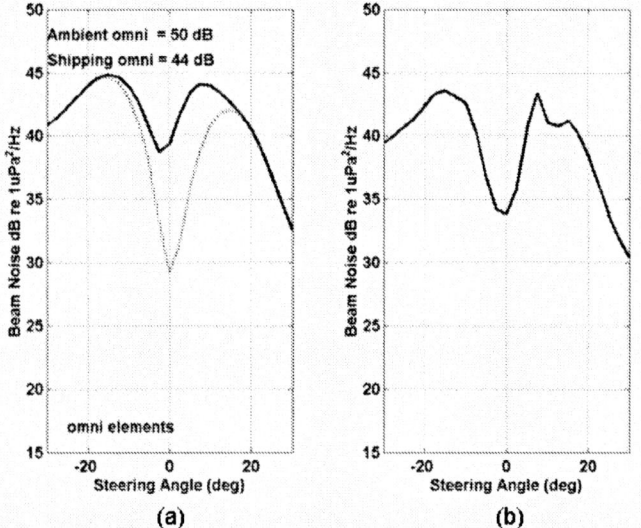

FIGURE 6. Ambient and shipping noise array response versus elevation angle at 8 kHz of the VLA (omni elements) assuming Kennedy monopole source directionality for ambient, (a) CBF processing and (b) MABF processing.

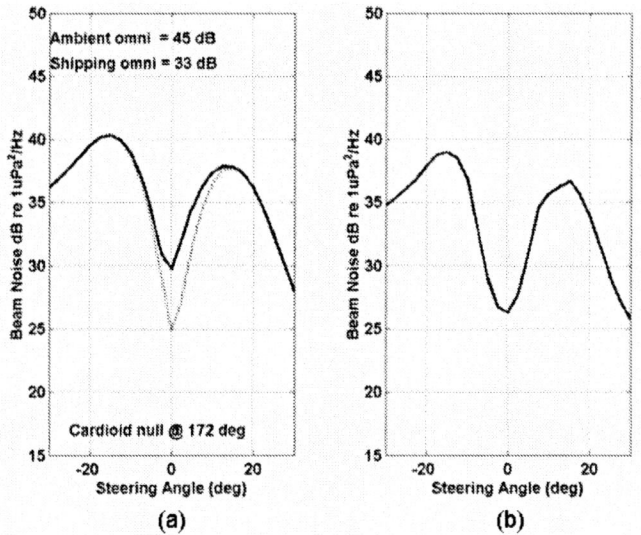

FIGURE 7. Ambient and shipping noise array response versus elevation angle at 8 kHz of the VLA (cardioid elements) assuming Kennedy source directionality for ambient, (a) CBF processing and (b) MABF processing.

Due to the low sidelobes provided by Chebychev shading, the 16 element VLA of omni elements is capable of theoretically measuring lower beam levels than the 72 element volumetric array in the notch with CBF, in the absence of shipping. But away from the notch, the beam levels rise rapidly. With shipping present, the notch gets filled in by 10 dB (Fig. 6(a)), although a shallow notch is still observed. The reason for this is that the VLA vertical beam width is narrow enough to steer into the 'shipping noise notch' and reject shipping noise in the vertical dimension. The volume scattering floor limits the level in this case. In the presence of shipping, MABF appears to provide 5 dB additional noise gain compared to CBF in the notch area. Note that the VLA has no ability to resolve ships in azimuth, hence there is only one curve shown. With the VLA of cardioids steered to null the nearest tanker at 172°T (Figure 7), the ambient level is reduced by 4.7 dB and the effective shipping level is reduced by 11 dB with CBF processing.

SUMMARY

A ray-based model has been developed to explore the sensitivities of wind-wave generated surface source directionality, volume scattering, element directionality and discrete shipping noise to the mid to high-frequency performance of a vertical line array and a volumetric array in a tactically relevant environment. Two surface source directionality models were shown in this paper to give different levels and structures to the ambient vertical noise directivity functions and corresponding array responses. Although based on empirical models found in the literature, future efforts by JHU/APL to develop physics-based surface directionality models will provide the necessary tools to validate the ambient noise model with measured data. In addition, the benefits of directional elements and adaptive methods were demonstrated to reduce the contamination of nearby ships in order to reach the ambient noise background.

ACKNOWLEDGEMENTS

The authors are grateful for the financial support from the IRAD program of the National Security Technology Department of the Johns Hopkins University / Applied Physics Laboratory.

REFERENCES

1. B. F. Cron and C. H. Sherman, "Spatial correlation functions for various noise models," *J. Acoust. Soc. Am.* **34** p. 1732 (1962)
2. W. S. Liggett and M. J. Jacobson, "Covariance of surface-generated noise in a deep ocean," *J. Acoust. Soc. Am.* **36** p. 303 (1965)
3. W. A. Kuperman and F. Ingenito, "Spatial correlation of surface generated noise in a stratified ocea," *J. Acoust. Soc. Am.* **67** p. 1988 (1980)
4. R. M. Hamson, "The theoretical responses of vertical and horizontal line arrays to wind induced noise in shallow water," *J. Acoust. Soc. Am.* **78** (5), p. 1702 (1985)
5. A.A. Aredov, N. N. Okhrimenko, and A.V. Furduev, "Anisotropy of the noise field in the ocean (experiment and calculations)," Sov. Phys. Acoust. **34** (2), p. 128 (1988)
6. R. M. Kennedy and T. K. Szlyk, "Modeling high-frequency vertical directional spectra," *J. Acoust. Soc. Am.* **89** (2), p. 673 (1991)
7. C. H. Harrison, "Formulas for ambient noise level and coherence," *J. Acoust. Soc. Am.* **99** (4), p. 2055 (1996)
8. C. H. Harrison, "Noise directionality for surface sources in range-dependent environments," *J. Acoust. Soc. Am.* **102**, p. 2655 (1997)
9. C. H. Harrison, R. Brind and A. Cowley, "Computation of noise directionality, coherence and array response in range dependent media with canary," J. Comp. Acoust. 9 (2), p. 327 (2001)
10. M. J. Buckingham and N. M. Carbone, "Source depth and the spatial coherence of ambient noise in the ocean," *J. Acoust. Soc. Am.* **102** (5), p. 2637 (1997)
11. G. B. Deane, M. J. Buckingham, and C. T. Tindle, "Vertical coherence of ambient noise in shallow water overlaying a fluid seabed," *J. Acoust. Soc. Am.* **102** (6), p. 3413 (1997)
12. "APL-UW High-Frequency Ocean Environmental Acoustic Models Handbook," APL/UW Technical Report 9407, (1994)
13. "Oceanic and Atmospheric Master Library Summary", Naval Oceanographic Office, Systems Integration Division, Stennis Space Center, MS 39522-5001, October 2002.
14. "Acoustic source levels of commercial vessels for use in sonar system modeling and analysis", Naval Undersea Systems Center, Tech. Memo 901157, August 1990.

Measurements and Predictions of High Frequency Ambient Noise

Andrew Holden

Dstl Winfrith, Dorchester, Dorset, UK

Abstract. A great deal has been published on ambient noise. Most of this has covered (a) omni directional levels, and (b) the vertical and horizontal directivity of shipping noise at low frequencies. There is some published material on the vertical directivity of wind generated noise at lower frequencies, but very little at higher frequencies. In order to study wind generated ambient noise at higher frequencies, work has recently started using a small planar array from QinetiQ Bincleaves. As well as measurements, a model called CANARY has been written to predict ambient noise vertical directivity and array responses to this noise. This paper contains some comparisons between CANARY predictions and (a) previous measured vertical directivity data at 4.5 kHz, (b) measured omni-directional data, and (c) initial analysis of the planar array measurements. The paper shows the nature of the ambient noise vertical structure at higher frequencies and that the CANARY predictions are in good agreement with the measurements.

INTRODUCTION

Ambient noise has been a big field of research from the 1960's. Initially, the emphasis was on its omni-directional properties, and then this shifted to the vertical and horizontal directivity of shipping noise at low frequencies. Since the mid 1980's near surface bubble layers have been an area of interest. An area that seems to have received little attention is the vertical directivity of wind generated noise, particularly at higher frequencies. Wind generated ambient noise in the ocean forms a noise field with a distinct vertical directivity. At low frequencies there are high noise levels at upward steer angles towards the sea surface, lower levels at angles towards the seabed, while there is a noise notch, a region of very low noise, in the horizontal. Recently there has been interest in studying the vertical directivity of ambient noise and determining the nature of the noise notch at higher frequencies of up to 50 kHz.

In order to study wind-generated ambient noise at frequencies of up to 50 kHz, work has recently started using a small planar array from QinetiQ Bincleaves. This array is being used to make measurements in a range of environments.

As well as measurements, predictions have been made using the CANARY (Coherence and Ambient Noise for ARraYs) model [1,2] Version 8.1. This model was written to predict the vertical directivity of ambient noise due to wind, rain and shipping, and to predict the array responses to ambient noise for arbitrary 3D arrays with directional hydophones. It is a range independent ray tracing model and was designed primarily for the 1 to 6 kHz frequency band. CANARY array response predictions have been found to be in good agreement with measured data in the 1 to 6 kHz band.

This paper contains previous measurements and predictions at 4.5 kHz, compares CANARY predictions with measured omni-directional data, and gives the initial analysis of the planar array measurements and corresponding CANARY predictions.

CANARY MODEL

This section gives brief details about the assumptions in CANARY model and how they relate to higher frequencies.

Wind generated surface noise is modelled by an infinite sheet of point sources at the sea surface, radiating sound with a dipole directionality. The wind source formula is an empirical fit to the derived source levels by Kuperman/Ferla [3]. Above 3.2 kHz the formula provides a 3.6 dB per Octave slope.

The sound absorption coefficient is calculated using the Thorp equation [4] that is only a function of frequency. A good fit can be obtained between this equation and Francois-Garrison equation [5] for particular combinations of frequency, salinity, PH and temperature. For other combinations there can be significant differences between the two equations.

Surface loss can be treated as either; (a) a look up table of loss versus grazing angle, or (b) using a formula that is limited to low grazing angles, low frequencies and moderate sea states. The formula can produce unrealistic surface loss values if any of these bounds are exceeded.

Bottom loss can be treated as either; (a) single layered solid with a look up table of loss versus grazing angle, or (b) as two layers, a sediment and a substrate. The two layered technique [6] uses the reflection coefficients at the water/sediment and sediment/substrate to derive an analytical expression for the overall three layer reflection coefficient. The technique in [6] is stated to be valid at high frequencies.

MEASURED AND PREDICTED AMBIENT NOISE

This section gives measurements and predictions of the vertical directivity of ambient noise. It should be noted that the vertical directivity cannot be seen directly, but only through the smoothing effect of an array beampattern.

In order to make the predictions, it is necessary to have good environmental data. Unfortunately, none of the measurements presented have a complete set of environmental data, hence some aspects have been assessed.

The first results to be shown come from previous work to assess CANARY. This work involved measuring and predicting the vertical array response of a line array in several environments and for a range of frequencies. The array was of the order of 1m high and was canted down at a $20°$ angle. The measurement shown in Figure 1 is at 4.5 kHz, in a deep water environment and for an array depth of about 100m. This measurement was typical of measurements obtained during the work. Two beamformer shading functions are applied, uniform and Dolph-Chebychev. The corresponding CANARY predictions are shown in Figure 2.

The vertical angle of 0° is the horizontal direction and –90° is directed straight down at the sea bed. The fall off in array response towards the endfire directions of the array is due to hydrophone directionality and the changing DI. This has the effect of making the array's view of the noise notch look less pronounced.

It can be noted that the array does not see the notch at 0°. This is due to the width of the beampattern's main lobe and the array geometry. It is also predicted that depending on the environment, the location of the notch can occur at angles slightly below 0°.

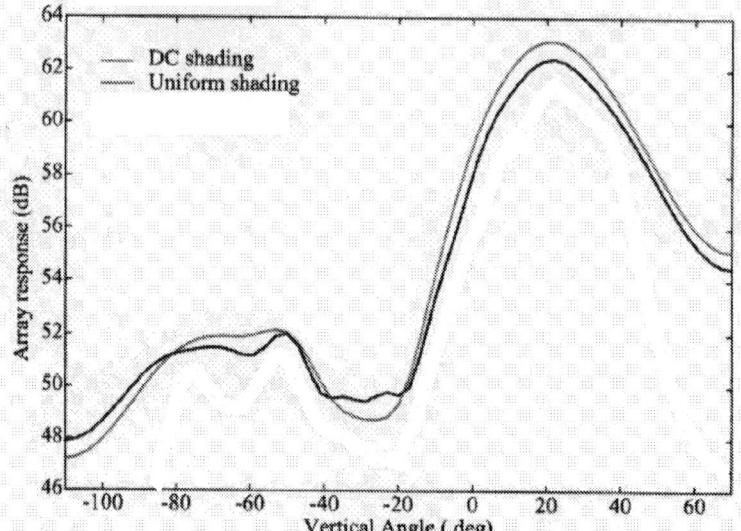

FIGURE 1. Measured vertical array response in deep water at 4.5 kHz.

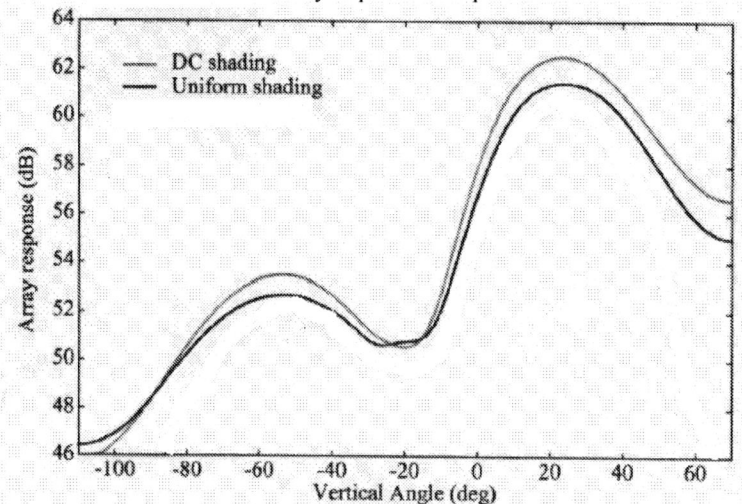

FIGURE 2. Predicted vertical array response in deep water at 4.5 kHz.

A wide range of measured omni directional data has been published. Some general upper and lower bounds of this data is shown in Figures 3 and 4. Some CANARY wind source only predictions for an omni directional hydrophone are also given for a possible high ambient noise environment (shallow water with a sandy sea bed) and a possible low ambient noise environment (deep water with a mud sea bed).

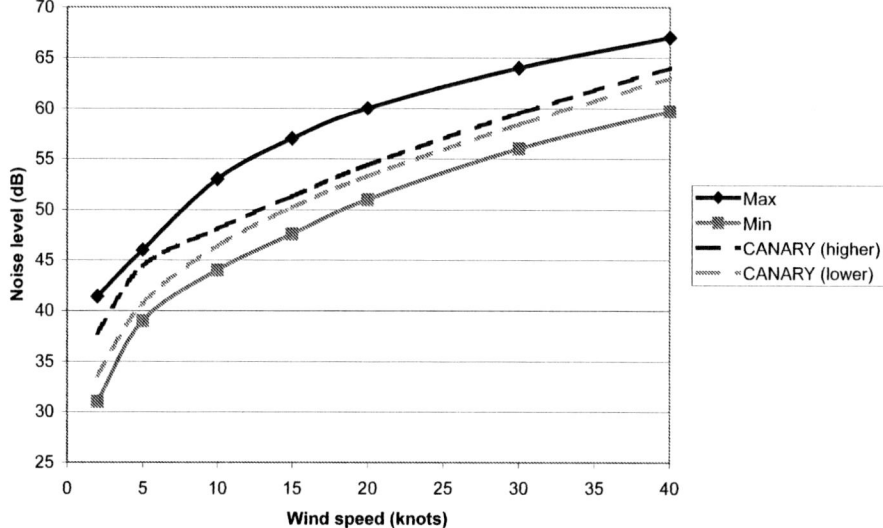

FIGURE 3. General bounds of measured omni directional noise and CANARY predictions 5 kHz.

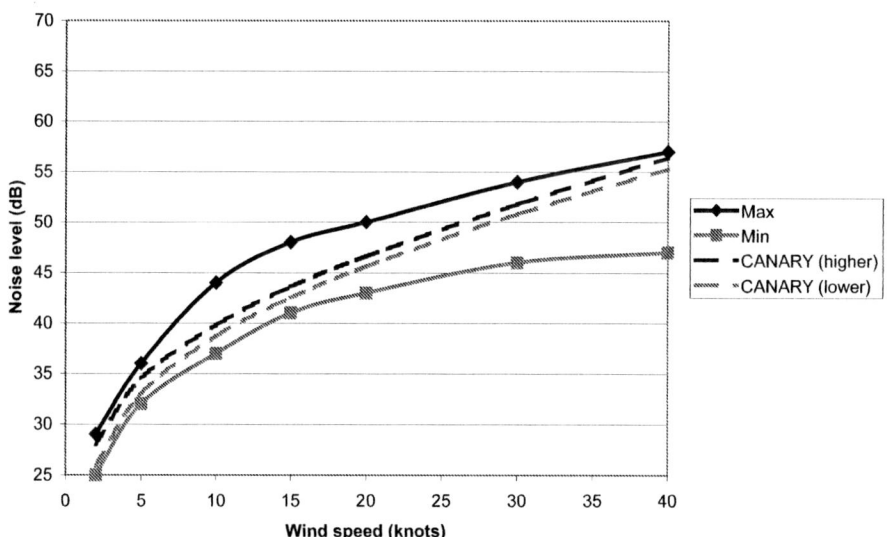

FIGURE 4. General bounds of measured omni directional noise and CANARY predictions 20 kHz.

The predictions show that the environment seems to make little impact on the ambient noise level, while the measurements show that the environment can make a significant difference. This needs further investigation. One explanation is that there could be ship noise in some of the measured data.

The current work involves measuring ambient noise using a small planar array in deep and shallow water for a range of frequencies. A measurement of ambient noise level vs vertical and horizontal angle at 20 kHz, an array depth of 90m, in deep water for a 20dB level scale is shown in Figure 5. The CANARY prediction is shown in Figure 6. High levels can be seen towards the sea surface and low level towards the sea bed. There is a reasonably good agreement between the measurement and predictions.

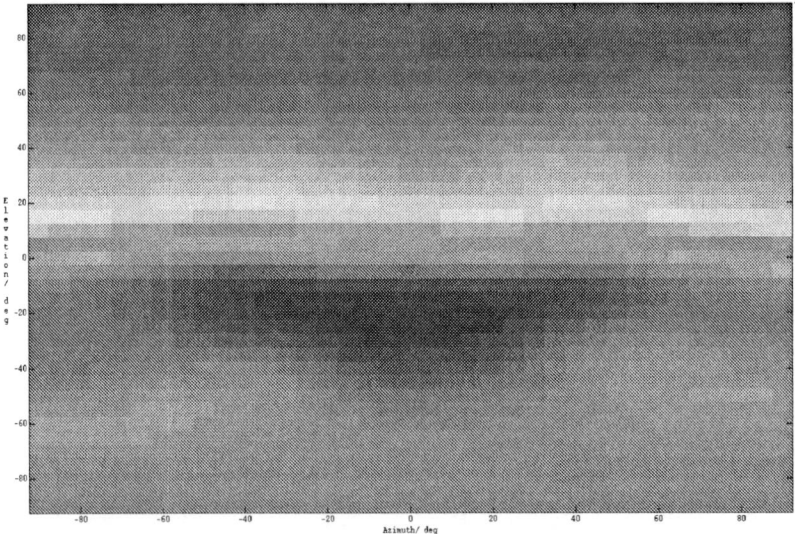

FIGURE 5. Measured ambient noise vs horizontal and vertical angle at 20kHz.

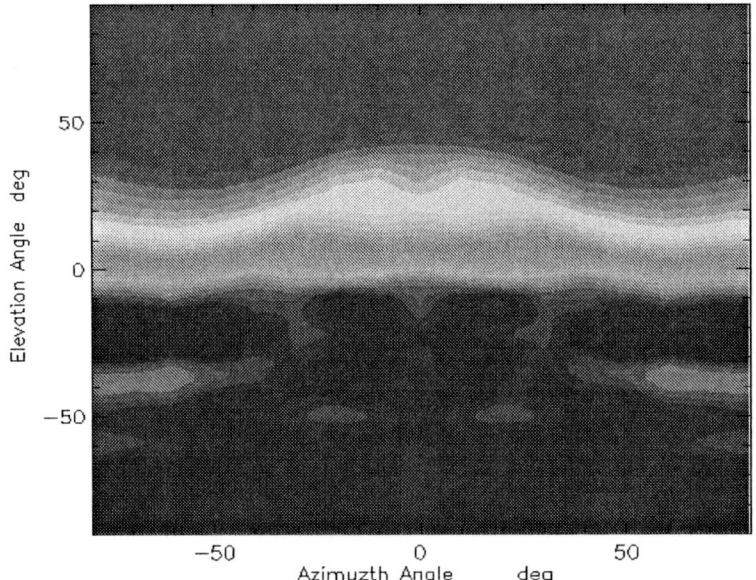

FIGURE 6. Predicted ambient noise vs horizontal and vertical angle at 20kHz.

CONCLUSIONS

This paper has investigated measurements and predictions of high frequency ambient noise. It has been shown that the ambient noise field has a distinct vertical structure at higher frequencies and that there is generally good agreement between the CANARY predictions and the measured data.

Further work needs to be done to explain the differences the environment brings to the variability of predictions and measurements.

ACKNOWLEDGMENTS

I would like to thank QinetiQ Bincleaves, Weymouth, UK for providing some of the measured data for this paper.

© British Crown Copyright 2004. Published with the permission of the Defence Science and Technology Laboratory on behalf of the controller of HSMO.

REFERENCES

1. Harrison C.H., 'CANARY – A Simple Model of Ambient Noise and Coherence', *Applied Acoustics* 51, pp. 289-315 (1997)
2. Harrison, C. H., 'Formulas for ambient noise level and coherence', *J. Acoust. Soc. Am.* 99(4), pp. 2055-2066 (1997)
3. Kuperman W.A. and Ferla C., 'A shallow water experiment to determine the source spectrum level of wind generated noise', *J. Acoust. Soc. Am.* 77, pp. 2067 (1985)
4. Thorp W. H., 'Analytic description of the low frequency attenuation coefficient', *J. Acoust. Soc. Am.* 42, pp. 270-271 (1967)
5. Francois R.E. and Garrison G.R., 'Sound Absorption based on ocean measurements', *J. Acoust. Soc. Am.* 72, pp. 1879-1890 (1982)
6. Ainslie M.A., 'Reflection and transmission coefficients for a layered fluid sediment overlying a uniform solid substrate', *J. Acoust. Soc. Am.* 99(2), pp. 893-902 (1996)

Ultrasonic Time Reversal Mirrors

Mathias Fink, Gabriel Montaldo, Mickael Tanter

Laboratoire Ondes et Acoustique, ESPCI, University Paris 7, 10 rue Vauquelin, Paris, France

Abstract. For more than ten years, time reversal techniques have been developed in many different fields of applications including detection of defects in solids, underwater acoustics, room acoustics and also ultrasound medical imaging and therapy. The essential property that makes time reversed acoustics possible is that the underlying physical process of wave propagation would be unchanged if time were reversed. In a non dissipative medium, the equations governing the waves guarantee that for every burst of sound that diverges from a source there exists in theory a set of waves that would precisely retrace the path of the sound back to the source. If the source is pointlike, this allows focusing back on the source whatever the medium complexity. For this reason, time reversal represents a very powerful adaptive focusing technique for complex media. The generation of this reconverging wave can be achieved by using Time Reversal Mirrors (TRM). It is made of arrays of ultrasonic reversible piezoelectric transducers that can record the wavefield coming from the sources and send back its time-reversed version in the medium. It relies on the use of fully programmable multi-channel electronics. In this paper we present some applications of iterative time reversal mirrors to target detection in medical applications.

THE TIME REVERSAL MIRROR IN PULSE ECHO MODE

One of the most promising areas for the application of TRMs is pulse-echo detection. In this domain, one is interested in detection, imaging and sometimes destruction of passive reflecting targets. A set of transducers first sends a short impulse and then detects the various echoes from the targets. One looks for calcifications, kidney or gallbladder stones, or tumors. As the acoustic detection quality depends on the availability of the sharpest possible ultrasonic beams to scan the medium of interest, the presence of an aberrating medium between the targets and the transducers can drastically change both the beam profiles and the detection capability. In medical imaging, a fat layer of varying thickness, bone tissues, or some muscular tissues may greatly degrade focusing.

For such applications, a TRM array can be controlled according to a three step sequence. One part of the array generates a brief pulse to illuminate the region of interest through any aberrating medium. If the region contains a point reflector, the reflected wavefront is selected by means of a temporal window and then the acquired information is time-reversed and reemitted. The reemitted wavefront refocuses on the target through the medium. It compensates also for unknown deformation of the mirror array. Although this self-focusing technique is highly effective, it requires the presence of a reflecting target in the medium [1].

In media containing several targets, the problem is more complicated and iterations of the TR operation may be used to select one target. Indeed, if the medium contains two targets of different reflectivities, the time-reversal of the echoes reflected from these targets generates two wavefronts focused on each target. The mirror produces the real acoustic images of the two reflectors on themselves. The highest amplitude wavefront illuminates the most reflective target, while the weakest wavefront illuminates the second target. In this case, the time-reversal process can be iterated. After the first time-reversed illumination, the weakest target is illuminated more weakly and reflects a fainter wavefront than the one coming from the strongest target. After some iterations, the process converges and produces a wavefront focused on the most reflective target. It converges if the target separation is sufficient to avoid the illumination of one target by the real acoustic image of the other one [2].

Application to lithotripsy

The method of choice to treat kidney and gall stones involves the use of large amplitude acoustic shock waves that are generated extracorporeally and focused onto a stone within the body. Lithotripters typically have a high focusing gain so that pressures are high at the stone but substantially lower in the surrounding tissue. The alignment of stone in the patient with the lithotripter focus is accomplished with fluoroscopy or ultrasonic imaging. Focusing is achieved geometrically, i.e., with ellipsoidal reflectors, concave focusing arrays of piezoelectric transducers, or acoustic lenses. Shock waves have amplitudes at the focus on the order of 1.000 bar and a duration of a few microseconds. They are typically fired at a 1 s pulse repetition rate [3,4]. The main problem to overcome in the field of lithotripsy is related to stone motion due to breathing. Indeed, the lateral dimension of the shock wave in current lithotripsy devices is less than 5 mm and the amplitude of the stone displacement can reach up to 20 mm from the initial position. Hence, in classical focusing technique, shock waves often miss the stone and subject neighboring tissues to unnecessary shocks that may cause local bleeding.

Different approaches have been investigated to overcome these limitations. Most of them are based on a trigger of the high power pulses when the stone goes through the focus of the shock wave generator. These approaches may reduce the number of shots needed to disintegrate the stone but increase considerably the time of treatment. The use of 2D arrays of piezoelectric transducers has opened the possibility of electronic steering and focusing the beam in biological tissues and a time-reversal piezoelectric generator has been developed [5] to move electronically the focus and track the stone during a lithotripsy treatment.

The goal is to locate and focus on a given reflecting target among others, for example, a stone in its surroundings: others stones and organ walls. Moreover, the stone is not a point like reflector but has dimensions up to ten times the wavelength. In the basic procedure that has been developed, the region of interest is first insonified by the transducer array. The reflected field is sensed on the whole array, time-reversed and retransmitted. As the process is iterated, the ultrasonic beam selects the target with the highest reflectivity. If the target is spatially extended, the process converges on one spot, whose dimensions depend only on geometry of the time-reversal mirror and the wavelength. High amplification during the last iteration can be used to produce a shock wave for stone destruction. However, two problems limit this technique. For human applications, it is necessary to use very short high power signals (bipolar and unipolar pulses) to prevent damages caused in the organs by cavitational gas bubbles, while the iterated pulses have long duration. Besides, a complete time-reversal electronic is expensive and the number of time-reversal channels must be limited. To solve these problems, another procedure has been developed. In the first step, only a subgroup of the array is used in a time-reversed mode. This step is conducted with low power ultrasound in order to remain in linear acoustics. After some iterations, a low power ultrasonic beam, generated by the array subgroup, is focused on the stone. The last set of received signals is used to deduce a time of flight profile on the subgroup. This time of flight profile is then interpolated to the whole array. The final step consists in the generation, by the whole array, of very short high power signals with the correct delays.

Different time reversal mirrors have been designed for lithotripsy. They use bidimensional transducer arrays working at a central frequency of 360 kHz, made of 121 prefocused piezoelectric transducer elements arranged on a spherical cup of 190 mm radius of curvature

Multiple target detection

Iterative time reversal technique is well adapted to focus on the target of higher reflectivity. However, in many cases it is also interesting to learn how to focus on the other reflectors. In order to achieve selective detection and focusing on each reflector inside an unknown multitarget medium, a matrix formalism approach, that extended time reversal analysis, was developed by Prada et al [6,7,8]. This method is derived from the theoretical analysis of iterative TRM and consists at the construction of the invariant of the time reversal process. This analysis consists at determining the possible transmitted waveforms that are invariant under the time reversal process. For these waveforms an iteration of the time reversal operation gives stationary results. Such waveforms can be determined through the calculation of the eigenvectors of the so called time reversal operator. Indeed, the echoes of a single target are an eigenvector of the time reversal operator K*K, where K=K(?) represents the inter-element response matrix.

Using this basic idea that each target is associated to an eigenvector of the time reversal operator, it is possible to record the whole time reversal operator and compute its eigenvectors decomposition. Thanks to this numerical eigenvector decomposition, a selective focusing on each target can be achieved. Using this technique, known as the DORT method, Chambers recently shown that the spectrum of the time reversal operator can be complex and very informative [9].

Nevertheless, the D.O.R.T method suffers several limitations as it requires the measurement of the NxN inter-element impulse responses and the computation of the eigenvector decomposition is quite time consuming and does not allow real time imaging.

A new real time technique has been recently proposed by G.Montaldo et al [10] for multitarget selective focusing that does not require the experimental acquisition of the time reversal operator. Actually, this technique achieves the operator decomposition simply by using a particular sequence of iterative wave illuminations instead of computational power. The general idea of this new approach is first to use the time reversal iterative process in order to estimate the signals focusing on the brightest target. These signals are then used to derive a cancellation filter allowing canceling target's echoes during the selection of the next brightest spot by iterative time reversal. This process can be extended to following multiple target detection using the cancellation filter that cleans up the targets already detected.

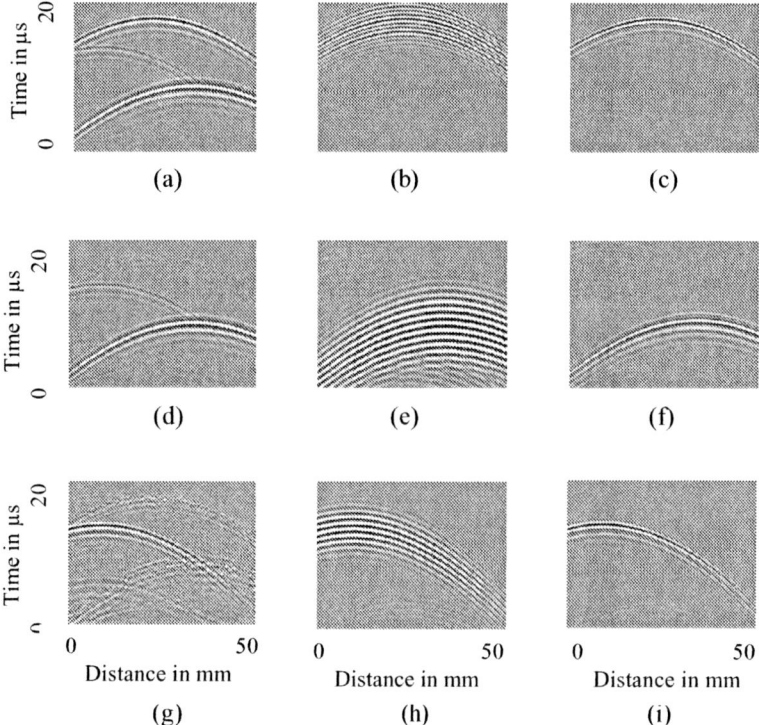

FIGURE 1. The iterative process. a) Echoes from the 3 scatterers after a plane wave emission. b) Detection of the strongest scatterer by iterative time reversal. c) Eigenpulse of the first scatterer. d) Echoes from the two scatterers after filtering the first one. e) Detection of the second scatterer by time reversal and filtering . f) Eigenvector of the second scatterer. g) Signal of the 3rd scatterer after filtering the first and second. h) Detection of the 3rd scatterer by time reversal and filtering. i) Eigenpulse of the 3rd scatterer.

A plane wave is first emitted in this medium. The backscattered signals are composed of three wavefronts of different amplitudes corresponding to each target (Fig. 1a). If these signals are time reversed and reemitted through the medium, the resulting wavefronts focus on each target and the brightest target is more illuminated than the others. Consequently, its contribution in the backscattered echoes is more important. After a few iterations this time-reversal process permits to select the most reflecting scatterer. However, at each iteration the signals are filtered by the limited bandwidth of the transducers and it results in a progressive temporal spreading of the emission signals. In Fig. 1b we can see the received signal after 8 iterations of the time-reversal process, the strongest scatterer was selected but the bandwidth was clearly reduced. The second step of the process allows for overcoming this problem.

The bandwidth narrowing suffered during the time reversal process is a real drawback in most applications. An easy solution consists of reconstructing wideband wavefront at each iteration by detecting the arrival time and amplitude law of the signals received on each transducer. This arrival time and amplitude law is then used to reemit a wideband pulsed signal identical on each transducer with the corresponding amplitudes and time delays on each transducer. It allows avoiding the bandwidth spreading of the signals during the iterative process. The arrival time and amplitude can be measured by using a simple maximum detection technique for each transducer (Fig. 2). Such a "pulsed" wavefront construction before each time reversal emission allows to correct the temporal spreading of the signals during the iterative process (Fig. 2c and 1c). In general, the use of a simple algorithm for the "pulsed wavefront" construction (for example a maximum detection) can induce some errors in the arrival time estimation. For example, if the signal is composed of several sinusoids as presented in Fig. 2.a, the maximum detection can be limited by a 2? uncertainty (Fig. 2 b). It results in the reemission of an incorrect wavefront at the next illumination. However, most of the energy of this incorrect wavefront is focused on the good location and the maximum detection on the next backscattered echoes becomes easier and more accurate. Thus, a few iterations of the time reversal process combined with the pulse compression allow obtaining a correct pulsed wave front or "eigenpulse" signal. Note that the duration of these combined steps is only limited by the waves travel time and the maximum detection hardware. As an example, for medical applications, the detection of a brightest target located at 50 mm depth in tissues achieved in 8 iterations of these combined steps could last less than a millisecond.

The basic idea for selecting a new scatterer is to filter the signals coming from the detected ones. This filter is built by subtracting the projection of the wave front from the signal each time. If we start with a backscattered signal containing the echoes of the three scatterers (Fig. 1.a), we obtain the filtered signals shown in (Fig. 1d). As one can notice, the echoes of the strongest scatterer have been cancelled. This new set of filtered signals is now used as initial illumination for the iterative time reversal process. The cancellation filter is applied at each step during the iterative time reversal process. Consequently, the second target generates the brightest echoes and is progressively selected by the iteration process (Fig. 1.e). The signals backscattered by the second target are temporally spread and can be "pulse compressed" (Fig. 1.f).

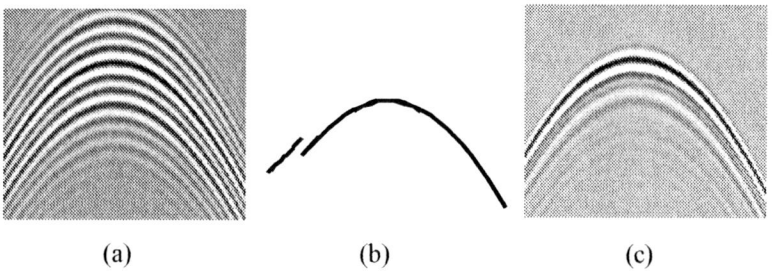

(a) (b) (c)

FIGURE 2. a) After the iteration of the time reversal the signal is enlarged. b) Selection of a wavefront only, this wavefront has a defect in the detection. c) After emitting this wavefront, we obtain a narrow pulse, in this narrow pulse thre are not defects in the identification of the wavefront.

Finally, the cancellation filter allows cancelling the first and second target and selecting the third target by iterating the time reversal process. Figures 1.g, 1.h and 1.i, describe the final eigenvector decomposition that was found for the third and weakest target.

As one can notice, the complete process does not require any fastidious calculation. The combination of a simple maximum detection with the iterative time reversal process was found sufficient in order to select multiple targets. The cancellation filter corresponds also to a simple signal subtraction. The main advantage is the simplicity of the procedures that can be implemented in hardware for real time selective focusing. As an example in medical imaging, the detection of 3 reflectors located at 50 mm depth in tissues could last less than 10 milliseconds!

This technique is also robust in speckle noise as it can be seen, for example, on a biological phantom containing nine wires embedded in a random distribution of unresolved scatterers. Fig. 3a shows the beamformed pulse-echo image of the phantom when it is illuminated with a plane wave. We can see the echoes of some target superimposed to the speckle noise. The iterative method is able to identify easily the 9 echoes as it is shown in Fig. 3b.

Compared to a conventional B scan image (Fig. 3c), our technique is able to perfectly recognize the 9 targets. Using the time delays, the position of each target can be estimated and in Fig. 3d, the positions of the targets are superimposed to the basic image. An interesting application of this technique is the identification of micro calcifications in the breast or other organs. This technique can also be implemented at lower frequency for real time mine detection.

CONCLUSION

Time-reversal shows startling applications in the field of ultrasound. Because ultrasonic time-reversal technology is now easily accessible to modern electronic technology, it is expected that applications in various areas will expand rapidly. Initial applications show promise in medical therapy. We have shown in this paper how passive reflecting targets embedded in the body (kidney stones, breast microcalcifications....) can be used as sources of time reversal waves. The very first, and perhaps most illustrative, application of time reversal concerns, real time tracking and destruction of moving kidney stones during lithotripsy treatment. Iterating the time reversal process leads to also interesting applications as it becomes possible in

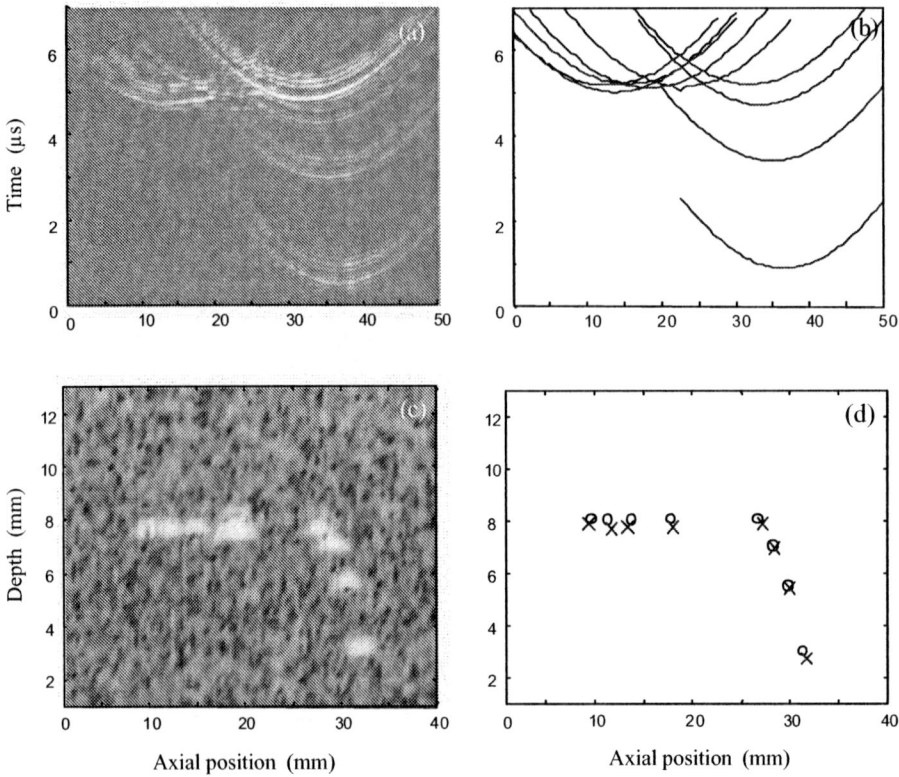

FIGURE 3. Target detection in a biological phantom. a) Signal received after a plane wave illumination, we can see some echoes of the targets with an important speckle noise from the phantom. b) Detection of the echoes of 9 targets. c) B scan image of the phantom. Due to the speckle noise, the smallest targets are difficult to resolve. d) Calculation of the targets positions from the detected echoes. The x are the measured positions and the circles are given by the furnisher of the phantom.

multiple target environments to select and focus in real time on each target of a medium. For this purpose, the ability of iterative time reversal to improve the detection of microcalcifications in the speckle noise of the breast was presented.

Time reversal is also a very powerful correction technique for distortions induced by sound velocity and density heterogeneities. Combined with absorption correction techniques, its potential is currently investigated for the local destruction of malignant brain tumors using trans-skull high intensity focused ultrasound.

Beyond these straightforward applications of time reversal to spatial focusing of waves through aberrating medium, time reversal techniques also allow us to revisit the complete concept of piezoelectric transducer designing. Contrary to conventional transducer technology avoiding unwanted reverberations in piezoelectric elements, time reversal can benefit from strongly reverberating media to create virtual transducers and thus to obtain a very high focusing quality with a small number of transducers. New generation of ultra-compact shock wave lithotripters can be implemented with this approach. The application of this breakthrough concept for 3D medical imaging is also being currently investigated.

All these applications of time reversal were discussed in the field of linear acoustics, but a very interesting point is that time reversal properties remain valid in the field of nonlinear acoustics. We are currently envisioning that time reversal techniques can also be very useful in nonlinear acoustics as it could enhance the image contrast in medical harmonic imaging.

REFERENCES

1. M. Fink, *IEEE Trans. Ultrason. Ferroelec. Freq. Contr*, 39 (5) 555-566 (1992)
2. C. Prada, F.Wu, M.Fink, *J.Acoust.Soc.Am*, 90 (2), 1119-1129 (1991).
3. M. Delius, *Ultrasound Med. Bio,* 26, Suppl. 1, S55–S58 (2000).
4. S. Nachef, D. Cathignol, and A. Birer, *J. Acoust. Soc. Am,* 92, 2292 (1992).
5. J.L Thomas, F. Wu, M. Fink, *Ultrasonic Imaging,* 18, 106-121, (1996)
6. C. Prada, M.Fink, *Wave Motion*, 20, 151-163 (1994).
7. C. Prada, J.L. Thomas, M. Fink, *J.Acoust.Soc.Am,* 97(1), 62-71, (1995).
8. C. Prada, S.Manneville, D. Spoliansly,M. Fink, *J.Acoust.Soc.Am,* 99, 2067-2076 (1996).
9. D. H. Chambers and A. K. Gautesen. *J. Acoust. Soc. Am.* 109 (6), 2616-2624, (2001).
10. G.Montaldo, M.Tanter, M. Fink, *J. Acoust. Soc.Am,* 115,776-784 (2004).

Time Reversal Ocean Acoustic Experiments At 3.5 kHz: Applications To Active Sonar And Undersea Communications

Heechun Song, P. Roux, T. Akal, G. Edelmann, W. Higley, W.S. Hodgkiss, W.A. Kuperman, K. Raghukumar, and M. Stevenson[†]

Marine Physical Laboratory/SIO, La Jolla, CA, USA
[†] *NATO SACLANT Undersea Research Centre, La Spezia, Italy*

Abstract. We have conducted a series of time-reversal experiments at a center frequency of 3.5 kHz with a 1 kHz bandwidth. These experiments and follow-up analysis suggest applications to active sonar and undersea communications. In the area of active sonar, time reversal physics points to procedures to minimize reverberation and therefore enhance the echo-to-reverberation ratio. These ideas have been confirmed under limited circumstances in our shallow-water acoustic experiments. For undersea communications, time reversal provides an opportunity to implement space-time multiplexing in complex environments. Our experiments indicate that vertical aperture provides a capability for implementing multiple input/multiple output (MIMO) communications. We also have demonstrated experimentally that a moving source and/or receiver can communicate by establishing a synthesized horizontal aperture time-reversal mirror.

INTRODUCTION

Over the last 40 years, time-reversal mirrors (TRMs) have been investigated for various applications [1-4]. This is a process that was first demonstrated in nonlinear optics, then in ultrasonic laboratory acoustic experiments, and most recently in ocean acoustics. A TRM takes advantage of reciprocity, a property of wave propagation in a static medium and a consequence of the linear wave equation invariance to time reversal. Therefore, phase conjugation in the frequency domain can be implemented in the time domain by a TRM.

Recently we have conducted a series of time-reversal experiments at a center frequency of 3.5 kHz with a 1 kHz bandwidth. These experiments and follow-up analysis suggest potential applications to active sonar and underwater communications. In the area of active sonar, a TRM focuses acoustic energy on a target enhancing the target echo while minimizing the reverberation from the boundaries below and above the focus, thereby resulting in echo-to-reverberation enhancement. These ideas have been confirmed under limited circumstances in our shallow-water acoustic experiments including active reverberation nulling [5-6].

For undersea communications, time-reversal provides an opportunity to implement space-time multiplexing in complex environments [7]. A recent time reversal experiment demonstrated that multiple foci can be projected from an array of sources to the same range but at different depths [8]. This Multiple Input/Multiple Output (MIMO) process potentially can improve the information data rate. We have also experimentally demonstrated that a moving source and/or receiver can communicate by establishing a synthetic, horizontal aperture, time-reversal mirror.

TRM AND ACTIVE SONAR

For active sonars, reverberation is defined as that portion of the received signal which is scattered by rough ocean boundaries or by volume inhomogeneities. Experiments have shown that when the focus is placed in the middle of water column, there is very little energy projected on the boundaries below and above the focus at the focal range (typically 20 dB down from the focal region). Hence, an echo return from a TRM focus will have a minimal reverberation at the echo range cell, resulting in an echo-to-reverberation enhancement.

Figure 1(a) shows the experimental configuration for reverberation measurements [5]. Figure 1(b) and (c) show the time-reversal (TR) focusing and the broadside (BS) transmission measured by the Vertical Receive Array (VRA). Broadside transmission is an excitation of the Source/Receive array (SRA) with equal amplitudes. Note that an enhancement in the ensonification level at the probe source (PS) location (60 m) by TR is approximately 5 dB as compared to the BS transmission which fills the water column.

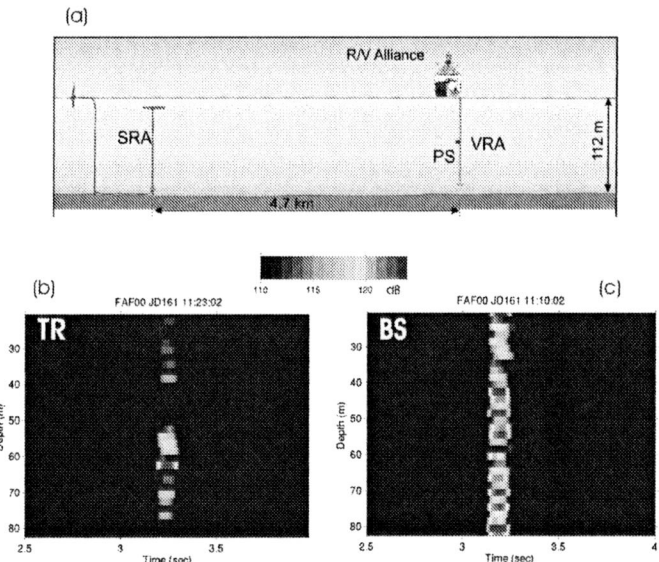

FIGURE 1. (a) Experimental configuration for reverberation measurements carried out north of Elba Island off the west coast of Italy. The PS was deployed from the R/V Alliance at 60 m depth and 4.7 km range away from the SRA. The PS pulse was a 100-ms long pulse at 3.75 kHz. (b) TR focusing recorded by the VRA near the PS. For comparison purposes, (c) shows a BS transmission received by the VRA which fills the water column.

The returning backscatter from these transmissions was recorded monostatically by the SRA. Figure 2 shows the measured reverberation fields: (a) BS and (b) TR transmission. The ambient noise level is also shown in (c) as a reference. The existence of a reverberation notch approximately 400-m wide and about 3 dB is evident. Note that the BS level is about 5 dB higher than that of TR due to the difference in transmitted level.

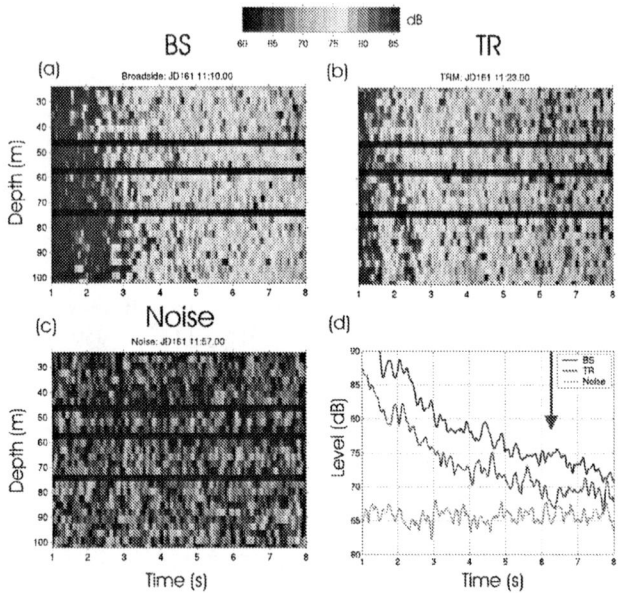

FIGURE 2. Measured backscattered field at the SRA: (a) BS transmission and (b) TR transmission. The ambient noise field is also displayed in (c) as a reference. (d) shows the corresponding reverberation level incoherently averaged across the upper SRA elements along with ambient noise level. The TR reverberation indicates about a 3 dB notch around 6.3 sec corresponding to the PS range of 4.7 km.

REVERBERATION NULLING

Backscattering from the rough water-bottom interface can serve as a surrogate probe source (PS) in time reversal. A time-gated portion of the reverberation then is refocused to the bottom interface at the corresponding range [9]. Here, reverberation nulling is investigated to enhance active target detection. The basic idea is to minimize the acoustic energy incident on the corresponding scattering interface by applying an excitation weight vector on the time-reversal mirror which is in the complementary subspace orthogonal to the focusing vector [6].

Figure 3 shows the reverberation nulling experiment at 3.5 kHz conducted in April 2003 near the Elba Island, Italy. The SRA was deployed in 105-meter water near Elba Island. Initially, we generated reverberation time series from 100-ms CW broadside transmission of the SRA (before). Due to the proximity of the SRA to the Island, the two prominent peaks around 4.25 km and 5.85 km result from the interaction with the

Island corresponding to the concentric circles denoted in the upper left panel. The peak at 2.5 km, however, is due to a seamount at the corresponding range which is visible in the bottom topography (small circle). Thus the peak at 2.5 km range provides a good candidate for reverberation nulling. The resulting reverberation nulling (after) is superimposed in the lower left panel, indicating the reduction of reverberation level to the background level by 2 dB. On the other hand, the reverberation return from the interaction with the island has increased at 5.8 km range.

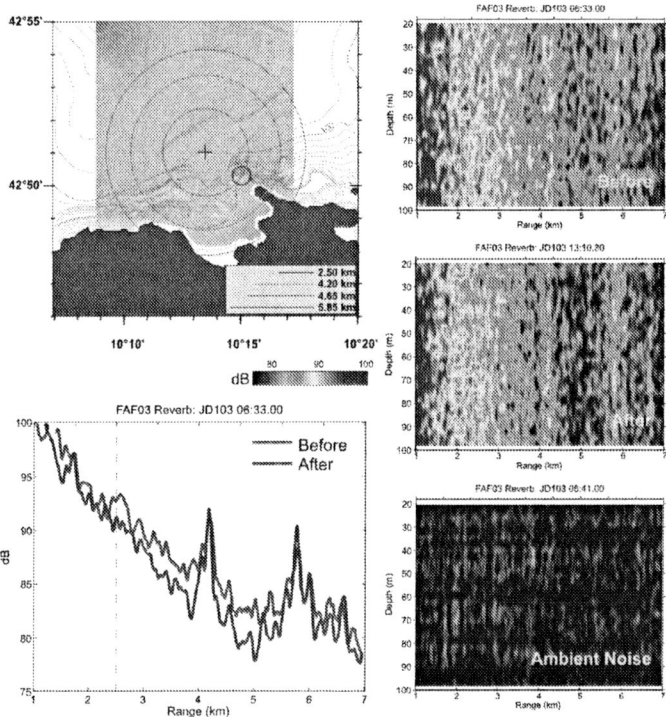

FIGURE 3. Experimental demonstration of reverberation nulling at 3.5 kHz. Note the reduction of reverberation level at the intended 2.5 km range as compared to the original reverberation from a broadside transmission (BT) of the SRA (bottom left). Right column: Reverberation level in time and depth along with ambient noise level.

MULTIPLE-INPUT MULTIPLE-OUTPUT (MIMO)

Time reversal can be implemented between a transmit and receive array without invoking reciprocity. This technique has been used in ultrasonic laboratory experiment, but never at sea. We refer to this approach as the "round robin" technique [8]. The process requires connectivity between the two arrays but does not require a Probe Source (PS) collocated with the receiver array as in a conventional TRM configuration as shown in Figure 1.

The procedure is illustrated in Fig. 4 in which a pulse is separately sent out from each SRA transducer and received on the VRA at a specific depth. This information is transferred to the SRA, and each respective pulse is synchronized, time reversed and sent out from the SRA simultaneously. Since the round robin procedure involved receiving all depths simultaneously on the VRA, the time reverse sequence for focusing at each depth is captured almost simultaneously.

Multiple time reversal focal spots then can be achieved simultaneously in the water column. Figure 5(a) shows the time-reversed focused field sequentially at every element of the VRA and (b) shows an example of simultaneous multiple focal spots at six different depths.

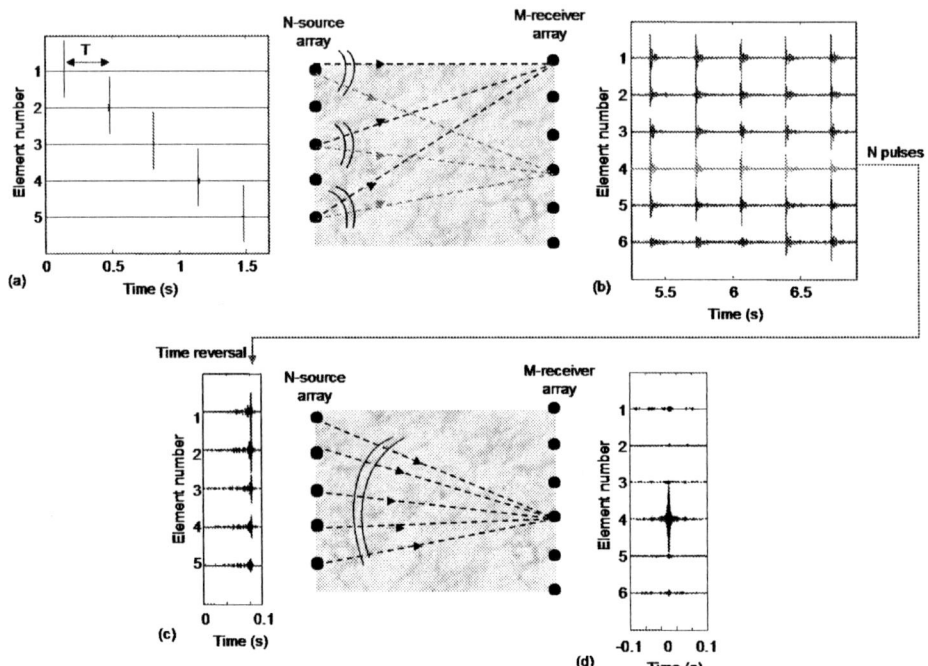

FIGURE 4. Schematic of a round-robin time-reversal implementation. A pulse is separately sent out from each SRA transducer and received on the VRA at a specific depth (e.g., element #4). This information is transferred to the SRA (lower plot), and each respective pulse is synchronized, time reversed and sent out from the SRA simultaneously.

UNDERWATER COMMUNICATIONS

With the at-sea, multiple-focal-spot demonstration with a TRM, spatial encoding of communication sequences is feasible. Here we demonstrate that different communication sequences can be simultaneously sent to and decoded at individual receivers on a vertical array using a simple binary Amplitude Shift Keying (ASK) modulation scheme. Although inefficient, the incoherent ASK modulation allows for initial feasibility study of MIMO communications.

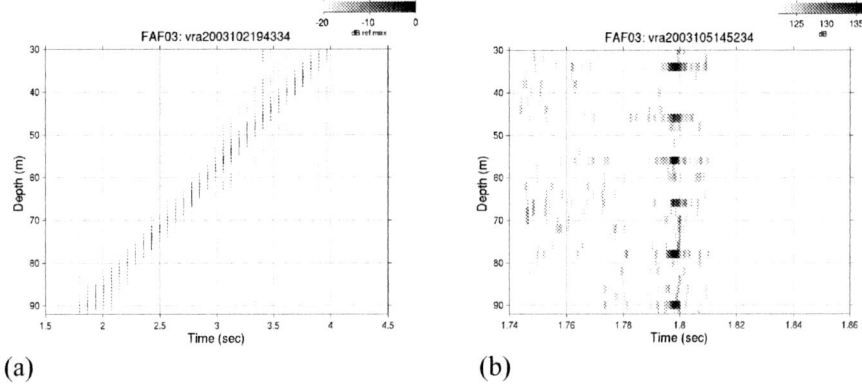

FIGURE 5. (a) Sequential focusing at every element of the VRA. (b) Simultaneous multiple focal spots at six depths.

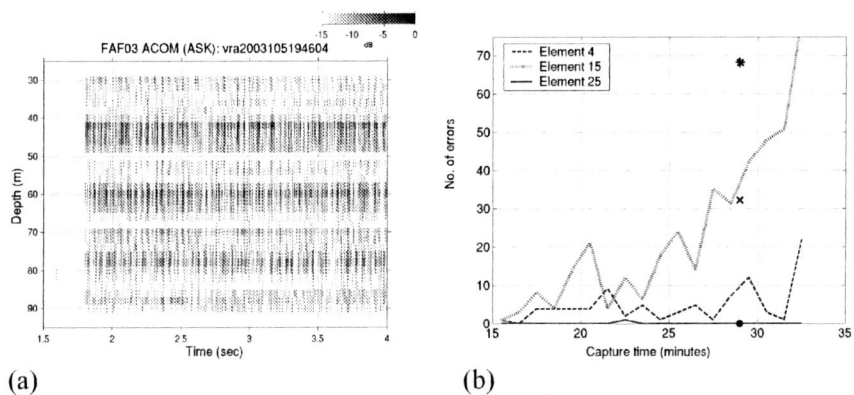

FIGURE 6. (a) The first 2 sec of measured binary ASK data to the three depths of VRA (42 m, 60 m and 78 m). (b) BER out of 4,900 bits as a function of time at three different focal depths using an incoherent binary ASK modulation. The results are from simulated ASK sequences based on 40-minute round-robin channel response data and decoded using passive time reversal. The dots denote the BER of active time reversal ASK communication data.

Figure 6(b) shows the Bit Error Rates (BER) for 4,900 bits transferred at three different depths as a function of time. These results are from simulated binary ASK sequences based on 40-minute round-robin channel response data and decoded using passive phase conjugation [10]. In comparison, the three dots at minute 29 denote the BER of active time reversal communication data. Simultaneous, multiple-depth, coherent communications currently are being investigated.

527

SYNTHETIC HORIZONTAL APERTURE TRM

An ultrasonic, synthetic aperture, endfire array has been constructed in our ultrasonic laboratory to study its time-reversal properties. The minimal hardware configuration of a synthetic endfire time-reversal array using only one transmitter and one receiver makes communications a viable application. In a recent TRM experiment, we investigated the synthetic aperture time-reversal communications.

The configuration is shown in Fig. 7. A 10-sec communication sequence is transmitted from the ITC towed source every 30 seconds at a 2-knot tow speed (1 m/s). The communication sequence consists of binary ASK coding with a 1-ms preamble and 199-ms spacing for synchronization and Doppler compensation processing. The bit length is 1 ms so each sequence contains about 9800 bits. Figure 7(a) displays the impulse response due to a single bit and (b) shows the reception from communication sequences convolved with the time reversed version of the impulse response. By combining the transmissions to produce a synthetic aperture array, the BER was drastically reduced. For a single transmission, the error was 3915/9800; for 14 contiguous transmissions (each 30 sec apart), the bit error was 114/9800; for 14 sparsely-spaced transmissions (each 150 sec apart), the error was 1/9800; for all 66 transmissions, the error was 0/9800.

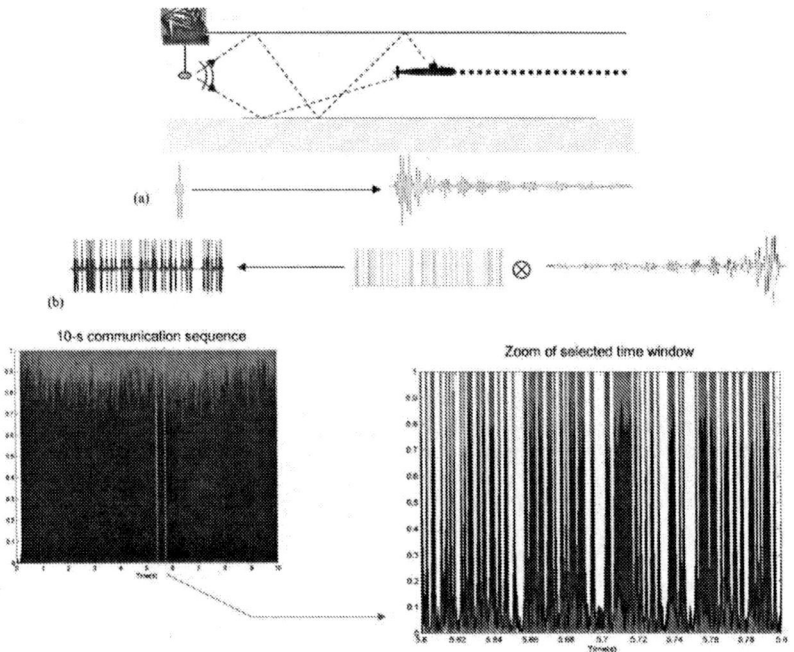

FIGURE 7. Synthetic, horizontal aperture communications. (a) Channel response due to a single bit. (b) Communication sequences convolved with a time-reversed version of the channel response. The bottom plots show decoded sequences after synchronization and Doppler compensation processing.

CONCLUSIONS

Recently a series of ocean acoustic experiments have been carried out confirming the robustness and potential utility of time reversal mirrors in underwater acoustics with applications in active sonar and undersea communications. In the area of active sonar, the echo-to-reverberation enhancement and reverberation nulling have been demonstrated using a time-reversal mirror in the 3-4 kHz band in shallow water.

For undersea communications, time reversal provides an opportunity to implement space-time multiplexing in complex environments. We demonstrated experimentally that multiple foci can be projected from an array of sources to the same range but at different depths. This MIMO process potentially can improve the information data rate. We also have demonstrated experimentally that a moving source and/or receiver can communicate by establishing a synthetic, horizontal aperture, time-reversal mirror.

ACKNOWLEDGMENTS

This work was supported by the Office of Naval Research, Contract No. N00014-01-D-0043-D06/D07.

REFERENCES

1. Parvelescu, A. and Clay, C., "Reproducibility of Signal Transmissions in the Ocean," *Radio Electron. Eng.* **29**, 223-228 (1965) and A. Parvelescu, "Matched-Signal ("Mess") Processing by the Ocean," *JASA* **98**, 943-960 (1995).
2. Fink, M., "Time-reversed Acoustics," in *Scientific American*, November, 91-97 (1999).
3. Jackson, D.R. and Dowling, D.R., "Phase Conjugation in Underwater Acoustics," *JASA* **91**, 3257-3277 (1991).
4. Kuperman, W.A., Hodgkiss, W.S., Song, H.C., Akal, T., Ferla, C., and Jackson, D.R., "Phase Conjugation in the Ocean: Experimental Demonstration of a Time Reversal Mirror," *JASA* **193**, 25-40 (1998).
5. Kim, S., Kuperman, W.A., Hodgkiss, W.S., Song, H.C., Edelmann, G., and Akal, T., "Echo-to-Reverberation Enhancement Using a Time Reversal Mirror," *JASA* **115**, 1525-1531 (2004).
6. Song, H.C., Kim, S., Hodgkiss, W.S., and Kuperman, W.A., "Environmentally Adaptive Reverberation Nulling Using a Time Reversal Mirror," *JASA* **116**, 763-768 (2004).
7. Edelmann, G.F., Akal, T., Hodgkiss, W.S., Kim, S., Kuperman, W.A., and Song, H.C., "An Initial Demonstration of Underwater Acoustic Communication Using Time Reversal," IEEE JOE **27**, 602-609 (2002).
8. Roux, P., Kuperman, W.A., Hodgkiss, W.S., Song, H.C., Akal, T., and Stevenson, M., "A Non-Reciprocal Implementation of Time Reversal in the Ocean," *JASA* **116**, 1009-1015 (2004).
9. Lingevitch, J.F., Song, H.C., and Kuperman, W.A., "Time Reversed Reverberation Focusing in a Waveguide," *JASA* **111**, 2609-2614 (2002).
10. Rouseff, D., Jackson, D.R., Fox, W.L., Jones, C.D., Ritcey, J.A., and Dowling, D.R., "Underwater Acoustic Communication by Passive Phase Conjugation: Theory and Experimental Results," *IEEE JOE* **26**, 821-831, 2001.

Time-Reversal and Spatial Diversity: Issues in a Time-Varying Geometry Test

S.M. Jesus* and A.J. Silva*

*SiPLAB - FCT, University of Algarve, Campus de Gambelas, 8005-139 Faro, Portugal

Abstract. Underwater acoustic communications in waveguides is known to be prone to severe multipath, which strongly limits practical transmission rates with actual channel equalization techniques. The time reversal principle uses the ocean waveguide response to a basic pulse shape to matched filter the received data sequence. Assuming the ocean response to be a version of the actual pulse shape ocean response corrupted by additive noise, the matched filter output remains a sum of four terms from which only one has the required data sequence in usable form. This paper analysis the ability of such a peculiar matched filter to reject the unwanted terms both with fixed and time-varying source-receiver geometries. One particular parameter with practical interest is the sensitivity of the matched filter performance to a change of the source-receiver range during the processing that induces a mismatch limiting factor. Simulation examples in realistic situations and results obtained with real data, collected during the INTIFANTE'00 sea trial, illustrate the theoretical assertions.

INTRODUCTION

Theoretical developments achieved in various areas of underwater acoustic signal processing such as time delay estimation, channel equalization and optimization, eventually contributed to the literal explosion of the air-wireless communications in the last 10 years. Conversely, recent developments in air-wireless communications are now becoming popular in underwater communications, where it is now frequent to see references to concepts such as digital links, multiple users, local area networks, etc. High-speed data communication requires digital modulation and coherent receivers which, due to the frequency-dependent sound absorption of the ocean channel, are clearly limited. Effective coherent receivers usually exploit spatial diversity and use powerful multichannel equalization algorithms to attain acceptable error rates [1]. Channel equalization is a process that attempts to undo the channel multipath with a digital filter, which coefficients are adapted on real time according to the variations of the acoustic channel propagation. Due to the high variability of the acoustic channel in practical situations, the usage of test sequences strongly limits the effective throughput of the acoustic link.

A different approach, known as acoustic Time-Reversal (TR), was originally proposed by [2, 3] and successfully tested at sea in [4]. In digital communications, the application of the TR principle consists in preceding each data packet by a probe signal that is later used as matched-filter to undo the channel multipath for each hydrophone, and then coherently summed over the receiving array channels. This process is known as passive phase conjugation [5, 6] or virtual Time-Reversal (vTR) [7, 8] and demonstrated with simulated data, in the underwater communication context, in [6, 9] and with real data in [10, 11]. As in acoustic TR, phase conjugation performance will depend upon the stability of the propagation channel and the ability of the receiving array to correctly sample the most important features of the acoustic field at the useful frequencies.

In practice, there are three main limiting factors in underwater communications: one is the stability of the oceanic characteristics that determine the channel acoustic response; the other is the noise, that can be of environmental nature - sea surface and microstructure induced - or due to electronics; the third, and most important, limiting factor is that due to the experimental geometry, where source and receiver motion relative to the sea surface and bottom are of paramount importance. Of course these limiting factors are not independent among them and often manifest themselves simultaneously. In addition to demonstrating the practical feasibility of TR in the ocean, the experiment by Kuperman et al., [4] also showed the remarkable temporal stability of this process, since pulses were successfully refocused up to one week after the original recordings. Through simulations Silva et al. [8], showed that the receiving array does not have to span all the water column or be extremely dense, but must intersect most of the energy propagating in the sound channel. Another important, and an often overlooked problem, is that of the choice of the probe signal window length to be recorded and used at a later time as matched-filter to the incoming data. It was shown in [11] that for each source-receiver range, there was an optimal probe signal window length, that minimized the empirical bit error probability.

It is common sense that a major limiting factor in real world applications is source-receiver motion in common tasks such as transmitting telemetry data from untethered measurement instruments and communicating with AUV's. Matched-Field Processing (MFP) represents an alternative for obtaining the channel impulse response required for vTR. Although it is recognized that an ocean response based TR receiver has a clear edge over MFP, it is also clear that, since source-receiver geometry appears explicitly on the MFP process, it can be easily compensated for by simply changing source/receiver range and/or depth. This paper addresses this problem first from a theoretical point of view, comparing the TR based receiver and the environmental based (MFP) receiver, then using simulations drawn from an experimental scenario and finally with a real data example.

THEORETICAL BACKGROUND

The Baseband Matched-Filter

In order to introduce the subject, let us consider that the system is working in baseband, that the bit sequence $\{a_n\}$ is memoryless and that the bit rate is such that each pulse is resolvable, i.e., the pulse duration is larger than the beamwidth of the channel response autocorrelation function. Let us assume that the transmitted signal is Pulse Amplitude Modulated (PAM) and can be written as

$$s(t) = \sum_{n=-\infty}^{+\infty} a_n p(t - nT_b), \tag{1}$$

where a_n is the symbol sequence, assumed white with power σ_a^2, T_b is the symbol period and $p(t)$ is the pulse shape function. Assuming the acoustic channel as a linear time-invariant system with impulse response $g(t,\theta)$, the received signal is

$$y(t,\theta) = g(t,\theta) * s(t) + w(t), \qquad (2)$$

where $w(t)$ is an additive zero mean white Gaussian noise with power σ_w^2, assumed to be uncorrelated with the signal and from sensor to sensor and where θ is a vector containing the environmental and model parameters, denoting the dependence of the channel response on the environment. In order to simplify the notation let us consider the discrete version of (2)

$$\begin{aligned} \mathbf{y}(\theta) &= \mathbf{G}(\theta)\mathbf{s} + \mathbf{w} & (3) \\ &= \mathbf{x}(\theta) + \mathbf{w}, & (4) \end{aligned}$$

where the observation vector $\mathbf{y}^T(\theta) = [y(0,\theta), y(1,\theta), \ldots, y(N-1,\theta)]$, is obtained over a set of N temporal samples at a sampling rate $T_s \leq 1/2f_{\max}$, where $f_{\max}$ is the maximum frequency in the observed signal $y(t)$. Considering that the channel impulse response and the signal $\mathbf{s}$, of dimension $M \times 1$, as causal signals, the matrix $\mathbf{G}(\theta)$, with dimension $N \times M$, is lower triangular with discrete and delayed replicas $\mathbf{g}(\theta)$ of the channel response $g(t,\theta)$, conditioned in the vector parameter θ. Using this notation and with the hypothesis stated above, both the noise and the observation will be additive white Gaussian distributed such that $\mathbf{w} : \mathcal{N}[\mathbf{0}, \sigma_w^2 \mathbf{I}]$ and $\mathbf{y} : \mathcal{N}[\mathbf{G}(\theta)\mathbf{s}, \sigma_w^2 \mathbf{I}]$.

The problem can be cast as the detection of signal $s(t)$, conditioned on the channel response $g(t)$ and on the parameter vector θ. This is a classical problem, which solution is given by the Neyman-Pearson (NP) theorem when the noise is Gaussian and by the (matched-)filter that maximizes the Signal-to-Noise Ratio (SNR) at its output, in the general case. The matched filter in this case is given by

$$h(n,\theta) = h_0 x(n_0 - n, \theta), \qquad (5)$$

where h_0 and n_0 are an arbitrary amplitude factor and a constant delay, respectively. The optimality of the matched-filter is attained when $n_0 = N - 1$ and the matched-filter output is sampled at $n = N - 1$, the duration of the observation record. The convolution of the observation $\mathbf{y}$ with filter (5) gives

$$z(n,\theta) = \mathbf{h}_n^T(\theta)\mathbf{y}(\theta) = z_o(\mathbf{n},\theta) + \mathbf{w}_o(\mathbf{n}), \qquad (6)$$

where

$$\mathbf{h}_n(\theta) = h_0 \mathbf{G}(\theta)\mathbf{s}|_n, \qquad (7)$$

and where $|_n$ index denotes a delay of n samples. Replacing (3) in (6) gives the signal and the noise terms as $z_o(n,\theta) = \mathbf{h}_n^T(\theta)\mathbf{G}(\theta)\mathbf{s}$ and $w_o = \mathbf{h}_n^T(\theta)\mathbf{w}$, respectively. The SNR at the output of the matched-filter can now be defined as

$$\rho(n,\theta) = \frac{|z_o(n,\theta)|^2}{E[w_o^2(n)]}. \qquad (8)$$

Replacing the values of $z_0(n,\theta)$ and $w_o(n)$ into the SNR expression gives

$$\rho(n,\theta) = \frac{\mathbf{s}^T \mathbf{G}^T(\theta)\mathbf{h}_n(\theta)\mathbf{h}_n^T(\theta)\mathbf{G}(\theta)\mathbf{s}}{\sigma_w^2 \mathbf{h}_n^T(\theta)\mathbf{h}_n(\theta)}, \qquad (9)$$

which maximum value is

$$\rho_{\max}(\theta) = \frac{\mathbf{s}^T \mathbf{G}^T(\theta)\mathbf{G}(\theta)\mathbf{s}}{\sigma_w^2}, \qquad (10)$$

obtained when the filter takes the form (7) and $n = N - 1$. Despite the apparent simplicity, the matched-filter concept is not always easy to implement in practice. This is due to the difficulties associated with (7) which, in the underwater communications context, deal with the fact that the channel response $\mathbf{G}(\theta)$ is unknown. Besides the classic solution of adaptive equalization, there at least three other possibilities used in practice:

- One is to ignore the channel response and use a receiving filter simply matched to emitted signal, such that $\mathbf{h}_n = h_0 \mathbf{s}$. This plain matched filter is commonly used in active sonar where the emitted signal is known and on classic underwater communication systems, where the receiving filter is only given by the pulse shape $p(t)$ as used on the signal coding expression (1).
- A second case became popular with the increasing usage of matched-field processing (MFP) techniques in the 80's, where the incoming signal is matched with a model-based replica $\hat{\mathbf{g}}(\hat{\theta})$ of the channel response $\mathbf{g}(\theta)$ whenever a sufficiently accurate approximation of the model parameter vector $\hat{\theta}$ of the true parameter vector θ is available. In that case an estimate $\hat{\mathbf{h}}_n(\hat{\theta}) = h_0 \hat{\mathbf{G}}(\hat{\theta})\mathbf{s}|_n$ is used, instead of $\mathbf{h}_n(\theta)$.
- The third situation arose more recently, with the appearance of the so-called time-reversal (TR) techniques that use the recording of the channel response to a single pulse shape (probe signal) during a given interval that is then used as matched-filter for the incoming signal at a later stage. This technique requires two important assumptions: one is that the source is cooperant with the receiver so it can, at regular intervals, transmit a channel probe so it can be used to calibrate the matched filter at the receiver and two, that in the time interval between probe signals the acoustic channel is sufficiently stable so the last probe, recorded a few seconds (or minutes!) before, can be used as the actual channel response, as required by the optimum receiver (7). Another important difference is that the recorded probe signal, apart from the signal itself, also contains noise which is implicitly used in the matched filter with, so far, unpredictable results.

Matched-Filter Relative Performance

Leaving aside the case of the plain matched-filter where the channel response is ignored, it is interesting to perform a comparison between the MFP version of the matched-filter and its TR counterpart. This comparison can be easily made using the matched-filter output SNR, since the SNR is a monotonic function of the probability of detection. Let us first consider the MFP case where the receiving filter is given by

$$\hat{\mathbf{h}}_n(\hat{\theta}) = h_0 \hat{\mathbf{G}}(\hat{\theta})\mathbf{s}|_n \tag{11}$$

conditioned on the acoustic model response $\hat{\mathbf{g}}$ and the assumed model environmental parameter vector $\hat{\theta}$. Replacing (11) as $\mathbf{h}(\theta)$ into SNR expression (9), allows to write

$$\rho_{\text{MFP}}(n, \theta) = \frac{|\mathbf{s}^T \mathbf{G}^T(\theta)\hat{\mathbf{G}}(\hat{\theta})\mathbf{s}|^2}{\sigma_w^2 \|\hat{\mathbf{G}}(\hat{\theta})\mathbf{s}\|^2}. \tag{12}$$

The maximum SNR performance of the MFP matched filter is obtained when $\hat{\mathbf{G}}(\hat{\theta}) = \mathbf{G}(\theta)$ and $n = N - 1$, *i.e.*, when the parameter vector and the channel response are perfectly modeled and the matched filter peak response is correctly chosen, such that

$$\rho_{\text{MFP}-\max} = \frac{\|\mathbf{G}(\theta)\mathbf{s}\|^2}{\sigma_w^2}. \tag{13}$$

In the TR case the matched-filter is given by

$$\mathbf{h}'_n(\theta) = \mathbf{G}'(\theta)\mathbf{s} + \mathbf{w}' \tag{14}$$

where $'$ denotes that the recording was made at a previous time such that the filter output can be written as[1]

$$z = [\mathbf{G}'\mathbf{s} + \mathbf{w}']^T [\mathbf{G}\mathbf{s} + \mathbf{w}] \tag{15}$$
$$= \mathbf{s}^T \mathbf{G}'^T \mathbf{G}\mathbf{s} + \mathbf{s}^T \mathbf{G}'^T \mathbf{w} + \mathbf{w}'^T \mathbf{G}\mathbf{s} + \mathbf{w}'^T \mathbf{w}, \tag{16}$$

where in the four terms obtained in (16), the first term corresponds to the desired signal, while the other three are noise components and treated as such in the evaluation of the output SNR. Thus, using definition (8), the output SNR is the ratio between the square module of the first term divided by the expectation of the square of the sum of the three last terms. Assuming that the noise is white and thus decorrelated between the time when the probe is recorded and when the actual matched-filtering takes place, greatly simplifies the final output SNR expression that can be written as

$$\rho_{\text{TR}} = \frac{|\mathbf{s}^T \mathbf{G}^T \mathbf{G}'\mathbf{s}|^2}{\sigma_w^2 \mathbf{s}^T [\mathbf{G}'^T \mathbf{G}' + \mathbf{G}^T \mathbf{G}]\mathbf{s}}. \tag{17}$$

This is a curious expression that shows an upper term that depends on the correlation of the channel responses at the probe time and at the actual receiving time, and a lower term that is the direct summation of the auto channel responses to the noise part. Due to the particular form of this filter an upper bound on its performance can not be obtained

---

[1] to simplify the notation the dependency on time index n and on the vector parameter θ will be omitted in the remaining of this section.

analytically. It is however interesting to perform a comparison between the two matched-filters making the ratio between (17) and (12) to obtain

$$\Lambda_{TR/MFP} = \frac{\|\hat{\mathbf{G}}(\hat{\theta})\mathbf{s}\|^2 |\mathbf{s}^T \mathbf{G}^T \mathbf{G}'\mathbf{s}|^2}{|\mathbf{s}^T \mathbf{G}^T(\theta)\hat{\mathbf{G}}(\hat{\theta})\mathbf{s}|^2 \mathbf{s}^T [\mathbf{G}'^T \mathbf{G}' + \mathbf{G}^T \mathbf{G}]\mathbf{s}}, \qquad (18)$$

where it can be easily remarked that if the same error is made in the TR than in the MFP, then $\hat{\mathbf{G}} = \mathbf{G}'$ and the performance ratio $\Lambda_{TR/MFP}$ is always ≤ 1, while in the no error case, $\hat{\mathbf{G}} = \mathbf{G}' = \mathbf{G}$, we have that $\Lambda_{TR/MFP} = 0.5$. In other terms, this means that if the same environmental mismatch exists in the two implementations, the TR approach will have a poorer performance than the MFP based approach, which is possibly explained by the fact that in the TR approach there is a re-injection of noise in the filter. Of course the question is how "close" can the model-based channel response be from the true channel response or equivalently how much can the channel response change in a given time interval between the probe signal and the actual message. The TR approach became popular with the experimental evidence that the acoustic channel was very stable relative to matched filtering a whole array of sensors in a shallow water environment [4]. There are, however, two aspects not covered by that experimental conclusions that are of major importance in practical underwater acoustic communications: One is the frequency band at which the experiment was performed that was significantly lower than that used in useful underwater communications systems; and the other is that both source and array where static in the water so no significant geometrical mismatch existed. An analysis of the relative performance of the two detectors in a realistic situation of a time-varying geometry environment encountered during the INTIFANTE'00 sea trial is accomplished via simulation in the next section.

SIMULATION AND REAL DATA RESULTS

This section gives a simulated example drawn from Event I of the INTIFANTE'00 sea trial that took place in the continental platform off the town of Setúbal (Portugal) in October 2000. This event involved an acoustic source suspended from a free drifting research vessel transmitting acoustic signals to a moored 16-hydrophone receiving Vertical Line Array (VLA). The source was at 70 m depth, steaming away from the VLA and making stations at various ranges. The environment was characterized by a nearly range independent 120 m-depth shallow water waveguide with a slightly downward refracting sound speed profile and a sandy bottom with 1,750 m/s sound speed, a density of 1.9 g/cm^3 and a compressional attenuation of 0.8 dB/λ. The transmitted signals were Differential Phase Shift Keying (DPSK) modulated sequences with a center frequency of 1.6 kHz and bandwidths between 100 and 350 Hz. A channel probe signal was transmitted at regular intervals of 20 seconds in between the data signals and was used to detect and synchronize each data sequence. DPSK was used in order to avoid dependence on individual pulse phase-offsets and to compensate the Doppler time compression/expansion an offline synchronization between the transmitted and the received signal was performed every second. The simulation was performed in the frequency domain using five frequencies equispaced in the band of 1,450 to 1,750 Hz. In order to test the relative performance of the matched-filters against time-geometry, the ratio Λ of equation (18) was estimated with $\hat{\mathbf{G}} = \mathbf{G}'$ and where the mismatch with $\mathbf{G}$ was due only to range variations during source drift. Realistic values of source drift were estimated

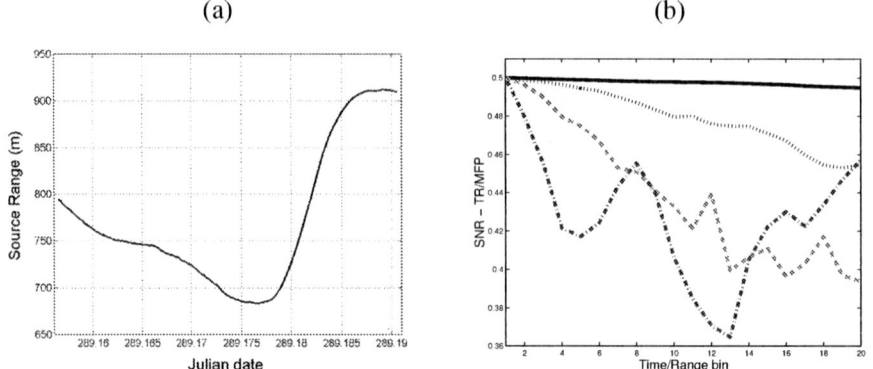

FIGURE 1. source-receiver range variation during station 1 of Event 1 (a) and Λ ratio (18) (b): 0.02 m/s (solid), 0.1 m/s(dot), 0.3 m/s (dash) and 1 m/s (dash-dot).

during station 1 using data shown in fig. 1(a). Figure 1(b) shows the results of Λ for one data window of 20 seconds for drift speeds of 0.02, 0.1, 0.3 and 1 m/s. It can be easily seen that, as expected, the relative performance of the two implementations starts at 0.5 and significantly decreases during the time/range window, depending on the drift speed. Generally a higher drift speed means a larger performance drop and more variable behavior. The variations of performance are related to the range variation relative to the signal mean wavelength ($\sim$1 m in this case). Unfortunately, due to VHF com-

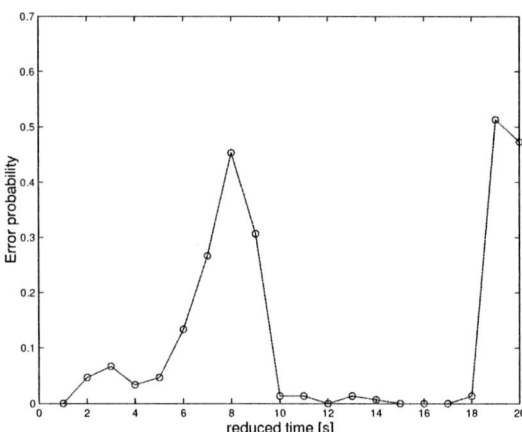

FIGURE 2. empirical probability of error during 20 seconds time window.

munication dropouts between the array and the ship, only very few complete 20 second long sequences were available and therefore the estimation of the error probability has a limited validity. In order to illustrate the effect of source-receiver range variation due to source drift during the transmissions, figure 2 shows the estimated empirical probability of error obtained with the TR based receiver over the 20 s window duration. It can be noticed that the performance degrades in the interval from 5 to 10 seconds, resumes to values close to zero between 10 and 18 seconds and degrades again at the end of the interval. Despite the sharp segments and zero error values due to the limited sample size, the overall performance pattern indicates a variation that is compatible with the predictions of figure 1(b), for a drift velocity of 1 m/s. It is anticipated that this behavior is due to successive phase alignment and misalignment due to range variation during the drift.

CONCLUSIONS

Unlike its airborne counterpart, underwater communications suffers from an extremely harsh propagation environment due to anisotropy, sound absorption and refraction. In shallow water the problem is particularly severe due to multiple reflections on the ocean bottom and surface, making it difficult to obtain a transmission rate of a few kbaud at ranges of several water depths. In any communication system the main component dealing with channel distortion is the receiving filter. Traditionally, the receiving filter is composed of a replica of the emitted pulse shape followed by a channel equalizer. This paper presents a theoretical comparison of two alternatives for the receiving filter: the time-reversal (TR) approach, that matched-filter the data with the channel response at a previous time, and the model-based MFP approach, that attempts to numerically compute the channel response from *a priori* environmental information. It is shown that, for the same channel distortion, the MFP based technique always outperforms (in terms of output SNR) the TR approach. Simulations drawn from a realistic example where channel mismatch is uniquely due to a drifting source-receiver geometry, confirm the theoretical assertions. An example obtained from the INTIFANTE'00 data set show an empirical bit error probability with a similar behavior to that seen in the simulations.

ACKNOWLEDGMENTS

The authors would like to thank the support of the NATO Undersea Research Centre and the Instituto Hidrográfico during the INTIFANTE'00 sea trial. This work was also partially supported by FUP/Ministry of Defence under project LOCAPASS, by FCT under project NUACE and by the HFi - High Frequency Initiative.

REFERENCES

1. Stojanovic, M., Catipovic, J., and Proakis, J., *J. Acoust. Soc. Am.*, **94**, 1621–1631 (1993).
2. Jackson, R., and Dowling, R., *J. Acoust. Soc. Am.*, **89**, 171–181 (1991).
3. Dowling, R., and Jackson, R., *J. Acoust. Soc. Am.*, **91**, 3257–3277 (1992).
4. Kuperman, W., Hodgkiss, W., Song, H. C., Akal, T., Ferla, C., and Jackson, D., *J. Acoust. Soc. Am.*, **103**, 25–40 (1998).
5. Dowling, R., *J. Acoust. Soc. America*, **95**, 1450–1458 (1994).

6. Rouseff, D., Fox, L., Jackson, D., and Jones, D., "Underwater Acoustic Communications Using Passive Phase Conjugation," in *Proc. of the MTS/IEEE Oceans 2001*, Honolulu, Hawai, USA, 2001, pp. 2227–2230.
7. Gomes, J., and Barroso, V., "A Matched Field Processing Approach to Underwater Acoustic Communication," in *Proc. of the MTS/IEEE Oceans 1999*, Seattle, USA, 1999, pp. 991–995.
8. Silva, A., Jesus, S., Gomes, J., and Barroso, V., "Underwater Acoustic Communications Using a "virtual" Electronic Time-Reversal Mirror Approach," in *5th European Conference on Underwater Acoustics*, edited by P. Chevret and M.Zakharia, Lyon, France, June, 2000, pp. 531–536.
9. Edelmann, G., Hodgkiss, W., Kim, S., Kuperman, W., Song., H., and Akal, T., "Underwater Acoustic Communications Using time-reversal," in *Proc. of the MTS/IEEE Oceans 2001*, Honolulu, Hawai, USA, 2001, pp. 2231–2235.
10. Jesus, S., and Silva, A., "Virtual Time Reversal in Underwater Acoustic Communications: Results on the INTIFANTE'00 Sea Trial," in *Proc. of Forum Acusticum*, Sevilla, Spain, 2002.
11. Silva, A., and Jesus, S., "Underwater Communications Using Virtual Time-Reversal in a Variable Geometry Channel," in *Proc. MTS/IEEE Oceans'2002*, Biloxi, USA, November, 2002, pp. 2416–2421.

A High-Frequency Active Underwater Acoustic Barrier Experiment Using a Time Reversal Mirror; Model-Data Comparison

Alessandra Tesei*, Hee Chun Song†, Piero Guerrini*, Philippe Roux†, William S. Hodgkiss†, Tuncay Akal†, Mark Stevenson* and William A. Kuperman†

*NATO Undersea Research Centre, La Spezia, Italy
†Marine Physical Laboratory, La Jolla, USA

Abstract. An underwater acoustic barrier based upon forward scattering in a Time-Reversal Mirror (TRM) was experimentally demonstrated for the first time in 2000 by Song et al. [1]. The barrier consisted of a TRM, a vertical receive array (VRA) and a co-located probe source working at 3,500 Hz in an ocean waveguide near the western coast of Italy. In April 2003 further barrier tests were performed by applying for the first time at sea a new method [2] that provided the capability to focus at different depths by using only the TRM transducers without the complication of additional probe sources. An echo repeater towed by R/V *Alliance* crossed the barrier, emulating the field forward scattered by a possible intruder insonified by the TRM. The presence of a target between the time reversal mirror and the focus can be detected if it significantly disturbs the quiescent region. A normal mode code is used to model the sound propagation in a waveguide. This is applied to predict the unperturbed and perturbed focused acoustic field measurements conducted at sea. Model-data comparison suggests that target detection performance is reasonably predictable using a numerical propagation model.

INTRODUCTION

The hypothesis under test in this study was whether mathematical modeling of an underwater acoustic barrier based on forward scattering in a time reversal mirror could accurately account for observations from a controlled at-sea experiment. The reason that a time reversal implementation is of interest is that a traditional forward-scatter acoustic barrier must detect forward scattered energy from a target arriving at a receiver together with the incident field, the so-called "looking into the sunlight" problem [3]. However, by focusing the transmitted blast in space and time on the downstream receive array, the signal of interest then becomes any temporal or spatial aberration on the receive array caused by the introduction of a scatterer into the propagation medium. As such, the implementation may lend itself to applications in an autonomous tripwire mode of operation.

Based upon earlier work by Song et al. [1], an underwater acoustic barrier experiment was conducted using a time reversal mirror in April 2003. In order to verify earlier work and plan future tests for a family of barrier systems under consideration, a data-model comparison was then conducted. The experiment was conducted using an omnidirec-

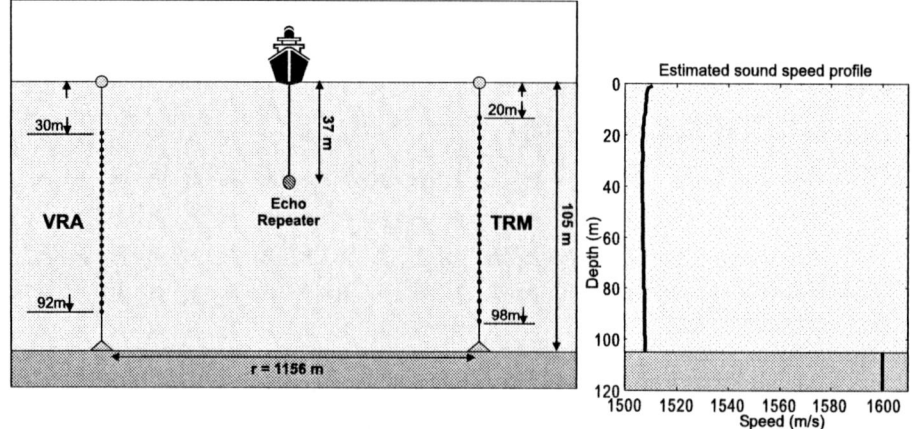

FIGURE 1. The experiment configuration, not drawn to scale. The SRA's 29 tx/rx transducers spanned 78 m of the water column. The VRA's 32 hydrophones covered 62 m of the water column. One VRA hydrophone did not function. The two arrays were linked to a common data acquisition point aboard the ship via RF telemetry.

tional echo-repeater as a target. We chose, for practical reasons, to study a simple case first. Also, a new method of focusing was implemented, the round-robin method described by Roux et al. [2].

We hoped to answer the question of whether one can mathematically model the observations provided from a simple barrier experiment conducted under highly controlled conditions. If one cannot model reality in a simple case, e.g., with an echo repeater as a target, then proceeding down this path in the future for real targets with their own radiation patterns is probably fruitless. However, if one can model reality for a simple case, then perhaps this line of research deserves further investigation. It may eventually lead to a predictive tool when designing forward-scatter barrier experiments using focused acoustic fields.

EXPERIMENT CONDUCT

The at-sea experiment was conducted off the western coast of Italy in a region described by Akal et al. [4]. Transmit (SRA) and receive (VRA) arrays spanning most of the water column were deployed 1,156 m apart, determined from acoustic travel time. Element spacing within the arrays was roughly six acoustic wavelengths. Any tilt in these arrays was disregarded. A ship, R/V *Alliance*, towed an echo-repeater while slowly drifting between the arrays. The hardware suite is described by Hodgkiss et al [5]. The experiment configuration is shown in Fig. 1. Note that the term SRA (Source Receive Array) is often used interchangeably with the term TRM (Time Reversal Mirror) in common parlance. Both terms refer to an array of transducers that are used both as receivers and transmitters.

The barrier experiment was performed at 3,500 Hz by applying for the first time at sea a new implementation of time reversal acoustics [2]. This new technique, sometimes called the "round-robin" technique, provided the capability to focus time reversed signals at any desired hydrophone on the VRA without the hardware complication of additional probe sources. However, it did require telemetry between the two arrays. An additional requirement of this method is that the propagation medium must remain adequately stationary during the procedure of forming a focus.

Briefly summarized, the procedure consisted of the following steps. A pulse was sent out separately from each transducer of the TRM and received at each hydrophone on the VRA at the other end of the barrier. We call this first step the forward propagation phase of focusing. The signals received by one VRA element at a selected depth are then transferred to the TRM, synchronized (or time-aligned), time reversed and sent out simultaneously, each signal from the respective transducer. We call this second step the back propagation phase of focusing.

Having established a focus at one hydrophone of the receive array, a target consisting of a towed Echo Repeater (ER) was introduced into the propagation medium. The signal of interest in this type of experiment is any temporal or spatial aberration observed on the receive array once a target is introduced.

We conducted the data-model comparison in two steps: (1) propagation from TRM to ER and (2) propagation from ER to VRA. The signal observed from TRM to VRA was then compared to the sum of steps (1) and (2). This was possible because the transmitted signal was captured at the ER in this controlled experiment, which would not normally be the case in an operational implementation of the technique. The echo-repeated signal was delayed in time due to the mechanics of the experimental set-up, so it could be separately added in post-processing. The data analyzed correspond to a nominal ship position at 568 m from the TRM and 645 m from the VRA (the sum is not exactly the straight-line distance between the two arrays, 1,156 m, because this was not the ship crossing point).

MATHEMATICAL MODELING

A numerical normal mode code (KRAKEN) was used to model the sound propagation in a waveguide. The KRAKEN normal mode program is a wave-theory model based on the expansion of the wave-equation solution into normal modes. KRAKEN only includes real-axis eigenvalues, and attenuation in the media is described by a perturbation theory. The continuous spectrum and attenuation in elastic media is not included. KRAKEN only includes the discrete set of normal modes, hence steep propagation angles at short ranges are not covered [6]. Approximate isospeed conditions were measured in the water column at 1,507 m/s. The average density of the sediment was estimated at 1.75 g/cm^3. The sediment propagation loss was estimated around 0.13 dB/λ, and the average sound speed of 1,610 m/s was determined by a tuning process.

Applying the model to this environment, Fig. 2 gives an indication of the expected fidelity of the data-model comparison. This figure shows a superposition of all the individual arrivals on the VRA, providing a depiction of the impulse response of the waveguide at the TRM if a point source were located at the VRA at depth 86 m. The

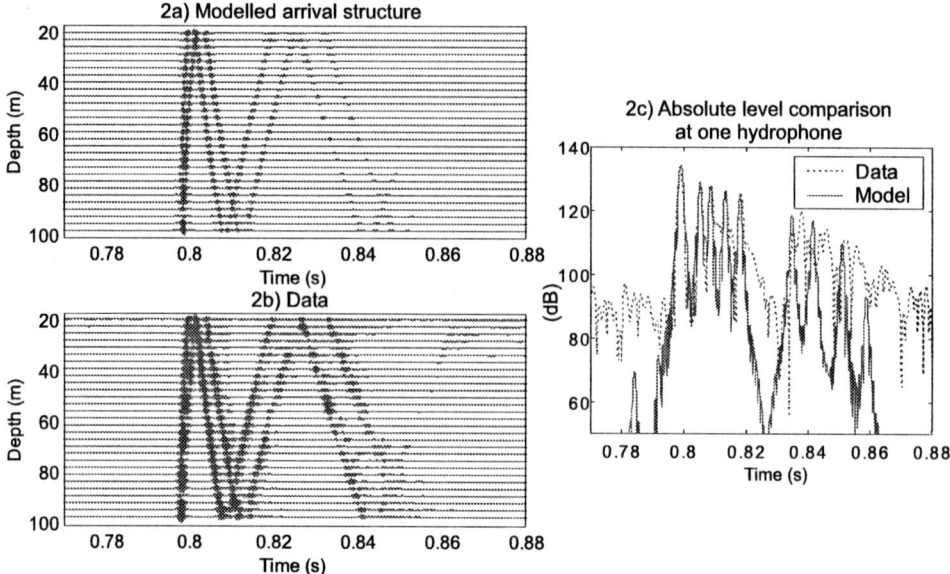

FIGURE 2. Modeled and measured impulse response of the waveguide given a point source at a depth of 86 m. The figure on the right shows the amplitude, in absolute level, of a representative arrival time series at one hydrophone (depth 67 m) on the VRA. The time envelope of pressure is shown. There is extremely good model-data fit for the first arrival and following four multipath echoes.

envelope of the raw pressure time series is presented. This figure also shows, in absolute decibel levels, that the model accurately accounts for the first arrival and following four multipath echoes, implying that knowledge of the channel impulse response is generally good. This is the required starting point for any data-model comparison. Discrepancy in the later arrivals when the number of multipaths (particularly of sea-surface bounces) becomes relatively high.

RESULTS

Modeling the Focus Unperturbed by a Target

Having achieved some confidence that the model could account adequately for the observed channel impulse response, we apply the model to the scenario wherein a focus is established but no target is present, i.e., the unpurturbed scenario. We ask the following questions: (1) How well does the model predict the observed received signal on a hydrophone at the focus? and (2) How well does the model account for the observed signal on a hydrophone in the quiescent region? Figure 3 shows the answers in absolute levels. Agreement is satisfactory. Note that the model is, of course, noise-free. However, the at-sea apparatus measured the actual noise floor present. Figure 4 shows the comparison at the focus in greater detail.

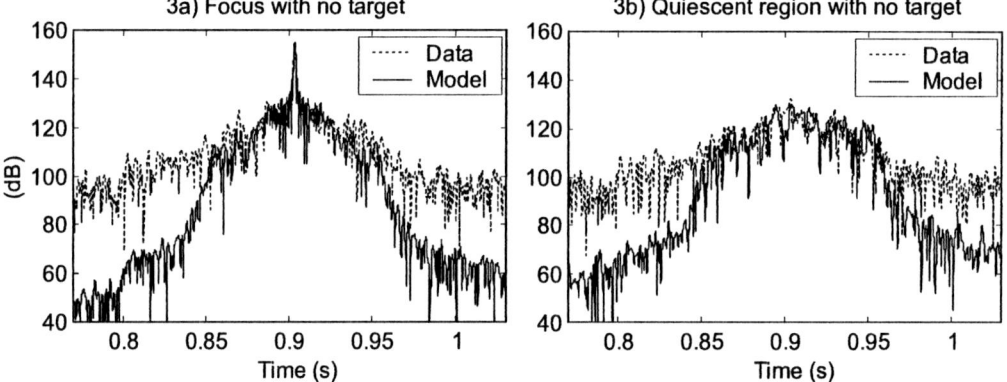

FIGURE 3. Comparison between data and model predictions for an unperturbed (no target present) focused field at two hydrophones on the VRA, one at the focal point at depth 86 m and the other in the quiescent region at depth 64 m.

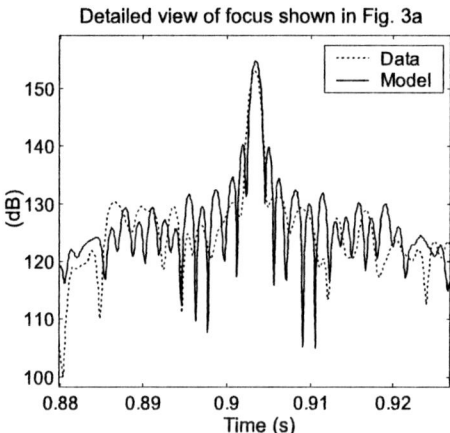

FIGURE 4. A more detailed view of Fig. 3a during a 50 ms time period spanning the focus. Model-data agreement is good to within a few dB at the focus.

Modeling the Received Field after Introducing a Target

The procedure described above was next repeated after introduction of a target into the field. The target was simulated, at sea, by an omnidirectional Echo Repeater (ER) with gain set at 60 dB. Here the data-model comparison process is conducted in two steps: propagation from SRA to ER, then from ER to VRA. One can initiate the SRA transmission with either the true signals, as derived from Fig. 2b or a model prediction (Fig. 2a). Likewise, the next link in the chain can be formed with either a modeled ER reception or the received signal. The significant difference is how noise is introduced and repropagated at each step. Figures 5 and 6 show how data and model compare. Of

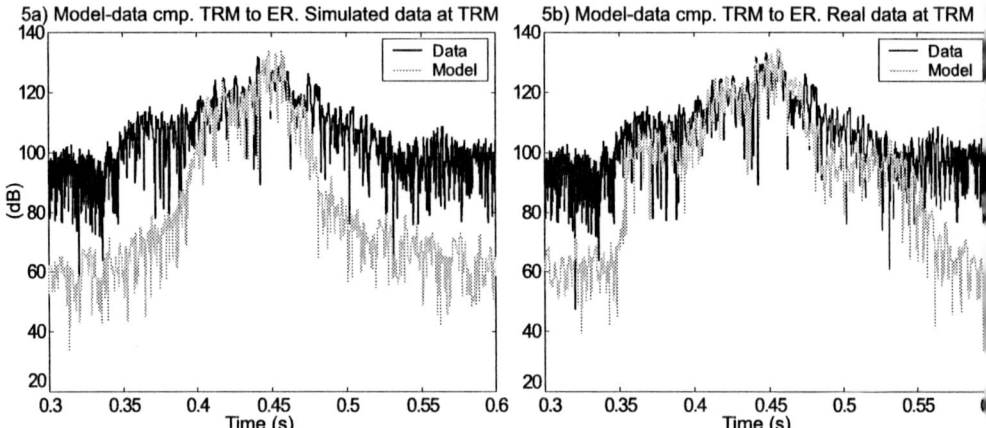

FIGURE 5. Model-data comparison of the field transmitted from the TRM and received by the echo repeater (ER). In the modeling result (Fig. 5a) the field transmitted is either simulated (Fig. 2a) or measured (Fig. 2b). The result using the measured signal includes noise recorded during the forward propagation phase, then time-reversed and retransmitted from the TRM. This noise contribution is not modeled.

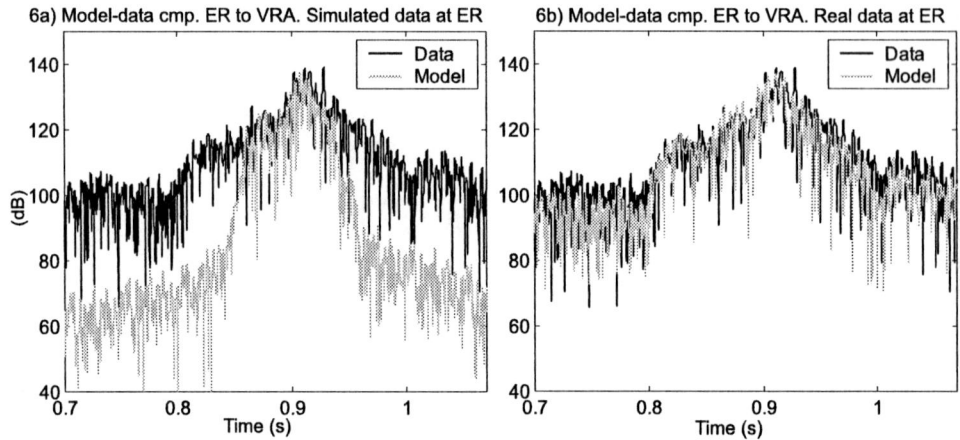

FIGURE 6. Model-data comparison of the field transmitted from ER to VRA, showing the received field that includes scattering from an insonified target. The ER gain simulated a target with an approximate TS of 60 dB. As in Fig.5, there is a noise contribution present when the measured signal is used that is not accounted for in the purely modeled prediction. However, in the 100 ms period spanning the focus, agreement is within a few dB. The data are from a VRA hydrophone at depth 64 m.

note is that the signal duration looks different because the result using a measured signal includes some noise recorded during the forward propagation phase, then time-reversed and transmitted back from the TRM. Noise is also amplified by the echo repeater when the actual received signal is used. This noise contribution is not modeled.

The end result of the experiment is shown in Fig 7. This figure shows a comparison

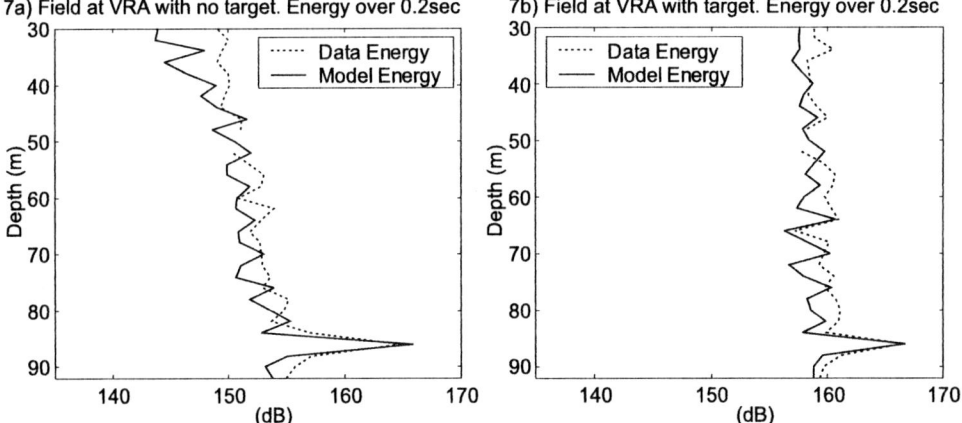

FIGURE 7. Comparison of energy received on the VRA over a 0.2 s time window centered at the signal's maximum peak. Left: model-data comparison with no target present. Right: model-data comparison with the target. Measured data energy is always higher than model prediction due to unmodeled noise, as in Figures 5 and 6.

of received energy at the VRA with and without a target present in the propagation medium. The integration time period is 0.2 s centered at the signal's maximum. Since the energy calculation includes an integration process, the result for data is always higher than model prediction due to noise, as explained above. The result for the case with the target was obtained by adding the unperturbed field to the field radiated by ER, which was possible because these signals were recorded independently. The comparison of the two plots in Fig. 7 suggests that the presence of a intruder with 60 dB of TS provides signal excess at the VRA of about 7-8 dB in the quiescent region. One VRA hydrophone was missing.

Calculated transmission loss (TL) from a rough sonar-equation-based formula [7] predicts TL of around 50 dB at 1,200 m of range. Measurements gave a value of 45 to 55 dB at about 1 km (with a source depth of 60 m). These two values agree reasonably. Calculated absolute received levels are also in general agreement with observation (SL - TL = 180 - 50 = 130 dB, which is roughly the value obtained through our model-data comparison of forward propagation; see Fig. 2c).

DISCUSSION

The echo repeater gain chosen for this study is a reasonably good representation of the foward-scatter TS of an object about the size of a small-medium submarine ensonified by a planewave in the free field (roughly at any incident angle, probably except endfire). Nevertheless, this study was not intended to be a conclusive test for the general performance of the technique in the real world. We chose, for practical reasons, to study a simple, totally repeatable, case first. The fact that we used an omni echo repeater with the gain set at 60 dB was a matter of convenience. The ER is omnidirectional and being the forward field by a target generally much changed by the insonification from many directions, as occurs in a waveguide with respect to a plane wave insonification in the free field. Also, the horizontal radiation pattern is generally smooth enough to work not only at the crossing point but within an azimuthal sector.

CONCLUSIONS

The reason that this study was conducted was to determine if observations made in the sea could be understood using mathematical modelling to provide some future predictive capability for a family of barrier systems. We hoped to answer the question of whether one can mathematically model the observations provided from one simple barrier experiment conducted under highly controlled conditions.

Model-data comparison suggests that target detection performance is reasonably predictable using a simple propagation modeling code and a simple geoacoustic field model of the waveguide. The main message of this study is that an available numerical model is a reasonable tool for prediction and design of experiments.

An echo repeater was used for its simplicity – a good thing in a pilot study like this one. It was not used as a representative surrogate submarine.

It appears that one can model reality for a simple case, so perhaps this line of research deserves further investigation. It may eventually lead to a useful predictive tool when designing forward-scatter barrier experiments using focused acoustic fields. The next step is to apply the method to more complicated targets.

ACKNOWLEDGMENTS

Work supported by NATO and the US Office of Naval Research. We would particularly like to acknowledge the efforts of SACLANTCEN in organizing and conducting the FAF-03 Sea Trial. This research would not have been possible without the contributions of Reginald Hollett, Pier Angiolo Boni, Roberto Lombardi, Bruno Miaschi, Jeff Skinner, Dave Ensberg, and Dick Harriss.

REFERENCES

1. Song, H. C., Kuperman, W. A., Hodgkiss, W. S., Akal, T., and Guerrini, P., *IEEE Journal of Oceanic Engineering*, **28** (2), 246 (2003).
2. Roux, P., Kuperman, W. A., Song, H. C., Hodgkiss, W. S., and Akal, T., *Journal of the Acoustical Society of America*, **114**, 2407 (2003).
3. Kuperman, W. A., Akal, T., Hodgkiss, W. S., Edelmann, G., Kim, S., and Song, H. C., *Journal of the Acoustical Society of America*, **108**, 2607 (2000).
4. Akal, T., Edelmann, G., Kim, S., Hodgkiss, W. S., Kuperman, W. A., and Song, H. C., "Low and high frequency ocean acoustic phase conjugation experiments," in *Procs of the Fifth European Conference on Underwater Acoustics*, 2000.
5. Hodgkiss, W. S., Skinner, J. S., Edmonds, G. E., Harriss, R. A., and Ensberg, D. E., "A high frequency phase conjugation array," in *Procs of IEEE OCEANS Conference*, 2001.
6. Porter, M. B., and Bucker, H. P., *Journal of the Acoustical Society of America*, **82**, 1349–1359 (1987).
7. Urick, R. J., *Principles of Underwater Sound*, McGraw Hill, New York, 1983, third edn.

AUTHOR INDEX

A

Abawi, A. T., 260
Akal, T., 522, 539
Arvelo, J., 497
Au, W. W. L., 247

B

Badiey, M., 214, 307, 322, 385
Bellettini, A., 489
Bradley, D., 204
Briggs, K. B., 12
Brown, B., 393, 402, 428, 438
Buckingham, M. J., 3
Burnett, D. S., 464

C

Calder, B., 307
Caruthers, J. W., 307, 366
Chandler, H., 393, 420
Commander, K. W., 141
Coviello, C. M., 237
Cox, H., 449
Culver, L., 204

D

Dahl, P. H., 194
de Moustier, C., 307, 478
D'Spain, G. L., 222
Duncan, A. J., 280

E

Edelmann, G., 522
Elam, W. T., 132
Ensberg, D. E., 222

F

Ferat, P., 497
Fialkowski, J., 157
Fink, M., 514
Fisher, C., 366
Fisher, R. A., 393, 402, 428
Flynn, J. A., 83
Forsythe, S. E., 322
Fox, W. L. J., 83, 307
Fromm, D., 149, 157

G

Gauss, R., 149, 157
Gendron, P. J., 98
Goodman, R. R., 296, 393, 413, 420
Gragg, R., 149
Green, D., 73
Guerrini, P., 539

H

Harrison, C., 22
Hayward, T. J., 114
Heaney, K. D., 47
Heitsenrether, R. M., 214
Henyey, F. S., 132
Higley, W., 522
Hillstrom, W. R., 296
Hodgkiss, W. S., 222, 307, 522, 539
Holden, A., 508
Hursky, P., 260, 307, 336, 350

I

Ioup, G. E., 288
Ioup, J. W., 288

J

Jackson, D. R., 125
Jensen, F. B., 385, 464

Jesus, S. M., 350

K

Kennedy, E. T., 393, 402, 413, 420, 428, 438
Kilfoyle, D., 385
Kraft, B. J., 307
Kuczaj, S., 296
Kuperman, W. A., 539
Kvadsheim, P. H., 272

L

Leighton, T. G., 180
LePage, K. D., 149, 157, 165
Lewis, J. K., 307, 373
Lim, R., 141
Lopes, J. L., 141
Lyons, A. P., 32

M

Maggi, A. L., 280
Malley, D., 393, 402, 428, 438
Marston, P. L., 472
Martin, S., 260
McCauley, R. D., 280
McDonald, B. E., 165
McDonald, V. K., 307, 336, 350, 385
Meredith, R. W., 366, 393, 402, 413, 420, 428, 438
Montaldo, G., 514
Munk, P., 489
Myers, V., 489

N

Nero, R., 157
Newcomb, J. J., 296
Norton, G. V., 413

O

Osler, J. C., 32

P

Pautet, L., 173, 385, 489
Pinkel, R., 230
Pinto, M., 489
Porter, M. B., 260, 307, 322, 358, 385
Pouliquen, E., 173
Preisig, J., 57

Q

Quiroz, E., 366

R

Raghukumar, K., 307, 522
Rajan, S., 307, 373
Ray, R., 393
Reynolds, S. A., 132
Rice, J., 73
Richardson, M. D., 12
Ritcey, J. A., 83
Roan, M. J., 237
Rouseff, D., 83, 307
Roux, P., 522, 539
Rudzinsky, J., 373
Ruppel, T. H., 413

S

Schmidt, H., 456, 464
Sevaldsen, E. M., 272
Sibul, L. H., 237
Siderius, M., 22, 307, 358, 385
Sidorovskaia, N. A., 288, 366
Silva, A. J., 350
Smith, J. A., 230
Song, H. C., 522, 539
Stanic, S., 393, 402, 413, 420, 428, 438
Stein, P. J., 307, 373
Stevenson, M., 522, 539
Stojanovic, M., 65

T

Tanter, M., 514
Tesei, A., 464, 539
Thames, R., 296

Thorsos, E. I., 132
Tiemann, C., 260

V

Vandiver, A., 373

W

Wang, L. S., 489
Williams, K. L., 132
Wilson, M. A., 413

Wright, A. J., 296

Y

Yang, T. C., 90, 98, 106, 114
Yang, W.-B., 106

Z

Zampolli, M., 464
Zoksimovski, A., 478